H. Haken H.C. Wolf

Atomic and Quantum Physics

An Introduction to the Fundamentals of Experiment and Theory

Translated by W. D. Brewer

Second Enlarged Edition

With 265 Figures

Springer-Verlag Berlin Heidelberg New York
London Paris Tokyo

Professor Dr. Dr. h.c. *Hermann Haken*

Institut für Theoretische Physik, Universität Stuttgart, Pfaffenwaldring 57,
D-7000 Stuttgart 80, Fed. Rep. of Germany

Professor Dr. *Hans Christoph Wolf*

Physikalisches Institut, Universität Stuttgart, Pfaffenwaldring 57,
D-7000 Stuttgart 80, Fed. Rep. of Germany

Translator:

Professor Dr. *William D. Brewer*

Freie Universität Berlin, Fachbereich Physik, Arnimallee 14,
D-1000 Berlin 33

Title of the german original edition:

H. Haken, H.C. Wolf: *Atom- und Quantenphysik.*
Eine Einführung in die experimentellen und theoretischen Grundlagen.
(Dritte, überarbeitete und erweiterte Auflage)

ISBN 3-540-17921-6 3. Auflage Springer-Verlag Berlin Heidelberg New York
ISBN 0-387-17921-6 3rd edition Springer-Verlag New York Berlin Heidelberg

---

ISBN 3-540-17702-7 2. Auflage Springer-Verlag Berlin Heidelberg New York
ISBN 0-387-17702-7 2nd edition Springer-Verlag New York Berlin Heidelberg

ISBN 3-540-13137-X 1. Auflage Springer-Verlag Berlin Heidelberg New York Tokyo
ISBN 0-387-13137-X 1st edition Springer-Verlag New York Berlin Heidelberg Tokyo

Library of Congress Cataloging-in-Publication Data. Haken, H. Atomic and quantum physics. Translation of: Atom- und Quantenphysik. Bibliography: p. Includes index. 1. Atoms. 2. Quantum theory. I. Wolf, H.C. (Hans Christoph), 1929-. II. Title. QC173.H17513 1987 539.7 87-9450

Typesetting: K+V Fotosatz, 6124 Beerfelden
Offset printing: Druckhaus Beltz, 6944 Hemsbach. Bookbinding: J. Schäffer GmbH & Co. KG, 6718 Grünstadt
2153/3150-543210

ÆDES CHRISTI
in Academia Oxoniensi

Preface to the Second Edition

The excellent critique and very positive response to the first edition of this book have encouraged us to prepare this second edition, in which we have tried to make improvements wherever possible. We have profited much from the suggestions of professors and students as well as from our own experience in teaching atomic and quantum physics at our university.

Following a widespread request, we have now included the solutions to the exercises and present these at the end of the book. Among the major new sections to be found in this second edition are the following:

We now include the derivation of the relativistic Klein-Gordon equation and of the Dirac equation because the latter, in particular, appears in atomic physics whenever relativistic effects must be taken into account. Our derivation of the Schrödinger equation allowed us to present this extension in a straightforward manner.

The high precision methods of modern spectroscopy allow the atomic physicist to measure extremely small but important shifts of the atomic lines. A very important effect of this kind is the Lamb shift, for which a detailed theoretical derivation is given in a new section. In order to put this in an adequate framework, the basic ideas of the quantization of the electromagnetic field as used in quantum electrodynamics are given. Again it turned out that all the concepts and methods needed to discuss these seemingly advanced theories had already been presented in previous chapters so that again the reader may easily follow these theoretical explanations.

The section on photoelectron spectroscopy has been enlarged and revised. Furthermore, the two-electron problem has been made more explicit by treating the difference between triplet and singlet states in detail. Finally, our previous presentation of nuclear spin resonance has been considerably enlarged because this method is finding widespread and very important applications, not only in chemistry but also in medicine, for instance in NMR tomography, which is an important new tool in medical diagnostics. This is only one example of the widespread and quite often unanticipated application of atomic and quantum physics in modern science and technology.

It goes without saying that we have not only corrected a number of misprints but have also tried to include the most recent developments in each area. This second English edition corresponds to the third German edition, which is published at about the same time. We wish to thank R. Seyfang, J.U. von Schütz and V. Weberruss for their help in preparing the second edition. It is again a pleasure for us to thank Springer-Verlag, in particular Dr. H. Lotsch and C.-D. Bachem for their always excellent cooperation.

Stuttgart, March 1987 *H. Haken H.C. Wolf*

Preface to the First Edition

A thorough knowledge of the physics of atoms and quanta is clearly a must for every student of physics but also for students of neighbouring disciplines such as chemistry and electrical engineering. What these students especially need is a coherent presentation of both the experimental and the theoretical aspects of atomic and quantum physics. Indeed, this field could evolve only through the intimate interaction between ingenious experiments and an equally ingenious development of bold new ideas.

It is well known that the study of the microworld of atoms caused a revolution of physical thought, and fundamental ideas of classical physics, such as those on measurability, had to be abandoned. But atomic and quantum physics is not only a fascinating field with respect to the development of far-reaching new physical ideas. It is also of enormous importance as a basis for other fields. For instance, it provides chemistry with a conceptual basis through the quantum theory of chemical bonding. Modern solid-state physics, with its numerous applications in communication and computer technology, rests on the fundamental concepts first developed in atomic and quantum physics. Among the many other important technical applications we mention just the laser, a now widely used light source which produces light whose physical nature is quite different from that of conventional lamps.

In this book we have tried to convey to the reader some of the fascination which atomic and quantum physics still gives a physicist studying this field. We have tried to elaborate on the fundamental facts and basic theoretical methods, leaving aside all superfluous material. The text emerged from lectures which the authors, an experimentalist and a theoretician, have given at the University of Stuttgart for many years. These lectures were matched with respect to their experimental and theoretical contents.

We have occasionally included in the text some more difficult theoretical sections, in order to give a student who wants to penetrate thoroughly into this field a self-contained presentation. The chapters which are more difficult to read are marked by an asterisk. They can be skipped on a first reading of this book. We have included chapters important for chemistry, such as the chapter on the quantum theory of the chemical bond, which may also serve as a starting point for studying solid-state physics. We have further included chapters on spin resonance. Though we explicitly deal with electron spins, similar ideas apply to nuclear spins. The methods of spin resonance play a fundamental role in modern physical, chemical and biological investigations as well as in medical diagnostics (nuclear spin tomography). Recent developments in atomic physics, such as studies on Rydberg atoms, are taken into account, and we elaborate the basic features of laser light and nonlinear spectroscopy. We hope that readers will find atomic and quantum physics just as fascinating as did the students of our lectures.

The present text is a translation of the second German edition *Atom- und Quantenphysik*. We wish to thank Prof. W. D. Brewer for the excellent translation and the most valuable suggestions he made for the improvement of the book. Our thanks also go to

Dr. J. v. Schütz and Mr. K. Zeile for the critical reading of the manuscript, to Ms. S. Schmiech and Dr. H. Ohno for the drawings, and to Mr. G. Haubs for the careful proof-reading. We would like to thank Mrs. U. Funke for her precious help in typing new chapters. Last, but not least, we wish to thank Springer-Verlag, and in particular H. Lotsch and G. M. Hayes, for their excellent cooperation.

Stuttgart, February 1984 *H. Haken H. C. Wolf*

Contents

List of the Most Important Symbols Used

The numbers of the equations in which the symbols are defined are given in parentheses; the numbers in square brackets refer to the section of the book. The Greek symbols are at the end of the list.

$\boldsymbol{A}$	Vector potential
A	Amplitude or constant
A	Mass number (2.2) or area
a	Interval factor or fine structure constant (12.28) and hyperfine splitting (20.10)
a_0	Bohr radius of the H atom in its ground state (8.8)
$\boldsymbol{B}$	Magnetic induction
b^+, b	Creation and annihilation operators for the harmonic oscillator
b	Constant, impact parameter
C	Constant
c	Velocity of light, series expansion coefficient
c.c.	Complex conjugate
$\boldsymbol{D}$	Dipole moment
d	Constant
dV	Infinitesimal volume element
$\boldsymbol{E}$	Electric field strength
E	Energy, total energy, energy eigenvalue
E_{kin}	Kinetic energy
E_{pot}	Potential energy
E_{tot}	Total energy
e	Proton charge
$-e$	Electron charge
e	Exponential function
$\boldsymbol{F}$	Electric field strength (14.1)
$\boldsymbol{F}, F$	Total angular momentum of an atom, including nuclear angular momentum and corresponding quantum number (20.6)
F	Amplitude of the magnetic induction [14.4, 14.5]
f	Spring constant
g	Landé g factor (12.10, 16, 21, 13.18, 20.13)
$\mathscr{H}$	Hamilton function, Hamiltonian operator
H_n	Hermite polynomial
h	Planck's constant
$\hbar$	$= h/2\pi$
$\boldsymbol{I}, I$	Nuclear angular momentum and corresponding quantum number (20.1)
I	Abbreviation for integrals [16.13] or intensity
i	Imaginary unit ($\mathrm{i} = \sqrt{-1}$)
$\boldsymbol{J}, J$	Total angular momentum of an electron shell and corresponding quantum number (17.5)
$\boldsymbol{j}, j$	Total angular momentum of an electron and corresponding quantum number [12.7]
$\hat{j}$	Operator for the total angular momentum
k	Boltzmann's constant, force constant
$\boldsymbol{k}$	Wavevector
$\boldsymbol{L}, L$	Resultant orbital angular momentum and corresponding quantum number (17.3)
L_n	Laguerre polynomial (10.81)
$\boldsymbol{l}, l$	Orbital angular momentum of an electron and corresponding quantum number
$\hat{l}$	Angular momentum operator
m, m_0	Mass
m	Magnetic quantum number
m_l	– for angular momentum
m_s	– for spin
m_j	Magnetic quantum number for total angular momentum
m_0	Rest mass, especially that of the electron

N, n Particle number, particle number density
N Normalisation factor
n Principal quantum number or number of photons or an integer
P Spectral radiation flux density (5.2) or probability
P_l^0 Legendre polynomial
P_l^m (m $\neq 0$) Associated Legendre function
$p, \bar{p}$ Momentum, expectation value of momentum
Q Nuclear quadrupole moment (20.20)
Q, q Charge
$R(r)$ Radial part of the hydrogen wavefunction
$\boldsymbol{r}$ Position coordinate (three-dimensional vector)
r Distance
$\boldsymbol{S}$ Resultant spin (17.4)
S Symbol for orbital angular momentum $L = 0$
$\boldsymbol{s}, s$ Electron spin and corresponding quantum number (12.15)
$\hat{\boldsymbol{s}}$ Spin operator $= (\hat{s}_x, \hat{s}_y, \hat{s}_z)$
T Absolute temperature
T_1 Longitudinal relaxation time
T_2 Transverse relaxation time
t Time
u Spectral energy density (5.2), atomic mass unit [2.2]
V Volume, potential, electric voltage
$\bar{V}$ Expectation value of the potential energy
$\boldsymbol{v}$ Velocity, particle velocity
x Particle coordinate (one-dimensional)
$\bar{x}$ Expectation value of position
$Y_{l,m}(\theta, \phi)$ Spherical harmonic functions (10.10, 48 – 50)
Z Nuclear charge
α Fine structure constant [8.10] or absorption coefficient (2.22)
β Constant
Γ Decay constant
γ Decay constant or linewidth gyromagnetic ratio (12.12)

∇^2 Laplace operator $= \partial^2/\partial x^2 + \partial^2/\partial y^2 + \partial^2/\partial z^2$
ΔE Energy uncertainty
Δk Wavenumber uncertainty
Δp Momentum uncertainty
Δt Time uncertainty (= finite measurement time)
ΔV Finite volume element
$\Delta\omega$ Uncertainty in the angular frequency
Δx Position uncertainty
$\delta(x)$ Dirac delta function (see mathematics appendix)
$\delta_{\mu,\nu}$ Kronecker delta symbol: $\delta_{\mu,\nu} = 1$ for $\mu = \nu$, $\delta_{\mu,\nu} = 0$ for $\mu \neq \nu$
ε Dimensionless energy (9.83)
$\varepsilon^{(n)}$ Energy contributions to perturbation theory
ε_0 Permittivity constant of vacuum
θ Angle coordinate (10.2)
κ Defined in (10.54)
λ Wavelength (exception: expansion parameter in [15.2.1, 2])
$\boldsymbol{\mu}, \mu$ Magnetic moment (12.1)
μ Reduced mass (8.15)
μ_B Bohr magneton (12.8)
μ_N Nuclear magneton (20.3)
ν Frequency [8.1]
$\bar{\nu}$ Wavenumber [8.1]
ξ Dimensionless coordinate (9.83)
ϱ Charge density, density of states, mass density; or dimensionless distance
σ Scattering coefficient, interaction cross section (2.16)
τ Torque (12.2)
Φ Phase
ϕ Phase angle, angle coordinate
$\phi(x)$ Wavefunction of a particle
$\phi_\uparrow, \phi_\downarrow, \phi$ Spin wavefunctions
ψ Wavefunction
Ψ Wavefunction of several electrons
$\hat{\Omega}$ Generalised quantum mechanical operator
Ω Frequency [14.4, 14.5, 15.3]
ω Angular frequency $2\pi\nu$, or eigenvalue [9.3.6]
$\triangleq$ means "corresponds to"

1. Introduction

1.1 Classical Physics and Quantum Mechanics

Atomic and quantum physics, which are introduced in this book, are essentially products of the first third of this century. The division of classical physics into branches such as mechanics, acoustics, thermodynamics, electricity, and optics had to be enlarged when – as a consequence of the increasing knowledge of the structure of matter – atoms and quanta became the objects of physical research. Thus, in the twentieth century, classical physics has been complemented by atomic physics and the physics of light or energy quanta. The goal of atomic physics is an understanding of the structure of atoms and their interactions with one another and with electric and magnetic fields. Atoms are made up of positively charged nuclei and negatively charged electrons. The electromagnetic forces through which these particles interact are well known in classical physics.

The physics of atomic nuclei cannot be understood on the basis of these forces alone. A new force – the nuclear or strong force – determines the structures of nuclei, and the typical binding energies are orders of magnitude larger than those of the electrons in atoms. The study of nuclei, of elementary particles, and the whole of high energy physics thus form their own branches of physics. They will not be treated in this book.

1.2 Short Historical Review

The word *atom* comes from the Greek and means "the indivisible", the smallest component of matter, which cannot be further divided. This concept was introduced in the 5th and 4th centuries B.C. by Greek natural philosophers. The first atomic theories of the structure of matter were those of *Democrites* (460 – 370 B.C.), *Plato* (429 – 348), and *Aristotle* (384 – 322). It required more than two millenia until this speculative atomism grew into an exact atomic physics in the modern sense.

The meaning of the word *atom* becomes less subject to misinterpretation if it is translated into Latin: an *individuum* is the smallest unit of a large set which possesses all the essential characteristics of the set. In this sense, an atom is in fact indivisible. One can, to be sure, split a hydrogen atom into a proton and an electron, but the hydrogen is destroyed in the process. For example, one can no longer observe the spectral lines characteristic of hydrogen in its optical spectrum.

Atomism as understood by modern science was first discovered for *matter*, then for *electricity*, and finally for *energy*.

The *atomism of matter*, the recognition of the fact that all the chemical elements are composed of atoms, followed from chemical investigations. The laws of constant and multiple proportions, formulated by *Proust* ca. 1799 and by *Dalton* ca. 1803, can be explained very simply by the atomic hypothesis:

The reaction equations

14 g nitrogen + 16 g oxygen yield 30 g NO and

14 g nitrogen + 32 g oxygen yield 46 g NO_2

mean: the atomic weights of nitrogen and oxygen are related as 14:16.

Only whole atoms can react with one another. The first atomic model (*Prout*, 1815) assumed that the atoms of all elements are put together out of hydrogen atoms. As a heuristic principle this hypothesis finally led to a scheme for ordering the elements based on their chemical properties, the *periodic system* of *L. Meyer* and *D. I. Mendeleev* (1869). More about this subject may be found in introductory textbooks on chemistry.

About the same time (1808), it was found by *Gay-Lussac* that not only the weights but also the *volumes* of gaseous reactants occur as ratios of small integers. In the above example,

1 volume nitrogen + 1 volume oxygen yield 2 volumes NO

1 volume nitrogen + 2 volumes oxygen yield 2 volumes NO_2.

Similar observations led to the *hypothesis of Avogadro* (1811): Equal volumes of gases under similar conditions (pressure, temperature) contain equal numbers of molecules.

Continued investigations of gases in the course of the 19th century led to the *atomism of heat*, that is, to the explanation of heat in general and of the thermodynamic laws in particular as consequences of atomic motion and collisions. In about 1870, the first theory to encompass a whole branch of physics, the *kinetic theory of gases*, was completed by the physicists *Clausius* and *Boltzmann*.

The *atomism of electricity* was discovered in 1833 by the English scientist *Michael Faraday*. Based on the quantitative evaluation of exceedingly careful measurements of the electrolysis of liquids, he formulated his famous laws:

> The quantity of an element which is separated is proportional to the quantity of charge transported in the process,

and

> various elements are separated into equivalent weights by the same quantity of charge.

From this, *Faraday* concluded:

> There are "atoms" of electricity – it was only after 70 years that their mass and charge could be determined –

and

> these "atoms" of electricity – the electrons – are bound to atoms of matter.

The discovery of the *atomism of energy* can be dated exactly: on December 14, 1900, *Planck* announced the derivation of his laws for black body radiation in a lecture before the Physical Society in Berlin. In order to derive these laws, he assumed that the energy of harmonic oscillators can only take on discrete values – quite contradictory to the classical view, in which the energy values form a continuum.

This date can be called the birth date of quantum theory. The further development of atomic and quantum physics is the subject of this book.

Our knowledge of the structure of atoms was influenced strongly by the investigation of optical spectra. After *Kirchhoff* and *Bunsen* had shown, about 1860, that optical spectra are characteristic of the elements which are emitting or absorbing the light, *Balmer* (1885) succeeded in finding an ordering principle in atomic spectra, expressed in the formula (8.1) which bears his name and which describes the spectral lines emitted from hydrogen atoms. As a result of the atomic model proposed by *Rutherford* (1911), *Bohr* was able, in 1913, to formulate the basic principles of the quantisation of electron orbits in atoms. These quantisation rules were considerably extended by *Sommerfeld*. A parallel development was the concept of matter waves, carried out by *De Broglie*. The actual breakthrough was attained by *Born, Heisenberg, Schrödinger, Pauli, Dirac,* and other researchers in the decade between 1920 and 1930.

The problems of atomic physics which are of current interest in research are:

- an increasingly detailed description of the structure of electronic shells of atoms and their excitations,
- the interactions between atoms and radiation fields, for example in view of their applications in optical pumping (Chap. 21) and in laser physics (Chap. 22),

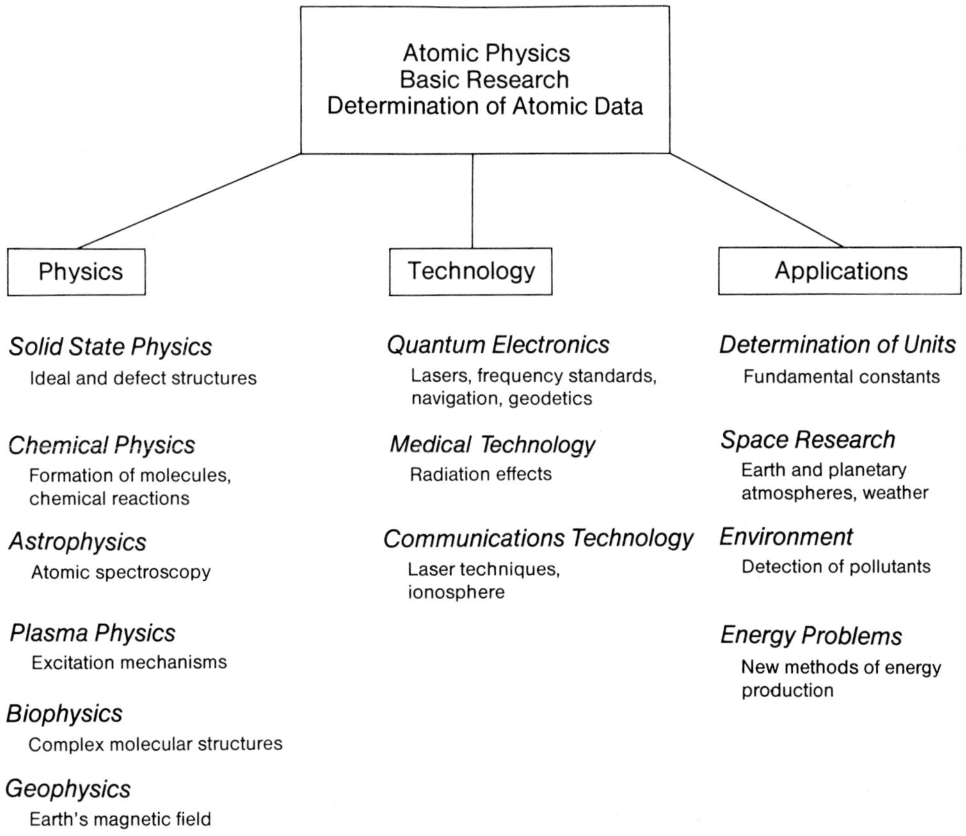

Fig. 1.1. The relevance of atomic physics for other disciplines of science and technology

- the interactions of atoms among themselves due to collisions in the gas phase and during the formation of molecules,
- the physical principles which lead to the formation of condensed phases from single atoms, and their properties.

Thus molecular and solid state physics are based on atomic physics, and chemistry as well makes constant use of its laws and principles.

Atomic physics is furthermore a basic science for many other disciplines of research, technology, and applications. A few examples are shown in Fig. 1.1.

The following chapters do not give a historical or a chronological presentation; they do, however, show the general line of developments and discoveries. An inductive approach is often used. It is of little use in physics to content oneself with the acquisition of factual knowledge. A physicist must learn to analyse, to explain, and to extract the essentials from experimental findings. In this way, one develops models for nature. In the process, it is important to recognise relationships to other experimental results and to be able to predict the outcome of new experiments. The predictions must then be experimentally tested. Because of this process, physics is not a dead, finalised science, but rather is in a constant state of development, since new experimental techniques open up new areas of research while, on the other hand, the process of developing physical concepts gives the impulse for ever newer experiments.

2. The Mass and Size of the Atom

2.1 What is an Atom?

An atom is the smallest unchangeable component of a chemical element. *Unchangeable* means in this case *by chemical means*; i.e., by reactions with acids or bases or the effect of moderate temperatures, atoms may only be slightly changed, namely, in their degree of ionisation. Moderate temperatures refers here to temperatures whose equivalent energy kT (k is Boltzmann's constant, T the temperature in K) is not larger than a few electron volts (eV) (see Table 8.1).

2.2 Determination of the Mass

Beginning with *Dalton*'s law of constant and multiple proportions, and *Avogadro*'s hypothesis, according to which equal volumes of gas contain the same number of molecules or atoms, we introduce *relative atomic masses* (also called atomic weights) A_{rel}. It was first discovered with the methods of chemistry that these atomic weights are approximately whole-number multiples of the atomic mass of the hydrogen atom. The relative atomic masses of nitrogen and oxygen are then $A_{\mathrm{rel}}(\mathrm{N}) \simeq 14$, $A_{\mathrm{rel}}(\mathrm{O}) \simeq 16$.

For this reason, an atomic mass unit has been defined, 1 u (abbreviation for unit, previously also referred to as 1 amu), which is approximately equal to the mass of a hydrogen atom. Since 1961 the unit of atomic mass has been based on the carbon atom $^{12}\mathrm{C}$ with $A_{\mathrm{rel}} = 12.00000$ u and is thus no longer exactly equal to the mass of the H atom. The use of C as base substance was found to be expedient for the experimental precision determination of atomic masses by chemical means. We have as definition

$$1\ \mathrm{u} = 1/12 \text{ of the mass of a neutral carbon atom with nuclear charge 6 and mass number 12, i.e., } {}^{12}_{6}\mathrm{C}\,. \tag{2.1}$$

Earlier scales were defined somewhat differently: the old "chemical" scale was based on oxygen in the naturally occurring isotope mixture:

$$1\ \mathrm{amu}_{\mathrm{chem}} = 1/16\ (\text{average mass of O atoms in the natural isotopic mixture})$$

and the old "physical" scale was based on the oxygen isotope $^{16}\mathrm{O}$:

$$1\ \mathrm{amu}_{^{16}\mathrm{O}} = 1/16\ (\text{mass of an } {}^{16}\mathrm{O} \text{ atom})\,.$$

The following conversion formulae hold:

$$\mathrm{amu}_{\mathrm{chem}} : \mathrm{amu}_{^{16}\mathrm{O}} : \mathrm{u}_{^{12}\mathrm{C}} = 0.99996 : 0.99968 : 1.00000 \tag{2.2}$$

and

$$A_{\text{rel, chem}}:A_{\text{rel},{}^{16}\text{O}}:A_{\text{rel},{}^{12}\text{C}} = 1.00004:1.00032:1.00000\,. \tag{2.3}$$

The absolute atomic masses can be obtained from the relative masses using the concept of the mole.

1 mole of a substance is, according to *Avogadro*, as many grams as the relative atomic weight (in the case of molecules, as the correspondingly defined relative molecular weight). Thus, 1 mole of the carbon isotope ${}^{12}_{6}C$ is 12 grams. 1 mole of any substance contains the same number (N_A) of atoms (molecules).

The number N_A which is defined in this way is called Avogadro's number (in the German literature, it is called the Loschmidt number after the Austrian physicist *Loschmidt* who determined it in 1865 by measurements on gases). Experimental methods for its determination will be discussed in the following section.

The absolute atomic mass m_{atom} can therefore be obtained by measuring Avogadro's number. We have:

$$\text{Mass of an atom} = \frac{\text{Mass of 1 mole of the substance}}{N_A}\,. \tag{2.4}$$

The determination of atomic masses is thus based on the determination of Avogadro's number; the size of the latter depends evidently on the choice of the base substance for the mole. N_A is currently defined as the number of carbon atoms in 12.000 g of isotopically pure ${}^{12}_{6}C$.

The present best value for N_A is

$$N_A = (6.022045 \pm 0.000005) \cdot 10^{23}\,\text{mole}^{-1}\,.$$

With this value, we can write (2.4) in the form

$$m_{\text{atom}} = \frac{A_{\text{rel},{}^{12}\text{C}}}{N_A}\,[\text{gram}]\,. \tag{2.5}$$

For the conversion of the mass unit u into other units the following relations hold:

$$1\,\text{u} = (1.660565 \pm 0.000005) \cdot 10^{-27}\,\text{kg} = 931.478\,\frac{\text{MeV}}{c^2}\,. \tag{2.6}$$

This last conversion results from the mass-energy equivalence $E = mc^2$. MeV is a measure of energy (see Table 8.1), c is the velocity of light. Numerical values for masses m, relative atomic masses A_{rel}, and the mass number A of a few atoms are shown in Table 2.1.

Table 2.1. Mass number, mass, and relative atomic mass of several atoms

	Mass number A	Mass m [kg]	A_{rel}
H atom	1	$1.67342 \cdot 10^{-27}$	1.007825
C atom	12	$19.92516 \cdot 10^{-27}$	12.000000
O atom	16	$26.5584 \cdot 10^{-27}$	15.99491

The mass number A of an atom is the integer which is closest to its relative atomic mass A_{rel}. It is found in nuclear physics that A is equal to the number of nucleons (protons and neutrons) in the atomic nucleus.

2.3 Methods for Determining Avogadro's Number

2.3.1 Electrolysis

In electrolytic decomposition of salts from a solution, the amount of salt decomposed is proportional to the charge which flows through the electrolyte. For one mole of a monovalent substance, a charge of 96485 As (ampere-seconds) is required. This is the Faraday constant F. Thus, since each ion carries one elementary charge e, we have the relation $N_{\mathrm{A}} = F/e$. The elementary charge e denotes the charge on a single electron (see Sect. 6.3). For example, in order to electrodeposit one mole or 63.5 g of copper from a solution of $CuSO_4$ in water, $2N_{\mathrm{A}}$ electrons are necessary, since the copper ion is doubly positively charged. By weighing the amount of material deposited and measuring the electric current as well as the time, one can obtain the constant N_{A}.

2.3.2 The Gas Constant and Boltzmann's Constant

The universal gas constant R and Boltzmann's constant k are related through the equation $k = R/N_{\mathrm{A}}$.

The gas constant can be determined by means of the ideal-gas law $pV = RT$; the Boltzmann constant, for example, from sedimentation equilibria (*Perrin*, 1908). In the latter method, the density distribution of small suspended particles in a liquid, determined by the simultaneous action of gravity and the Brownian molecular motion, is given by the equation

$$n_h = n_0\, \mathrm{e}^{-mgh/kT}, \tag{2.7}$$

where n_h is the number of particles in a unit volume at a height h, n_0 the number of particles in a unit volume at height $h = 0$, m the mass of the particles, g the acceleration of gravity, k the Boltzmann constant, and T the absolute temperature. In Fig. 2.1, a model

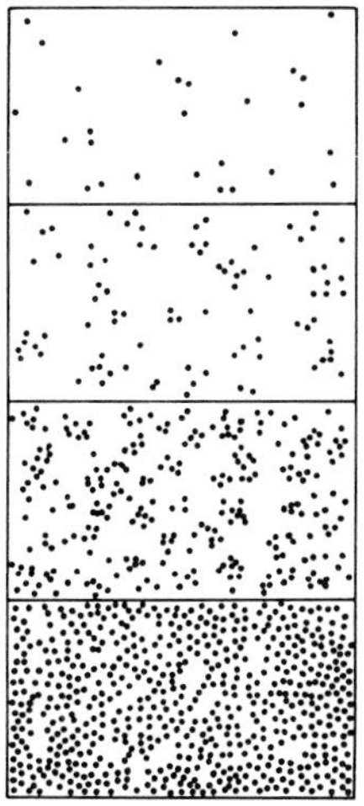

Fig. 2.1. Sedimentation equilibrium: distribution of suspended mastix spheres of 0.6 μm diameter at four different heights in the field of view of a microscope (after *Perrin*)

experiment demonstrating sedimentation is shown. The formula given by (2.7) is a special case of the famous Boltzmann distribution. Since we will use this distribution in numerous places in this book, we will give an explicit general formula for it here. The exact derivation may be found in texts on statistical physics. According to the Boltzmann distribution, the number of particles with energies in the interval $E \ldots E+dE$ in thermal equilibrium is given by

$$n_E dE = NZ(T)\,e^{-E/kT} g(E)\,dE\,, \tag{2.8}$$

where, generalising (2.7), the following quantities appear: N is the total number of particles, $Z(T)$ is the *partition function*. The latter ensures that, integrating over the whole energy range, one obtains the total particle number N, i.e., $\int n_E dE = N$. It is therefore given by $Z(T)^{-1} = \int \exp(-E/kT)\,g(E)\,dE$. Finally, $g(E)$ is the *density of states*; it is necessary since, for example, particles with the same energy can be moving in different directions, i.e., there can be more than one state with the energy E.

A completely satisfactory definition of $g(E)$ only becomes possible with the help of quantum mechanics. Using quantum numbers, of which we will later encounter a number of examples, one can count the number of "states" in the interval $E \ldots E+dE$.

2.3.3 X-Ray Diffraction in Crystals

With x-radiation of a known wavelength, one can determine the lattice constant, or the volume of an atom or molecule in a crystal. The volume of a mole V_{mol} is then N_A times the atomic volume. For one mole one thus has

$$N_A V_{atom} = V_{mol} = M/\varrho\,, \tag{2.9}$$

where M denotes the molar mass and ϱ the mass density.

Figure 2.2 illustrates the principle; it shows a section of a NaCl lattice. NaCl crystallises in the face-centred cubic structure. The NaCl lattice can be built up from two face-centred cubic lattices containing the Na^+ and the Cl^- ions. These ions occupy the corners of cubes of side $a/2$, where a is the edge length of the Na^+ or Cl^- unit cell. The unit cell is the smallest unit of a crystal, in the sense that the crystal structure consists of a repetition of this element in each of the three dimensions of space.

The size of a can be determined by x-ray diffraction if the x-ray wavelength is known (Sect. 2.4.5). In a cube of volume $(a/2)^3$, there are $4/8 = 1/2$ NaCl molecules, since each ion belongs to 8 cubes. The number of molecules per unit volume is therefore

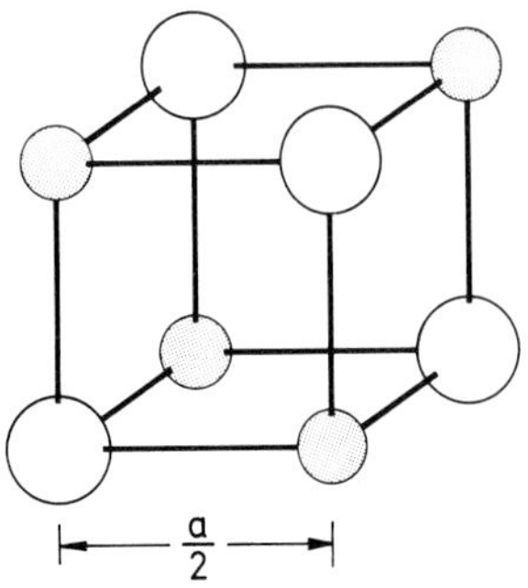

Fig. 2.2. Section of a NaCl lattice. The unit cell of the face-centred cubic lattice is a cube of side a. It contains one face-centred cubic cell each of Na^+ and of Cl^- ions

$$n = (1/2)(2/a)^3 . \quad (2.10)$$

When we set this equal to the quotient $N_A/V_{mol} = N_A \varrho/M$, we obtain

$$N_A = \frac{4M}{a^3 \varrho} = \frac{4 \cdot 58.4}{(5.63)^3 \cdot 10^{-24} \cdot 2.16} = 6.05 \cdot 10^{23}\,\mathrm{mol}^{-1} \quad \text{or} \quad (2.11)$$

$$6.05 \cdot 10^{26}\,(\mathrm{kmol})^{-1} \quad \text{with} \quad a = 5.63 \cdot 10^{-8}\,\mathrm{cm} \quad \text{and} \quad \varrho = 2.16\,\mathrm{gcm}^{-3} .$$

The accuracy of a measurement of N_A by this method is about $5 \cdot 10^{-6}$ (relative uncertainty). The density ϱ cannot, up to now, be determined with greater accuracy. However, the lattice constant a can be obtained with much greater accuracy using an x-ray interferometer of Si single crystals, resulting in a relative error of $6 \cdot 10^{-8}$. This method becomes an absolute technique for determining N_A if the measurement of the x-ray wavelength is made using a mechanically ruled grating and can thus be related to the meter directly. This becomes possible with the method of grazing-incidence diffraction; in the normal-incidence method, the mechanically prepared rulings are too wide relative to the wavelength.

2.3.4 Determination Using Radioactive Decay

Among the many other methods with which N_A has been determined, we will only mention here that of *Rutherford* and *Royds* from the year 1909.

In the experimental setup shown in Fig. 2.3, a radon source is contained in the interior of the glass tube A. The α particles which are emitted by this source can pass through the thin walls of tube A. In the second, thick-walled tube B, the α particles,

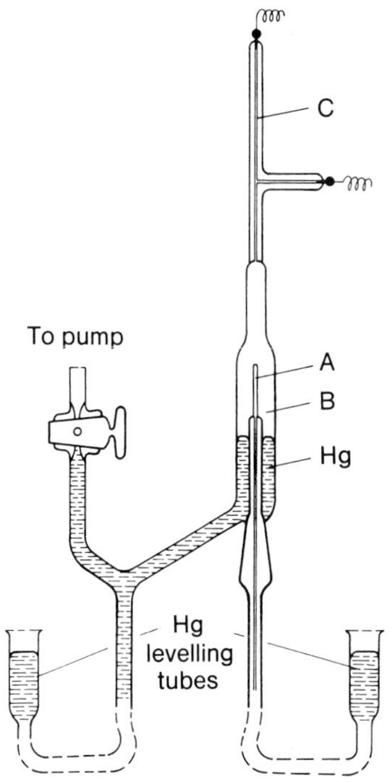

Fig. 2.3. Experimental arrangement of *Rutherford* and *Royds*: Phil. Mag. **17**, 281 (1909). The thin-walled glass tube A contains the α-active gas radon, $^{222}_{86}Rn$. The helium atoms which collect after some days in the evacuated space B are compressed into the capillary C and detected in the spectrum of a gas discharge. The mercury levelling vessels serve to compress the gases

which are the atomic nuclei of helium atoms, collect as He gas. Through ionisation processes in the source, in the glass walls, and in the gas, electrons are set free which can combine with the α particles and make them into He atoms. Ignition of a gas discharge in tube C excites these atoms and causes them to emit light; spectral analysis shows that the gas is, in fact, helium.

In this manner, it was demonstrated that α particles are helium nuclei. If one measures the quantity of gas which is formed in a certain time, and knows the decay rate of the source (e.g., by counting with a Geiger counter or scintillation detector), one can determine the number of atoms per unit volume and thus N_A.

2.4 Determination of the Size of the Atom

2.4.1 Application of the Kinetic Theory of Gases

The kinetic theory of gases describes the macroscopic state variables of gases such as pressure and temperature on an atomic basis. Its application to the explanation of the macroscopically measurable quantities relevant to gases also leads to a determination of the size of the atoms. To understand this, we must first recall the arguments which provide convincing evidence for the correctness of the kinetic theory.

The ideal-gas law states

$$pV = nRT\,, \tag{2.12}$$

where p is the pressure, V the volume, n the number of moles, R the universal gas constant, and T the temperature.

At constant temperature, this is Boyle's law. Equation (2.12) can also be derived kinetically. To do this, one calculates the number of particles in a given volume which collide with a unit surface of the walls per unit time and thereby transfer momentum to the walls (this is the number of particles contained in the so-called Maxwellian cylinder of length v). The pressure which is exerted by the gas on the walls is then given by

$$p = (1/3)Nm\overline{v^2}\,, \tag{2.13}$$

where m is the mass of the particles (gas atoms or molecules), $\overline{v^2}$ is their mean-square velocity, and N is the number of particles per unit volume. Since the mean kinetic energy $m\overline{v^2}/2$ of a free particle in thermal equilibrium is equal to $(3/2)\,kT$, (2.13) becomes $p = NkT$. This equation is identical to the ideal-gas law, as one immediately recognises upon multiplication by the molar volume V_{mol}:

$$pV_{mol} = NV_{mol}kT = N_AkT = RT\,. \tag{2.14}$$

The demonstration that the kinetic theory gives a good description of the physical behaviour of gases is provided by experimental testing of the predictions of the theory. For example, the distribution of the molecular velocities in a gas which can be derived from the kinetic theory (Maxwell distribution) has been experimentally verified with great accuracy. This distribution is again a special case of the Boltzmann distribution [cf. (2.8)]. Here the energy of a particle is $E = mv^2/2$. We wish to calculate the number of particles, $n(v)dv$, whose *absolute* velocity, independent of direction, lies in the interval $v \ldots v+dv$. Thus we must recalculate the density function $g(E)$ in terms of a new density function $\tilde{g}(v)$, using the condition

$$g(E)\,dE = \tilde{g}(v)\,dv\,.$$

Since the calculations yield no physical insights, we will only give the end result, the Maxwellian velocity distribution:

$$n(v)\,dv = n_0 v^2 \sqrt{\frac{2}{\pi}} \left(\frac{m}{kT}\right)^{3/2} \mathrm{e}^{-mv^2/2kT} dv \qquad (2.15)$$

with $n(v)\,dv$ being the number of particles with a velocity in the interval $v \ldots v+dv$ and n_0 the total number of particles. In the experimental test of the velocity distribution, the relative number of gas atoms with a given velocity v is measured.

2.4.2 The Interaction Cross Section

The size of an atom in a gas may be measured from the *interaction cross section* with which the atom collides with other atoms. The derivation of the concept *interaction cross section* is illustrated in Fig. 2.4. A beam of atoms of type 1 (beam cross-sectional area A, particle radius r_1, particle number density N_0) strikes a layer made of atoms of type 2 (layer thickness Δx, particle radius r_2, particle number density n). We ask, "How many atoms of type 1 collide with those of type 2 and are deflected from their course, so that they do not pass undisturbed through the layer?" This interaction cross section is thus frequently referred to in physical language as a *scattering cross section*.

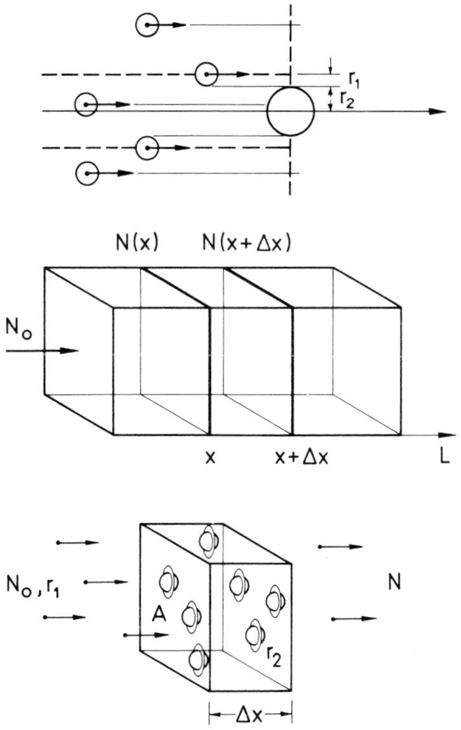

Fig. 2.4. Definition and measurement of interaction cross sections. *Upper part:* The interaction cross section of particles with radius r_1 with those of radius r_2 is found to be $(r_1+r_2)^2\pi$. *Centre part:* Out of N_0 particles which reach the volume element of thickness Δx from the left, N pass through undeflected. In this manner, the interaction cross section may be experimentally determined. *Lower part*: Derivation of (2.20). The radii of particles 1 and 2 are combined into the radius (r_1+r_2)

The problem corresponds roughly to the following macroscopic situation: someone is standing in front of a forest and shoots a bullet from a rifle. The probability that the bullet will pass through the forest undeflected is larger, the smaller the thickness and the density of the trees. If one shoots repeatedly and counts the number of undeflected bullets relative to the total number fired, it is possible to determine the thickness of the trees (that is, their interaction cross section). In order to do so, one must know their density. Naturally, one would not use such a method for a real forest. In atomic physics, it is, however, necessary, since it is not possible to put a meter stick up to an atom as it is to a tree. The same problem occurs in a number of contexts in physics: for example, in nuclear physics, a collision between two particles can be used to determine the interaction cross section for a nuclear or particle reaction. In atomic physics, we shall see that the interaction between a light quantum and an atom is described by a cross section. Because of its wide application in many areas of physics, the concept of the interaction cross section will be treated in some detail here.

A collision between atoms of radii r_1 and r_2 leads to a deflection of the atoms out of their initial directions when it occurs within an area $\sigma = (r_1 + r_2)^2 \pi$ (see Fig. 2.4). We may thus combine the deflection of both colliding particles into a common cross section. The probability of a collision is then given as the quotient of the number of favorable to the number of possible cases:

$$W = \frac{\text{Area of all the interaction cross sections in the volume of the beam}}{\text{Total area } A} .$$

This is valid under the assumption that the areas πr^2 of various particles which are located behind one another do not overlap. This is fulfilled for a sufficiently small layer thickness. In order to calculate the number of deflected atoms in a finite layer of thickness L, we first divide up the layer into thin layers of thickness Δx. If N atoms enter a thin layer at the position x (see Fig. 2.4), a number ΔN is deflected out of the beam after passing through the distance Δx:

$$\Delta N = -WN = -\frac{\text{Total number of atoms in the volume} \cdot \sigma}{\text{Total area}} \cdot N . \qquad (2.16)$$

Since the total number of atoms in a given volume is given by the product of particle number density n with the area A and the layer thickness Δx, we obtain from (2.16)

$$\Delta N = -\frac{nA\,\Delta x\,\sigma}{A} N . \qquad (2.17)$$

If we replace differences by the corresponding infinitesimal quantities, we have

$$dN/N = -n\sigma\,dx . \qquad (2.18)$$

To obtain the number of atoms which are deflected (or not deflected, respectively) along the entire length x, we integrate (2.18):

$$\ln N = -n\sigma x + \ln N_0 . \qquad (2.19)$$

Here, $\ln N_0$ is a constant of integration, with N_0 being the number of particles which are incident at the point $x = 0$. From this relation we obtain immediately

$N = N_0 \exp(-n\sigma x)$ as the number of particles which are still present after a distance x, or, after passing through a total length L

$$N = N_0 \mathrm{e}^{-n\sigma L} . \tag{2.20}$$

The number of deflected atoms is correspondingly

$$N_{\mathrm{scatt}} = N_0(1 - \mathrm{e}^{-n\sigma L}) . \tag{2.21}$$

The product $n\sigma = \alpha$ is also denoted as the (macroscopic) *scattering coefficient* and σ as the (microscopic) total interaction cross section.

From a measurement of σ follows, according to $\sigma = (r_1 + r_2)^2\pi$, the quantity $(r_1 + r_2)$. In the case of identical atoms with $r = r_1 = r_2$, we have thus determined r, i.e., the *size* of the atoms.

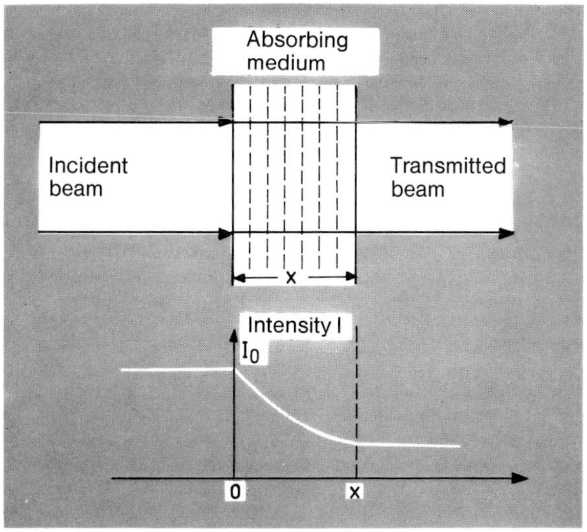

Fig. 2.5. Attenuation of a light beam on passing through an absorbing medium

We will frequently encounter the concept of the interaction cross section, which we have defined here, in later sections of this book. Figure 2.5 shows the dependence of the *intensity* of a light beam on the thickness of absorbing medium through which the beam has passed, as described by (2.20). For the absorption of *light* by atoms or molecules, the Lambert-Beers law is valid:

$$I = I_0 \mathrm{e}^{-n\alpha x}, \tag{2.22}$$

where I is the transmitted intensity, I_0 the incident intensity, and α the absorption coefficient *per absorbing particle.* n is again the number density of atoms or molecules in the absorbing medium.

2.4.3 Experimental Determination of Interaction Cross Sections

Interaction cross sections can be directly measured by collision experiments using an atomic beam and a gas target. An apparatus for such measurements is shown in Fig. 2.6.

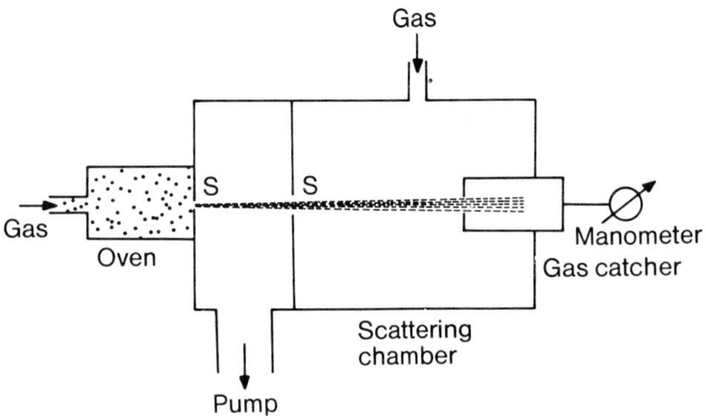

Fig. 2.6. Setup for measuring scattering cross sections of atoms by atoms. A beam of gas atoms enters the scattering chamber through the slits S. Scattering by the gas atoms in the chamber leads to an attenuation of the beam which arrives at the catcher

Frequently, however, interaction cross sections or atomic sizes are determined indirectly. For example, one measures the mean free path λ, which we define with the help of (2.20): λ is the distance L or x, after which the initial density N_0 has been reduced to N_0/e. Thus, with (2.20), where n is again the particle number density, and taking $r_1 = r_2 = r$,

$$\lambda = \frac{1}{4\pi r^2 n} = \frac{1}{\sigma n}. \qquad (2.23)$$

Up to now, we have assumed that the target atoms are at rest. If they are also in motion, the expression for λ must be modified somewhat. We give the result without derivation:

$$\lambda = \frac{1}{4\pi \sqrt{2} r^2 n}. \qquad (2.24)$$

The mean free path thus defined enters into macroscopically measurable quantities, for example the viscosity η. The viscosity is in fact a measure of the momentum transfer between atoms or molecules in gases or liquids and therefore also depends on the frequency of collisions between the particles. The mean free path can thus also be macroscopically determined. The detailed relation is (without derivation)

$$\eta = \tfrac{1}{3} \varrho \lambda \bar{v}, \qquad (2.25)$$

where η is the viscosity, ϱ the density, λ the mean free path, and $\bar{v}$ the mean velocity of the particles. The quantity η can be measured, e.g., from the flow velocity through a capillary.

Another method for measuring λ results from thermal conductivity in gases. The latter also depends on the frequency of collisions between the particles and the energy transfer which thus occurs. Here we have – likewise without derivation – a relation between the heat flow dQ/dt and the thermal gradient dT/dx which produces it:

$$dQ/dt = -\lambda_{TC} \cdot A \cdot dT/dx \,, \tag{2.26}$$

where dQ is the differential quantity of heat, λ_{TC} the thermal conductivity, and A the cross-sectional area of the heat transport medium. The thermal conductivity λ_{TC} depends upon the mean free path according to the relation

$$\lambda_{TC} = \frac{n}{2} k \bar{v} \lambda \,. \tag{2.27}$$

Table 2.2 on p. 20 contains some values for atomic radii. Further details and the derivations which we have passed over here may be found in standard texts on experimental physics.

2.4.4 Determining the Atomic Size from the Covolume

The Van der Waals equation for one mole of a real gas states

$$(P + a/V^2)(V - b) = RT \,. \tag{2.28}$$

Here the expression a/V^2 denotes the "internal pressure" which adds to the external pressure P and is due to the forces between the particles. Another correction due to the internal forces is the reduction of the free volume V of the gas by the volume b (the so-called covolume). This quantity b, which can be experimentally determined by measuring the P-V diagram of the equation of state, is equal to the fourfold volume of the particles. We thus have

$$b = 4 \cdot \frac{4\pi}{3} \cdot r^3 \cdot N_A \,. \tag{2.29}$$

2.4.5 Atomic Sizes from X-Ray Diffraction Measurements on Crystals

The famous experiment of *von Laue, Friedrich*, and *Knipping* in 1912 on the diffraction of x-radiation in crystals yielded:
- the final proof that crystals are built up of atoms,
- the wavelength of x-radiation,
- the lattice constant of crystals – and with it, information on the size of the atoms in the crystal.

Figure 2.7 shows the experimental set-up schematically. For an exact derivation of the interference conditions, one would have to treat the interference from a three-dimensional lattice. Here we will use the simplified method of *Bragg* (1913) to illustrate how the lattice constants can be determined.

X-ray diffraction may be regarded as a reflection of x-radiation by the so-called lattice planes at certain specular angles. A lattice plane is a plane in a crystal which is occupied by atoms. In a crystal there is a large number of families of parallel and equi-

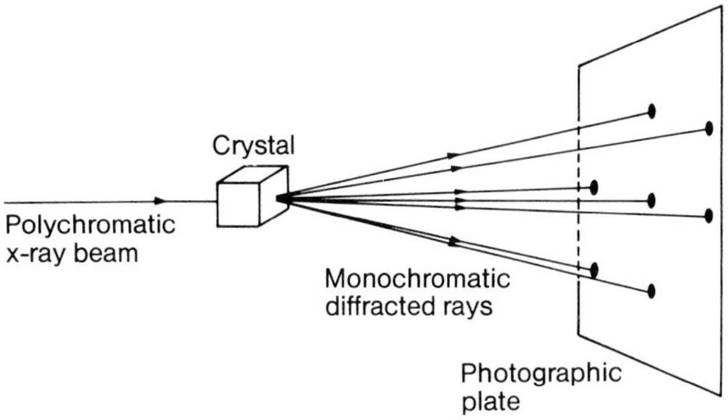

Fig. 2.7. X-ray diffraction from a single crystal after *von Laue*; schematic of the experimental arrangement. X-radiation with a continuous distribution of wavelengths (polychromatic or white x-radiation) is diffracted by a single crystal. The conditions for interference from a three-dimensional lattice yield constructive interference at particular directions in space and at particular wavelengths. One thus observes interference maxima, which correspond to certain discrete wavelengths (monochromatic x-radiation)

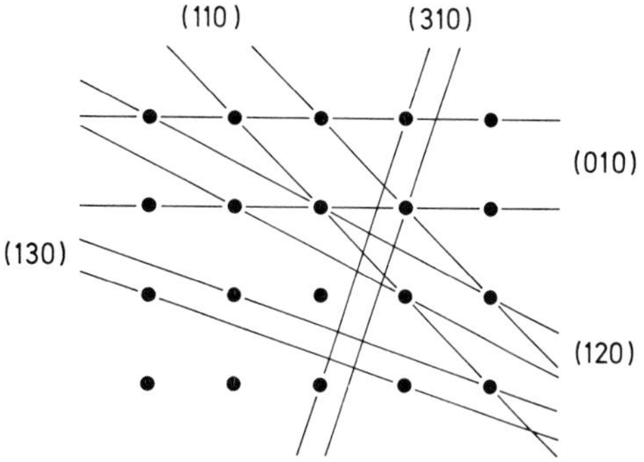

Fig. 2.8. Simple cubic lattice with several lattice planes. These are characterised by the *Miller Indices*. The spacing between two parallel lattice planes decreases with increasing Miller indices

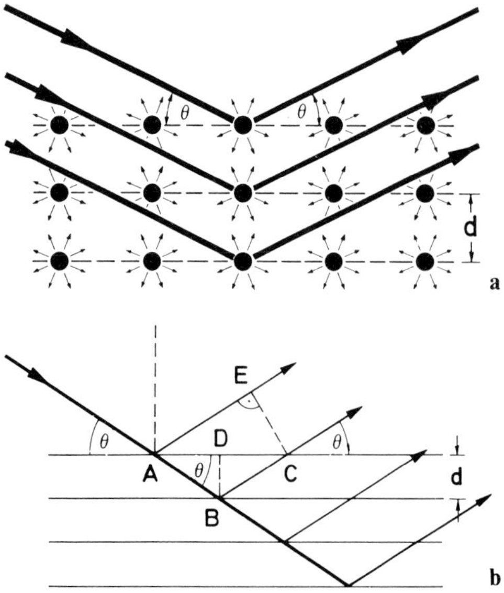

Fig. 2.9a, b. Derivation of the Bragg Law of Reflection. The horizontal lines symbolise lattice planes, from which the incident x-radiation arriving at angle θ is scattered. **a)** Each atom of a lattice plane acts as a scattering centre. **b)** The derivation of the Bragg condition for the reflection of x-radiation from a lattice plane

distant lattice planes. They are distinguished from one another by their spacing, by the density of atoms within the planes, and by their orientations within the crystal lattice (see Fig. 2.8). According to Huygens' principle, each atom which is struck by the incident x-radiation acts as the source point for a new elementary wave (Fig. 2.9a). These elementary waves produce constructive interferences at certain angles. The reflection condition is derived as follows: amplification occurs when the path difference Δ between two adjacent beams corresponds to a whole multiple of the wavelength, $n\lambda$. For the path difference Δ we have, according to Fig. 2.9b,

$$\Delta = \mathrm{AB} + \mathrm{BC} - \mathrm{AE} = 2\mathrm{AB} - \mathrm{AE} = \frac{2d}{\sin\theta} - 2\mathrm{AD}\cos\theta\,. \tag{2.30}$$

With the relation $\mathrm{AD} = d/\tan\theta$, one obtains from (2.30)

$$\Delta = 2\frac{d}{\sin\theta}(1-\cos^2\theta)\,,$$

or, finally, the condition for constructive interference

$$\Delta = 2d\sin\theta = n\lambda\,. \tag{2.31}$$

The various methods of observing x-ray diffraction from crystals which are used in practice differ in the following ways:

- In the *Laue* method one uses a single crystal, a particular value of the angle of incidence, and x-radiation with a continuous spectrum ("polychromatic" x-rays). The condition for constructive interference is fulfilled for individual *points* in the plane of observation for particular wavelengths.
- In the *Bragg* rotating-crystal method one also uses a single crystal, but monochromatic x-rays. The crystal is rotated, so that the angle of incidence varies through a continuous range of values. The condition for constructive interference is fulfilled for various lattice planes successively.
- In the *Debye-Scherrer* method (Figs. 2.10, 11), the sample is polycrystalline or powdered. The x-rays are monochromatic. Since each lattice plane occurs in all possible orientations relative to the incident beam, one obtains interference *cones* whose intersection with the plane of observation gives interference rings.

Equation (2.31) relates the wavelength of the x-rays to the lattice constant or the spacing of the lattice planes. The x-ray wavelength can be measured by other means than with crystal interferences. Its measurement can be directly correlated to the meter by utilising x-ray interference at grazing incidence from a diffraction grating. Since it is not possible to manufacture diffraction gratings with a grating constant of the order of x-ray wavelengths, one uses coarse gratings, for example with 50 lines/mm, and lets the x-radiation strike the grating at a grazing angle of less than 1°. The effective grating constant is then the projection of the actual line spacing at this angle. It is sufficiently small to permit the measurement of the x-ray wavelength.

We make two additional remarks concerning x-ray diffraction.

- In practice, x-ray diffraction is much more complicated than indicated above. The exact intensity distribution in the diffraction pattern must be carefully determined, and account must be taken of the fact that the scattering centres are not points, but instead are electronic shells with a finite extension. A complete, quantitative ana-

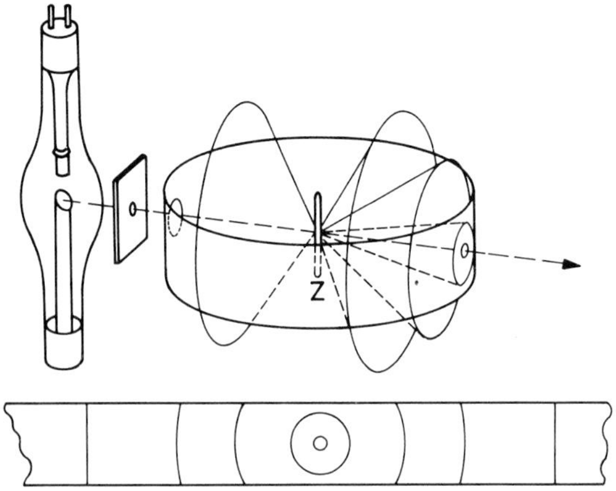

Fig. 2.10. Debye-Scherrer method: x-ray diffraction of monochromatic x-radiation by a polycrystalline sample Z. On the film, the intersections of the diffraction cones from the various families of lattice planes appear as rings

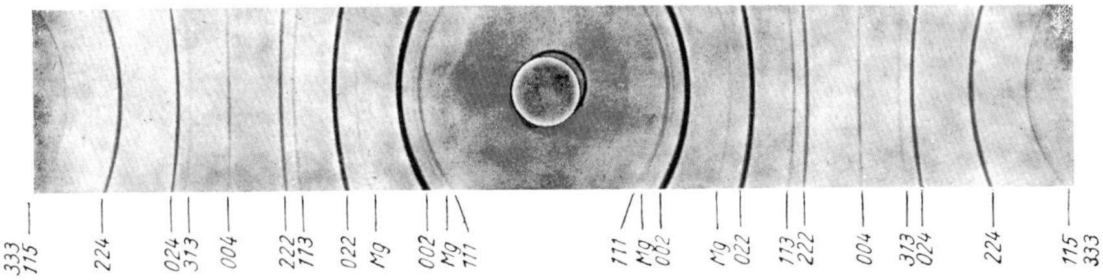

Fig. 2.11. Debye-Scherrer diagram of MgO [from Gerthsen, Kneser, Vogel: *Physik,* 13th ed. (Springer, Berlin, Heidelberg, New York 1978) Fig. 12.37]

lysis of x-ray diffraction patterns leads finally to an exact determination of the electron density with the sample crystal lattice. From it, we obtain not only the spacing between the atoms in the lattice, but also their sizes and even their shapes. Figures 2.12 and 2.13 illustrate experimentally determined electron density distributions in crystals. A contour map of this type raises the question, "Where does an atom have its boundary?", and this leads in turn to the question,

"What do we really mean by the *size* of an atom?"

– In the case of hard spheres, the size can be defined exactly. For atoms, the concept "size" cannot be defined without reference to the method of measurement. Various methods are sensitive to different properties of the atom, which depend on the "size of the atom" in differing ways.

Let us consider the methods of investigation described above once more in light of this remark.

From the viscosity η one obtains a measure of the interatomic distance in the presence of thermal motion. Because the atoms are not perfectly hard spheres, the radius determined in this manner will, however, be a function of the velocity. Furthermore, the results depend on the shape of the atom; the spatial extension of the electronic shells of atoms and molecules deviates more or less strongly from a spherical shape, in general.

The covolume b in the real-gas law is derived under the assumption that the atoms are elastic spheres. The lattice plane spacing d measures an equilibrium distance between the particles in the crystal lattice.

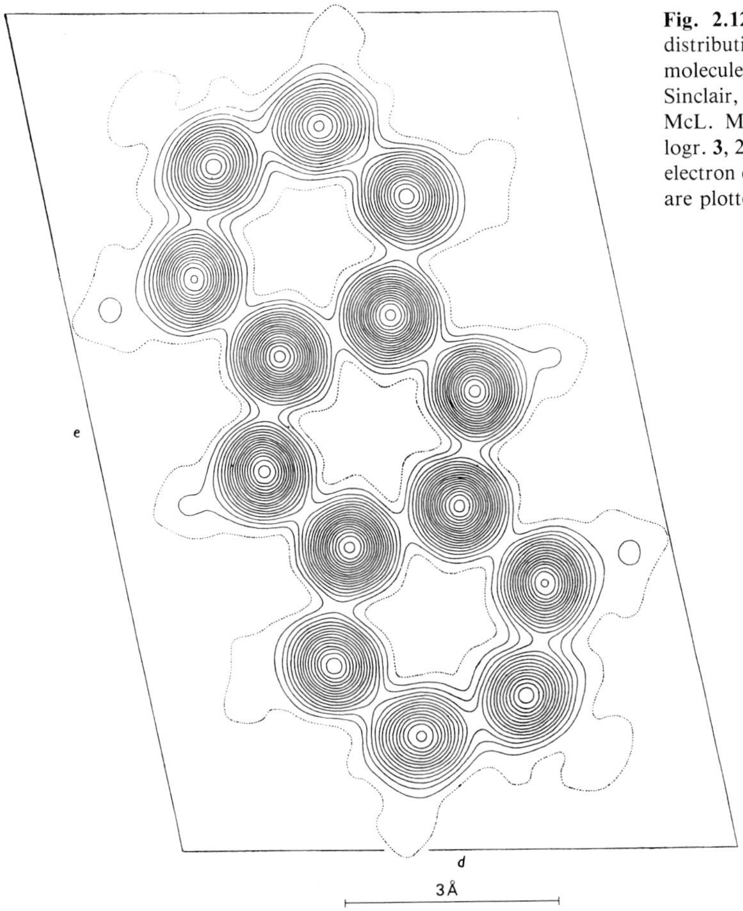

Fig. 2.12. The electron density distribution in the anthracene molecule determined by V. L. Sinclair, J. M. Robertson, A. McL. Mathieson: Acta Crystallogr. **3**, 254 (1950). Lines of equal electron density ("contour lines") are plotted

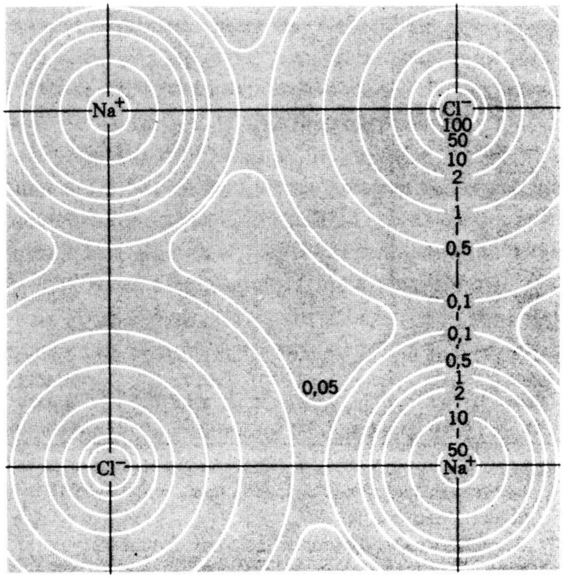

Fig. 2.13. Distribution of the electron density in the basal plane of NaCl from x-ray analysis by Schoknecht: Z. Naturforsch. **12a**, 983 (1957). The solid curves are again lines of equal electron density

It should therefore not be surprising that the values of atomic radii measured by the various methods deviate somewhat from each other – the order of magnitude is, however, always the same – a few Ångstroms. Table 2.2 shows a comparison of the measured values.

Table 2.2. Atomic radii [Å], measured by various methods (1 Å = 0.1 nm)

	from η	from d	from b
Neon	1.18	1.60	1.2
Argon	1.44	1.90	1.48
Krypton	1.58	1.97	1.58
Xenon	1.75	2.20	1.72

2.4.6 Can Individual Atoms Be Seen?

The *resolving power* of a microscope is defined as the smallest spacing between two structures in an object which can still be imaged separately. According to Abbé's theory of image formation, the resolving power is limited by diffraction in the opening of the lens which forms the image. In texts on optics, the condition that – in addition to the zeroth order – at least one additional diffraction maximum is necessary in order to form an image, is used to derive the equation for resolving power,

$$d = \frac{\lambda}{n \sin \alpha}, \tag{2.32}$$

where d is the resolving power, λ the wavelength, n the index of refraction, and α the angular opening of the lens. For visible light, one obtains a resolution of ca. 5000 Å or 500 nm.

For other types of electromagnetic radiation, the theoretical resolving power cannot be reached. For x-rays, it is not possible to construct suitable lenses, since the index of refraction of all substances for x-radiation is approximately equal to 1. Electrons may be deflected by electric and by magnetic fields; thus, they may be used to construct lenses for electrons and to form images. Because of the unavoidable "lens aberrations", however, it is only possible to work with beams of very small divergence in electron microscopes. Table 2.3 gives an overview of the resolving powers of various methods of image formation.

In recent years, success in the effort to form images of individual atoms has been obtained with two special types of electron microscopes: with the *field emission microscope* it has been possible to visualize single atoms or large molecules on the tips of fine metal points (Fig. 2.14), and with the *scanning electron microscope* it has proved possible to form images of atoms and molecules. Here the resolution is about 5 Å or 0.5 nm. With high-voltage electron microscopes, one can now obtain a resolution of 0.2 nm. This makes it possible to image the individual atoms in molecules and in crystals. An example is shown in Fig. 2.15.

Table 2.3. Resolving powers for various wavelengths

	Resolving Power [Å]		Remarks
	theory	practice	
Light	ca. 5000	ca. 5000	
Dark field	ca. 500	ca. 500	No image formation, only diffraction pattern
X-rays ($\lambda = 1$ Å)	1	several 100	No lenses
Electrons (100000 Volt $\triangleq \lambda = 0.037$ Å)	0.04	2 – 5	Lens aberrations

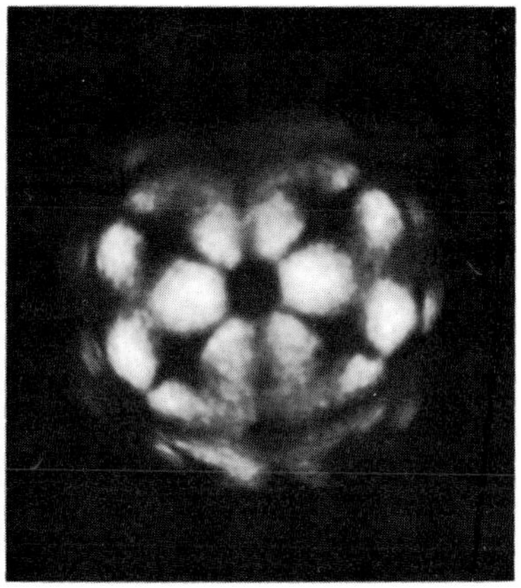

Fig. 2.14. Image of the point of a tungsten needle with a field emission microscope, developed by *F. W. Müller*. The picture was provided by the Leybold-Heraeus Co. in Cologne. The image can easily be produced as a lecture demonstration. The various lattice planes which intersect the tip of the needle have differing emission probabilities for the field emission of electrons; this produces the geometric pattern of light and dark regions. This pattern reflects the crystal lattice geometry. Barium atoms were then vapour-deposited onto the tungsten needle. Where they are present, the emission probability is increased, so that they appear as bright points in the image. During the vapour deposition one can observe them falling onto the point like snowflakes. It can be shown in this manner that individual atoms become visible

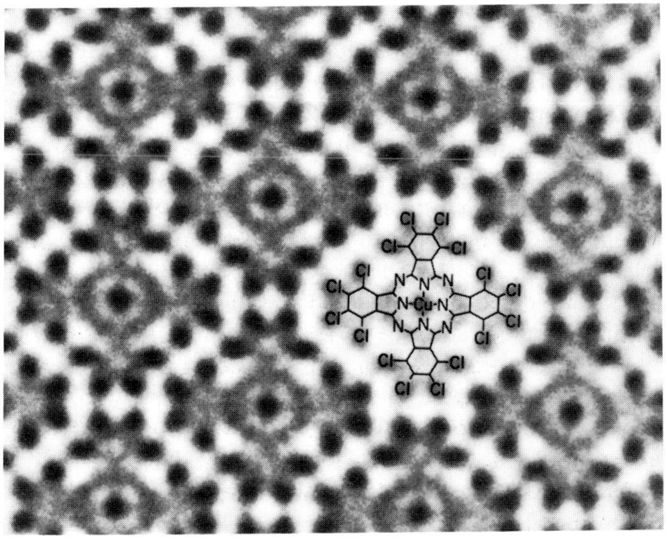

Fig. 2.15. An electron microscope picture of hexa-deca-chloro-copper-phthalocyanin molecules. The molecules form a thin crystalline growth-layer on the alkali halide crystal which serves as substrate. The image formation and processing were done with a 500 kV high-resolution electron microscope and with a special image enhancement technique. The central copper atoms are especially clear, as are the 16 peripheral chlorine atoms. (The picture was kindly placed at our disposal by Prof. *N. Uyeda*, Kyoto University)

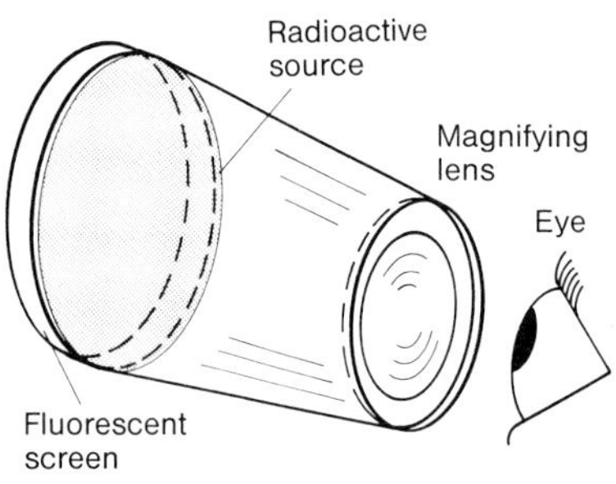

Fig. 2.16. Spinthariscope, schematic illustration. The fluorescent screen scintillates due to the irradiation from the radioactive source. The scintillation processes may be observed through the magnifying lens

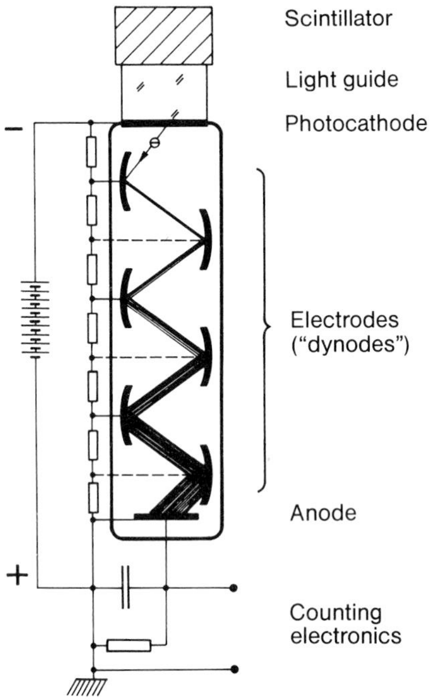

Fig. 2.17. Schematic diagram of a scintillation detector. The light flashes which are produced in the scintillator by the incident radiation pass through the light guide to the photocathode of an electron multiplier tube. The photoelectrons which are released by the cathode are amplified in a series of steps and registered at the anode as a current pulse

Although it is possible at present only in special cases to make individual atoms visible, there are experimental methods for observing processes in which only single atoms take part.

Single atomic processes (nuclear decays) were made visible in the first years of research into radioactive decay by using the "spinthariscope" (Fig. 2.16). This is nothing more than a fluorescent screen, which produces light flashes upon bombardment with decay products from radioactive material and which may be observed with a magnifying lens. With this instrument, single atomic events – decays – were counted in Rutherford's laboratory at the beginning of this century. Today, scintillation detectors or semiconductor detectors are used for this purpose. When radiation from a radioactive decay falls on a NaI crystal, the crystal produces light flashes, which can be amplified in a photomultiplier tube (PMT) (Fig. 2.17). In this way, individual events can be conveniently registered. For example: one electron with an energy of 10000 eV produces ca. 200 light quanta in the scintillator (it requires on average about 50 eV per light quantum). Each light quantum creates one photoelectron at the photocathode of the PMT. The PMT amplifies each of these electrons about 10^5-fold, so that per light quantum, about 10^5 electrons are released from the anode. This results in a charge per incident electron (beta particle) of $3 \cdot 10^{-12}$ C, which can easily be measured.

An arrangement which played an especially important role in the early period of modern atomic physics and which is still in use today for the excitation of atoms and for producing particle beams is the *gas discharge tube*, Fig. 2.18. It can be employed both for exciting the emission of light from the atoms of the gas inside the tube and for the production of cathode and canal rays. *Plücker* described *cathode rays* for the first time in 1859. They were given that name because they could be observed through a hole in the anode and seemed to emanate from the cathode. In fact, they are generated in the gas volume. The rays which strike the fluorescent screen in front of the hole in the

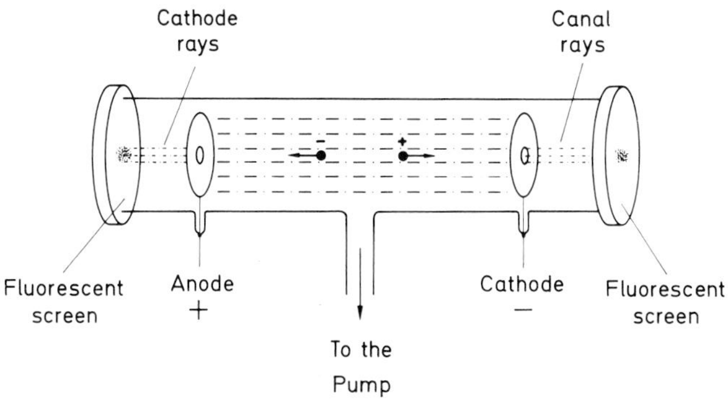

Fig. 2.18. A gas discharge tube for producing cathode rays. Between the cathode and the anode is a potential difference of several thousand volts. The fluorescence which appears in the tube and the formation of cathode rays depend strongly on the gas pressure. In the field-free region between the cathode and the fluorescent screen one observes the canal beam, which consists of positive ions. The cathode and canal rays produced in this manner were particularly important in the early period of atomic physics for the investigation of charge, mass, and scattering cross sections of electrons and ions

cathode, the *canal rays*, were discovered in 1886 by *Goldstein*. In 1897, *Thomson* showed that the cathode rays consists of negatively charged particles – the *electrons*. *Wien* demonstrated in 1900 that the canal rays are electrically charged atoms, that is, ions.

Atoms as the basic particles of matter have been thus theoretically and experimentally detected and made visible in the course of the past century.

Problems

2.1 a) Colloidal particles are dispersed in a liquid. Show that at equilibrium, the number of particles per cm^3 is given by a Boltzmann distribution:

$$n(h) = n_0 \exp \frac{N_A}{RT} V(\varrho - \varrho')gh ,$$

where N_A is Avogadro's number, R is the gas constant, T is the absolute temperature, V is the volume of the particles, ϱ is the density of the particles, ϱ' is the density of the liquid, g is the acceleration due to gravity, h is the height and n_0 is the number of particles at the height $h = 0$.

b) Determine Avogadro's number, using the above relation and the following experimental data:

$n_0 = 134$ particles/cm^3, $n\,(h = 0.0030$ cm$) = 67$ particles/cm^3, $\varrho = 1.23$ g/cm^3, $\varrho' = 1.00$ g/cm^3, $T = 293$ K, particle diameter $= 4.24 \cdot 10^{-5}$ cm.

Hint: To derive an expression for $n(h)$, use the barometric altitude formula: $dp = -\varrho(h)g\,dh$. Treat the particles as heavy, non-interacting molecules of an ideal gas and use the ideal gas equation $pV = RT$ to obtain the relation between dp and $d\varrho$ or dn.

2.2 Liquid helium (atomic weight 4.003) has a density $\varrho = 0.13$ g/cm^3. Estimate the radius of a He atom, assuming that the atoms are packed in the densest possible configuration, which fills 74% of the space.

2.3 Canal rays, i.e., positive ion rays are generated in a gas discharge tube. How often does an ion ($r = 0.05$ nm) collide with an atom of the ideal filler gas ($r = 0.1$ nm) if it travels 1 m in a straight path through the discharge tube and if the pressure in the tube is 1 mbar? 10^{-2} mbar? 10^{-4} mbar and the temperature $T = 300$ K?

2.4 The covolume of helium gas was determined from pressure-volume diagrams to be $b = 0.0237$ litre/mole. The covolume of mercury is 0.01696 litre/mole. What is the size of the atoms in the two gases?

2.5 Why are monochromatic x-rays used for the Debye-Scherrer method, and how are they produced? Does the diffraction cone with the smallest apex angle represent the smallest or the largest lattice plane spacing? How large is this spacing if a first-order angle $\alpha = 5°$ is measured between the surface of the cone and the undiffracted beam? (Assume that the quantum energy of the x-rays is 50 keV, $E_{\text{x-ray}} = 50$ keV).

2.6 Monochromatic x-rays ($\lambda = 0.5$ Å) fall on a sample of KCl powder. A flat photographic plate is set up 1 m from the powder and perpendicular to the beam. Calculate the radii of the sections of the Bragg diffraction cone (Fig. 2.10) for first- and second-order diffraction, assuming a lattice-plane spacing of 3.14 Å.

2.7 A tight bunch of slow neutrons (2 eV), which is produced in a nuclear reactor, lands on a crystal with a lattice spacing of 1.60 Å. Determine the Bragg angle for first-order diffraction.

Hint: Use (7.1) for the wavelength of the neutrons.

3. Isotopes

3.1 The Periodic System of the Elements

One of the early significant achievements of atomic physics in the past century – or rather, of chemists working together with the physicists – was the explanation of the periodic system of the chemical elements on the basis of atomic structure.

This system (Table 3.1) is constructed by listing the atoms according to increasing nuclear charge number (or atomic number), Z. In the process, the chemical properties of the atoms are taken into account, so that chemically similar atoms are placed under each other in columns. With this procedure, we find eight vertical columns with subgroups and seven horizontal rows or periods. Each position is occupied by an atom which belongs there because of its chemical properties. To be sure, in this system all fourteen rare earths would have to be placed in the same position, i.e., at $Z = 57$, and all the actinides in position $Z = 89$. Both the periodicity and the above mentioned discrepancies will be explained on the basis of the electronic structure of the atoms near the end of the book in Chap. 19. Using the heavy-ion accelerator in Darmstadt (F.R. Germany) additional elements having the atomic numbers 107, 108, and 109 were artificially produced in the period from 1981 to 1984. Their atomic nuclei are unstable and decay within a few milliseconds.

The periodic system is an ordering of the elements according to periodically recurring chemical as well as physical properties. As an example of the latter we show here the atomic volumes and the ionisation energies as functions of the nuclear charge Z (Fig. 3.1). Chemical properties which periodically repeat themselves are, for example, the monovalence of the alkali atoms or the lack of reactivity of the rare gases. These empirical regularities indicate corresponding regularities in the atomic structure.

A first attempt at an explanation was the hypothesis of *Prout* (1815): all atoms are made up of hydrogen atoms. This picture was refined and modified as further ele-

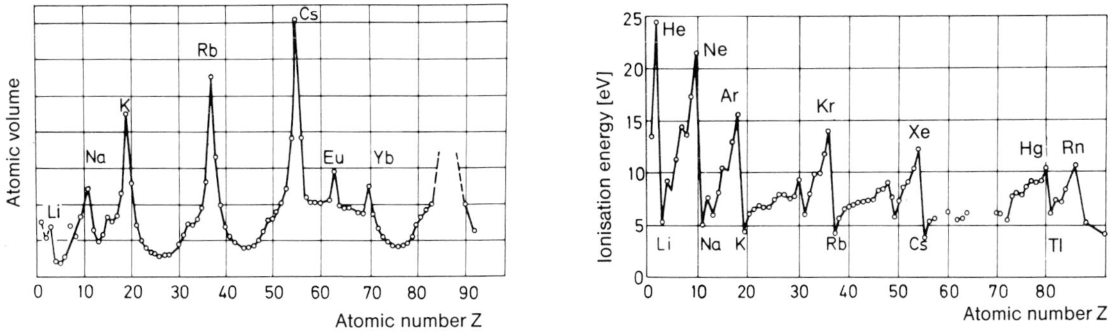

Fig. 3.1. Atomic volumes and ionisation energies as functions of the position in the periodic system of the elements. Particularly noticeable are the (relatively) large atomic volumes of the alkali metal atoms and the large ionisation energies of the noble gas atoms

Table 3.1. Periodic system of the elements

1*s*	**H** 1 1.008 1																	**He** 2 4.0026 2	1*s*
2*s* 2*p*	**Li** 3 6.939 1 —	**Be** 4 9.012 2 —											**B** 5 10.81 2 1	**C** 6 12.01 2 2	**N** 7 14.01 2 3	**O** 8 16.00 2 4	**F** 9 19.00 2 5	**Ne** 10 20.18 2 6	2*s* 2*p*
3*s* 3*p*	**Na** 11 23.00 1 —	**Mg** 12 24.31 2 —											**Al** 13 26.98 2 1	**Si** 14 28.09 2 2	**P** 15 30.97 2 3	**S** 16 32.06 2 4	**Cl** 17 35.45 2 5	**Ar** 18 39.95 2 6	3*s* 3*p*
3*d* 4*s* 4*p*	**K** 19 39.10 — 1 —	**Ca** 20 40.08 — 2 —	**Sc** 21 44.96 1 2 —	**Ti** 22 47.90 2 2 —	**V** 23 50.94 3 2 —	**Cr** 24 52.00 5 1 —	**Mn** 25 54.94 5 2 —	**Fe** 26 55.85 6 2 —	**Co** 27 58.93 7 2 —	**Ni** 28 58.71 8 2 —	**Cu** 29 63.55 10 1 —	**Zn** 30 65.38 10 2 —	**Ga** 31 69.72 10 2 1	**Ge** 32 72.59 10 2 2	**As** 33 74.92 10 2 3	**Se** 34 78.96 10 2 4	**Br** 35 79.90 10 2 5	**Kr** 36 83.80 10 2 6	3*d* 4*s* 4*p*
4*d* 5*s* 5*p*	**Rb** 37 85.47 — 1 —	**Sr** 38 87.62 — 2 —	**Y** 39 88.91 1 2 —	**Zr** 40 91.22 2 2 —	**Nb** 41 92.91 4 1 —	**Mo** 42 95.94 5 1 —	**T** 43 98.91 6 1 —	**Ru** 44 101.07 7 1 —	**Rh** 45 102.9 8 1 —	**Pd** 46 106.4 10 — —	**Ag** 47 107.9 10 1 —	**Cd** 48 112.4 10 2 —	**In** 49 114.8 10 2 1	**Sn** 50 118.7 10 2 2	**Sb** 51 121.8 10 2 3	**Te** 52 127.6 10 2 4	**I** 53 126.9 10 2 5	**Xe** 54 131.3 10 2 6	4*d* 5*s* 5*p*
5*d* 6*s* 6*p*	**Cs** 55 132.9 — 1 —	**Ba** 56 137.3 — 2 —	**La** 57 138.9 1 2 —	**Hf** 72 178.5 2 2 —	**Ta** 73 181.0 3 2 —	**W** 74 183.9 4 2 —	**Re** 75 186.2 5 2 —	**Os** 76 190.2 6 2 —	**Ir** 77 192.2 7 2 —	**Pt** 78 195.1 9 1 —	**Au** 79 197.0 10 1 —	**Hg** 80 200.6 10 2 —	**Tl** 81 204.4 10 2 1	**Pb** 82 207.2 10 2 2	**Bi** 83 209.0 10 2 3	**Po** 84 (210) 10 2 4	**At** 85 (210) 10 2 5	**Rn** 86 (222) 10 2 6	5*d* 6*s* 6*p*
6*d* 7*s* 7*p*	**Fr** 87 (223) — 1 —	**Ra** 88 (226) — 2 —	**Ac** 89 (227) 1 2 —	**Ku** 104 (258) 2? 2? —	**Ha** 105 (260) 3? 2? —														

4*f* 5*d* 6*s*	**Ce** 58 140.1 2 — 2	**Pr** 59 140.9 3 — 2	**Nd** 60 144.2 4 — 2	**Pm** 61 (145) 5 — 2	**Sm** 62 150.4 6 — 2	**Eu** 63 152.0 7 — 2	**Gd** 64 157.3 7 1 2	**Tb** 65 158.9 8 1 2	**Dy** 66 162.5 10 — 2	**Ho** 67 164.9 11 — 2	**Er** 68 167.3 12 — 2	**Tm** 69 168.9 13 — 2	**Yb** 70 173.0 14 — 2	**Lu** 71 175.0 14 1 2	4*f* 5*d* 6*s*
5*f* 6*d* 7*s*	**Th** 90 232.0 — 2 2	**Pa** 91 231.0 2 1 2	**U** 92 238.0 3 1 2	**Np** 93 237.0 5 — 2	**Pu** 94 239.1 6 — 2	**Am** 95 (243) 7 — 2	**Cm** 96 (247) 7 1 2	**Bk** 97 (247) 9 — 2	**Cf** 98 (251) 10 — 2	**Es** 99 (252) 11 — 2	**Fm** 100 (257) 12 — 2	**Md** 101 (258) 13 — 2	**No** 102 (259) 14 — 2	**Lr** 103 (260) 14 1 2	5*f* 6*d* 7*s*

Key:

	Fe 26	—Element and atomic number
	55.85	—Atomic mass in u; for some unstable elements, the mass number of the most stable isotope is given in parentheses.
3*d* 4*s* 4*p*	6 2 —	Electron configuration; the filled shells of the previous periods are not shown. For example, the complete electron configuration of Fe is $1s^22s^22p^63s^23p^63d^64s^2$

mentary particles were discovered, first the electron, then the proton. Only after 1932 did it become clear that the atomic nucleus consists of neutrons as well as protons. The number of electrons of an atom is smaller than the mass number, since the nucleus contains just as many protons as the electronic shells have electrons, but it also contains neutrons.

The relative atomic masses A_{rel} could originally only be measured by chemical methods. By these means, it was determined that the addition of hydrogen atoms alone cannot explain the observed "atomic weights" without contradictions. If the model of Prout is correct, then the atomic weights must be integers. For the most part, they *are* integers to a good approximation; A and A_{rel} are nearly equal. However, there are counter examples: the relative atomic mass – the atomic weight – of chlorine, for example, is $A_{rel} = 35.5$ in the naturally occurring element. Furthermore, it was determined that lead from various ores had differing atomic weights. Today we know that this is due to the lead having been produced as the end product from different radioactive decay chains.

These observations were the starting point for investigations which led to the discovery of the *isotopes*. This term denotes the fact that atoms with differing mass numbers may belong to the same position in the periodic table, i.e., they may have the same nuclear charge number Z. The differing mass numbers result from the different numbers of neutrons in the atomic nuclei. The concept of isotopes will be treated in the following sections. The existence of isotopes was discovered and thoroughly investigated with the aid of *mass spectroscopy*.

3.2 Mass Spectroscopy

3.2.1 Parabola Method

The physical techniques for exact measurement of atomic masses and for separating atoms with differing masses are mostly methods for determining the ratio e/m, i.e., the ratio of charge to mass. For this purpose one uses the deflection of ionized atoms moving through electric fields $\boldsymbol{E}$ and magnetic fields $\boldsymbol{B}$.

The oldest and most easily understood method is the parabola method of *Thomson* (1913). An ion beam from a gas discharge passes through the electric field of a condenser and a magnetic field $\boldsymbol{B}$ which is oriented parallel to the electric field (Fig. 3.2). In the plane of observation, particles of the same charge and mass, but having different velocities v, are distributed along a parabola whose origin is in the point where the undeflected beam would pass. This can be shown in the following manner: The homogeneous electric field $\boldsymbol{E}$, which is applied in the y direction, causes a deflection in this direction. The y coordinate of the particles changes according to the equation for the acceleration:

$$\ddot{y} = (e/m) \cdot E\,. \tag{3.1}$$

The y coordinate itself is given by the solution of (3.1),

$$y = \frac{1}{2}\,\frac{eE}{m}\,t^2 = \frac{e}{2m}\,E \cdot \frac{l^2}{v^2}\,, \tag{3.2}$$

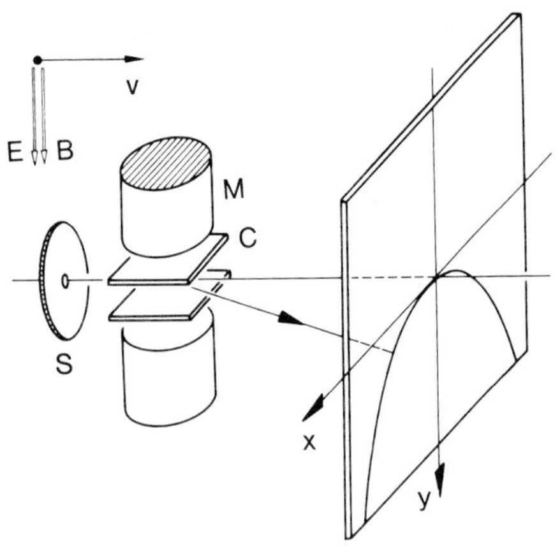

Fig. 3.2. Schematic representation of the parabola method. The ion beam, collimated by the slit S, is deflected by the magnet M and the condenser C in the x and y directions. Equation (3.5) describes the path of the particles on a catcher screen immediately after exiting from the magnet and the condenser. If the screen is placed at a greater distance, a corresponding distortion of the parabolas due to projection is seen

where the last equation is found by expressing the time spent by the particles in the electric field in terms of the velocity v and the length l of the condenser. Since the deflection of the particles in the y direction is inversely proportional to the kinetic energy $mv^2/2$, the condenser is referred to as an *energy filter.*

The homogeneous $\boldsymbol{B}$ field, which is also applied in the y direction, produces a deflection in the x direction. This deflection can be calculated as follows:

The particles which enter the homogeneous $\boldsymbol{B}$ field are forced to follow circular orbits in a plane perpendicular to the direction of the field (y direction). Since, however, the $\boldsymbol{B}$ field is limited in spatial extent (Fig. 3.2), the particles pass through only a segment of this circular orbit and then move on in a straight line. The resulting deflection in the x direction may be derived by means of the radius of curvature of the circular orbit, which is obtained by setting equal the magnitudes of the Lorentz force in the magnetic field, $\boldsymbol{F} = e(\boldsymbol{v} \times \boldsymbol{B})$, and of the centrifugal force $\boldsymbol{F}_c = mv^2\boldsymbol{r}/r^2$:

$$r = mv/eB . \tag{3.3}$$

For the centrifugal acceleration $a_c = v^2/r$ we obtain [by inserting (3.3) for the radius] the following relation:

$$a_c = eBv/m .$$

Since the particle only moves through a relatively short segment of the circle, we may replace its acceleration in the x direction with the centrifugal acceleration a_c. The total deflection in the time t is given by

$$x = a_c t^2/2 .$$

In this equation, we replace a_c by eBv/m and the time of flight t by the quotient l/v, where l is the distance traveled in the field. We then obtain for the deflection in the x direction

$$x = \frac{eBl^2}{2mv} . \tag{3.4}$$

Fig. 3.3. Separation of a mixture of hydrocarbon ions with the Thomson parabola method. For calibration, one uses ions of known mass. The intensities of the individual parabolic sections correspond to the relative amounts within the mixture of the ions which produced them. [Photo after *Conrad* from W. Finkelnburg: *Einführung in die Atomphysik,* 11, 12th ed. (Springer, Berlin, Heidelberg, New York 1976) Fig. 12]

The x deflection is inversely proportional to the momentum mv of the particles. For this reason, one often calls the magnet causing the deflection a *momentum filter*. From the expressions for x and y we can eliminate v, so that we obtain the equation for the orbit of deflection of the particles:

$$y = \frac{2E}{l^2 B^2} \cdot \frac{m}{e} x^2 . \tag{3.5}$$

This is the equation of a parabola $x^2 = 2py$ with the parameter $p = el^2 B^2/4mE$. This parameter has the same value for ions with the same ratio m/e but with differing velocities v. An example of a measurement is shown in Fig. 3.3.

The total intensity of the partial beam which produces a particular parabola is a measure of the relative abundance of the corresponding ion or isotope. Since the ions in general have differing velocities due to their preparation in an oven or a gas discharge tube, those ions having the same values of m/e will be distributed over the entire length of a particular segment of a parabola.

Aston used this method in 1920 to investigate the composition of naturally occurring neon, which consists of 3 types of atoms with the mass numbers 20, 21, and 22; this was the first exact demonstration of the existence of isotopes by means of mass spectroscopy (Table 3.2).

In any case, the most important result of the measurements with the parabola method was the following: many elements consist of several *isotopes*, that is atoms with the same nuclear charge number Z and differing mass numbers A. Nuclei with particular values of A and Z are referred to as *nuclides*.

Table 3.2. Isotopic composition of neon. The values of A_{rel} given were not determined with the parabola method, but rather, with the precision quoted, by the use of a double-focussing mass spectrometer

$^{20}_{10}Ne$	90.92%	$A_{rel} = 19.99244$
$^{21}_{10}Ne$	0.26%	$A_{rel} = 20.99385$
$^{22}_{10}Ne$	8.82%	$A_{rel} = 21.99138$

3.2.2 Improved Mass Spectrometers

The first essential improvement of *Thomson*'s mass spectrograph was achieved in 1919 by *Aston*, namely the introduction of *velocity focussing*. He did not use parallel electric and magnetic fields as in the parabola method, but rather perpendicular fields $\boldsymbol{E}$ and $\boldsymbol{B}$. The $\boldsymbol{E}$ field splits up the incident particle beam according to m/e, but also according to different velocities. By proper adjustment of the field strengths, one may however ensure that the $\boldsymbol{B}$ field brings all the particles with differing velocities together at a particular point in space, while particle beams with different m/e ratios remain separated. Particles with the same m/e ratio are collected at one point by the detector and not along a parabolic segment as in the parabola method (Fig. 3.4).

An apparatus with velocity focussing thus has a higher transmission for the ions than one which uses the simple parabola method, i.e., one can detect smaller amounts of ions and so, by closing down the slits, obtain a better mass resolution. The resolution attained by *Aston* (1919) was about 130 for the ratio $m/\Delta m$, that is, for the mass divided by the separable mass difference Δm.

The second major improvement was the introduction of directional focussing (first done by *Dempster* in 1918). By means of properly dimensioned sector fields (Fig. 3.5), it can be ensured that ions with the same m/e ratio but with somewhat differing angles of incidence, which are therefore deflected by differing amounts, are again collected at a point.

In modern high-resolution mass spectrographs, both methods – velocity and directional focussing – are used, leading to what is called *double focussing*. The precision attainable today for the relative atomic masses A_{rel} is down to 10^{-7} u. The same criteria apply as for optical spectrographs: by using narrow slits one obtains high resolution,

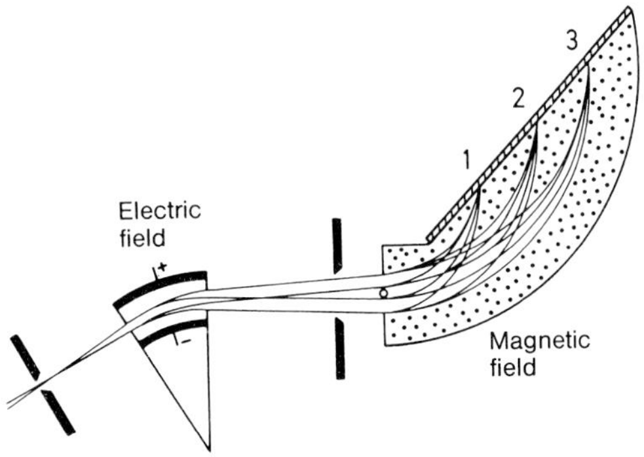

Fig. 3.4. A focussing mass spectrograph as designed by *Aston*. The points 1, 2, and 3 denote the points at which three types of particles with three different values of m/e are collected

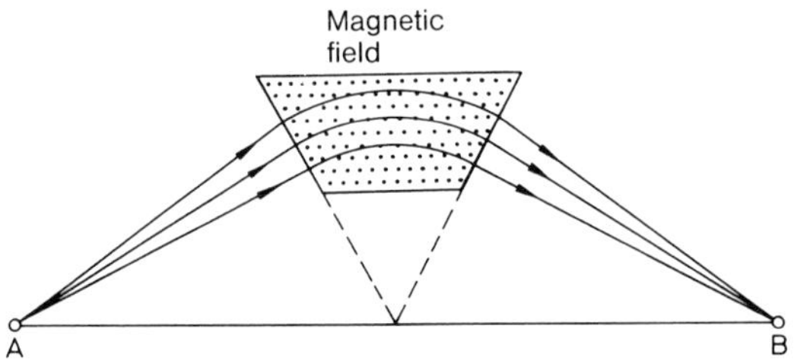

Fig. 3.5. Directional focussing in a magnetic sector field, schematically illustrated. Particles which pass a longer distance through the region of magnetic field are more strongly deflected

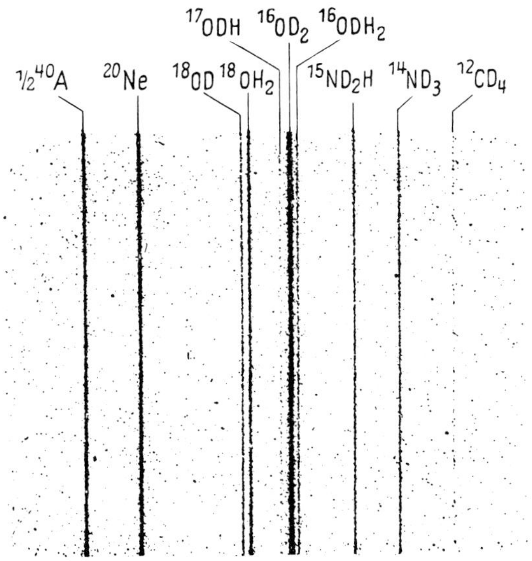

Fig. 3.6. An example of high-resolution mass spectroscopy: separation of 10 different ions with mass number 20, whose atomic or molecular weights lie between 19.9878 and 20.0628. The picture was made with the double-focussing mass spectrometer of *Mattauch* and coworkers [from W. Finkelnburg: Einführung in die Atomphysik, 11, 12th ed. (Springer, Berlin, Heidelberg, New York 1976) Fig. 15]

but at the cost of intensity. This represents the principal problem for the experimentalist. The high resolution is mainly needed for nuclear physics problems, e.g., for the measurement of the so-called mass defect, but also for problems in analysis and structure determination in chemistry, Sect. 3.2.4. The resolution $m/\Delta m$ which can be attained at present, i.e., the possibility of separating two masses with the values m and $m+\Delta m$, is more than 100000. An example is shown in Fig. 3.6.

3.2.3 Results of Mass Spectrometry

In atomic physics, mass spectrometers are primarily of interest as instruments for analysing the isotopic composition of chemical elements.

An element often has several isotopes, for example chlorine: an isotope with mass number 35 occurs with an abundance of 75.4%; the other stable isotope with mass number $A = 37$ has an abundance of 24.6%. The resulting relative atomic mass of the isotope mixture is $A_{rel} = 35.457$. There are elements with only one stable isotope, for example

$$^{9}_{4}\mathrm{Be},\ ^{27}_{13}\mathrm{Al},\ ^{127}_{53}\mathrm{I},$$

and others with two stable isotopes, e.g.,

$$\begin{array}{ll} ^{1}_{1}\mathrm{H} & ^{2}_{1}\mathrm{H} \\ 99.986\% & 0.014\%, \end{array}$$

and finally there are elements with many stable isotopes. For example, mercury, $_{80}\mathrm{Hg}$, has 7 stable isotopes with A between 196 and 204. A few further examples are contained in Table 3.3.

Table 3.3. Some examples of isotopes

	Mass number	Rel. atomic weight	Abs. atomic weight [10^{-27} kg]
^{1}H	1	1.007825	1.67342
^{2}H	2	2.014102	3.34427
^{12}C	12	12.000000	19.9251
^{16}O	16	15.99415	26.5584
^{35}Cl	35	34.96851	58.0628
^{37}Cl	37	36.965898	61.37929

3.2.4 Modern Applications of the Mass Spectrometer

Aside from precision measurements in atomic and nuclear physics, mass spectrometers with limited mass resolution are utilised today in many applications in science and technology.

In chemistry, simplified double-focussing spectrometers are used for analytical purposes. The molecular fragments which result from electron or ion bombardment of molecules are identified; from their distribution, a clue to the identification of the original molecules is obtained.

In physics, chemistry, and technology, simple, compact spectrometers are used to analyse residual gases in vacuum systems. For this purpose, a mass resolution of $m/\Delta m = 100$ is usually sufficient.

A further application of these relatively simple spectrometers is the production of pure atomic and molecular beams. Recently, high-frequency mass spectrometers have been applied for this purpose. In these so-called time-of-flight spectrometers, charged particles are differently accelerated by high-frequency electromagnetic fields depending on their specific charges, and pass through the spectrometer with different velocities. The different times of flight (through the spectrometer) are a measure of the ratio e/m.

In a quadrupole mass filter, the superposition of direct and alternating potentials on the four cross-connected, parabolically shaped electrodes results in an inhomogeneous high-frequency field in the interior of the electrode assembly. A static field is superposed on the high-frequency field. Only particles with a particular mass and energy can pass through a filter with a given geometry and frequency (Fig. 3.7).

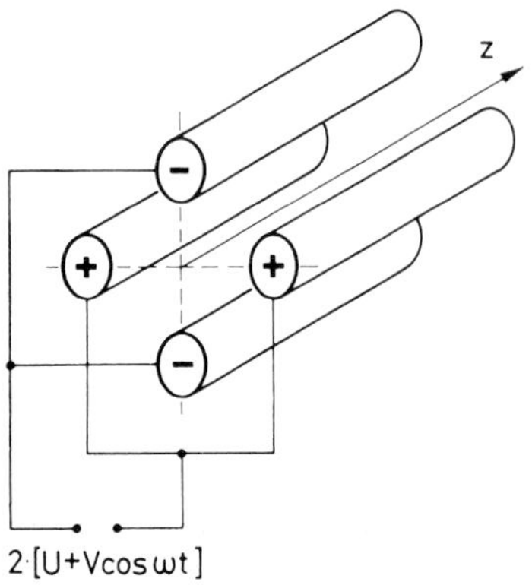

Fig. 3.7. Schematic of a quadrupole mass filter. The ion beam, moving in the $+z$ direction, is deflected by a high-frequency alternating potential. In order for the beam to pass through the filter, a relation between e/m, the frequency ω, and the deflection voltages U and V must be fulfilled

3.2.5 Isotope Separation

The separation of isotopes is more a problem of technology and nuclear physics than of atomic physics, which is the main topic of this book. For this reason, we will only briefly treat the subject here.

In principle, any method which can distinguish particles on the basis of a physical property depending on the mass may be used to separate isotopes. Which one is employed in a particular application depends on questions of economics and the state of the technology. The requirements are rather varied.

Separation of the two hydrogen isotopes ^1_1H and ^2_1H with a mass difference of 100% is relatively easy, while by contrast the separation of $^{235}_{92}U$ and $^{238}_{92}U$ is considerably more difficult. In the latter case, the masses differ by only 1.25%. In the following, the most important methods will be briefly described.

Electromagnetic separation with mass spectrographs is usually expensive and slow. The yields which can be obtained are of the order of 1 mg per hour at a current of 10^{-4} A. For example, 35 g Cl as singly charged ions corresponds to 96500 As transported charge. At a current of 10^{-4} A, 35 g of Cl will be deposited in a time

$$\frac{9.65 \cdot 10^4\,\mathrm{As}}{10^{-4}\,\mathrm{A}} = 9.65 \cdot 10^8\,\mathrm{s} = 30\ \mathrm{years}\,.$$

Nevertheless, this technique is applied on a large scale for isotope separation, for example for separating uranium isotopes, initially for the manufacture of uranium fission bombs. The necessary investment of technology and energy is enormous.

Mass separation by means of *diffusion* through porous membranes is based on the fact that in a gas, particles of differing masses m_1 and m_2 have different velocities v_1 and v_2 at a given temperature. The following relation holds:

$$v_1/v_2 = \sqrt{m_2/m_1}\,, \quad \text{since} \quad m_1 v_1^2 = m_2 v_2^2\,,$$

that is, the mean kinetic energy for both types of particle is the same. Light atoms are therefore on the average faster and diffuse more quickly. To obtain efficient isotope separation, many diffusion layers must be connected in series. This method is today the most important technology for uranium separation: the gaseous compound UF_6 is employed to enrich the uranium isotope $^{235}_{92}U$ relative to $^{238}_{92}U$.

The *gas centrifuge* is also applied on a large scale for uranium separation. Here, the heavier isotope is acted upon by a stronger centrifugal force. The lighter isotope is enriched in the region of the centrifuge axle. For effective separation, many stages must be employed one after another. The most serious technical problem is the strength of the materials used in view of the extreme accelerations necessary.

The *separation tube* utilises thermodiffusion: it is based on the principle that a temperature gradient in a mixture of gases leads to a separation of the mixture; the effect is increased by convection. Along the axis of a long tube, a heater wire is suspended. The lighter isotope is enriched in the middle and at the top, the heavier isotope at the outer wall and at the bottom of the tube.

Fractional distillation in repeated steps uses the fact that the heavier isotope in general has the higher boiling point. For example, the boiling point of heavy water (D_2O) lies 1.42 degrees above that of H_2O.

In *electrolysis*, molecules with the heavier isotope are less easily decomposed than those with the lighter isotope. This technique is used for large-scale separation of heavy and light hydrogen.

There are also *chemical reactions* in which molecules with differing isotopic compositions react with different rates. In such cases, isotope separation can be achieved through chemical reaction. Since the availability of narrow-band, tunable light sources in the form of dye lasers, (see Chap. 21), *laser photochemistry* can also be used for isotope separation. In this method, certain isotopes in a mixture of molecules composed of various isotopes can be selectively photoexcited, leading to photochemical reactions of the selected molecules. Some interesting new techniques for isotope separation have been developed in recent years based on this principle.

Problems

3.1 Show that a transverse homogeneous magnetic field can be used to sort charged particles according to their momenta, and to sort monoenergetic particles according to their masses. All the particles have the same charge.

3.2 An ion beam containing $^1H^+$, $^2H^+$, and $^3H^+$ is accelerated through a voltage of 1000 V and is directed perpendicular to the field lines of a 0.05 tesla magnetic field. How far apart are the component beams when they have travelled 5 cm through the homogeneous magnetic field and are measured at a distance of 25 cm from the beginning of the magnetic field?

3.3 A beam of positive ions traverses for a distance $l = 4$ cm an electric field $|\boldsymbol{E}| = 5000$ V/m and a parallel magnetic field $|\boldsymbol{B}| = 0.1$ tesla. The ions travel perpendicular to the direction of the two fields (parabola method). They then cross a field-free region $l' = 18$ cm and land on a flat fluorescent screen. What are the parameters of the parabolas on the screen if the beam consists essentially of singly charged hydrogen ions and hydrogen molecules with a velocity corresponding to an accelerating voltage between 1000 V and 4000 V? What does the image on the screen look like if both positively and negatively charged ions are directed at it?

3.4 The isotopic abundance of ^{235}U and ^{238}U in naturally occurring uranium is 0.72% and 99.28%, respectively. If the isotopes are separated by diffusion, the isotopic mixture after one separation step is 0.754% ^{235}U. How many separations are needed to enrich the ^{235}U to 50%? to 99%?

Hint: The separation coefficient $\alpha =$ (abundance before separation)/(abundance after separation) is independent of the composition of the isotopic mixture.

4. The Nucleus of the Atom

4.1 Passage of Electrons Through Matter

Beginning in the 1890s, *Lenard* investigated the attenuation of electron beams passing through matter. This attenuation can have two causes: the electrons can lose their energy by exciting or ionising atoms, or they can be elastically scattered and so change their directions and leave the beam. *Lenard* produced the beam by means of a cathode ray tube. Today, one would use thermionic emission as the electron source.

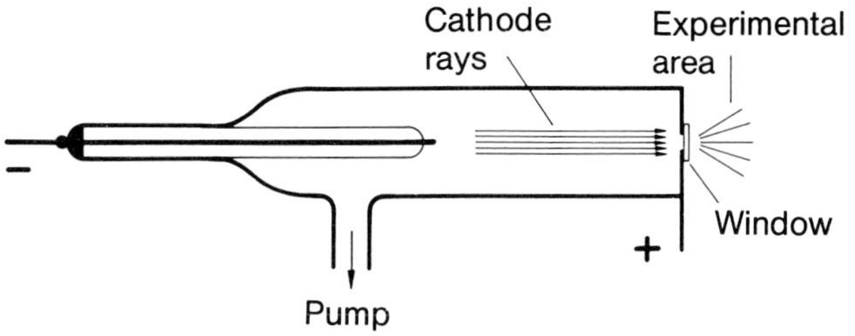

Fig. 4.1. Schematic of *Lenard*'s cathode ray tube. The cathode rays pass through a thin foil – the Lenard window – into the surrounding air and excite it, producing fluorescence

An important result may be obtained from the qualitative experiment illustrated in Fig. 4.1: if the gas discharge tube used for producing the cathode rays is closed with an extremely thin aluminium foil, it may be observed that electrons from the cathode ray beam pass through the foil. They excite the air for a distance of several centimeters outside the tube, yielding a bluish-red fluorescence light, and can be detected several cm away from the end of the tube by using a fluorescent screen. An aluminium window of this type, with a thickness of ca. $5 \cdot 10^{-4}$ cm, is called a *Lenard window*. The experiment offers visible proof that the electrons can pass through some 10000 atomic layers as well as several cm of air at NTP. Under the assumption that atoms were impenetrable for electrons, the scattering of electrons by air would take place over a length of the order of the gas kinetic mean free path, i.e., in a range of about 10^{-5} cm.

From such qualitative experiments, it follows that the interaction cross section for collisions of an electron from a gas discharge tube with atoms is small compared to the cross section for collisions between two atoms.

For the quantitative determination of the interaction cross section between electrons and atoms, one may employ a setup analogous to that shown in Fig. 4.1, where, however, the cathode ray beam passes through the Lenard window into a scattering chamber. In the chamber, the electron current is measured after the beam has passed through a gas atmosphere of known composition and density. The collisions of the electrons with the atoms in the foil can also be investigated; for this purpose, the experimental parameters (foil thickness, foil material, pressure and composition of the gas, and distance between foil and electron detector) may all be varied. The interaction cross section is obtained from the ratio of incident (I_0) and transmitted (I) electron intensities by means of the equation derived above (2.22):

$$I(x) = I_0 e^{-\alpha x}, \tag{4.1}$$

where x indicates the thickness of the scattering layer.

It may be shown that:

- The *absorption* or *scattering coefficient* α is proportional to the pressure in the scattering chamber. This is in agreement with the definition of the total interaction cross section given earlier as being equal to the sum of the partial cross sections, $\alpha = \sum_{i=1}^{n} \sigma_i$, since, for identical scattering particles, $\alpha = \sigma n$ is then the sum of all the interaction cross sections per unit volume, where n gives the number of particles per unit volume and is proportional to the gas pressure.
- In foils and in gases, *independent of the phase of matter and of the particular properties of the material*, for a given electron velocity it is found that $\alpha/\varrho = \text{const}$, i.e., the interaction cross section is proportional to the density ϱ of the scattering material.
- With increasing electron velocity, the ratio α/ϱ decreases strongly (Fig. 4.2).

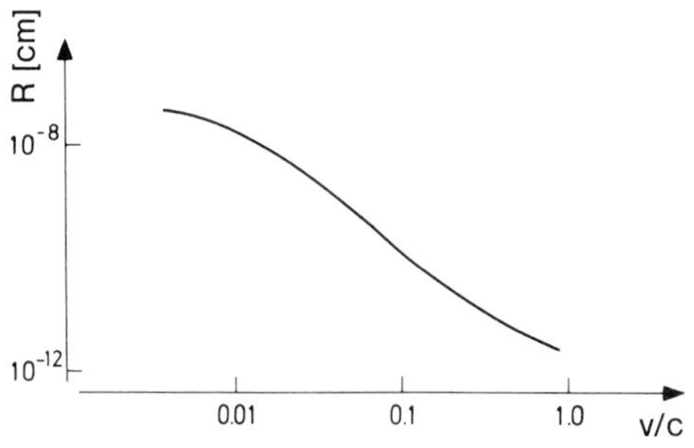

Fig. 4.2. Qualitative behaviour of the interaction cross section for gas atoms and electrons as a function of the electron velocity. The collision radius R, which is connected with the cross section σ by means of $\pi R^2 = \sigma$ (Sect. 2.4.2), is plotted against the ratio v/c of the electron velocity to the velocity of light

In Fig. 4.2, the collision radii calculated from the measured interaction cross sections α are plotted as a function of the electron velocity. For very fast electrons, atoms are thus a factor of 10^8 more penetrable than for slow electrons. The experiments lead to the following conclusion: only a small portion of the atomic volume is impenetrable for fast electrons, or, as expressed by *Lenard* – the inside of an atom is as empty as interplanetary space.

At first, the only general conclusion which could be drawn from this result was that the mass and charge in an atom are distributed in a "lumpy" fashion, rather than being evenly spread throughout the atomic volume. *Lenard* spoke of nuclei and force fields. The analogy with a solar system was tempting. Today, we know that slow electrons are scattered by the atomic electron cloud, while fast electrons are scattered by the nucleus only.

The realisation that there is *one* small nucleus, which contains the entire positive charge and almost the entire mass of the atom, is due to the investigations of *Rutherford*, who utilised the scattering of alpha particles by matter.

4.2 Passage of Alpha Particles Through Matter (Rutherford Scattering)

4.2.1 Some Properties of Alpha Particles

Alpha particles are emitted by some radioactive nuclei. They consist of doubly ionised helium nuclei, ${}_2^4He^{2+}$, with high kinetic energies (several MeV). They can, for example, be detected by means of their ability to ionise air in a cloud chamber; alpha particles with an energy of 5 MeV have a range of about 3.5 cm in air at NTP. In this distance, they lose their initial kinetic energy to the air molecules through ionisation and excitation processes. Since the mean free path for atoms or molecules as calculated by the kinetic theory of gases amounts to about 10^{-5} cm, we see that alpha particles can penetrate and pass through thousands of atoms ($3.5\ \mathrm{cm}/10^{-5} \triangleq 3.5 \cdot 10^5$ atoms) without being noticeably deflected from a straight path. Cloud chamber pictures show that the paths of the alpha particles are for the most part straight; only near the ends of the tracks, when the particles have lost most of their kinetic energy and are moving slowly, do we observe large deflections from straight-line paths (Fig. 4.3). Another possibility for observing the paths of alpha particles is offered by the spinthariscope or the scintillation detector (Figs. 2.16 and 2.17). Using scintillation detectors, *Geiger* and *Marsden* investigated the scattering of alpha particles in matter, which we will now treat in detail.

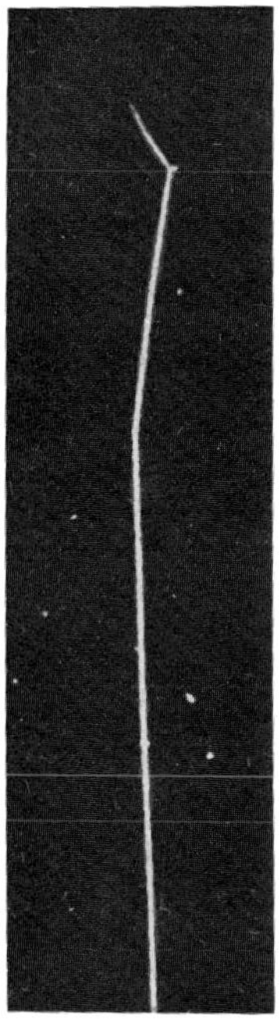

Fig. 4.3. Cloud chamber photograph of the track of an alpha particle, by *Wilson*. The particle passes through several cm of air without noticeable deflection. At the end of the track, we see two deflections; at the second, we can also see the short track of the target nucleus, which was accelerated to the right by the collision. [From W. Finkelnburg: *Einführung in die Atomphysik,* 11,12th ed. (Springer, Berlin, Heidelberg, New York 1976) Fig. 3]

4.2.2 Scattering of Alpha Particles by a Foil

In order to investigate the interaction cross section for collisions between alpha particles and atoms quantitatively, *Rutherford* and coworkers utilised the following experimental setup (Fig. 4.4):

The alpha particles, which are emitted by naturally radioactive material R, pass through a collimator and strike a thin metal foil F. The transmitted alpha intensity is determined by means of a scintillation screen S, observed through the lens L. In contrast to the determinations of interaction cross sections described above, in *Rutherford*'s experiments the directly transmitted alpha intensity was not the main object of the investigation; instead, the dependence of the scattered intensity on scattering angle θ was determined. θ is the angle between the directions of the deflected and the incident particle beams (Fig. 4.7). Scattering experiments of this type have become one of the most important tools in nuclear physics. A typical experiment yields a result like the one shown in Fig. 4.5.

The scattered intensity decreases strongly with increasing scattering angle. The angular dependence is well described by the inverse fourth power of the sine of half the scattering angle. At large scattering angles, deviations from this dependence are seen; we will treat this so-called anomalous Rutherford scattering in Sect. 4.2.4.

It is further observed that scattering occurs even at very large angles. It can be concluded that this is not due to multiple scattering processes; in scattering of alpha particles by helium atoms in a cloud chamber, large deflection angles, namely 90°, can be seen directly. An example is shown in the cloud chamber photograph in Fig. 4.6.

A quantitative explanation of these results may be given with the help of the Rutherford atomic model (1911). The model states that:

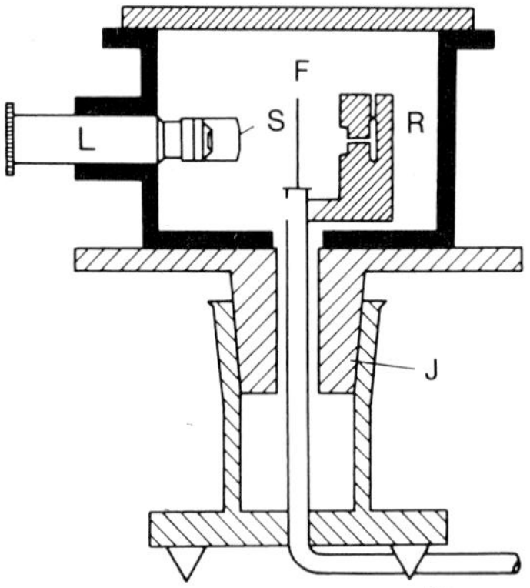

Fig. 4.4. Experimental setup for the investigation of Rutherford scattering: alpha particles from the radioactive source R are scattered by the foil F. The light flashes which are produced by the scintillation screen S are observed through the observation telescope L. The chamber can be evacuated and the observation lens L may be turned around the foil axis by means of the ground-glass joint J

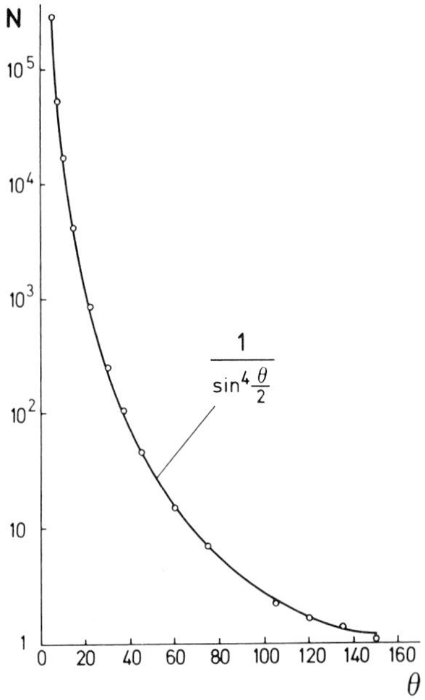

Fig. 4.5. Graphical representation of the experimental results of *Geiger* and *Marsden* for the Rutherford scattering of alpha particles by a gold foil. The scattering rate N is plotted as a function of the scattering angle θ. The solid curve represents the theoretical function for Coulomb scattering

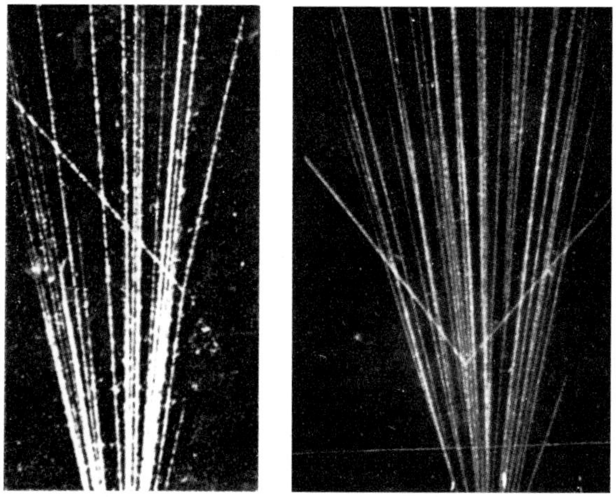

Fig. 4.6. Cloud chamber photographs of alpha particles. Collision processes with the gas in the chamber can be seen; left, the chamber gas is hydrogen, right, it is helium. In hydrogen, the alpha particle is only slightly deflected from a straight-line track, while the hydrogen target nucleus recoils sharply off to the left. In helium, the angle between the tracks of the alpha particle and the recoiling nucleus after the collision is 90°, since the two particles have the same mass. [From K. H. Hellwege: *Einführung in die Physik der Atome*, Heidelberger Taschenbücher, Vol. 2, 4th ed. (Springer, Berlin, Heidelberg, New York 1974) Fig. 4]

- Atoms have nuclei with a radius R of about 10^{-12} cm. The nucleus contains nearly the entire mass of the atom. A collision between an alpha particle and a much lighter atomic electron produces no measurable deflection in the alpha particle's path.
- The atomic nucleus has a positive charge Ze, where Z is the position of the element in the periodic table.
- Around the positively charged nucleus is a Coulomb field given by (at distance r)

$$E = (1/4\pi\varepsilon_0)\frac{Ze}{r^2}\frac{\boldsymbol{r}}{r}. \tag{4.2}$$

4.2.3 Derivation of the Rutherford Scattering Formula

The above model leads to the Rutherford scattering formula (4.20) if we take into account only the Coulomb repulsion between the nuclear charge and the charge of the alpha particle. We will use the model to calculate the dependence of the scattering probability on the deflection angle in two steps: first, for a single scattering event we determine the dependence of the deflection angle on the *impact parameter* b, which is the distance of closest approach of the alpha particle to the target nucleus, assuming no deflection occurs (see Fig. 4.7). We shall see that a unique relation between the impact parameter b and the deflection angle θ exists. Secondly, we will average over all possible impact parameters, since we cannot follow a single alpha particle on its path through the target foil, but rather observe the scattering of many alpha particles. Multiple scattering will not be considered; for the experiment, this means that the target foil must be sufficiently thin that each alpha particle is only scattered once on passing through the target.

In order to calculate the path of the particle we recall the motion of a planet under the influence of an attractive gravitational field. The effective force is proportional to $1/r^2$ where r is the sun-planet distance. The orbits which one finds in this case are known to be either elliptical, parabolic, of hyperbolic.

Since the Coulomb force has the same dependence upon distance r as the gravitational force, the orbital calculations from celestial mechanics can be utilised directly. Admittedly, since the Coulomb force is here repulsive, only the hyperbolic orbits represent possible solutions when we are dealing with charges of the same signs.

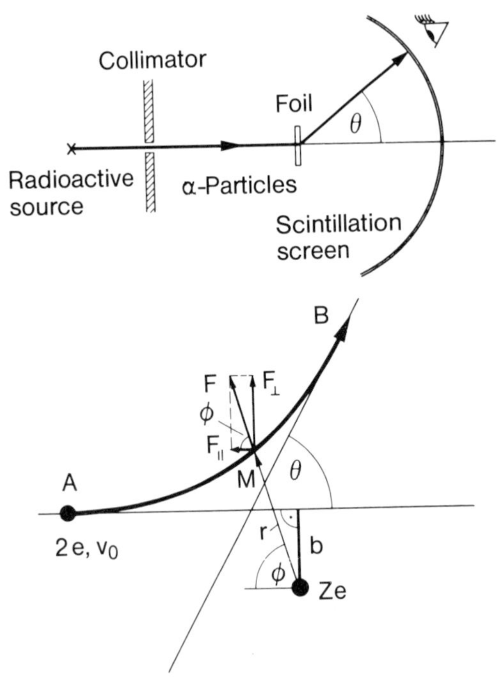

Fig. 4.7. Rutherford scattering. *Upper part:* Schematic illustration of the experimental setup. *Lower part:* The geometry of the model calculation. The alpha particle is deflected from A to B through scattering by the nucleus *Ze*

We now wish to determine the relation between the scattering angle θ and the impact parameter b (Fig. 4.7). The particle arrives at point A, still distant from the nucleus, with a velocity v_0. If it were not deflected, it would pass the nucleus at a distance b. Between the alpha particle and the nucleus, there is a repulsive Coulomb force F

$$F = \frac{2Ze^2}{4\pi\varepsilon_0 r^2}\frac{r}{r} \tag{4.2a}$$

with the nuclear charge Ze, the elementary charge e, the permittivity constant ε_0, and a distance r between the nucleus and the alpha particle.

We assume that the particle has reached point M in its orbit and express the force which acts there in terms of two components:

$$F_\perp = F\sin\phi \quad \text{perpendicular to the original direction}, \tag{4.3}$$

and

$$F_{||} = F\cos\phi \quad \text{antiparallel to the original direction}. \tag{4.4}$$

ϕ is the angle between the horizontal (i.e., the direction of the incident beam) and the radius vector $\boldsymbol{r}$ to the momentary position of the particle.

We now apply the law of conservation of angular momentum, placing the origin of the coordinate axes at the centre of the atomic nucleus. Since the force which acts here is radial (4.2a), it produces no torque and the angular momentum is constant; in particular, the angular momenta at the points A and M are the same, or, mathematically,

$$(mv_0b)_A = (mr^2\dot{\phi})_M \tag{4.5}$$

in which we have used polar coordinates (r, ϕ). Solving for $1/r^2$ yields

$$1/r^2 = \dot{\phi}/v_0b \,. \tag{4.6}$$

If we consider only the motion perpendicular to the original beam direction, Newton's equation of motion reads

$$m\frac{dv_\perp}{dt} = F_\perp = \frac{2Ze^2}{4\pi\varepsilon_0}\,\frac{1}{r^2}\sin\phi \,. \tag{4.7}$$

If we replace $1/r^2$ in this equation with the right-hand side of (4.6) and integrate over time, using the abbreviation $k = 2Ze^2/4\pi\varepsilon_0$, we obtain

$$\int_{t_A}^{t_B}\frac{dv_\perp}{dt}dt = \frac{k}{mv_0b}\int_A^B \sin\phi\frac{d\phi}{dt}dt \,. \tag{4.8}$$

In order to determine the limits of the integral, we imagine the point A to be infinitely distant from the nucleus. Since now no Coulomb force acts, we have $v_\perp = 0$, and the angle $\phi = 0$.

To determine the scattering angle θ between the incident direction and the direction of the particle after scattering, we let point B (see Fig. 4.7) move away to infinity. Then the angle ϕ is seen to be related to θ through the expression $\phi = 180° - \theta$. Because of conservation of energy, the final velocity at the point B is equal to the initial velocity v_0 at point A, since the potential energy vanishes at a sufficiently large distance from the nucleus. The component $v_\perp$ has, using $\phi = 180° - \theta$, the value $v_\perp = v_0 \sin\theta$. Then the integral equation (4.8) becomes, using

$$\frac{dv_\perp}{dt}dt = dv_\perp \quad \text{and} \quad \frac{d\phi}{dt}dt = d\phi \,,$$

the following equation:

$$\int_0^{v_0\sin\theta} dv_\perp = \frac{k}{mv_0b}\int_0^{\pi-\theta}\sin\phi\, d\phi. \tag{4.9}$$

Upon integration, we obtain

$$v_0 \sin\theta = \frac{k}{mv_0b}(1+\cos\theta) \,. \tag{4.10}$$

With the trigonometric identity

$$\frac{1+\cos\theta}{\sin\theta} = \cot(\theta/2) \tag{4.11}$$

we obtain the relation between the impact parameter and the deflection angle which we are seeking:

$$b = \frac{k}{mv_0^2} \cot(\theta/2) . \qquad (4.12)$$

In an actual experiment, we cannot measure the number of scattered particles arriving at a particular angle θ, but rather we have to consider the finite range of angles between θ and $\theta + d\theta$; these correspond to impact parameters in the range b to $b + db$. Then, by differentiating (4.12), we obtain the relation between db and $d\theta$:

$$db = -\frac{k}{2mv_0^2} \frac{1}{\sin^2(\theta/2)} d\theta . \qquad (4.13)$$

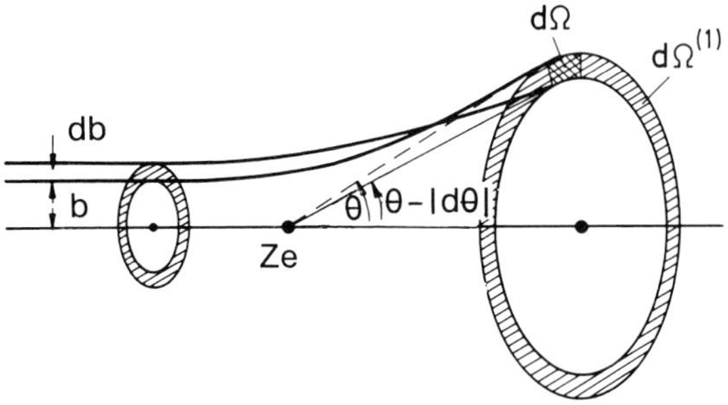

Fig. 4.8. Rutherford scattering. The incident alpha particles with impact parameters in the range b to $b + db$ are deflected into the range of angles θ to $\theta - |d\theta|$

Finally, we have to consider that the whole problem has rotational symmetry around an axis through the target nucleus and parallel to the direction of the incident beam (Fig. 4.8). Therefore, we have to consider a circular ring with radii $r_1 = b$ and $r_2 = b + db$, through which the incident beam enters and is scattered into the angular region from $\theta - |d\theta|$ to θ. [We note that with increasing impact parameter b the angle θ becomes smaller, see (4.12)]. This range of angles corresponds to an "effective area", the *differential cross section* da:

$$da = 2\pi b\, db . \qquad (4.14)$$

If we shoot alpha particles through a thin foil with thickness D and area A, containing N atoms/cm^3, the "effective area" of all the atoms is

$$dA = 2\pi b\, db\, N D A \qquad (4.15)$$

with the condition that the "effective areas" of the atoms do not overlap one another, which is a good assumption in a thin foil (up to 10000 atomic layers).

The probability that an incident alpha particle strikes the "effective area" of an atom in the foil is given by

$$W = \frac{\text{"effective area"}}{\text{total area}} = dA/A = 2\pi N D\, b\, db . \qquad (4.16)$$

With a total of n alpha particles, the number dn' of the particles which strike the "effective area" and thus are deflected into the angle range θ to $\theta - |d\theta|$ is given by

$$dn' = n \cdot 2\pi N D\, b\, db\,. \tag{4.17}$$

These particles pass through the unit sphere around the target foil on a ring of area $d\Omega^{(1)} = 2\pi \sin\theta |d\theta|$. In the following, it is convenient to use the half-angle $\theta/2$; doing so, we obtain

$$d\Omega^{(1)} = 4\pi \sin(\theta/2)\cos(\theta/2)|d\theta|\,. \tag{4.18}$$

The detector which is used in the measurement cuts out a segment $d\Omega$ from this ring-shaped area. This surface element on the unit sphere is called a *solid angle*. The number of particles actually measured is therefore smaller than the number dn' by the ratio $d\Omega/d\Omega^{(1)}$. If the detector subtends a solid angle of $d\Omega$, the number of particles observed at angle θ is given by

$$dn = dn' \cdot d\Omega/d\Omega^{(1)}\,. \tag{4.19}$$

Inserting b and db from (4.12) and (4.13), we obtain the full Rutherford formula:

$$\frac{dn(\theta, d\theta)}{n} = \frac{Z^2 e^2 D N}{(4\pi\varepsilon_0)^2 m^2 v_0^4 \sin^4(\theta/2)}\, d\Omega \tag{4.20}$$

with n the number of incident particles, dn the number of particles scattered at an angle θ into the solid angle $d\Omega$, Z the (target) nuclear charge, e the elementary charge, D the target foil thickness, N the number of atoms/cm^3 in the target foil, $d\Omega$ the solid angle subtended by the alpha particle detector, ε_0 the permittivity constant of vacuum, m the mass of the scattered (alpha) particles, v_0 the velocity of the incident particles, and θ the angle of deflection.

This formula tells us how many particles dn out of the incident number n are scattered at a particular angle θ into a particular solid angle $d\Omega$, when target properties and incident particle velocity are known. Corresponding to (4.20), we find for the differential cross section (4.14)

$$da = \frac{Z^2 e^4}{(4\pi\varepsilon_0)^2 m^2 v_0^4 \sin^4(\theta/2)}\, d\Omega^{(1)}\,. \tag{4.21}$$

Furthermore, it is useful to define the *macroscopic (differential) cross section* $N da$, which is equivalent to the "effective area" dA per unit volume. By integration of (4.21) over $\Omega^{(1)}$, we can obtain the *total interaction cross section* a; the latter, however, diverges in the present case of a pure (unscreened) Coulomb potential, since (4.21) diverges for $\theta \to 0$. In the Rutherford scattering formula (4.20) for scattering by a foil, the limiting case $\theta \to 0$ is in principle not physically relevant: this is a result of the model, since $\theta = 0$ means that $b = \infty$. An infinite value of the impact parameter is, however, unreasonable given the assumed dense packing of the target atoms; the largest possible impact parameter is equal to half the distance between target atoms in the foil. For $\theta = \pi$, dn/n shows a minimum. This corresponds to $b = 0$. For very small impact parameters, there are deviations between the results of the calculation using the scattering formula (4.20) and the experiments. This occurs because the model of a deflection of the alpha particles by the Coulomb field of the nuclei alone is insufficient. From the values of the impact parameter b for which these deviations become important, we can determine the nuclear radius R. This will be discussed in the following section.

4.2.4 Experimental Results

The Rutherford formula has been experimentally tested with great care. Keeping the solid angle $d\Omega$ constant, the $\sin^{-4}(\theta/2)$-law is found to be excellently reproduced in the counting rate (Fig. 4.5). Even with alpha particles of energy 5 MeV and scattering angles of 150°, no deviations from the Rutherford formula are found; this corresponds to an impact parameter of $6 \cdot 10^{-15}$ m. In this region, only the Coulomb potential of the nucleus has a measurable effect on the alpha particles.

The experimental tests of the Rutherford scattering formula can be summarised as follows:

The Coulomb law is obeyed well even at very small impact parameters, since the Rutherford formula is still valid. The nuclear radius is thus

$$R < 6 \cdot 10^{-15}\,\mathrm{m}\,.$$

From experiments with different foil materials, the nuclear charge Z can be determined. The experiments of *Chadwick* (1920) verified that Z is identical with the position of the element in the periodic table.

The nucleus was originally assumed to be constructed from A protons and $(A-Z)$ electrons where A is the mass number defined on p. 6. After 1932 it was known that this model is not correct; $(A-Z)$ is rather the number of neutrons and Z is the number of protons in the nucleus.

We come now to the so-called *anomalous* Rutherford scattering. In the scattering of very fast alpha particles ($E > 6$ MeV) at large angles θ, i.e., with small impact parameters b – nearly central collisions – one observes clear *deviations* from the Rutherford formula. Here the Coulomb law is apparently no longer obeyed. The alpha particles approach the nuclei so closely that another, short-range interaction force becomes effective: the nuclear force. From the values b and θ at which deviations from the Rutherford formula, i.e., from the Coulomb law begin to occur, a nuclear size of $R \cong 10^{-15}$ m is obtained. This means that the density of the nucleus is about 10^{15} times larger than the density of the atom as a whole. These deviations from the scattering behaviour expected on the basis of the Rutherford formula are called *anomalous Rutherford scattering.*

The Rutherford model may be developed further. Negative electrons orbit around the positively charged nucleus with nuclear charge Z. This represents a dynamic equilibrium: without the motion of the electrons, no stability would be possible. If deflections of alpha particles through large angles are possible without causing a noticeable energy loss on the part of the alpha particles, then the mass of the target nucleus must be large compared to that of the alpha particle. On the other hand, observations with cloud chambers filled with helium gas, in which the target and the projectile, i.e., a He atom and an alpha particle, have virtually the same mass, show deflections of about 90°. From such experiments one can show that the nucleus must contain nearly the whole mass of the atom.

By contrast, momentum conservation requires that in a collision between an alpha particle and an electron, due to the small electron mass only very little momentum can be transferred. With the electron/alpha particle mass ratio, the deflection of the alpha particles can be no larger than 28″.

For very large impact parameters (small deflection angles), the Rutherford formula is likewise no longer exactly valid. The Coulomb potential of the nucleus is perturbed by the atomic electrons. These effects occur for $b \geqq 10^{-10}$ cm (deflection angles of a few seconds of arc) and are very difficult to detect experimentally. Completely analogous scattering formulae and scattering problems occur in the scattering of protons by atomic nuclei. The angular dependence of the scattering is related to the scattering potential; the latter can thus be determined from experiment. Scattering processes play an important rôle in nuclear and elementary particle physics, in the investigation of the internal structure of nuclei and of certain elementary particles. For example, *Hofstadter* was granted the Nobel Prize in 1961 for his scattering experiments using fast electrons (10^9 eV) on protons and neutrons. From the angular dependence of the scattering intensity, he was able to obtain information about the inner structure of the proton and of the neutron.

4.2.5 What is Meant by Nuclear Radius?

We can summarise our considerations in the above sections as follows: an alpha particle, which approaches a nucleus from outside the atom, is acted on at first only by the repulsive Coulomb potential. If it approaches the nucleus sufficiently closely, it will also be acted upon by the attractive nuclear force. The nuclear radius is defined as the distance at which the effect of the nuclear potential is comparable to that of the Coulomb potential (Fig. 4.9). For such investigations, alpha particles of high kinetic energies are used, so that they can approach the nucleus sufficiently closely.

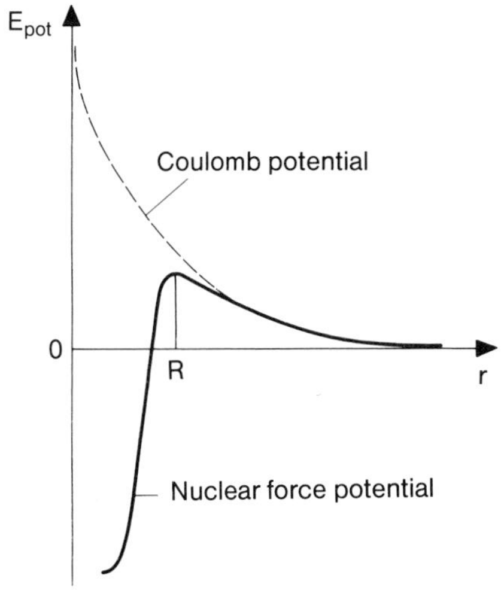

Fig. 4.9. Nuclear force and Coulomb potentials, used for defining the nuclear radius *R*

The empirical result of such measurements on nuclei with the mass number A is found to be

$$R = (1.3 \pm 0.1) A^{1/3} \cdot 10^{-15}\,\mathrm{m}\,.$$

Numerical examples for $A = 12$ and $A = 208$ are:

$$R(^{12}_{6}\mathrm{C}) = 2.7 \cdot 10^{-15}\,\mathrm{m}\,, \qquad R(^{208}_{82}\mathrm{Pb}) = 7.1 \cdot 10^{-15}\,\mathrm{m}\,.$$

This relationship between the nuclear mass and the nuclear radius implies that the density of nuclear matter is constant and independent of the size of the nucleus. This is one of the experimental results underlying the liquid-drop model for nuclei.

Problems

4.1 An aluminium foil scatters 10^3 α particles per second in a given direction and solid angle. How many α particles will be scattered per second in the same direction and solid angle if the aluminium foil is replaced by a gold foil of the same thickness?

4.2 The number of α particles scattered from a foil into a counter is $10^6\,\mathrm{s}^{-1}$ for a scattering angle of 10°. Calculate from this the number of α particles which will be scattered into this counter as it is moved on a circular path from 10° to 180°. Show your results for $N(\theta)$ graphically.

4.3 Determine the distance of the closest approach of protons to gold nuclei in head-on collisions in which the protons have kinetic energies of (a) 10 MeV and (b) 80 MeV, and compare the results with the nuclear radius. In which case would the proton "touch" the nucleus? Determine the kinetic energy of the proton when it "touches" the nucleus.

4.4 Through what angle is a 4 MeV α particle scattered when it approaches a gold nucleus with a collision parameter of 2.6×10^{-13} m?

4.5 How large is the collision parameter of an α particle with 4 MeV kinetic energy which is scattered through the angle $\theta = 15°$ by collision with a gold nucleus ($Z = 79$)?

4.6 A beam of α particles with 12.75 MeV kinetic energy is scattered off a thin aluminium foil ($Z = 13$). It is observed that the number of particles which are scattered in a certain direction begins to deviate from the value calculated for pure Coulomb scattering at the deflection angle $\theta = 54°$. How large is the radius of the Al nucleus if one assumes that the α particle has a radius $R_\alpha = 2 \times 10^{-15}$ m?

Hint: Calculate the orbit according to (4.8 and 9) up to $\phi_0 = (180 - \theta)/2$, the point of closest approach, and determine $r(\phi_0)$.

4.7 A tight bunch of protons with uniform energy strikes a 4 μm thick gold foil perpendicular to the direction of flight. Of these protons, the fraction $\eta = 1.35 \times 10^{-3}$ is scattered through the angle $\theta = 60°$ in the angular interval $d\theta$.

a) What is the kinetic energy of the colliding protons?
b) Calculate the differential effective cross section $da(\theta)/d\Omega$ of the gold nucleus.
c) What is the collision parameter b?

Hint: Use (4.20) and the expression

$$\frac{dn/n}{d\Omega} = ND\frac{da(\theta)}{d\Omega}\,.$$

5. The Photon

5.1 Wave Character of Light

The fact that light can be regarded as a wave phenomenon was experimentally shown in the 17th and 18th centuries by the Dutch physicist *Huygens* and the English physician *Young* with the aid of interference experiments. In the 19th century, the physical nature of these waves came to light: they are electromagnetic waves, described by Maxwell's equations. They are characterised by the field vectors $\boldsymbol{E}$ and $\boldsymbol{B}$ of the electric and the magnetic field and exhibit a periodicity with the frequency ω.

In the year 1885, the theory of electromagnetic phenomena was completed with the formulation of the Maxwell equations. Two years later (1887), *Hertz* succeeded in demonstrating that such waves can be produced in the laboratory as emissions from an oscillating dipole. According to *Maxwell*, an electric and a magnetic field propagate away from an accelerated charge with the velocity of light. The accelerated charge radiates energy. The emission of light in the oscillator model is a result of a high frequency oscillation carried out by a charged particle. In absorption and in scattering of light, the incident electromagnetic wave excites the oscillator to forced oscillations. This classical Maxwell theory permits the precise calculation of the electromagnetic waves which are emitted by radio and radar antennas. Furthermore, it completely describes all of the wave properties of electromagnetic radiation, for example interference and diffraction.

Electromagnetic waves may be produced over an extremely wide range of frequencies (see Fig. 8.1); for this purpose, a number of different processes may be used.

Besides the oscillating dipole, some other examples are:

- the emission of light by the electrically charged particles in particle accelerators. Here, the *synchrotron radiation* is particularly noteworthy. The circulating particles in a circular electron accelerator emit radiation with a continuous spectrum. This radiation is utilised – for example at the German Electron Synchrotron (DESY) in Hamburg – as an intense, continuous, polarised light source for spectroscopy in the near, mid-, and far ultraviolet spectral regions. Figure 5.1 shows a schematic illustration of the accelerator in Hamburg. In Fig. 5.2, the spectrum of the synchrotron radiation is indicated. At relativistic particle energies, i.e., when the particle velocity is no longer small compared to the velocity of light, the emitted synchrotron radiation energy is a considerable fraction of the total energy which must be expended to operate the accelerator.
- A radiation emission which is produced in a similar manner and which is also called synchrotron radiation occurs when charged particles become trapped in the magnetic field of the earth. This phenomenon also occurs in distant regions of space, for example in the famous Crab nebula. Various astronomical objects are

known to emit radiation in frequency regions from the far ultraviolet down to radio frequencies.

- A negative acceleration of electrons – for example a slowing down in the field of an atomic nucleus – leads to the emission of x-rays, the so-called bremsstrahlung.
- The thermal radiation of the sun is the energy source for all life on the earth.

While the wave character of light must be considered to be an experimentally and theoretically well-established fact, especially because of diffraction and interference phenomena, there are, on the other hand, experiments in which light behaves as particles; these are called *light quanta* or *photons.*

Before we describe the experiments which demonstrate the particle nature of light, we will summarise the most important physical properties of photons.

Photon

Energy	$h\nu$
Velocity	c
Rest mass	$m = 0$
Momentum	$p = h\nu/c = h/\lambda$
Intrinsic angular momentum (spin)	$h/2\pi$

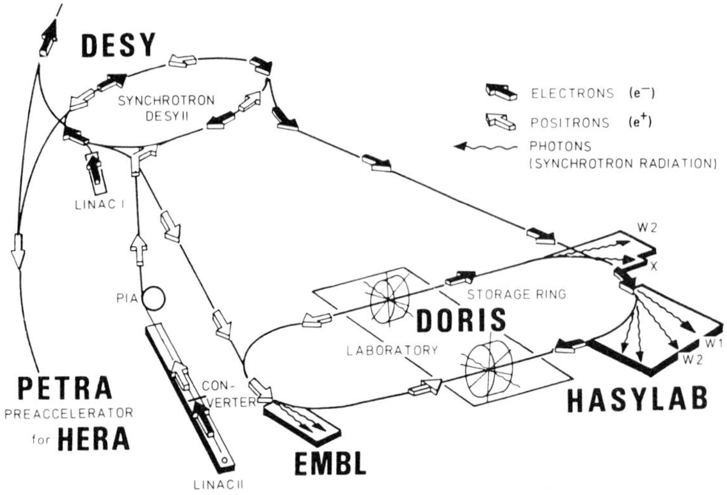

Fig. 5.1. A schematic representation of the electron synchrotron DESY and the storage ring DORIS in Hamburg. Electrons are preaccelerated by linear accelerators; positrons are generated in a converter target and stored in the accumulator storage ring PIA. Then both kinds of particles are further accelerated in the synchrotron DESY II and injected into the DORIS ring, where they can be stored for several hours. Alternatively, the particles may be accelerated to still higher energies in the former storage ring PETRA, and then injected into the electron-proton storage ring HERA (not shown in the drawing), where they are used for experiments in high-energy physics. The DORIS storage ring is used both for high-energy physics and as a large-scale source of photons for experiments with synchrotron radiation. In the facilities of the Hamburg Synchrotron Radiation Laboratory (HASYLAB) and the European Molecular Biology Laboratory (EMBL), about 30 experimental stations have been constructed for synchrotron radiation experiments. (Provided by C. Kunz, Hamburg University)

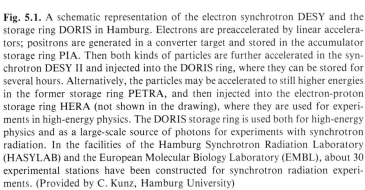

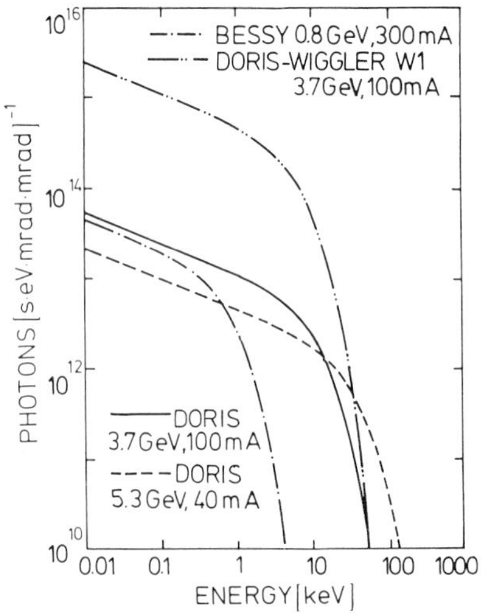

Fig. 5.2. The spectral intensity distribution of the synchrotron radiation from various electron synchrotrons and storage rings (photons/s eV mrad^{-2}). The radiation is continuous from the visible into the x-ray regions. In the far-ultraviolet and soft x-ray regions, an electron or positron storage ring is currently the best radiation source available. In addition to the spectrum emitted from the bending magnets of BESSY (in Berlin) and HASYLAB (in Hamburg), we show the considerably higher photon flux from the 32-pole "wiggler" W1 at HASYLAB. (Provided by C. Kunz, Hamburg University)

For converting from the quantum energy of a photon $E = h\nu$, which is often expressed in eV, to the vacuum wavelength λ_{vac} of the light wave, the following relation holds:

$$E\,[\text{eV}] = 12398/\lambda\,[\text{Å}]\,. \tag{5.1}$$

In the next sections we will describe three experiments which can only be understood by assuming the existence of photons.

5.2 Thermal Radiation

5.2.1 Spectral Distribution of Black Body Radiation

The quantisation of energy in the interaction of light with matter was postulated for the first time by *Planck* in the year 1900 in his theoretical analysis of the spectral distribution of the light emitted by a black body radiator (defined below), which had been experimentally determined. This light is referred to as thermal or black body radiation.

Hot objects emit light radiation as a result of their temperature. This is an everyday experience. It is well known that the colour which we see in a thermal radiator (for example a furnace) changes from dark red to bright red to yellow to white as the temperature of the furnace is increased. The determination of the colour in the interior of a furnace is used as a measure of its temperature; this technical application is called *pyrometry*.

At temperatures below a few hundred kelvins, the radiation emitted is for the most part infrared light, also called heat radiation. This infrared radiation is responsible for the fact that a thermally isolated object eventually reaches the same temperature as its surroundings. If one wishes to carry out experiments at very low temperatures (e.g., 4.2 K and below), the experimental region must therefore be screened from the thermal radiation of the laboratory, which is at room temperature, by using cooled thermal shields.

The laws governing the spectral intensity distribution of thermal radiation are obtained by the experimental analysis of the *black body radiator*. This is a cavity which emits radiation being in thermal equilibrium with its walls; the material of the walls emits and absorbs thermal radiation.

Experimentally, a source of black body radiation is most easily obtained by making a small hole in the wall of a cavity which is held at constant temperature. The hole is so small that neither the radiation which enters the cavity from outside, nor that which escapes to the outside, is sufficient to alter the thermal equilibrium in the cavity (Fig. 5.3). The energy density $u(\nu, T)$ of the radiation field within the cavity can be determined by measuring the radiative power (energy per unit time) $N(\nu, T)$ connected with the spectral radiative flux density $P(\nu, T)$ which passes out of the hole, using a spectrometer. The spectral energy density $u(\nu, T)$ is defined as

$$u(\nu, T)\,d\nu = \frac{\text{radiation energy in the frequency range } \nu \ldots \nu + d\nu}{\text{volume}}$$

and the spectral radiative flux density $2P(\nu, T)$ as

$$2P(\nu, T)\,d\nu = \frac{\text{radiative power in the frequency range } \nu \ldots \nu + d\nu}{\text{area} \cdot \text{solid angle}}\,.$$

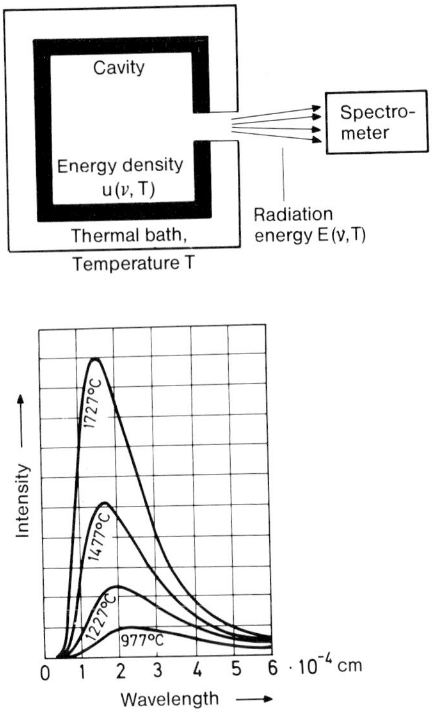

Fig. 5.3. Thermal radiation. *Upper part:* Schematic illustration of a cavity radiator. *Lower part:* Typical measured curves of the intensity distribution in the black body radiation at various temperatures

$2P(\nu, T)\,d\nu$ is the quantity of energy within the frequency interval $\nu \ldots \nu + d\nu$ which is radiated per unit time through a unit area into the solid angle 1 sterad normal to the surface. The factor 2 in the definition comes from the fact that the radiation can be decomposed into two components with polarisation directions perpendicular to each other. $P(\nu, T)\,d\nu$ for one component, i.e., for linearly polarised radiation, is independent of the polarisation direction for a black body radiator. The energy density u in the cavity and the power measured in the radiation leaving the hole are related by the equation

$$u(\nu, T)\,d\nu = (8\pi/c)\,\frac{N(\nu, T)\,d\nu}{A\cos\theta \cdot \Delta\Omega}\,, \qquad (5.2)$$

where A is the cross-sectional area of the hole, θ the angle between the normal to A and the radiation direction, and $\Delta\Omega$ is the solid angle subtended by the radiation detector.

Typical measured curves are shown in Fig. 5.3. The radiation has a continuous spectrum with a clearly visible maximum, which lies in the infrared region at room temperature.

The following results are important:

- For a given temperature the energy distribution is the same, regardless of the form and material of the cavity walls.
- The total power (radiated into a half-sphere) per unit area of the surface of the cavity is a simple function of the temperature:

$$S = 2\pi \int_0^\infty P\,d\nu = \sigma T^4\,. \qquad (5.3)$$

This is the Stefan-Boltzmann Law, with $\sigma = 5.670 \cdot 10^{-8}\ \mathrm{Wm^{-2}K^{-4}}$.

- The Wien Displacement Law holds for the wavelength at which the maximum intensity λ_{max} occurs in the emitted spectrum, as a function of temperature:

$$\lambda_{max} T = \text{const} = 0.29\ \text{cm K}\,. \tag{5.4}$$

As an example we can take the solar radiation: the surface temperature of the sun is 6000 K; the wavelength at maximum intensity is $\lambda_{max} = 480$ nm.

- The law derived by *Rayleigh* and *Jeans* from classical electrodynamics,

$$P = \frac{\nu^2}{c^2} kT \tag{5.5}$$

describes the radiative flux density per polarisation direction very well for *low* frequencies. However, at *high* frequencies, this law cannot be correct: if it is integrated over all frequencies, it yields an infinitely large energy density – we encounter the so-called ultraviolet catastrophe. Within classical electrodynamics and thermodynamics it was not possible to find an expression for P which agreed with experiment at high frequencies. This was accomplished for the first time by the Planck formula.

5.2.2 Planck's Radiation Formula

According to *Planck*, the experimentally determined spectral energy density of the radiation per unit volume, in the frequency interval $\nu \ldots \nu + d\nu$, can be represented by

$$u(\nu, T)\, d\nu = \frac{8\pi h \nu^3}{c^3} \frac{1}{\mathrm{e}^{h\nu/kT} - 1} d\nu\,. \tag{5.6}$$

From this we obtain for the radiative flux density per polarisation direction

$$P(\nu, T)\, d\nu = \frac{h\nu^3}{c^2(\mathrm{e}^{h\nu/kT} - 1)} d\nu\,.$$

These radiation formulae may be derived by making the following assumptions:

1) The atoms in the walls of the cavity behave like small electromagnetic oscillators, with each having its characteristic oscillation frequency ν. They radiate electromagnetic radiation out into the cavity and absorb radiation from the cavity; a thermal equilibrium is established between the radiation in the cavity and the atoms in the cavity walls. The excitation of the oscillators depends on the temperature.
2) The oscillators cannot – like classical oscillators – take on all possible values of energy, but rather only discrete values described by

$$E_n = n \cdot h \cdot \nu\,, \tag{5.7}$$

where n is an integer $n = 0, 1, 2, \ldots$ and h is Planck's constant ($h = 6.626176 \cdot 10^{-34}$ Js $= 4.14 \cdot 10^{-15}$ eVs). Today we know that this *quantisation* is more correctly described by the equation

$$E_n = (n + 1/2)\, h\nu \,. \tag{5.8}$$

The quantity $h\nu/2$ is the *zero-point energy* of the oscillator. We will derive (5.8) using a quantum mechanical treatment in Chap. 9.

3) As long as the oscillator is not emitting or absorbing (radiation) energy, it remains in its quantum state, characterised by the quantum number n.
4) The number of possible states of oscillation of the electromagnetic field in the cavity of volume V between ν and $\nu + d\nu$ for both polarisation directions is given by

$$dZ = \frac{8\pi V \nu^2}{c^3}\, d\nu \,, \tag{5.9}$$

which can be derived in classical electrodynamics.

The existence of discrete energy values represents a contradiction to the experience of classical physics, where energy always seems to occur with continuous values. The reason that quantised energy steps are not observed in classical physics is the smallness of the Planck constant h. We give a numerical example to make this clear:

A mass-and-spring harmonic oscillator with a mass $m = 1$ kg and a spring constant $f = 20\ \mathrm{Nm}^{-1}$ is oscillating with an amplitude $x_0 = 10^{-2}$ m. Its characteristic frequency is then given by

$$\nu = (1/2\pi)\sqrt{f/m} = 0.71\ \mathrm{s}^{-1} \,.$$

The energy of the oscillator is

$$E = f x_0^2/2 = 1.0 \cdot 10^{-3}\ \mathrm{J} \,.$$

This energy corresponds to n energy quanta of the frequency ν:

$$n = E/h\nu = \frac{10^{-3}\ \mathrm{J}}{6.6 \cdot 10^{-34}\ \mathrm{Js} \cdot 0.7\ \mathrm{s}^{-1}} = 2.1 \cdot 10^{30} \,.$$

If n now changes to $n \pm 1$, this produces a relatively energy change of

$$\Delta E/E = \frac{h\nu}{n \cdot h\nu} \cong 10^{-30} \,.$$

The relative change is thus extremely small.

From this we conclude that an energy quantisation for macroscopic systems in the realm of classical physics cannot be detected, due to the extremely large quantum numbers which occur.

5.2.3 Einstein's Derivation of Planck's Formula

The derivation of the Planck radiation formula due to *Einstein* (1917) is an interesting example of the combination of optics, thermodynamics, and statistics. *Einstein* assumed that light consists of particles, the so-called light quanta or photons. Each light wave of frequency ν corresponds to a number of photons. Furthermore, in this theory the existence of discrete atomic energy levels is already assumed. The justification for this latter assumption in terms of the Bohr atomic model will be treated in Chap. 8 and 9; we thus anticipate it here.

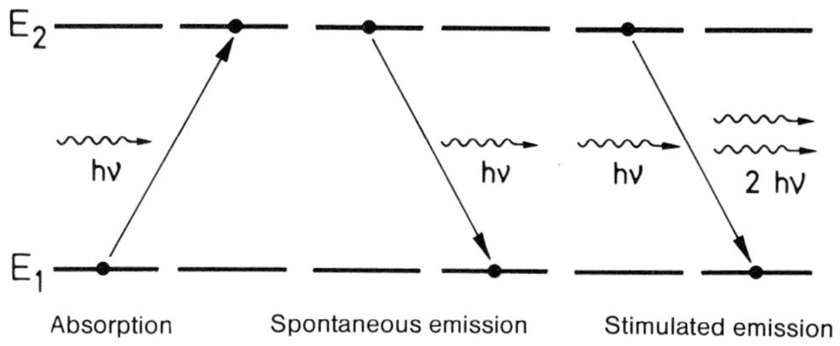

Fig. 5.4. Absorption, spontaneous and stimulated emission of radiation with the quantum energy $h\nu$ between two energy levels E_1 and E_2

An atom with two energy levels E_1, E_2 may, according to *Einstein*, interact with electromagnetic radiation in three different ways:

- *Absorption* of a light quantum takes the atom from the lower level E_1 into the energetically higher level E_2. In the process, a light quantum of energy $\Delta E = E_2 - E_1 = h\nu$ is removed from the radiation field.
- *Emission* occurs from the level E_2 *spontaneously* within a time known as the mean lifetime. A light quantum of energy ΔE is thereby released into the radiation field.
- Just as light quanta can be absorbed, light quanta in the radiation field can also stimulate the emission of further quanta when the atom is in the higher level E_2. For this stimulated or *induced emission*, primary light quanta are thus necessary. Another light quantum joins those which were already present in the radiation field.

These processes are shown schematically in Fig. 5.4.

In order to derive *Planck*'s formula, we consider, following *Einstein*, a system of N atoms. The numbers of atoms in the level E_1 or E_2 are denoted by N_1 and N_2, respectively. The system is taken to be in thermal equilibrium with its surroundings. Interactions with the radiation field are only possible in the form of emission or absorption of radiation as discrete energy quanta $h\nu = E_2 - E_1$.

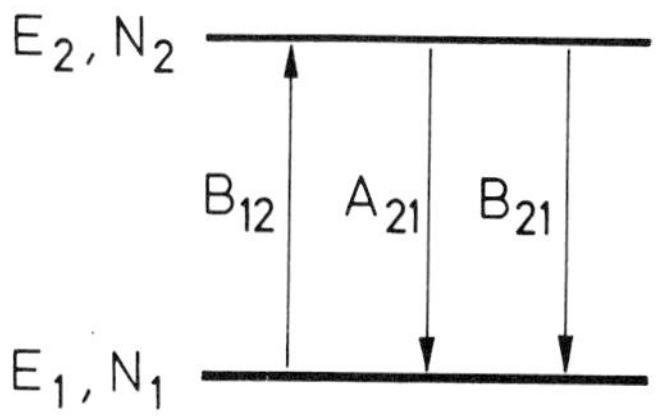

Fig. 5.5. Derivation of the Planck formula: two energy levels E_1 and E_2 with occupation numbers N_1 and N_2 are connected by transitions with the probabilities B_{12}, B_{21} and A_{21}

The radiation field is taken to have the radiation energy density $u(\nu, T)$, which we will denote in the following simply as $u(\nu)$. It then produces, per unit time, the following transitions (Fig. 5.5):

Absorption from 1 to 2. The number of processes in time dt is proportional to the occupation number N_1 of level 1 and to the radiation energy density $u(\nu)$:

$$dN_{12} = B_{12} u(\nu) N_1 dt\,. \tag{5.10}$$

The proportionality constant B_{12} is called the Einstein coefficient and is a measure of the transition probability per unit time and radiation density.
Transitions from 2 to 1 are composed of two contributions, as seen in Fig. 5.5: the first is *spontaneous emission from 2 to 1*. The number of such processes per unit time is proportional to the occupation number N_2. We have:

$$dN'_{21} = A_{21} N_2 dt\,. \tag{5.11}$$

A_{21} is also an Einstein coefficient and is a measure for the transition probability per unit time. Furthermore, we have *induced emission from 2 to 1*. It is, analogously to (5.10), proportional to the occupation number N_2 and to the radiation density $u(\nu)$. The result is:

$$dN''_{21} = B_{21} u(\nu) N_2 dt\,. \tag{5.12}$$

B_{21} is defined in an analogous way to the Einstein coefficient B_{12} in (5.10).

In equilibrium, an equal number of transitions occurs in each direction. We must therefore have

$$dN_{12} = dN'_{21} + dN''_{21}\,. \tag{5.13}$$

Setting (5.10) and (5.11, 12) equal leads to the following ratio of the occupation numbers:

$$N_2/N_1 = \frac{B_{12} u(\nu)}{A_{21} + B_{21} u(\nu)}\,. \tag{5.14}$$

Since the system is in thermal equilibrium, the ratio of the occupation numbers of the two energy levels can also be calculated according to the Boltzmann distribution. It must then be true that

$$N_2/N_1 = \mathrm{e}^{-E_2/kT}/\mathrm{e}^{-E_1/kT}\,. \tag{5.15}$$

From these two equations follows

$$\frac{B_{12} u(\nu)}{A_{21} + B_{21} u(\nu)} = \mathrm{e}^{-E_2/kT}/\mathrm{e}^{-E_1/kT} \tag{5.16}$$

and

$$u(\nu) = \frac{A_{21}}{B_{12}\mathrm{e}^{h\nu/kT} - B_{21}} \tag{5.17}$$

with the abbreviation $E_2 - E_1 = h\nu$.

To determine the coefficients A and B, we use the limiting condition that $u(\nu)$ must go to infinity when $T \to \infty$, i.e., the denominator of (5.17) must go to zero. Then we have

$$B_{12} = B_{21}\,. \tag{5.18}$$

From this follows

$$u(\nu) = \frac{A_{21}}{B_{12}(e^{h\nu/kT} - 1)} . \tag{5.19}$$

Furthermore, the experimentally verified Rayleigh-Jeans law must hold for small frequencies, i.e., for $h\nu \ll kT$, see (5.5),

$$u(\nu) = \frac{8\pi\nu^2 kT}{c^3} . \tag{5.20}$$

For small values of the exponent $(h\nu/kT)$ we can use a series expansion for the exponential function: $\exp(h\nu/kT) = 1 + h\nu/kT + \dots$. Inserting this in (5.19) yields

$$u(\nu) = \frac{A_{21}}{B_{12}} \frac{kT}{h\nu} \quad (\text{for } h\nu \ll kT) , \tag{5.21}$$

which, combining with the Rayleigh-Jeans law (5.20), leads to

$$\frac{A_{21}}{B_{12}} = \frac{8\pi h\nu^3}{c^3} \quad (\text{holds generally}) . \tag{5.22}$$

Finally, inserting in (5.19),

$$u(\nu) = \frac{8\pi h\nu^3}{c^3} \frac{1}{e^{h\nu/kT} - 1} . \tag{5.23}$$

Equation (5.23) is the Planck formula.

Rearranging (5.22), we find for the relation between the Einstein coefficients for transitions between levels 2 and 1,

$$A_{21} = \frac{8\pi h\nu^3}{c^3} B_{12} . \tag{5.24}$$

This corresponds to the Kirchhoff relation, according to which the probabilities for spontaneous emission and absorption are proportional.

Equation (5.18) is, furthermore, an expression of the fact that the radiation field takes up and gives out radiation quanta in like fashion; absorption and stimulated emission are fully complementary physical processes.

Einstein's derivation of the Planck formula lends strong support to the existence of light quanta of energy $h\nu$. From the equation $E = h\nu$ and the equivalence of mass and energy, $E = mc^2$, it follows that a mass can also be ascribed to the photon, having the value $m_{ph} = h\nu/c^2$. However, the rest mass of the photon is in fact zero.

5.3 The Photoelectric Effect

In the year 1888, *Hallwachs* measured for the first time the laws governing the release of electrons from metals by light, the *photoelectric effect*, following earlier observations by *Hertz*. The results of his experiments were explained in 1905 by *Einstein* using the hypothesis of light quanta.

The photoeffect may be simply demonstrated, qualitatively, using the setup shown in Fig. 5.6. A zinc plate is activated on its surface by rubbing with mercury and is mounted in an electrically insulated holder. If it is negatively charged and then illuminated with ultraviolet light, it rapidly discharges. A positively charged plate cannot be discharged by light.

From these experiments we see that light sets electrons from the plate free. The negatively charged plate releases these electrons to the surrounding air; the positively charged plate retains them due to Coulomb attraction. These experiments may be made more quantitative by replacing the electrometer with a so-called dropping electrometer; the quantity of charge released from the plate can then be measured as a function of the intensity and energy of the light.

If one also wishes to measure the kinetic energy of the electrons, the counter-field method may be used: one measures the maximum voltage V_{max} which is just sufficient

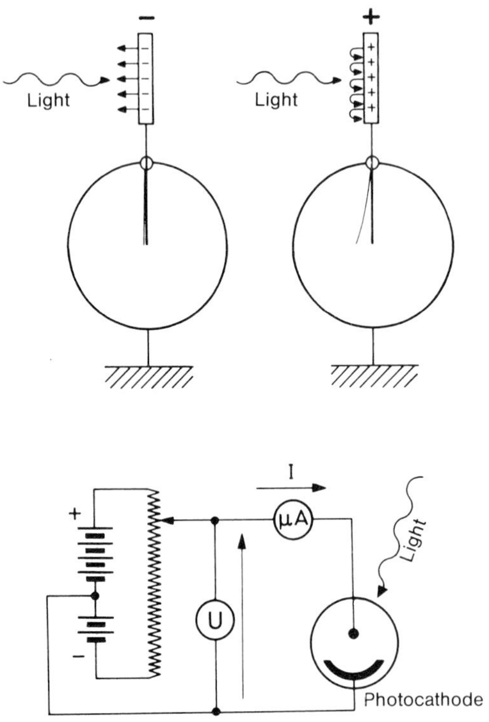

Fig. 5.6. Photoeffect. *Upper part:* A negatively charged electrometer is discharged upon illumination of the electrode, a positively charged one is not. *Lower part:* Arrangement for quantitative study of the photoeffect (voltage U, current I)

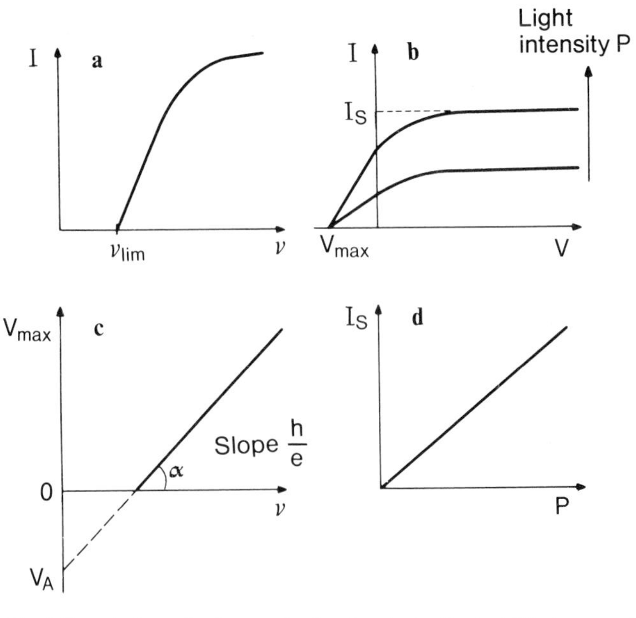

Fig. 5.7 a – d. Quantitative results for the photoeffect. **a)** Photocurrent I as a function of the frequency ν of the light. Below the limiting frequency ν_{lim} there is no longer a photocurrent. **b)** Photocurrent I as a function of the applied voltage V. Positive values of the voltage here mean that the irradiated electrode is the *cathode*. The largest negative voltage which can still be overcome by the photoelectrons (when the irradiated electrode is the *anode*) is V_{max}. The saturation current I_S is a function of the light intensity P. **c)** Maximum braking voltage V_{max} as a function of the light frequency ν; measurement of the ratio h/e and of the work function V_A as the slope and intercept of the straight line according to (5.28). **d)** Saturation current I_S as a function of the light intensity P. The current increases with the intensity

to keep the electrons from leaving the plate. For this purpose, the setup shown in the lower part of Fig. 5.6 may be used, with the illuminated electrode however attached to the *positive* pole of the voltage source.

The results of such an experiment are shown in Fig. 5.7: the electron current I as a function of the frequency ν of the light begins at a limiting frequency ν_{lim} which is characteristic for the electrode material (Fig. 5.7a). The maximum kinetic energy of the electrons follows from the current-voltage characteristic curve of the apparatus (Fig. 5.7b). If the counter voltage – the braking potential – reaches a certain value V_{max}, dependent on the frequency of the light, the photocurrent drops to zero and remains so. The electrons which are emitted no longer have sufficient energy to overcome the braking potential. The expression $eV_{\mathrm{max}} = mv^2/2$ gives the velocity of the electrons. If the maximum braking potential V_{max} is plotted against the frequency of the light, a straight line is found (Fig. 5.7c).

To understand these experiments, we need the *light quantum* hypothesis. Classically, it would be expected that the electric field E of the light, which is proportional to the square root of the light intensity, is responsible for the acceleration and release of the electrons from the electrode. The energy of the photoelectrons should increase with increasing light intensity. We find, however, that the energy of the photoelectrons does not depend on the light intensity (and thus on the radiation power), but only on the frequency of the light.

On the other hand, the number N of emitted electrons is proportional to the intensity P of the light (Fig. 5.7d).

Photoelectrons are only emitted when the light frequency is larger than a characteristic value ν_{lim} which depends on the electrode material. The following relation must hold:

$$h\nu \geqq h\nu_{\mathrm{lim}} = eV_{\mathrm{A}}\,. \tag{5.25}$$

Clearly, a part of the light energy $h\nu$ is used to release an electron from the metal electrode. For this purpose an amount of energy equal to eV_{A}, called the work function, is required. This work function is specific for the electrode material. The remainder of the energy of the light quantum is available to the electron as kinetic energy. The total energy of the light quantum is thus transferred in an elementary process to the electron. The energy balance is given by

$$mv^2/2 \qquad = h\nu \qquad - eV_{\mathrm{A}}\,. \tag{5.26}$$

kinetic energy of the photoelectron = quantum energy of the light − work function of the photoelectron

The kinetic energy of the photoelectrons is equal to the energy eV_{max}; thus we can write (5.26) in the form

$$eV_{\mathrm{max}} = h\nu - eV_{\mathrm{A}} \tag{5.27}$$

or

$$V_{\mathrm{max}} = h\nu/e - V_{\mathrm{A}}\,. \tag{5.28}$$

The slope of the straight line which is obtained by plotting V_{max} against the frequency ν of the exciting light (Fig. 5.7c) can be used for a precision measurement of the ratio h/e. For the angle α of the slope we have

$$\tan \alpha = h/e\,. \tag{5.29}$$

Table 5.1 gives some examples for the work functions of various metals. The alkali metals are notable for their especially small work functions.

Table 5.1. Work functions eV_A and limiting wavelengths λ_{lim} of several metals

Metal	eV_A [eV]	λ_{lim} [nm]
Li	2.46	504
Na	2.28	543
K	2.25	551
Rb	2.13	582
Cs	1.94	639
Cu	4.48	277
Pt	5.36	231

An arrangement in which the electrons released from an electrode (photocathode) complete a circuit, open in the absence of light, between the photocathode and another electrode is called a *photocell.* Photocells are used in numerous applications in measurement and control technology.

Apart from the so-called external photoeffect, which we have discussed above, the same phenomenon is met again in many other areas of physics. In solid state physics, the release of normally bound charge carriers by the action of light is called the *internal photoeffect*. In this case, an increase in the electrical conductivity of semiconductors or insulators upon illumination may be observed. In nuclear physics, atomic nuclei can be excited and caused to emit nucleons, i.e., the particles of which the nucleus is composed, upon absorption of very short wavelength radiation (x-rays or gamma rays). This is termed the *nuclear photoeffect*.

5.4 The Compton Effect

5.4.1 Experiments

The Compton effect is the name given to the scattering of light by weakly bound or free electrons. This effect occurs in particular in the x-ray region of the electromagnetic spectrum. The incident light wave (x-radiation) excites electrons in the target atoms so that they oscillate. The oscillating electrons in the field of the positively charged nuclei may be considered as classical oscillators; they emit radiation themselves, with the same frequency as that with which they are driven. This radiation is called Rayleigh-scattered radiation. The theory of Rayleigh scattering was first developed for visible light; it explains the blue colour of the sky. Light of short wavelength (blue) is more strongly scattered than light of long wavelength (red light). The scattered radiation has the same wavelength as the primary radiation, and is polarised. In 1909 it was shown by *Barkla* that this type of scattering also occurs with x-radiation.

In 1921, *Compton* observed that in addition to the spectrally unshifted scattered radiation, a spectrally displaced component appeared (Fig. 5.8). There is a simple

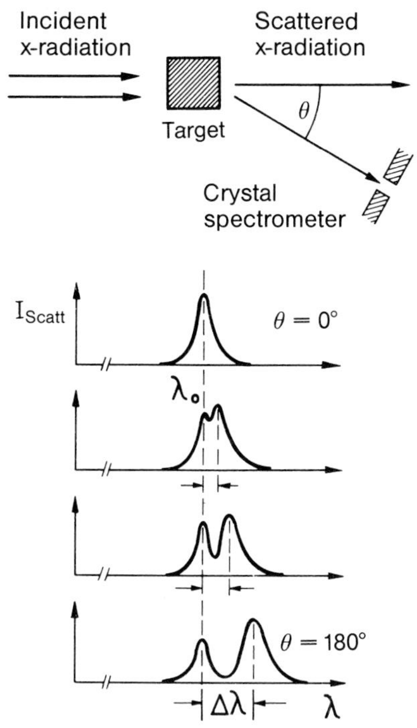

Fig. 5.8. Compton Effect. *Upper part:* Experimental setup. The x-radiation which is scattered by the target (e.g., a graphite block) is measured as a function of the scattering angle θ, which is defined to be 0° for undeflected radiation and 180° for radiation reflected back towards the source. *Lower part:* Measured scattered radiation as a function of wavelength for various scattering angles. The unshifted Rayleigh-scattered radiation as well as the shifted Compton-scattered radiation are seen

relation between the shift in the wavelength and the scattering angle: independently of the target material, it is found that

$$\Delta\lambda = \lambda_c(1 - \cos\theta) \tag{5.30}$$

with the *Compton wavelength* $\lambda_c = 0.024$ Å. The wavelength shift $\Delta\lambda$ is also completely independent of the primary wavelength. Only the *intensity* of the Compton scattering depends on the target material: it is especially large for light materials (Z small), as compared to the x-ray absorption which increases approximately with Z^3 (18.5).

Two numerical examples may serve to illustrate Compton scattering; the wavelength shift is a maximum at $\theta = 180°$. At an energy of 1000 eV for the light quanta before scattering, the energy of the 180°-scattered radiation is 996 eV; at a primary energy of 1 MeV, the 180°-scattered radiation has an energy 200 keV. In the former case, the energy is reduced by 4 eV or 0.4%, in the latter by 800 keV or 80%. In both cases, the corresponding shift in the *wavelength* is about $\Delta\lambda = 0.050$ Å.

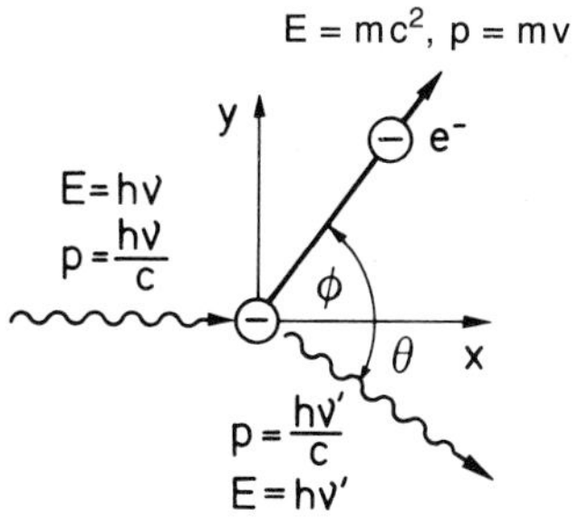

Fig. 5.9. Explanation of the Compton effect: the incident x-ray quantum with energy $E = h\nu$ and momentum $p = h\nu/c$ collides with an electron. In the collision, it transfers energy and momentum to the electron; the scattered x-ray quantum thus has a reduced energy $h\nu'$ and a reduced momentum $h\nu'/c$

The explanation of these experiments was not possible within the wave picture of light. Using the hypothesis of light quanta, the effect can be represented as a collision between two particles, a photon and an electron (Fig. 5.9). In the collision, energy and momentum are transferred.

More precisely, we are dealing with an elastic collision between radiation quanta and electrons that are weakly bound in the outer shells of atoms with initial velocities $v_0 \cong 0$. The binding energy of the electrons is assumed to be so small that it can be neglected in comparison with the photon energy in the following derivation.

5.4.2 Derivation of the Compton Shift

We consider the Compton effect as an elastic collision between a photon and an electron. Energy and momentum conservation must both be fulfilled. The momentum and the kinetic energy of the electron before the collision are practically zero. The calculation must be done relativistically, leading to the following equations (see Fig. 5.9).

Energy before and after the collision is conserved, so that

$$h\nu + m_0 c^2 = h\nu' + mc^2 . \tag{5.31}$$

Here m_0 is the rest mass of the electron and m its mass after the collision, and ν and ν' are the frequencies of the radiation before and after the collision.

For the momentum in the y direction before and after the collision we have

$$0 = \frac{h\nu'}{c}\sin\theta - mv\sin\phi \tag{5.32}$$

and for the momentum in the x direction

$$\frac{h\nu}{c} = \frac{h\nu'}{c}\cos\theta + mv\cos\phi . \tag{5.33}$$

In (5.31), we move $h\nu'$ to the left side and abbreviate $\nu - \nu' = \Delta\nu$. We then express the mass of the moving electron in terms of the rest mass, using the relativistic mass formula $m = m_0(1 - v^2/c^2)^{-1/2}$. If we now square (5.31) and rearrange somewhat, we obtain

$$h^2(\Delta\nu)^2 + 2m_0c^2 h\Delta\nu = m_0^2c^4\frac{v^2}{c^2 - v^2} . \tag{5.34}$$

In order to eliminate the angle ϕ from (5.32) and (5.33), we solve these equations for $\sin\phi$ and $\cos\phi$ and use the identity $\sin^2\phi + \cos^2\phi = 1$. When we rearrange, we obtain

$$h^2[(\Delta\nu)^2 + 2\nu(\nu - \Delta\nu)(1 - \cos\theta)] = m_0^2c^4\frac{v^2}{c^2 - v^2} . \tag{5.35}$$

Since the right-hand sides of (5.34) and (5.35) are identical, we can set the left-hand sides equal to one another, obtaining

$$m_0c^2h\Delta\nu = h^2\nu(\nu - \Delta\nu)(1 - \cos\theta) . \tag{5.36}$$

We wish to express the result in terms of wavelengths instead of frequencies. Using $c = \lambda\nu$, we obtain

$$|\Delta\lambda| = \left|\frac{c}{\nu} - \frac{c}{\nu - \Delta\nu}\right| = \frac{c\Delta\nu}{\nu(\nu - \Delta\nu)} . \tag{5.37}$$

Inserting (5.37) into (5.36), we have finally

$$|\Delta\lambda| = \frac{h}{m_0 c}(1 - \cos\theta) = \lambda_c(1 - \cos\theta) , \tag{5.38}$$

where we have introduced the abbreviation $\lambda_c = h(m_0 c)^{-1}$ (the "Compton wavelength"). Incidentally, the quantum energy of radiation which has the Compton wavelength λ_c is just equal to the rest energy of an electron:

$$h\nu = hc/\lambda_c = m_0 c^2 = 511 \text{ keV} . \tag{5.39}$$

The energy and momentum of the recoil electrons may also be calculated with these equations. The energy received by the electrons is relatively small, but their paths can still be seen in a cloud chamber and be quantitatively determined. This was shown in 1925 by *Compton* and *Simon*.

Another experiment done by *Bothe* and *Geiger* in 1925 shows that the electrons and the photons are, in fact, "emitted" simultaneously in the Compton effect (Fig. 5.10). A scattering target is set up in the centre between symmetrically placed electron and photon detectors. The number of simultaneous counts in the two detectors is measured with a coincidence circuit, with the result that the number of coincidences observed is far greater than the number which would be expected by chance from uncorrelated events.

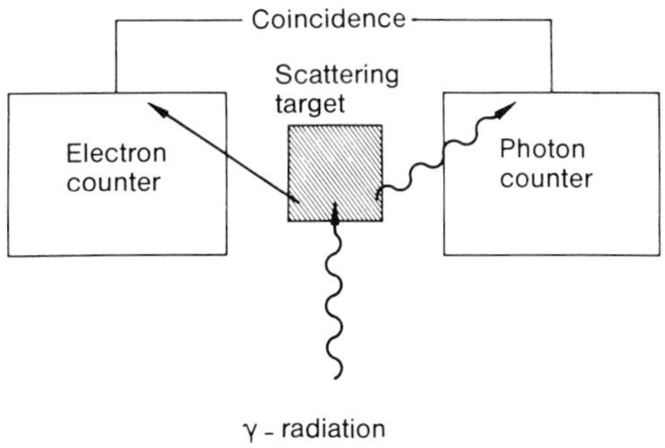

Fig. 5.10. Apparatus to detect coincidences between scattered x-ray quanta and recoil electrons from the Compton effect

The following remarks may be useful for a deeper understanding of the Compton effect:

- Compton scattering is relatively weak in the case of strongly bound electrons, that is in heavy atoms. When the electron's binding energy is large compared to the quantum energy $h\nu$ of the photon, no momentum transfer is possible.

- In certain energy ranges – in particular for medium-hard x-rays – the Compton effect is the principal cause of scattering and attenuation of radiation in matter.
- In Compton scattering, the incident and the scattered radiation are *incoherent* relative to one another (if $h\nu \neq h\nu'$).

We see that the photoeffect and the Compton effect can only be understood by assuming that light (radiation) consists of individual particles with a momentum p. On the other hand, we know from interference and diffraction experiments that light behaves as a wave, characterised by a wavelength λ and a frequency ν. The particle aspects of light which we summarised at the beginning of this chapter find their confirmation in the photoelectric and Compton effects.

How can we reconcile the wave and the particle character of the same phenomenon? To clarify this question we imagine the following experiment (Fig. 7.5): we allow a beam of light to pass through a small hole in an opaque wall and to fall on a screen. On this screen we observe, following the laws of wave optics, a diffraction pattern. We could, however, detect the light falling on the screen by means of the photoeffect or the Compton effect. If we now make the intensity of the light weaker and weaker, we find, using the photoeffect as detector, that locally, at particular points, single photoelectrons are set free by the light. The particle character of the radiation is here predominant. If this experiment is now repeated a number of times and the relative abundance of photoelectron releases is kept track of as a function of position, a distribution curve will be obtained which is in exact agreement with the diffraction pattern.

This thought experiment, which could actually be carried out as a real experiment, gives us the key to explaining the nature of light. Light carries – virtually, so to speak – both properties – wave and particle – within itself. Depending on the experiment we carry out, it shows us the one or the other aspect of its character. In order to combine both aspects, which at first appear contradictory, we have to apply statistical considerations. Thus in the case just described, if we perform an experiment which is intended to detect the diffraction pattern, but then inquire about the particle character, we find that we cannot predict the point at which the light particle will strike the screen with certainty. Instead, we can only give the probability that it will arrive at a particular point. The resulting probability distribution is then identical with the diffraction pattern which we would calculate according to the laws of classical physics. This statistical way of considering things is, as we shall see again and again, fundamental to the quantum mechanical interpretation of physical phenomena (Chap. 7.2).

Problems

5.1 Express the relativistic mass of a photon in terms of h, λ and c.

5.2 What is the momentum of a photon with 1 eV energy? Give the corresponding wavelength in Ångstroms.

5.3 How much mass does a 100 W light bulb lose in one year due to light emission?

5.4 A photon with 2 MeV of energy is converted into a positron-electron pair. What is the kinetic energy of the positron and electron, if the energy is equally distributed between the two particles and the electrostatic interaction between the two is ignored? ($m_{e^-} = m_{e^+} \triangleq 0.511$ MeV). How large is their velocity?

5.5 In the upper atmosphere, molecular oxygen is split into two oxygen atoms by photons from the sun. The longest wavelength of photons which can do this have $\lambda = 1.75 \times 10^{-7}$ m. What is the binding energy of O_2?

5.6 A person can perceive yellow light with the naked eye when the power being delivered to the retina is 1.8×10^{-18} W. The wavelength of yellow light is about 6000 Å. At this power, how many photons fall on the retina each second?

5.7 A monochromatic beam of electromagnetic radiation has an intensity of 1 W/m^2. What is the average number N of photons per m^2 and second for a) 1 kHz radio waves and b) 10 MeV gamma rays?

5.8 Calculate the radiation pressure of sunlight when the incoming energy/(s m^2) is 1.4×10^3 W/m^2 and the radiation is completely absorbed. Compare this value with the atmospheric pressure. What is the force on a surface of 1 m^2? What is the force if the light is completely reflected?

Hint: The radiation pressure is the momentum transferred per unit of time and surface area.

5.9 A photon which is emitted by an atom imparts an equal and opposite momentum to the atom.

a) What is the kinetic energy transferred to the atom if the frequency of the photon is ν and the mass of the atom is M?
b) How much energy is transferred to the Hg atom in the emission of the mercury spectral line $\lambda = 2357$ Å? ($M_{\text{Hg}} \approx 200.6$ u).
c) What is the corresponding reaction energy in the emission of γ quanta with 1.33 MeV energy by ^{60}Ni? ($M_{\text{Ni}} \approx 58.7$ u).

Compare these values with the energy uncertainty due to the lifetime according to (7.29) ($\tau_{\text{Hg}} \simeq 10^{-8}$ s, $\tau_{\text{Ni}} \simeq 10^{-14}$ s).

5.10 What is the temperature of a black sphere with a diameter of 10 cm which is emitting a total of 100 W thermal radiation?

Hint: Use the numerical values from (5.3).

5.11 Calculate the temperature of the sun and the energy density of the radiation in its interior, assuming that the sun is a spherical black body with a radius R of 7×10^8 m. The intensity of the solar radiation at the surface of the earth (which is 1.5×10^{11} m from the sun) is 1.4×10^3 W/m^2. Assume that the energy density in the interior of the sun is homogeneous. Is this realistic?

5.12 What is the wavelength of the spectral maximum of the radiation from a black body at 300 K (room temperature)? Calculate the monochromatic energy density at this frequency.

5.13 A photon releases a photoelectron with an energy of 2 eV from a metal which has a work function of 2 eV. What is the smallest possible value for the energy of this photon?

5.14 The work function for the photoeffect in potassium is 2.25 eV. For the case when light with a wavelength of 3.6×10^{-7} m falls on the potassium, calculate a) the braking potential U_{max} of the photoelectrons and b) the kinetic energy and the velocity of the fastest of the emitted electrons.

5.15 A homogeneous monochromatic light beam, wavelength 4.0×10^{-7} m, falls perpendicularly on a material with a work function of 2.0 eV. The beam intensity is 3.0×10^{-9} W/m^2. Calculate a) the number of electrons emitted per m^2 and second, b) the energy absorbed per m^2 and second, and c) the kinetic energy of the photoelectrons.

5.16 A metal surface is irradiated with light of various wavelengths λ. The braking potentials V indicated in the table are measured for the photoelectrons.

λ [10^{-7} m]	V [V]	λ [10^{-7} m]	V [V]
3.66	1.48	4.92	0.62
4.05	1.15	5.46	0.36
4.36	0.93	5.79	0.24

Plot the braking potential along the ordinate versus the frequency of the light along the abscissa. Calculate from the curve a) the threshold frequency, b) the photoelectric work function of the metal, and c) the quotient h/e.

5.17 The yellow D lines of sodium appear when sodium vapour is irradiated with electrons which have been accelerated through a potential difference of 2.11 V. Calculate the value of h/e.

5.18 In a Compton effect experiment, the scattered light quantum is observed at an angle of 60° to the direction of the incident light. After the collision, the scattered electron moves in a circular path with a radius $R = 1.5$ cm in a magnetic field $|\boldsymbol{B}| = 0.02$ Vs/m^2 which is perpendicular to the plane of the electron path. What is the energy and the wavelength of the incident light quantum?

Hint: Use (6.7) with $|\boldsymbol{E}| = 0$ to calculate the electron's path.

5.19 A photon with 10^4 eV energy collides with a free electron at rest and is scattered through an angle of 60°. Calculate a) the change in energy, frequency and wavelength of the photon, and b) the kinetic energy, momentum and direction of the electron after the collision.

5.20 X-rays with a wavelength of 1 Å are scattered on graphite. The scattered radiation is observed perpendicular to the direction of the incident x-rays.

a) How large is the Compton shift $\Delta\lambda$?
b) How large is the kinetic energy of the ejected electron?
c) What fraction of its original energy does the photon lose?
d) How large is the corresponding fraction of energy lost by a photon with a wavelength $\lambda = 0.1$ Å if it is deflected through 90° by Compton scattering?

The electron should be considered at rest before the collision, and the binding energy should be neglected.

6. The Electron

6.1 Production of Free Electrons

The name "electron", which is derived from the Greek word for amber, was coined by the English physicist *Stoney* in 1894.

In the early days of atomic physics, free electrons were usually produced as cathode rays from gas dicharges. Now, however, they are most often obtained using thermionic emission from wires. This process has the advantage that the electrons can easily be focussed and accelerated. Free electrons can also be produced by utilising the photoeffect (Chap. 5) or in the form of emissions from radioactive nuclei.

6.2 Size of the Electron

The electron is just as invisible as the atom; indeed, as a component of the latter, it must be smaller. We first arbitrarily define a parameter, called the classical electron radius, by making the following assumptions:

- the electron is a sphere with radius r_{el} and surface charge $-e$,
- the energy of the rest mass $E = m_0c^2$ is equal to the potential electrostatic energy of the surface charge.

We use the formulae of classical electrostatics to calculate the electrostatic energy. The capacitance of a spherical surface of radius r is

$$C = 4\pi\varepsilon_0 r\,. \tag{6.1}$$

The work required to add a charge q to a capacitor with capacitance C is

$$W = \tfrac{1}{2}q^2/C\,. \tag{6.2}$$

Therefore the potential energy of a spherical capacitor, i.e. the energy of its electrostatic field, is

$$E_{pot} = \frac{e^2}{8\pi\varepsilon_0 r}\,. \tag{6.3}$$

The condition that $E_{pot} = m_0c^2$ (m_0 is the rest mass of the electron and c is the velocity of light) determines the radius $r = r_{el}$:

$$r_{el} = \frac{e^2}{2\cdot 4\pi\varepsilon_0 m_0 c^2}\,. \tag{6.4}$$

Other assumptions regarding the distribution of charge (e.g., continuous distribution of charge throughout the volume instead of a surface charge) lead to somewhat different numerical values. The "classical" electron radius is, finally, defined as

$$r_{el} = \frac{e^2}{4\pi\varepsilon_0 m_0 c^2} = 2.8 \cdot 10^{-15}\,\mathrm{m}\,. \tag{6.5}$$

It can be seen from this derivation that the parameter is purely conceptual. Is it possible to measure the electron's radius? In principle, yes. For example, the scattering cross section can be determined by irradiating with x-rays, exactly as the scattering cross sections of gas atoms may be determined (Chap. 4). The result of such experiments is a cross section $\sigma = \pi \cdot r_{el}^2$, and the parameter r_{el} is found to be of the same order of magnitude as the classical electron radius defined above.

Experiments in which electrons are scattered by electrons reveal no deviations from the Coulomb law, even at very small collision distances. All results up to now have thus yielded only the information that the electron is a structureless, point-like particle.

6.3 The Charge of the Electron

As mentioned earlier, the charge of the electron, $-e$, can be derived from the Faraday constant F:

$$e = F/N_A\,. \tag{6.6}$$

However, since we wish to use this equation to determine Avagadro's number N_A, we require an independent method of measuring the elementary charge e.

This was accomplished by *Millikan*'s experiment (1911), in which the charge on small drops of oil is determined from their motion in the electric field of a capacitor. It remains the best method for determining e.

The principle of the method is illustrated in Fig. 6.1. Figure 6.2 shows the entire experimental setup. The rising or falling velocity of a charged oil droplet in the homogeneous electric field of a capacitor is determined by the resultant of all the forces acting on the droplet: electrostatic force, gravitational force, buoyancy in the air, and friction with the air. We will pass over the details of the measurement procedure; we mention only, as a curiosity, the fact that an incorrect value of the quantity e was used for some years due to an error in the determination of the viscosity of the air. The latter quantity is needed to calculate the frictional force acting on the droplet which is rising or falling. The best value of the elementary charge is currently $e = (1.6021917 \pm 0.0000070) \cdot 10^{-19}\,\mathrm{C}$.

The question has been repeatedly raised, whether there are smaller amounts of charge than the so-called elementary charge e. Up to the present time, no smaller charges have been detected unambiguously.

To explain the structure of elementary particles, the existence of more fundamental elementary particles, the "quarks", has been suggested; they would have charges of $e/3$ or $2e/3$. Some experiments in high energy physics can, in fact, be interpreted by the assumption that such particles exist, but are bound to each other and/or to their anti-

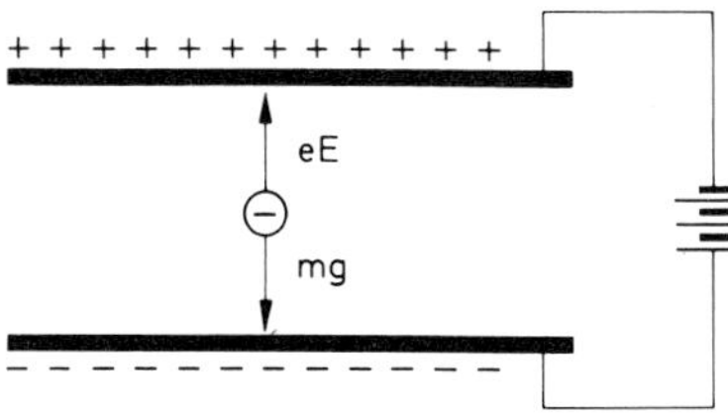

Fig. 6.1. Principle of the Millikan oil-drop experiment for measuring the charge of the electron. A negatively charged oil droplet experiences the force neE, where n is the number of elementary charges on the droplet; the graviational force mg acts in the opposite direction

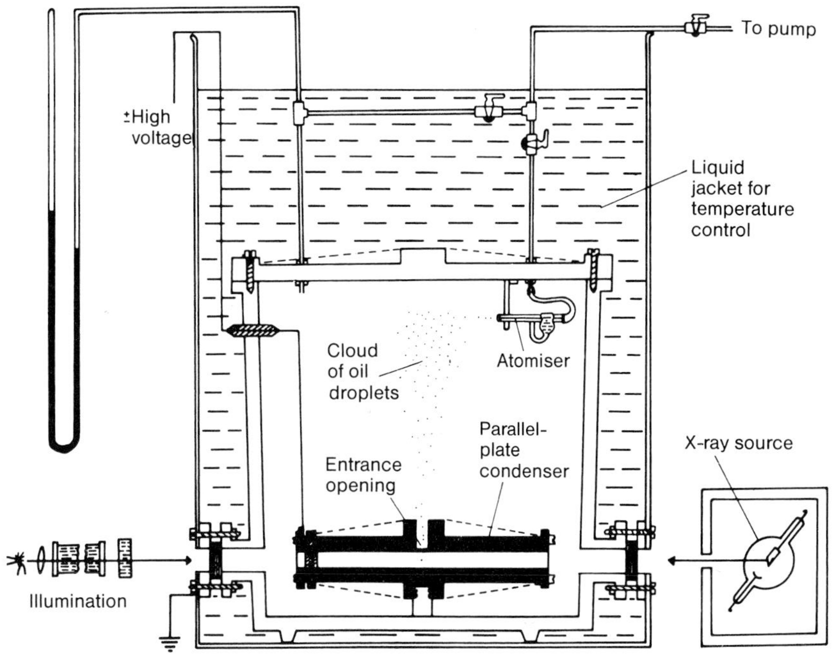

Fig. 6.2. Experimental arrangement of *Millikan* from Phys. Rev. **2**, 109 (1913). The oil droplets, which are formed by the atomiser, can be charged or discharged by irradiation with x-rays

particles. On the other hand, there has so far been no convincing proof of the existence of free quarks, and theoreticians have even developed a theory of "confinement". According to this theory, the forces between quarks become so large that they can never appear as individual particles.

6.4 The Specific Charge e/m of the Electron

The mass of the electron is determined by measuring the deflection of electrons in electric and magnetic fields. The motion is determined by the ratio of charge to mass e/m according to the equation

$$\boldsymbol{F} = m\frac{d\boldsymbol{v}}{dt} = -e[\boldsymbol{E}+(\boldsymbol{v}\times\boldsymbol{B})] . \tag{6.7}$$

Following the first e/m measurement by *Thompson* (1897), many methods for measuring this quantity were developed in the next 50 years, but all of them were basically variations of the same principle which we have already discussed for the parabolic method in mass spectroscopy.

The method of *Classen* (1907) is particularly elegant (Fig. 6.3). The $\boldsymbol{E}$ field between the cathode and the film imparts a uniform, known velocity to the electrons. The kinetic energy, to a non-relativistic approximation, is

$$\frac{m}{2}v^2 = eV , \tag{6.8}$$

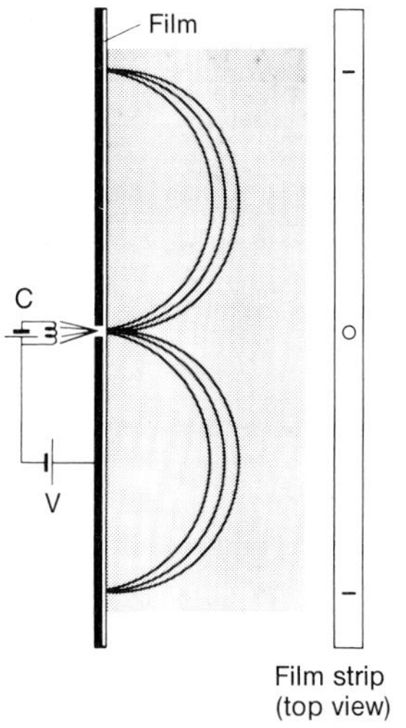

Fig. 6.3. Experimental arrangement for measurement of the specific mass m/e of the electron. The electrons are generated at the cathode C and accelerated by the voltage V. They are deflected into circular paths by a magnetic field perpendicular to the plane of the figure and recorded on a film. The direction of the deflection is reversed by reversing the poles of the magnet

where V is the accelerating voltage. Rearranging,

$$v = \sqrt{\frac{2eV}{m}} \, . \tag{6.9}$$

The $\boldsymbol{B}$ field deflects the electrons into a circular path with radius r. Equating the Lorentz and the centrifugal forces yields

$$\frac{mv^2}{r} = evB \, . \tag{6.10}$$

The desired ratio of charge to mass follows from (6.9) and (6.10):

$$\frac{e}{m} = \frac{2V}{r^2 B^2} \, . \tag{6.11}$$

One can thus obtain the ratio e/m by measurement of a voltage, a magnetic field strength and a distance. Over the years, various other methods were used to measure the specific mass m/e. They differ primarily in the relative positions of the electric and magnetic fields. Figure 6.4 shows one of these other experimental arrangements, one which corresponds in principle to the Aston mass spectrograph.

The dependence of the mass on the velocity was examined quite early with the help of these experiments. Table 6.1 lists some measured values for e/m.

The limiting value of m as the kinetic energy of the electron went to zero was found to be $m_0 = 1.7588 \cdot 10^{-31}$ kg or $(5.485930 \pm 0.000034) \cdot 10^{-4}$ u. 1 u is thus 1822.84 m_0.

The dependence of the ratio e/m on the particle velocity was found experimentally in 1901 (4 years before *Einstein*'s theory of relativity) by *Kaufmann*. *Kaufmann* used the method known in mass spectroscopy as the Thomson parabola method (Figs. 3.2, 3). The particles are deflected in transverse $\boldsymbol{E}$ and $\boldsymbol{B}$ fields. The electrons studied by *Kaufmann* were β particles from radioactive sources, because his experiments were

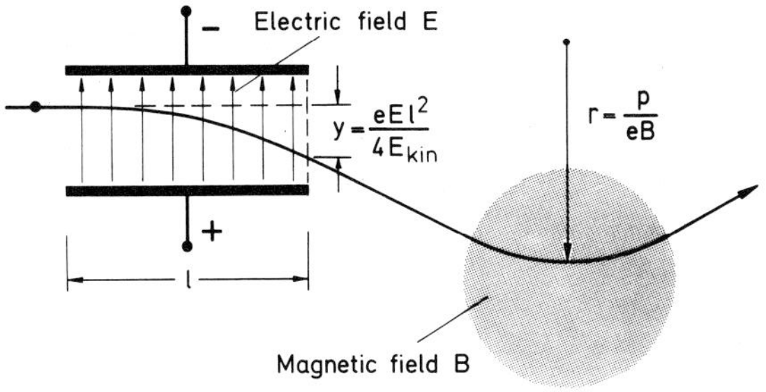

Fig. 6.4. Arrangement for measurement of the ratio e/m for electrons. Here the electrons are deflected first by an electric field and then by a magnetic field

Table 6.1. Specific charge of the electron at different accelerating voltage

Accelerating voltage V [kV]	0	500	1000	1500
Measured specific charge e/m $[10^{11}$ C/kg]	1.76	0.88	0.56	0.44

meant to clarify the physical nature of β rays. However, the photographic record of the particles did not yield the segment of a parabola which would be expected if the ratio of e/m were constant at different velocities of the particles.

Kaufmann's curves can be understood in the following way: For a uniform value of e/m, each point of the parabola correspondings to a particular value of the velocity v. From the fact that a parabolic segment is actually observed for lower velocities (larger deflections), it can be concluded that the slower particles have a continuous velocity distribution at a constant mass. However, at high velocities v the mass appears to increase steadily. The curve therefore passes through points on each of a series of adjacent parabolas, corresponding to successively higher masses m.

These measurements were the first to substantiate the dependence of mass on velocity. They fit the Lorentz equation

$$m = m_0 \frac{1}{\sqrt{1 - v^2/c^2}} . \qquad (6.12)$$

It was later shown that this equation can also be derived from the theory of relativity, if the validity and Lorentz invariance of the conservation of energy is assumed. The equation is equivalent to the principle $E = mc^2$. Figure 6.5 shows experimental values of the dependence of the mass on the velocity.

The following qualitative argument was advanced in an attempt to understand the change in mass with velocity: When the electron is accelerated, part of the energy is absorbed in the creation of the magnetic field of the moving electron – which is of course an electric current. Thus an "electromagnetic mass" is added to the inertial

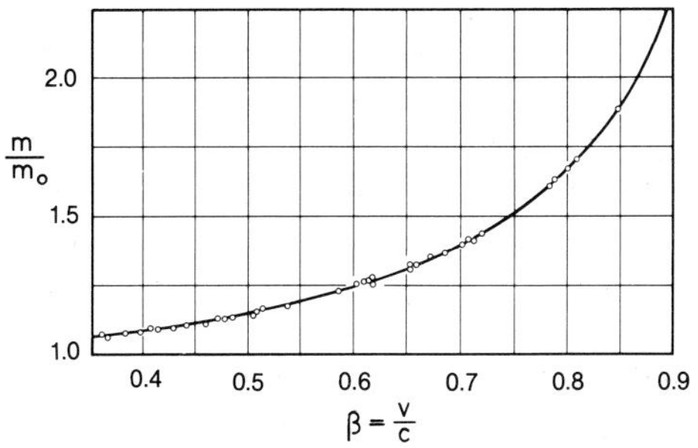

Fig. 6.5. Experimental values for the mass of the electron as a function of its velocity (test of the Lorentz formula). The mass m, in units of the rest mass m_0, is plotted against the velocity in units of the velocity of light, v/c

mass. This argument fails, however, because the mass increase is also observed with neutral particles, e.g. neutrons.

As time went on, the Lorentz equation was tested by many highly precise measurements. It was possible to confirm it to within 1.5% by measurement of electron energies under 1 MeV. In modern electron accelerators, it has been confirmed to far better precision, as shown in Fig. 6.5, and is now a well established tenet of physics.

It is instructive to consider the error one would make by neglecting the relativistic mass increase. From conservation of energy and (6.12) we find for a kinetic energy $E_{kin} = 1$ keV a velocity $v/c = 0.063$, and for 1 MeV, $v/c = 0.942$. The relativistic mass increase is then according to (6.12) at 1 keV $4 \cdot 10^{-3}$ times the rest mass m_0, but at 1 MeV it is already 2 times m_0, i.e. $(m - m_0)/m_0 = 2$.

6.5 Wave Character of Electrons

The motion of electrons in electric and magnetic fields may initially be understood as a particle motion. We have treated it as such in explaining various experiments up to now. There are, however, a number of different experiments in which electrons and other particles show interference and diffraction phenomena, that is they exhibit wave character. In this section, we will discuss the experimental grounds for assuming the existence of *matter waves*.

Experiment 1: The Ramsauer Effect (1921)

The measurement of the interaction cross section for collisions of extremely slow electrons with gas atoms yielded very small values, much smaller than those found in the kinetic theory of gases; at somewhat higher electron energies, the values were found to be much larger (Fig. 6.6). This type of minimum in the velocity dependence of the interaction cross section could be explained as the result of diffraction by particles whose size was comparable with the wavelength of the electrons.

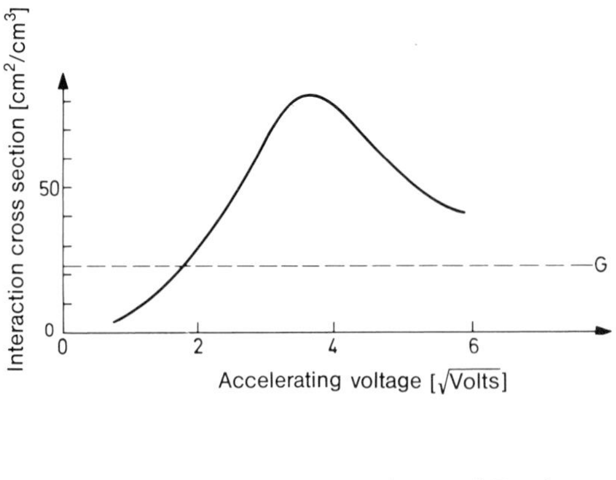

Fig. 6.6. Scattering coefficient $\alpha = n\sigma$ for electrons on gas atoms at various electron velocities (Ramsauer Effect). The dashed line represents the gas kinetic cross section G of the atoms. For the relation between scattering coefficient and cross section see (2.21)

Experiment 2: The Investigations of Davisson and Germer (1919);
Their Explanation (1927)

On reflecting slow electrons from crystals, *Davisson* and *Germer* observed interference effects, i.e. maxima and minima in the intensities of the reflected electrons, which were

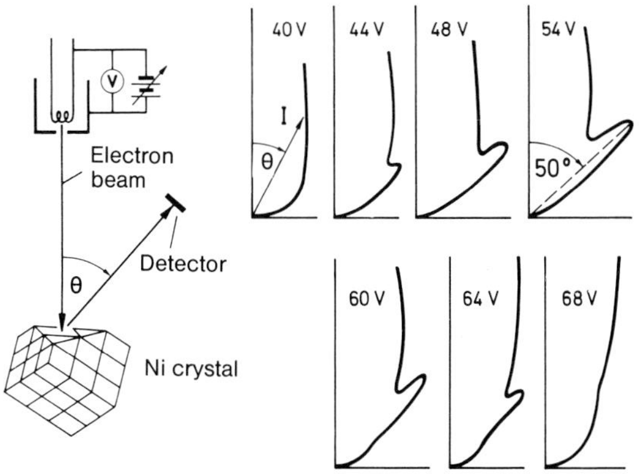

Fig. 6.7. Electron interference experiment of *Davisson* and *Germer*. *Left:* Diagram of the apparatus. The electrons accelerated through the voltage V are reflected from selected surfaces of a single crystal of nickel and recorded as a function of the reflection angle θ. *Right:* Angular distribution of reflected electrons at various accelerating voltages. The diagrams shown are polar plots in which the distance from the zero point to the curve corresponds to the intensity of reflection at the corresponding angle. There is a maximum at a scattering or reflection angle of about 50°, due to interference

uniquely determined by the electron velocities, the crystal orientation, and the angle of observation. Their experimental setup and the results are shown schematically in Fig. 6.7. The interference maxima and minima come about in a manner similar to x-ray diffraction from the lattice planes of a crystal (Bragg reflections, see Sect. 2.4.5). The occurrence of interference means that the motion of the electrons must be connected with a wave phenomenon. Indeed, *de Broglie* put forth the suggestion that just as light can possess a particle character, electrons must also have a wave character; he assumed the validity of the fundamental relation $p = h/\lambda$ between the momentum and the wavelength.

If we express the momentum by means of the mass and the velocity, i.e. $p = m_0 v$, and set $v = \sqrt{2E_{\mathrm{kin}}/m_0}$ for non-relativistic velocities, we find

$$\lambda = h/\sqrt{2m_0 E_{\mathrm{kin}}} \, . \tag{6.13}$$

It follows that for electrons which have been accelerated by a voltage V,

$$\lambda = \frac{12.3}{\sqrt{V}} \, [\text{Å}] \, .$$

The wavelength is measured in angstroms and the kinetic energy is converted to eV, because the electrons acquire their kinetic energy by traversing a voltage V. An accelerating voltage of 54 V, for example, produces $\lambda = 1.67$ Å.

De Broglie's hypothesis applies to all particles, not only electrons. The values given in Table 6.2 are for electrons.

Table 6.2. Wavelength of electrons in Å corresponding to various energies [eV], according to *de Broglie*

E_{kin}	[eV]	10	100	10^3	10^4	10^5	10^6	10^7	10^8
λ	[Å]	3.9	1.2	0.39	0.12	$3.7 \cdot 10^{-3}$	$8.7 \cdot 10^{-3}$	$1.2 \cdot 10^{-3}$	$1.2 \cdot 10^{-4}$

Many other experiments were suggested and could now be understood.

Experiment 3: Fresnel Diffraction from a Sharp Boundary, Boersch (1956)

One of the basic diffraction experiments in optics is the diffraction by a semi-infinite plane. Like light, electrons can be diffracted from a sharp boundary. In this experiment they are diffracted from the edge of an extremely thin foil of Al_2O_3 (Fig. 6.8).

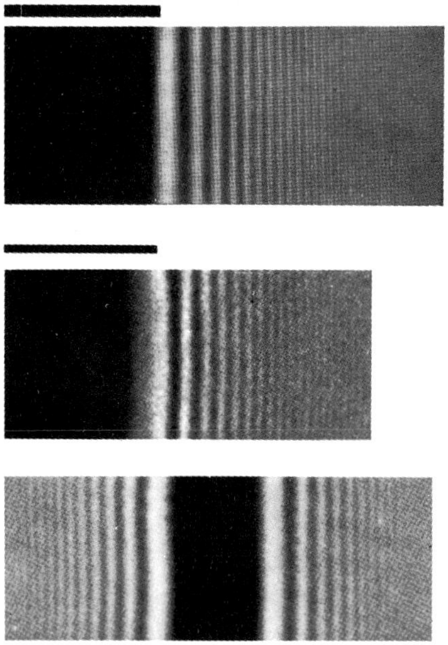

Fig. 6.8. *Above:* Diffraction lines of filtered red light at the geometric shadow boundary of a semi-infinite plane. *Middle:* Diffraction of electrons from the edge of an Al_2O_3 foil in a semi-infinite plane. $\lambda = 5 \cdot 10^{-12}$ m, corresponding to an electron energy of $3.4 \cdot 10^4$ eV. *Below:* Diffraction of electrons from a 2 μm thick gold-coated wire. Electron energy $1.94 \cdot 10^4$ eV. All photos from R. W. Pohl, *Optik und Atomphysik*, 11, 12th ed. (Springer, Berlin, Heidelberg, New York 1967) Figs. 202, 522, 523

Experiment 4: Diffraction from a Fresnel Double Prism, Möllenstedt (1956)

The Fresnel double prism experiment of classical optics was carried out with electrons. In this experiment, an electrically charged quartz fibre acts a double prism for electrons. Electrons from the two virtual electron sources interfere (Fig. 6.9). Measurement of the resulting interference lines in the image plane confirmed the de Broglie relationship to within 0.5% (Fig. 6.10).

Experiment 5: Atoms as Waves (1931)

The wave character of particles other than electrons was also demonstrated by interference experiments. *Stern*, *Frisch* and *Estermann* (1931) observed the diffraction of beams of helium atoms from the surface of a LiF crystal. The wavelength of the helium atoms is derived from the temperature $T = 400$ K and the mean kinetic energy $\bar{\varepsilon} = 3\,kT/2$ of the atomic beam. It lies in the angstrom range:

$$\lambda_{\mathrm{He}} = h/\sqrt{2mE} = h/\sqrt{3\,m_{\mathrm{He}}kT}\ .$$

This relationship has also been experimentally confirmed.

Interference and diffraction with particle beams are now included among the routine physical methods. Electrons can be used in the same interference experiments as x-rays.

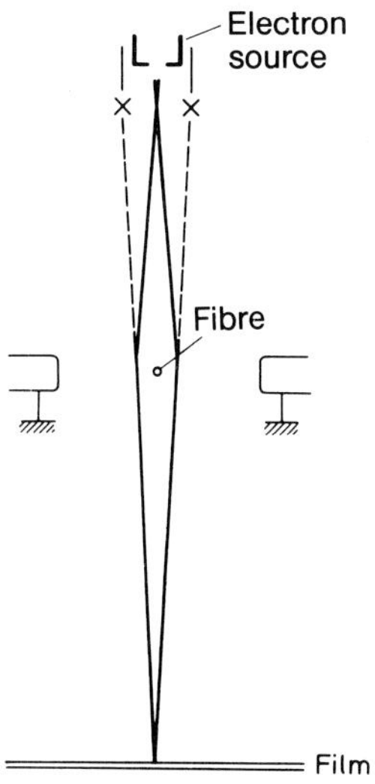

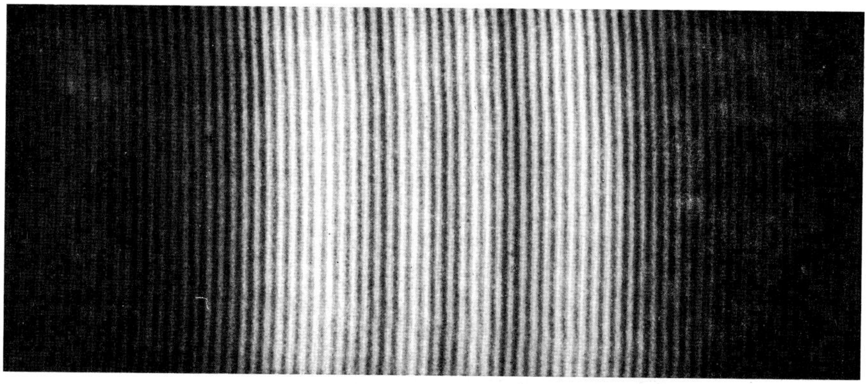

Fig. 6.10. Electron interference from the electrostatic double prism according to *Möllenstedt* and *Düker*. Data from Gerthsen, Kneser, Vogel: *Physik*, 13th ed. (Springer, Berlin, Heidelberg, New York 1977) Fig. 10.69

Fig. 6.9. Electron interference with an electrostatic double prism according to *Möllenstedt* and *Düker*. There is a voltage between the fibre and the counterelectrodes. The electrons are deflected by the resulting inhomogeneous field as shown. From Gerthsen, Kneser, Vogel: *Physik*, 13th ed. (Springer, Berlin, Heidelberg, New York 1977) Fig. 10.68

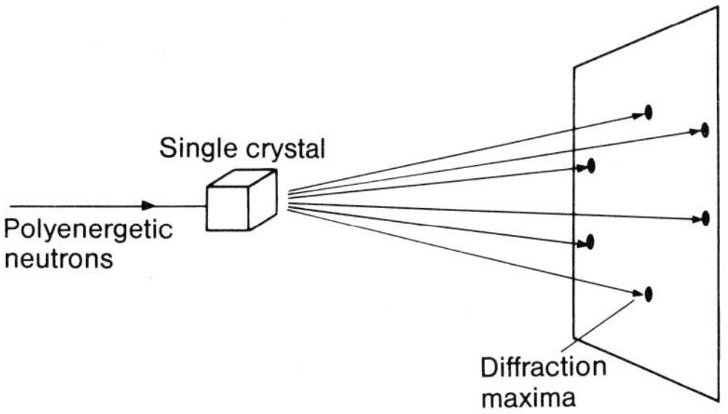

Fig. 6.11. Neutron diffraction from a single crystal, von Laue arrangement. With polyenergetic neutrons, one obtains Laue diagrams from scattering on the single crystal

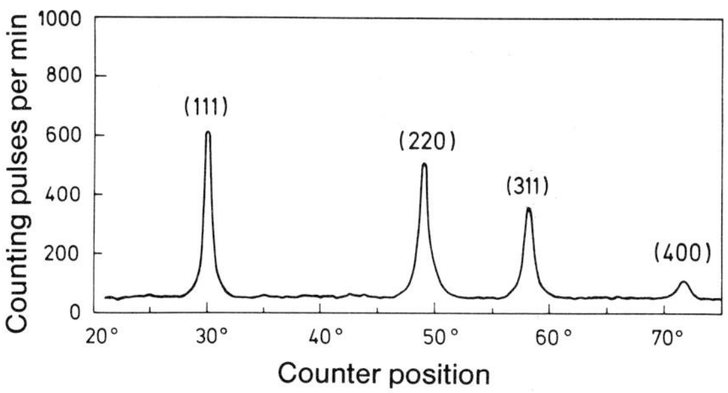

Fig. 6.13. Neutron diffraction from diamond powder, according to *G. Bacon*. Diffraction maxima occur from four families of lattice planes, which are denoted by the crystallographic indices (111), (220), (311) and (400)

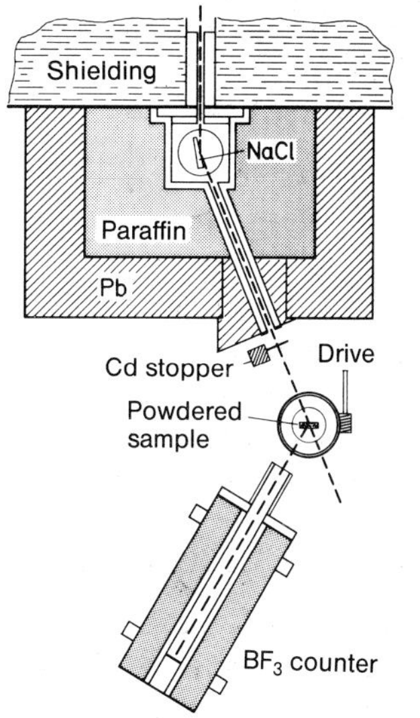

Fig. 6.12. Neutron spectrometer according to E. V. Wollan, C. G. Shull, Phys. Rev. **73**, 830 (1948). The neutrons are monochromatised by reflection from a NaCl crystal. When diffracted from a polycrystalline sample, they generate interference rings according to *Debye-Scherrer*. They are measured by means of a BF_3 counter

Neutron diffraction methods have become important in solid state physics. They are among the most useful means of determining crystal structures, analysing magnetically ordered systems (due to the existence of a neutron magnetic moment), and detection of lattice oscillation spectra in crystals.

Figure 6.11 shows schematically an arrangement for measurement of neutron diffraction from single crystals according to *von Laue*. It is completely analogous to the x-ray arrangement (Fig. 2.7). High intensity neutron beams are most conveniently obtained from a nuclear reactor. They are produced by nuclear fission with a velocity distribution which is continuous, within certain limits. If one wishes to work with particles of a single wavelength, or according to *de Broglie*, with uniform velocity, one must monochromatise the neutron beam.

This can be done, for example as shown in Fig. 6.12, by reflection from a single crystal (here NaCl). If these monochromatic neutrons are directed onto a polycrystalline or powdered sample, one obtains the same interference patterns as with the Debye-Scherrer x-ray technique (Fig. 2.10). An example of such a pattern is shown in Fig. 6.13; it is the result of diffraction of a neutron beam from diamond powder.

Problems

6.1 In the Millikan experiment to determine the elementary charge, a voltage of $V = 50$ V is applied between the plates of a capacitor, which are 1 cm apart. What must the diameter $2r$ of the oil droplets be, in order that a droplet with a single charge is held in suspension? What is the velocity $\boldsymbol{v}$ of the droplets as they fall when the direction of the electric field is reversed? The coefficient of viscosity of air is $\eta = 1.84 \times 10^{-4}$ poise, and the density of the oil $\varrho = 0.9\ \mathrm{g/cm^3}$.

Hint: The frictional force F_R is given by Stokes' Law, $\boldsymbol{F}_R = -6\pi\eta r\boldsymbol{v}$ (1 poise = 1 dyn s/cm^2 = 10^{-1} Ns/m^2).

6.2 A singly charged particle with 0.12 MeV of kinetic energy moves through a transverse homogeneous magnetic field with field strength $B = 100$ G (1 G = 10^{-4} T). Calculate the mass of the particle, given that the deflection from the original direction of travel is 3 mm in a 10 cm flight path.

6.3 To measure their specific charge, electrons are accelerated through a voltage V. Then they pass through the transverse fields of two small plate capacitors, which are placed at a distance l from one another. Both capacitors are connected to a single frequency generator (frequency ν). When the frequency is suitably adjusted, the electrons leave the second capacitor in their original direction of travel.

a) Under what condition is this possible?
Derive a relation between e/m and the experimental data.
b) What is the minimum frequency required from the generator if $V = 500$ V and $l = 10$ cm?
c) Sketch the apparatus.

6.4 If the kinetic energy of an electron is equivalent to its rest mass, what is its velocity?

6.5 The rest energy of the electron is 0.5 MeV. Give the ratio of inertial mass to rest mass for an electron with a kinetic energy of 1 MeV.

6.6 Calculate the de Broglie wavelength of an electron with the velocity $v = 0.8\,c$, using relativistic relationships.

6.7 Calculate the de Broglie wavelength of an electron with kinetic energy of 1 eV, 100 eV, 1000 eV, 100 keV. Which wavelengths will be noticeably diffracted in a nickel crystal, in which the atomic spacing is about 2.15 Å? Calculate the energy of those electrons which are scattered through angles of less than 30°.

6.8 What is the average kinetic energy and the corresponding de Broglie wavelength of thermal neutrons, i.e., neutrons which are in thermal equilibrium with matter at 25 °C? According to the Bragg formula, what is the angle of incidence at which the first interference maximum occurs when these neutrons are reflected from a NaCl crystal in which the lattice spacing d is 2.82 Å? The mass of the neutron is 1.675×10^{-27} kg.

6.9 Consider an electron which is far away from a proton and at rest. It is attracted to the proton. Calculate the magnitude of the wavelength of the electron when it has approached within a) 1 m and b) 0.5×10^{-10} m of the proton. (The latter distance is of the same order as the orbital radius of the electron in the ground state of the hydrogen atom.)

7. Some Basic Properties of Matter Waves

7.1 Wave Packets

In the two preceding chapters it was shown that light, electrons and other elementary particles can have both wave and particle characteristics. In this chapter we will examine more closely how the wave properties of matter can be understood and described mathematically.

For both light and material particles there are basic relationships between energy and frequency, and between momentum and wavelength, which are summarised in the following formulae:

Light	Matter
$E = h\nu$	$E = h\nu = \hbar\omega$
$p = \dfrac{h\nu}{c}$	$p = \dfrac{h}{\lambda} = \hbar k\,.$

(7.1)

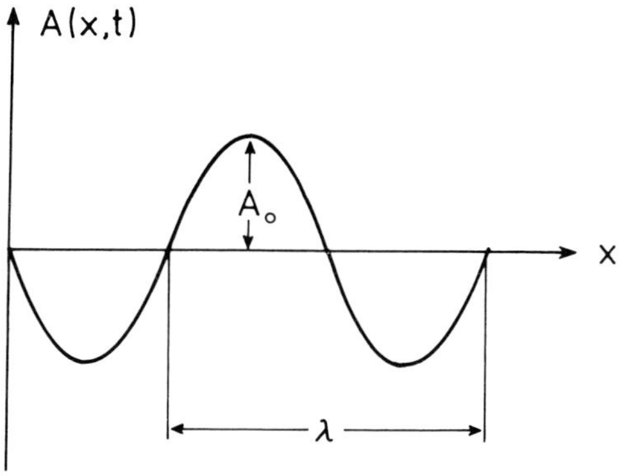

Fig. 7.1. Instantaneous view of a wave with amplitude A_0 and wavelength λ

We now wish to expand these relationships into a more exact theory. We are familiar with descriptions of wave motion from the study of light. If we consider a plane monochromatic wave (Fig. 7.1) travelling in the x direction, the wave amplitude A at time t and point x is $A(x,t) = A_0\cos(kx-\omega t)$. The wave number k is related to the wavelength λ by $k = 2\pi/\lambda$. The circular frequency ω is related to the frequency by $\omega = 2\pi\nu$. In many cases it is more useful to use complex notation, in which we express the cosine by exponential functions according to the formula

$$\cos\alpha = \tfrac{1}{2}(\mathrm{e}^{\mathrm{i}\alpha}+\mathrm{e}^{-\mathrm{i}\alpha})\,. \tag{7.2}$$

We accordingly expand $A(x,t)$:

$$A(x,t) = A_0 \tfrac{1}{2}[\exp(\mathrm{i}kx - \mathrm{i}\,\omega t) + \exp(-\mathrm{i}kx + \mathrm{i}\,\omega t)] \,. \tag{7.3}$$

Applying the relations (7.1), we obtain

$$\exp(\mathrm{i}kx - \mathrm{i}\,\omega t) = \exp\left[\frac{\mathrm{i}}{\hbar}(px - Et)\right] . \tag{7.4}$$

The wave represented by (7.4) is an infinitely long wave train.

On the other hand, since we ordinarily assume that particles ("point masses") are localised, we must consider whether we can, by superposing a sufficient number of suitable wave trains, arrive at some spatially concentrated sort of "wave". We are tempted to form what are called *wave packets*, in which the amplitude is localised in a certain region of space. In order to get an idea of how such wave packets can be built up, we first imagine that two wave trains of slightly differing frequencies and wavenumbers are superposed. We then obtain from the two amplitudes $A_1(x,t)$ and $A_2(x,t)$ a new amplitude $A(x,t)$ according to

$$A(x,t) = A_1(x,t) + A_2(x,t) \,, \tag{7.5}$$

or, using cosine waves of the same amplitude for A_1 and A_2,

$$A(x,t) = A_0[\cos(k_1 x - \omega_1 t) + \cos(k_2 x - \omega_2 t)] \,. \tag{7.6}$$

As we know from elementary mathematics, the right-hand side of (7.6) may be expressed as

$$2A_0 \cos(kx - \omega t)\cos(\Delta k x - \Delta\omega t) \,, \tag{7.7}$$

where

$$k = \tfrac{1}{2}(k_1 + k_2) \,, \qquad \omega = \tfrac{1}{2}(\omega_1 + \omega_2) \,,$$

and

$$\Delta k = \tfrac{1}{2}(k_1 - k_2) \,, \qquad \Delta\omega = \tfrac{1}{2}(\omega_1 - \omega_2) \,.$$

The resulting wave is sketched in Fig. 7.2. The wave is clearly amplified in some regions of space and attenuated in others. This suggests that we might produce a more and more complete localisation by superposing more and more cosine waves. This is, in fact, the case. To see how, we use the complex representation. We superpose waves of the form (7.4) for various wavenumbers k and assume that the wavenumbers form a continuous distribution. Thus, we form the integral

$$\int_{k_0-\Delta k}^{k_0+\Delta k} a \exp[\mathrm{i}(kx - \omega t)]\, dk = \psi(x,t) \,, \tag{7.8}$$

where a is taken to be a constant amplitude.

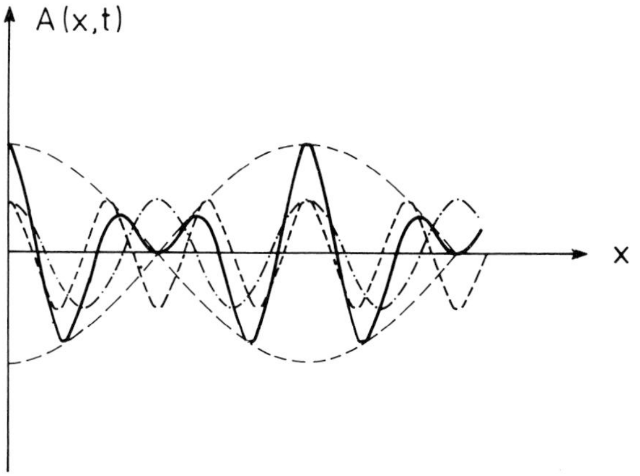

Fig. 7.2. Superposition of two waves of the same amplitude. Fundamental wave 1: (– · –), fundamental wave 2: (– –) same amplitude as 1. Resulting wave A: (——). The envelope $\cos(\Delta kx - \Delta\omega t)$ for constant t is also shown as a dashed curve

In taking this integral, we must notice that ω and k are related to one another, since the energy and the momentum of an electron are connected by the relation $E = p^2/(2m_0)$, and this in turn means that ω and k are related according to (7.1). To evaluate the integral we set

$$k = k_0 + (k - k_0) \tag{7.9}$$

and expand ω about the value k_0 using a Taylor series in $(k - k_0)$, which we terminate after the second term:

$$\omega = \omega_0 + \left(\frac{d\omega}{dk}\right)(k - k_0) + \ldots \,. \tag{7.10}$$

In the following, we abbreviate $d\omega/dk$ as ω'. Inserting (7.9) and (7.10) in (7.8), we obtain

$$\psi(x, t) = a\exp[-\mathrm{i}(\omega_0 t - k_0 x)] \int_{-\Delta k}^{\Delta k} \exp[-\mathrm{i}(\omega' t - x)\xi]\, d\xi\,, \tag{7.11}$$

where we have set $(k - k_0) = \xi$. The remaining integral may be evaluated in an elementary manner and (7.11) finally takes the form

$$\psi(x, t) = a\exp(-\mathrm{i}\,\omega_0 t + \mathrm{i}\,k_0 x) \cdot 2\,\frac{\sin[(\omega' t - x)\Delta k]}{\omega' t - x}\,. \tag{7.12}$$

The real part of ψ is shown in Fig. 7.3.

We can draw two important conclusions from (7.12):

1) The wave packet represented by ψ is strongly localised in the region of $x = \omega' t$. The maximum amplitude moves with a velocity $\omega' \equiv d\omega/dk$. With the help of (7.1), we can express ω and k in terms of E and p, obtaining $\omega' = \partial E/\partial p$, or, if we use the standard relation $E = p^2/2m_0$, finally $\omega' = p/m_0 = v_{\text{particle}}$. In order to understand this result, we recall the concepts of phase velocity and group velocity.

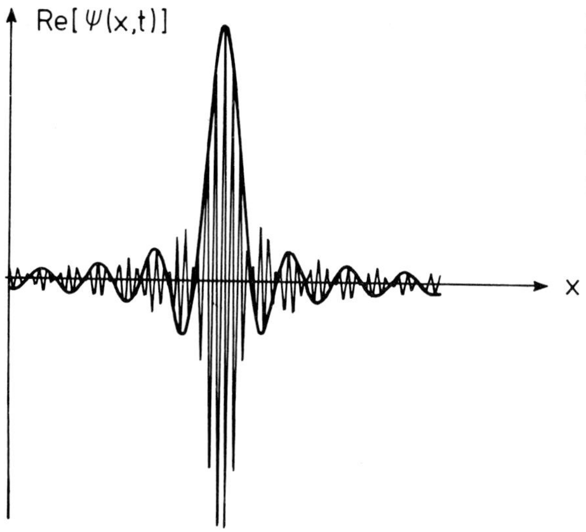

Fig. 7.3. The real part of $\psi(x,t)$ as a function of the position x. The rapid oscillations are described by $\cos(k_0x-\omega_0t)$ with t fixed. The envelope is given by $\sin[(\omega' t-x)\Delta k]/(\omega' t-x)$ with t fixed. Note that the scale of the x axis has been greatly reduced in comparison to Fig. 7.2

If we let the time variable increase in the wavefunction $\cos(kx-\omega t)$, then the position x_{max} at which a particular wave maximum is to be found moves according to the relation $kx_{max}-\omega t=0$, i.e. $x_{max}=(\omega/k)t$. The position x_{max} thus moves with the *phase velocity* $v_{phase}=\omega/k$.

If we replace ω by E and k by p according to (7.1), we find immediately that this v does *not* equal the particle velocity. On the other hand, we have just seen that the maximum of a *wavepacket* moves with the velocity $v_G=d\omega/dk$. This velocity of a wave group (wavepacket) is called the *group velocity*. Thus the group velocity of the de Broglie waves (matter waves) is identical with the particle velocity.

We could be tempted to unify the wave and particle pictures by using wave packets to describe the motion of particles. This is unfortunately not possible, because in general, wave packets change their shapes and flow apart with time. We are therefore compelled to adopt a quite different approach, as will be shown below.

2) A second implication of the result (7.12) is the following: The width of a wave packet is roughly the distance between the first two zero points to the left and right of

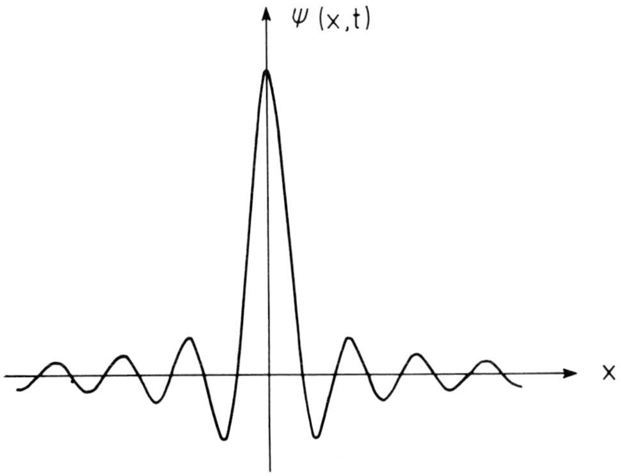

Fig. 7.4. The envelope of the real part of the wave packet (7.12) (Fig. 7.3). The first node is at $x_0=\pi/\Delta k$

the maximum (Fig. 7.4). Since the first zero point is at $x_0 = \pi/\Delta k$, the width of the wave packet would be $\Delta x = 2\pi/\Delta k$. The more we wish to concentrate the wave packet, i.e., the smaller we make Δx, the larger we must make the k-region, or Δk.

In order to clarify the relationship between the particle and the wave descriptions, we shall consider the experiment described in the following section as we have already for light.

7.2 Probabilistic Interpretation

We wish to illustrate, using the electron as an example, how one can unify the wave and particle descriptions. To determine the position of an electron in the x direction (Fig. 7.5), we allow an electron beam to pass through a slit with a width Δx. We can thus ensure that the electron coming from the left must have passed through this position. Now, however, the wave properties come into play, and the electron is accordingly diffracted by the slit. A diffraction pattern is produced on the screen S (Fig. 7.5). According to wave theory, the intensity of the diffraction pattern is proportional to the square of the amplitude. When we consider the electron as a wave, and take ψ as its wave amplitude, we obtain the intensity $I = |\psi(x,t)|^2$ at time t and position x on the observation screen. It is better, for both mathematical and physical reasons, not to speak of the intensity at a *point* in space, but rather of the intensity in the three-dimensional region dx, dy, dz around the point x, y, z.

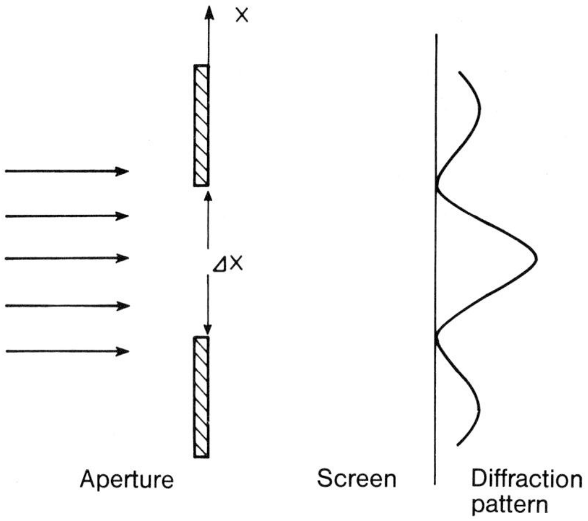

Fig. 7.5. An electron beam (arrows at left) passes through an aperture and generates a diffraction pattern on a screen. The intensity distribution on the screen is shown schematically on the right

Therefore, in the following we shall consider the intensity in a volume element $dV = dx\,dy\,dz$:

$$I\,dx\,dy\,dz = |\psi(x,y,z,t)|^2 dx\,dy\,dz\,. \tag{7.13}$$

(Compare this to the one-dimensional example in Fig. 7.6.)

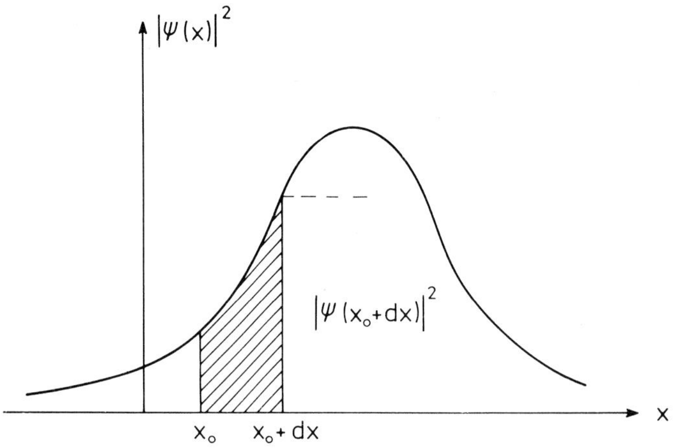

Fig. 7.6. $|\psi(x)|^2$ as a function of x at a given time t. The shaded area corresponds to the probability that the electron is located in the interval x_0 to x_0+dx

Now comes the essential point. The screen can also be considered as an apparatus which detects electrons individually as particles. A fluorescent screen flashes at the point of impact each time an electron hits it. The electron is thus highly localised, and there is no diffraction pattern. If we repeat the experiment, we observe other flashes of light, and in general these are at different points on the screen. Only when we carry out many experiments, or allow many electrons to pass through, do we obtain a diffraction pattern of the form described by (7.13). This is the key to the explanation of the phenomenon of "wave-particle duality". On the one hand, the intensity of the diffraction pattern in a volume ΔV is proportional to the absolute square of the amplitude,

$$|\psi|^2 \Delta V, \tag{7.14}$$

and on the other, it is proportional to the probability of finding the electron in ΔV. $|\psi|^2 \Delta V$ is thus itself proportional to the frequency of finding the electron in ΔV. $|\psi(x,y,z,t)|^2 dx\,dy\,dz$ must therefore be seen as the *probability* of finding the electron in a volume element dV about the point x,y,z.

Because the statistical interpretation of quantum mechanics will be mentioned frequently, and is absolutely necessary to an understanding of the subject, we shall spend a bit more time on the concept of probability. Let us compare a quantum mechanical experiment with a game of dice. Since a die has six different numbers on its faces, it has, so to speak, six different experimental values. We cannot say in advance, however, which face, i.e. which experimental value we will obtain in any given throw. We can only give the probability P_n of obtaining the value n. In the case of a die, P_n is very easy to determine. According to a basic postulate of probability theory, the sum of all probabilities P_n must be one (i.e., one face must come up on each throw):

$$\sum_n P_n = 1 \,. \tag{7.15}$$

Since all the numbers $n = 1, 2, \ldots 6$ are equally probable, the six values of P_n must be equal, so $P_n = 1/6$.

It is not so easy to determine $|\psi|^2 dx\,dy\,dz$. We can infer from the above, however, that there must be a normalisation condition for $|\psi|^2 dx\,dy\,dz$. If we integrate over all points in space, the particle must be found *somewhere*, so the total probability must therefore be equal to 1. We thereby obtain the basic normalisation condition

$$\int |\psi(x,y,z)|^2 dx\,dy\,dz = 1 \,. \tag{7.16}$$

We shall illustrate the use of this normalisation condition with two examples.

1) We assume that the electron is enclosed in a box with volume V. The integral (7.16) must then extend only over this volume. If we use for ψ the wavefunction

$$\psi = A_0 \exp(\mathrm{i}\boldsymbol{k}\cdot\boldsymbol{x} - \mathrm{i}\omega t)\,, \tag{7.17}$$

where $\boldsymbol{k}\cdot\boldsymbol{x} = k_x x + k_y y + k_z z$, then A_0 must be

$$A_0 = V^{-1/2}\,. \tag{7.18}$$

2) If the space extends to infinity, there is a difficulty, because here $A_0 = 0$ if we simply allow V to go to infinity in (7.18). It can be shown, however, that a generalised normalisation condition can still be derived. In one dimension the normalised wavefunction is

$$\psi_k(x,t) = (1/\sqrt{2\pi})\exp(\mathrm{i}kx - \mathrm{i}\omega t)\,, \tag{7.19}$$

and the normalisation condition is

$$\int \psi_k^*(x,t)\,\psi_{k'}(x,t)\,dx = \delta(k-k')\,. \tag{7.20}$$

Here $\delta(k-k')$ is the Dirac δ function (see Appendix A).

The probabilistic interpretation of the wavefunction is also necessary for the following reason: if the impact of an electron on the screen were to cause it to flash at more than one point, this would mean that the electron had divided itself. All experiments have shown, however, that the electron is *not* divisible. The determination of $|\psi|^2 dV$ allows us only to predict the probability of finding the electron in that volume. If we have found it at one position (localised it), we are certain that it is not somewhere else as well. This is evidently a "yes-no" statement and leaves no ambiguity for an individual electron. If we consider the reflection of electrons in this way, and observe that 5% are reflected, it means this: if we carry out a very large number of experiments, 5% of all the electrons would be reflected. It would be completely false, however, to say that 5% of a single electron had been reflected.

7.3 The Heisenberg Uncertainty Relation

We now consider some of the implications of the fact that the electron sometimes acts as a particle and sometimes as a wave. As we calculated earlier, the one-dimensional distribution of the wave packet is

$$\psi(x) \sim \frac{\sin(x\Delta k)}{x}\,. \tag{7.21}$$

If we take the position of the first zero point as a measure of the uncertainty in the position, we obtain from (7.21) (Fig. 7.4) the relation

$$\frac{\Delta x}{2}=\frac{\pi}{\Delta k}\,. \tag{7.22}$$

The uncertainty in the position is clearly connected with the uncertainty in the wavenumbers k. But the wavenumber is related to the momentum by the equation

$$p=\hbar k\,. \tag{7.23}$$

If we insert this in (7.22), we obtain the basic Heisenberg uncertainty relation

$$\Delta x\Delta p \geqq h\,. \tag{7.24}$$

This relation states that it is impossible to measure the position and the momentum of an electron exactly at the same time. A lower bound to the simultaneous measurability is given by (7.24). Indeed, if we wished to let Δx go to zero (exact determination of the position), we would have to allow Δp to become infinite, and *vice versa*. The fact that we notice nothing of this uncertainty relation in daily life is a result of the smallness of Planck's constant h. If, on the other hand, we consider the microscopic world, then we can only understand the results of experiments if we take the finite size of the constant h into account. We will clarify the meaning of (7.24) with the example of an experiment.

An electron is moving in a horizontal direction (y). We wish to determine its coordinate in the perpendicular (x) direction. For this purpose, we set up a collimator perpendicular to the direction of motion with a slit of width $d=\Delta x$. If the electron passes through this slit, then we know that it was at that position with the uncertainty Δx. Now, however, we must take into account the wave nature of the electron. From the theory of diffraction we know that a wave produces a diffraction pattern on the obser-

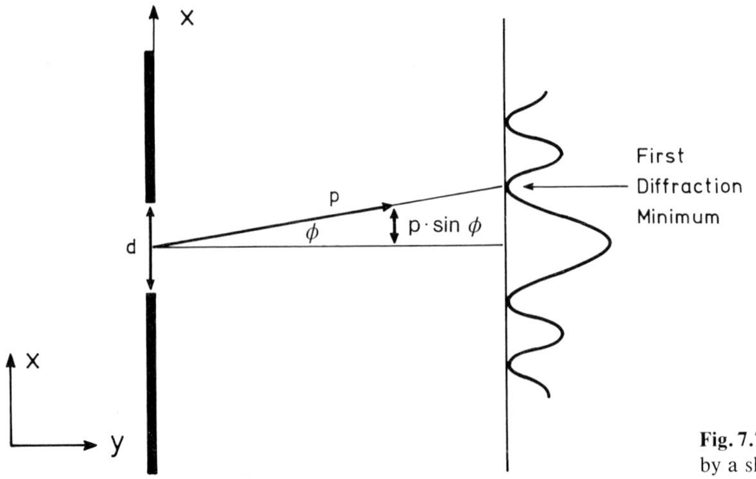

Fig. 7.7. Diffraction of a wave by a slit

vation screen after passing through a slit of width d (Fig. 7.7). The angle ϕ at which the first diffraction minimum occurs is given by

$$\sin\phi = \lambda/d\,. \tag{7.25}$$

If we denote the total momentum of the electron by $\boldsymbol{p}$, the projection of the momentum on the x axis is $p\sin\phi$. This x component in the momentum is produced by diffraction of the electron wave at the slit; the resulting uncertainty in the momentum is then

$$\Delta p_x = p\sin\phi\,. \tag{7.26}$$

If we once again use the relation

$$p = h/\lambda \tag{7.27}$$

and insert (7.26) and (7.27) in (7.25), we again obtain the Heisenberg uncertainty relation (7.24).

This example shows clearly that a measurement of one quantity, here the position, immediately produces a perturbation of the complementary quantity, namely the momentum. Before we set up the collimator with its slit, we could have determined the momentum of the electron. The result would have been that the electron was moving exactly in the y direction, i.e., that its momentum component in the x direction was exactly equal to zero. In the above experiment, we were able to determine the position with a certain accuracy, but we had to accept the fact that the momentum thereby became uncertain in the x direction. There is also a relation between energy and time which is analogous to (7.24).

7.4 The Energy-Time Uncertainty Relation

In the wavefunction $\sim\exp(\mathrm{i}kx - \mathrm{i}\omega t)$, which was the starting point of this chapter, the position x and the time t occur in a symmetric fashion. Just as we could form wave packets which exhibited a certain concentration in space, we can also construct wave packets which have a concentration about a time t with an uncertainty Δt. Instead of the relation $\Delta x\Delta k \geqq 2\pi$, we then have

$$\Delta t\Delta\omega \geqq 2\pi\,. \tag{7.28}$$

Utilising the relation $E = \hbar\omega$, we find from this that

$$\Delta E\Delta t \geqq h\,. \tag{7.29}$$

This relation, which we shall discuss in more detail at a later point in the book, states among other things, that one must carry out a measurement for a sufficiently long time, in order to measure an energy with good accuracy in quantum mechanics.

7.5 Some Consequences of the Uncertainty Relations for Bound States

In the preceding sections of this chapter we have explicitly considered *free* electrons. In the next chapters we shall be concerned with the experimental and theoretical questions associated with *bound* electrons, for example in the hydrogen atom. In this section we shall to some extent anticipate the presentation in the rest of the book. The reader will recognise even in this section that wave mechanics will play a fundamental role in the treatment of bound states.

We will consider the hydrogen atom as the simplest case of a bound state. We assume that the electron travels around the nucleus in an orbit, as a planet around the sun. Why the electronic shells of the atom have a finite extent – why there is a smallest electron orbit – was an insoluble problem in classical physics.

The energy of an electron is equal to the sum of the kinetic and the potential energy,

$$E_{\text{class}} = E_{\text{kin}} + E_{\text{pot}} . \tag{7.30}$$

If we express the kinetic energy of a particle $E_{\text{kin}} = (m_0/2)\, v^2$ in terms of the momentum p, and substitute the Coulomb potential energy $-e^2/(4\pi\varepsilon_0 r)$ for E_{pot}, the expression for E is

$$E = \frac{p^2}{2m_0} - \frac{e^2}{4\pi\varepsilon_0 r} . \tag{7.31}$$

Here r is the distance of the electron from the nucleus.

It can be shown in classical mechanics that $E_{\text{class}} = -e^2/(2 \cdot 4\pi\varepsilon_0 r)$.

If we allow r to go to zero, the energy naturally goes to $-\infty$. In other words, the energy decreases continually and there is no smallest orbital radius. Let us now consider the expression (7.31) from a "naive" quantum mechanical point of view. Then "orbit" would mean that we have the electron concentrated at a distance of approximately r from the nucleus. The positional uncertainty would therefore be of the order of r. This, however, would entail uncertainty in the momentum p of the order of h/r, which in turn establishes a minimum for the order of magnitude of the kinetic energy (Fig. 7.8). If we therefore substitute

$$p \approx \frac{h}{r} \tag{7.32}$$

in (7.31), we realise that the minimum of the energy expression

$$E = \frac{1}{2m_0} \frac{\mathrm{h}^2}{r^2} - \frac{e^2}{4\pi\varepsilon_0 r} = \text{Min} \tag{7.33}$$

is no longer at $r = 0$. If we let r go to zero, the kinetic energy would increase very rapidly. We shall leave the determination of the minimum of (7.33) to the reader as an simple exercise in differential calculus and give the result immediately. The radius is

$$r = \frac{h^2 4\pi\varepsilon_0}{m_0 e^2} . \tag{7.34}$$

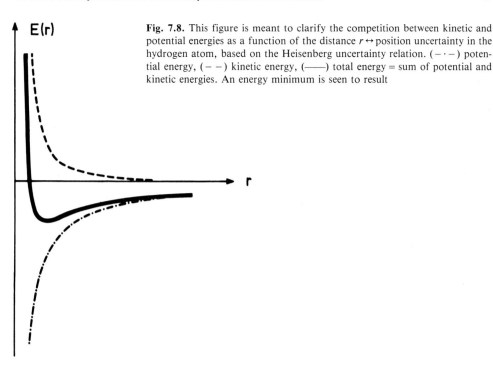

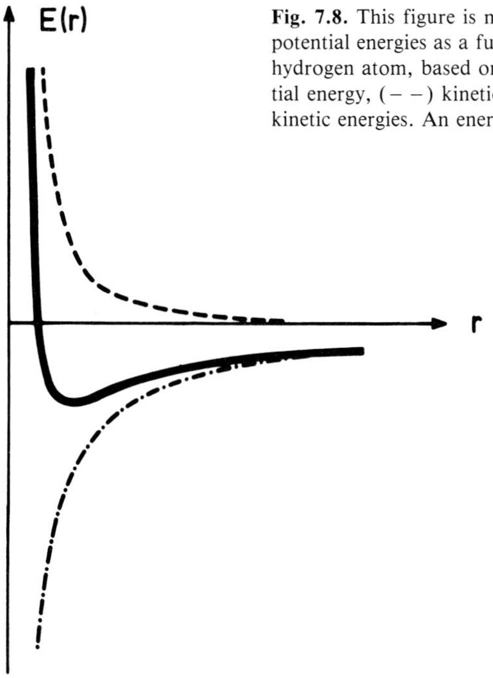

Fig. 7.8. This figure is meant to clarify the competition between kinetic and potential energies as a function of the distance $r \leftrightarrow$ position uncertainty in the hydrogen atom, based on the Heisenberg uncertainty relation. (– · –) potential energy, (– –) kinetic energy, (——) total energy = sum of potential and kinetic energies. An energy minimum is seen to result

If we substitute this r in (7.33), the corresponding energy is

$$E = -\frac{1}{2}\frac{e^4 m_0}{(4\pi\varepsilon_0)^2 h^2}. \tag{7.35}$$

When we substitute the known numerical values for Planck's constant, and the mass and charge of the electron, we obtain a radius of about 10^{-8} cm, which is the right order of magnitude for the hydrogen atom. As we shall see later, the exact quantum mechanical calculation of the energy yields

$$E = -\frac{1}{2}\frac{e^4 m_0}{(4\pi\varepsilon_0)^2 \hbar^2}. \tag{7.36}$$

The only difference between (7.35) and (7.36) is the factor $\hbar^2 \equiv (h/2\pi)^2$ which replaces h^2.

The Heisenberg uncertainty principle also allows us to calculate the so-called *zero-point energy of a harmonic oscillator.* Here we consider the motion of a particle elastically bound by a spring with a spring constant f. Since the elastic energy increases quadratically with the displacement x and the kinetic energy again has the form $p^2/2m_0$, the total energy is

$$E = \frac{p^2}{2m_0} + \frac{f}{2}x^2. \tag{7.37}$$

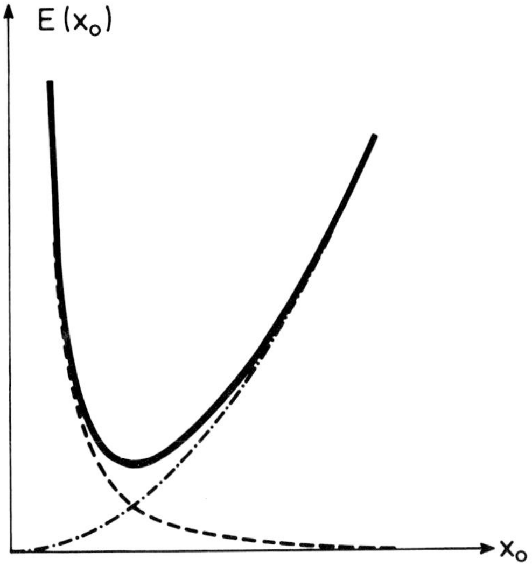

Fig. 7.9. Illustrating the competition between kinetic and potential energies as a function of the displacement ↔ position uncertainty in the harmonic oscillator. (– · –) potential energy, (– –) kinetic energy, (——) total energy. The classical energy minimum at $x_0 = 0$ is shifted to a finite value

In classical physics this energy is at a minimum when both the momentum and the position are zero, i.e., the particle is at rest. However, since according to the Heisenberg relation an exact position is associated with an infinite momentum, we allow a positional uncertainty of the same magnitude as the oscillation amplitude x_0 and have the corresponding momentum uncertainty according to (7.24), where x_0 assumes the role of r (Fig. 7.9). We again require that the total energy is minimised by the appropriate choice of x_0:

$$E = \frac{h^2}{2m_0 x_0^2} + \frac{f}{2} x_0^2 = \text{Min}\,. \tag{7.38}$$

Solving this equation for x_0 yields the amplitude of the harmonic oscillator,

$$x_0 = \sqrt[4]{\frac{h^2}{m_0 f}}\,. \tag{7.39}$$

The energy then has the form

$$E = h\,\omega\,. \tag{7.40}$$

As we shall see later, an exact quantum mechanical calculation yields the relations

$$E = \tfrac{1}{2}\hbar\,\omega \tag{7.41}$$

and

$$x_0 = \sqrt{\frac{\hbar}{2m_0\omega}}\,. \tag{7.42}$$

It follows from these considerations that atomic, elastically bound particles are fundamentally incapable of being at rest. Such elastically coupled particles occur, for example, in crystal lattices. The quantum theory predicts that these atoms will constantly carry out zero-point oscillations.

Problems

7.1 Normalise the wave packet

$$\psi(x,t) = N \int\limits_{-\infty}^{+\infty} \exp\left[-\frac{k^2}{2(\Delta k)^2}\right] e^{i[kx-\omega(k)t]}dk$$

for $t = 0$. Then calculate $\psi(x,t)$ for a free particle of mass m_0 for $t > 0$. Does the normalisation hold for $t > 0$? On the basis of the occupation probability, decide whether the wave packet falls apart. What is the significance of

$$\exp\left(-\frac{k^2}{2(\Delta k)^2}\right)?$$

Hint: Use the relation

$$\int\limits_{-\infty}^{+\infty} e^{-a\xi^2-b\xi}d\xi = e^{b^2/(4a)} \int\limits_{-\infty}^{+\infty} e^{-a[\xi+b/(2a)]^2}d\xi$$

(completing the square!)

The second integral can be converted to the Gaussian integral by changing the coordinates.

7.2 By the appropriate choice of Δk in Problem 7.1, let the probability of locating the wave packet outside $\Delta x = 10^{-8}$ cm be zero. How long would it take Δx to attain the size of the distance between the earth and sun ($\simeq$150 million km)?

Hint: Choose Δx so that $\psi(\Delta x, 0) = 1/e$ [$e \equiv \exp(1)$!]

7.3 Treat Problems 7.1 and 7.2 in three dimensions.

8. Bohr's Model of the Hydrogen Atom

8.1 Basic Principles of Spectroscopy

In the following chapters we shall take up the detailed analysis of the spectra of atoms in every wavelength region. The most important sources of information about the electronic structure and composition of atoms are spectra in the visible, infrared, ultraviolet, x-ray, microwave and radio frequency ranges. Figure 8.1 summarises these spectral regions.

Optical spectra are further categorised as line, band and continuous spectra. Continuous spectra are emitted by radiant solids or high-density gases. Band spectra consist of groups of large numbers of spectral lines which are very close to one another. They are generally associated with molecules. Line spectra, on the other hand, are typical of atoms. They consist of single lines, which can be ordered in characteristic series.

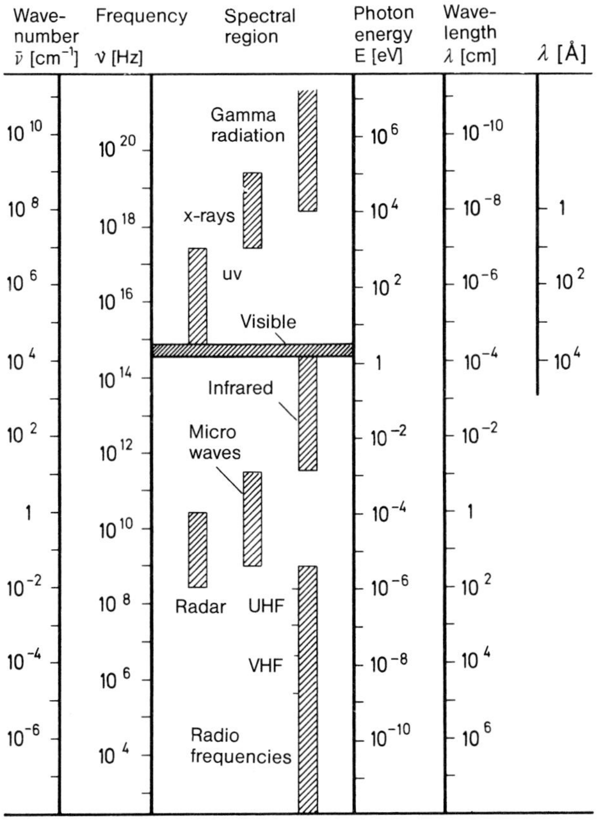

Fig. 8.1. The electromagnetic spectrum. Regions and units

Optical spectra can be observed either by emission or by absorption. The former mode requires that the substance to be examined be made to emit light; this can be achieved by transferring energy to the atoms by means of light, electron collisions, x-ray excitation or some other process. If a substance re-emits the light it has absorbed, the process is called resonance fluorescence. The best known example of this is the resonance fluorescence of sodium vapour (Fig. 8.2).

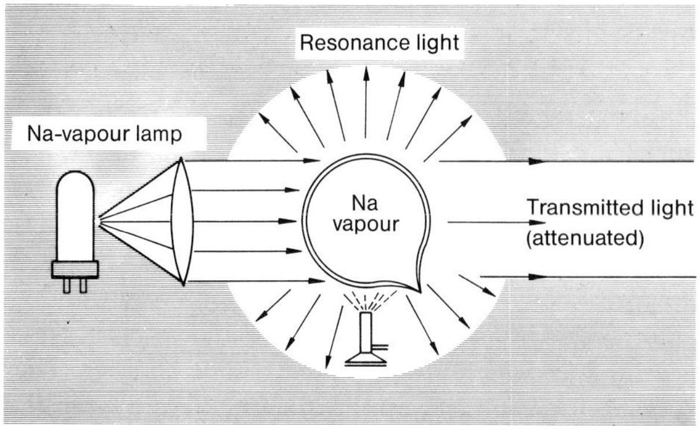

Fig. 8.2. Resonance fluorescence of sodium vapour. Sodium metal is heated in an evacuated glass sphere. The resulting sodium vapour absorbs the light of a sodium vapour lamp and emits the same light as resonance fluorescence in all directions

Details of apparatus will be discussed in the following sections in connection with particular problems.

Spectra are traditionally measured in several different units, due to the features of the apparatus or for practical reasons:

- In *wavelength* units. These can be determined absolutely with a diffraction grating. Usually, however, one uses a calibrated comparison spectrum, which allows greater accuracy.
  One wavelength standard is the yellow ^{86}Kr line, that is a yellow line in the spectrum of the ^{86}Kr atom. For this line,

$$\lambda_{\text{vac}} = 6057.80211\ \text{Å} \mathrel{\hat{=}} \bar{\nu} = 16507.6373\ \text{cm}^{-1} \qquad \text{(see below)}\ .$$

In general the wavelengths are referred to vacuum. The corresponding wavelength in air is somewhat smaller, because the index of refraction of air is somewhat greater than 1, and the velocity of light in air is thus somewhat less than in a vacuum. To convert wavelengths measured in air ("normal" air, 15 °C, 760 Torr), the formula is

$$\lambda_{\text{air}} = \lambda_{\text{vac}}/n\ .$$

The refractive index of air is a function of the wavelength. At 6000 Å, $n = 1.0002762$.

For the yellow ^{86}Kr line in normal air,

$$\lambda_{\text{air}} = 6056.12941\ \text{Å}\ .$$

- Specifying the *frequency* is more general, since it is not dependent on the medium. We have:

$$\nu = c/\lambda_{\text{vac}} = c/(n\lambda_{\text{air}})\ .$$

– A frequently cited quantity is the *wavenumber*:

$$\bar{\nu} = \nu/c = 1/\lambda_{\text{vac}} = 1/(n\lambda_{\text{air}}) .$$

The wavenumber is, like the frequency, a quantity proportional to the energy; conversion may be made according to the equation

$$E = \bar{\nu}hc .$$

– Finally, the unit *electron volt* (eV) is often used as a measure of the energy.

Several units which are important and practical in atomic physics as well as conversion factors are set out in Table 8.1 and in Fig. 8.1.

Table 8.1. Frequently used units and conversion factors (see also the table on the inner side of the front cover)

Quantity	Unit and conversion factor
Wavelength λ	$1\ \text{Å} = 10^{-10}\ \text{m} = 0.1\ \text{nm}$
Wavenumber $\bar{\nu}$	$1\ \text{cm}^{-1}$ (=1 kayser)
$\bar{\nu} = 1/\lambda$	$\bar{\nu} = 8066\ E(\text{eV})\ \text{cm}^{-1}$
	$1\ \text{cm}^{-1} = 29.979\ \text{GHz}$
Energy E	$1\ \text{electron volt} = 1.602 \cdot 10^{-19}\ \text{J} = 1.96 \cdot 10^{-6}\ m_0c^2$
	$E = h\nu = hc/\lambda = hc\bar{\nu}$
	$1\ \text{eV} \triangleq 2.418 \cdot 10^{14}\ \text{Hz} \triangleq 8066\ \text{cm}^{-1}$
	$E(\text{eV}) = 1.24 \cdot 10^{-4} \dfrac{\bar{\nu}}{\text{cm}}$
Mass m_0	$1\ \text{electron mass} = 9.11 \cdot 10^{-31}\ \text{kg} = 511\ \text{keV}/c^2$
Charge e	$1\ \text{elementary charge} = 1.6 \cdot 10^{-19}\ \text{C}$
Planck's constant h	$h = 4.14 \cdot 10^{-15}\ \text{eV s}$
	$\hbar = h/2\pi = 6.58 \cdot 10^{-16}\ \text{eV s}$

8.2 The Optical Spectrum of the Hydrogen Atom

Kirchhoff and *Bunsen*, the founders of spectroscopic analysis, were the first to discover in the mid-19th century that each element possesses its own characteristic spectrum. Hydrogen is the lightest element, and the hydrogen atom is the simplest atom, consisting of a proton and an electron. The spectra of the hydrogen atom have played an important rôle again and again over the last 90 years in the development of our understanding of the laws of atomic structure and of the structure of matter.

The emission spectrum of atomic hydrogen shows three characteristic lines in the visible region at 6563, 4861, and 4340 Å ($H_{\alpha,\beta,\gamma}$). The most intense of these lines was discovered in 1853 by Ångström; it is now called the H_α line. In the near ultraviolet region, these three lines are followed by a whole series of further lines, which fall closer and closer together in a regular way as they approach a short-wavelength limit (H_∞) (Fig. 8.3).

Balmer found in 1885 that the wavelengths of these lines could be extremely well reproduced by a relation of the form

$$\lambda = \left(\frac{n_1^2}{n_1^2 - 4}\right) G . \tag{8.1}$$

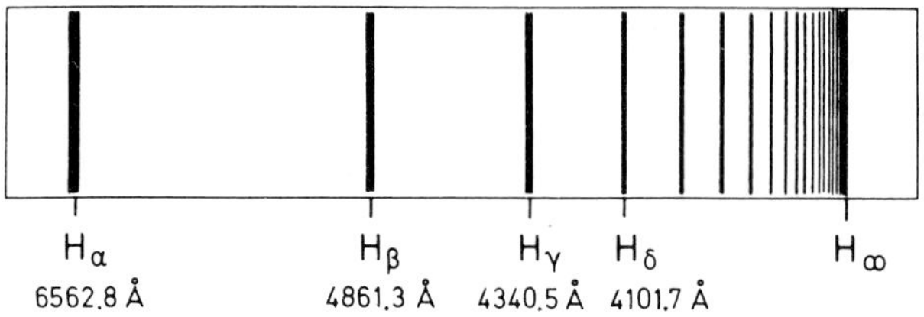

Fig. 8.3. Balmer series in the hydrogen emission spectrum. The convergence of the lines to the series limit H_∞ is clearly seen

Here n_1 is an integer, $n_1 = 3, 4, \ldots$ and G is an empirical constant. Today, we write the Balmer formula somewhat differently. For the wavenumbers of the lines we write

$$\bar{\nu} = 1/\lambda = R_\mathrm{H}\left(\frac{1}{2^2} - \frac{1}{n^2}\right), \qquad n \text{ an integer } > 2 \tag{8.2}$$

The quantity $R_\mathrm{H}(=4/G)$ is called the *Rydberg constant* and has the numerical value

$$R_\mathrm{H} = 109677.5810\,\mathrm{cm}^{-1}.$$

The series limit is found for $n \to \infty$ to be

$$\bar{\nu}_\infty = R_\mathrm{H}/4 .$$

For the further investigation of the hydrogen spectrum, astrophysical observations have played an important rôle. In the spectra of stars, photographically recorded as early as 1881 by *Huggins*, a large number of lines from the hydrogen spectrum are seen.

Table 8.2. The first 20 lines of the Balmer series of hydrogen. The numbers quoted are wavelengths in air, the wavenumbers in vacuum, and the values calculated from the Balmer formula

n	λ_air [Å]	$\bar{\nu}_\mathrm{vac}$ [cm^{-1}]	$R_\mathrm{H}\left(\frac{1}{2^2} - \frac{1}{n^2}\right)$
H_α 3	6562.79	15233.21	15233.00
H_β 4	4861.33	20564.77	20564.55
H_γ 5	4340.46	23032.54	23032.29
H_δ 6	4101.73	24373.07	24372.80
H_ε 7	3970.07	25181.33	25181.08
H_ζ 8	3889.06	25705.84	25705.68
H_η 9	3835.40	26065.53	26065.35
H_ϑ 10	3797.91	26322.80	26322.62
H_ι 11	3770.63	26513.21	26512.97
H_κ 12	3750.15	26658.01	26657.75
H_λ 13	3734.37	26770.65	26770.42
H_μ 14	3721.95	26860.01	26859.82
H_ν 15	3711.98	26932.14	26931.94
H_ξ 16	3703.86	26991.18	26990.97
H_o 17	3697.15	27040.17	27039.89
H_π 18	3691.55	27081.18	27080.88
H_ϱ 19	3686.83	27115.85	27115.58
H_σ 20	3682.82	27145.37	27145.20

Using modern radio-astronomical techniques, transitions between states with extremely large n-values have been found; levels with n between 90 and 350 could be identified.

The reason that many lines were discovered first in astrophysical observations and not by experiments on the earth is connected with the difficulty of preparing pure atomic hydrogen in the laboratory. Gas discharges, in which H_2 gas is decomposed into atomic hydrogen and excited to fluorescence, always contain fluorescing hydrogen molecules as well, whose spectrum overlaps the atomic-hydrogen spectrum.

Above the series limit we observe the so-called series-limit continuum, a region in which the spectrum shows no more lines, but is, instead, continuous.

A comparison of the calculated spectral lines obtained from the Balmer formula (8.2) with the observed lines (Table 8.2) shows that the formula is not just a good approximation: the series is described with great precision. The whole spectrum of the H atom is represented by equations of the form

$$\bar{\nu} = R_{\mathrm{H}}\left(\frac{1}{n'^2} - \frac{1}{n^2}\right) \quad \text{with } n' < n \text{ being integers} . \tag{8.3}$$

The numbers n and n' are called *principal quantum numbers*. Table 8.3 contains some of the lines from the first four series.

Table 8.3. The wavelengths of some lines of the various spectral series in hydrogen. The series with $n' = 5$ was observed in 1924 by *Pfund*; it begins with a line of $\lambda = 74000$ Å, but is not shown in the table

n \ n'	1 *Lyman*	2 *Balmer*	3 *Paschen*	4 *Brackett*
2	1216 Å ≙ 82257 cm^{-1}			
3	1026 Å ≙ 97466 cm^{-1}	6563 Å ≙ 15233 cm^{-1}		
4	973 Å ≙ 102807 cm^{-1}	4861 Å ≙ 20565 cm^{-1}	18751 Å ≙ 5333 cm^{-1}	
5	950 Å ≙ 105263 cm^{-1}	4340 Å ≙ 23033 cm^{-1}	12818 Å ≙ 7801 cm^{-1}	40500 Å ≙ 2467 cm^{-1}
Year of discovery	1906	1885	1908	1922

The relation (8.3) was formulated first by *Rydberg* in 1889. He found, "to his great joy", that the Balmer formula (8.1) is a special case of the Rydberg formula (8.3). Table 8.3 also illustrates the Ritz Combination Principle, which was found empirically in 1898. It states:

The difference of the frequencies of two lines in a spectral series is equal to the frequency of a spectral line which actually occurs in another series from the same atomic spectrum. For example, the frequency difference of the first two terms in the Lyman series is equal to the frequency of the first line of the Balmer series, as can be seen from the wavenumber entries in Table 8.3.

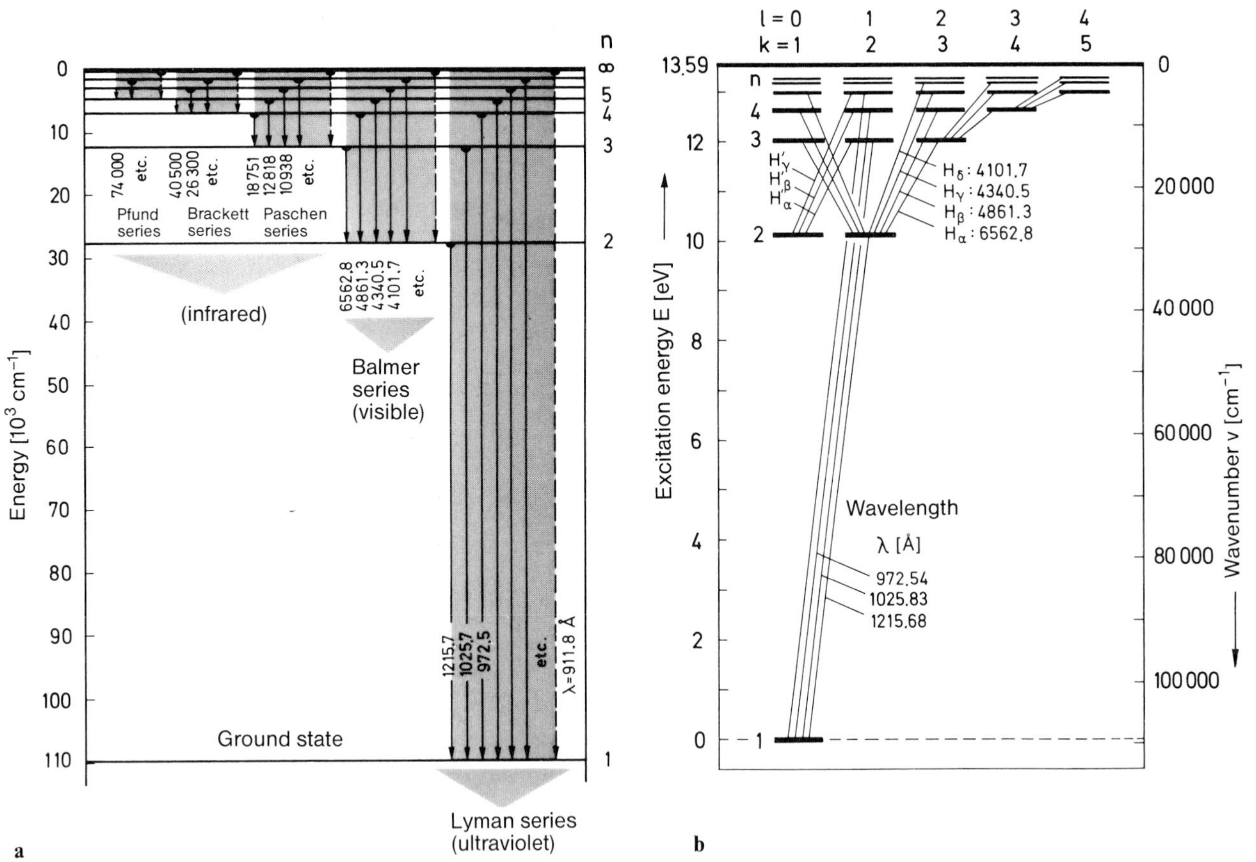

Fig. 8.4. a) Term diagram of the lines of the hydrogen spectrum and series classification. The wavelengths of the transitions are given in Å. The energies can be given as (negative) binding energies, with the zero of energy being the ionisation limit, or they can be given as excitation energies, beginning with the ground state, so that the energy of the term n_∞ is equal to the ionisation energy **b)** This represents the lines of the hydrogen spectrum in the term scheme of *Grotrian* [Struktur der Materie VII (Springer, Berlin 1928)]. The symbols l and k appearing in the upper margin of the figure will be explained later (Sect. 8.9)

We can conclude from observation and inductive reasoning that the frequencies (or wavenumbers) of all the spectral lines can be represented as differences of two *terms* of the form R/n^2. As we shall see in the following, these are just the energy levels of the electron in a hydrogen atom. The spectral lines of the hydrogen atom can be graphically pictured as transitions between the energy levels (terms), leading to a spectral *energy level diagram* (Fig. 8.4).

8.3 Bohr's Postulates

In the early years of this century, various models were suggested to explain the relationship between atomic structure and the spectral lines. The most successful of these is due to *Bohr* (1913). Following the Rutherford model, he assumed that the electrons move around the nucleus in circular orbits of radius r with velocity v, much as the planets move around the sun in the Solar System. A dynamic equilibrium between the centrifugal force and the Coulomb attraction of the electrons to the nucleus is assumed to exist. Thus, for the hydrogen atom, one has

$$\frac{e^2}{4\pi\varepsilon_0 r^2} = m_0 r\omega^2 . \tag{8.4}$$

The corresponding energy is the sum of the kinetic and the potential energies of the electrons:

$$E = E_{\mathrm{kin}} + E_{\mathrm{pot}} ,$$

where the kinetic energy, as usual, is given by $m_0 v^2/2$ or $m_0 r^2 \omega^2/2$. The potential energy is defined as the work which one obtains on allowing the electron to approach the nucleus under the influence of the Coulomb force from infinity to a distance *r*. Since the work is defined as the product of force and distance, and the Coulomb force changes continuously with the distance from the nucleus, we must integrate the contributions to the work along a differential path *dr*; this gives

$$E_{\mathrm{pot}} = \int_{\infty}^{r} \frac{e^2}{4\pi\varepsilon_0 r'^2}\, dr' = -\frac{e^2}{4\pi\varepsilon_0 r} . \tag{8.5}$$

E_{pot}, as a binding energy, may be seen to be negative, with the zero point being the state of complete ionisation. The total energy is thus found to be

$$E = \frac{1}{2} m_0 r^2 \omega^2 - \frac{e^2}{4\pi\varepsilon_0 r} . \tag{8.6}$$

Thus far, the model corresponds to that of *Rutherford*.

We may rewrite (8.6) by using (8.4):

$$E = -\frac{e^2}{2\cdot 4\pi\varepsilon_0 r} = -\frac{1}{2(4\pi\varepsilon_0)^{2/3}} (e^4 m_0 \omega^2)^{1/3} . \tag{8.7}$$

If, however, one attempts to understand the emission and absorption of light using this model and the known laws of classical electrodynamics, one encounters fundamental difficulties. *Classically*, orbits of arbitrary radius and thus a *continuous* series of energy values for the electron in the field of the nucleus should be allowed. But on identifying the energy levels which are implied by the spectral series with the values of the electron's energy, one is forced to assume that only *discrete* energy values are possible. Furthermore, electrons moving in circular orbits are accelerated charges, and as such, they should radiate electromagnetic waves with frequencies equal to their orbital frequencies, $\nu = \omega/2\pi$. They would thus lose energy continuously, i.e. their orbits are unstable and they would spiral into the nucleus. Their orbital frequencies would change continuously during this process. Therefore, the radiation emitted would include a continuous range of frequencies.

In order to avoid this discrepancy with the laws of classical physics, *Bohr* formulated three *postulates* which describe the deviations from classical behavior for the electrons in an atom. These postulates proved to be an extremely important step towards quantum mechanics. They are:

- The classical equations of motion are valid for electrons in atoms. However, only certain *discrete* orbits with the energies E_n are allowed. These are the energy levels of the atom.
- The motion of the electrons in these quantised orbits is *radiationless*. An electron can be transferred from an orbit with lower (negative) binding energy E_n (i.e. larger r) to an orbit with higher (negative) binding energy $E_{n'}$ (smaller r), emitting radiation in the process. The frequency of the emitted radiation is given by

$$E_n - E_{n'} = h\nu . \tag{8.8}$$

Light absorption is the reverse process.

By comparing (8.8) and (8.3), *Bohr* identified the energy terms $E_{n'}$ and E_n as

$$E_n = -\frac{Rhc}{n^2}, \qquad E_{n'} = -\frac{Rhc}{n'^2}, \tag{8.9}$$

where the minus sign again implies that we are dealing with binding energies.
- Finally, for the calculation of the Rydberg constant R in (8.9) from atomic quantities, *Bohr* used the comparison of the orbital frequencies of the electrons with the frequency of the emitted or absorbed radiation. In classical physics, these frequencies would be equal, as mentioned above. However, using (8.4), one can easily calculate that this is not at all the case in the hydrogen atom for small orbital radii r.

Bohr's decisive idea was then to postulate that with increasing orbital radius r, the laws of quantum atomic physics become identical with those of classical physics. The application of this *"Correspondence Principle"* to the hydrogen atom allows the determination of the discrete stable orbits.

We consider the emission of light according to the first two postulates for a transition between neighboring orbits, i.e. for $(n-n')=1$, and for large n. From (8.3) we have for the frequency ν, with $n-n'=\tau$

$$\nu = Rc\left(\frac{1}{n'^2} - \frac{1}{n^2}\right) = Rc\left(\frac{1}{(n-\tau)^2} - \frac{1}{n^2}\right) \tag{8.10}$$

$$Rc\frac{1}{n^2}\left(\frac{1}{(1-\tau/n)^2} - 1\right) \cong Rc\,\frac{2\tau}{n^3}$$

or, with $\tau = 1$,

$$\nu = \frac{2Rc}{n^3}. \tag{8.11}$$

This frequency is now set equal to the classical orbital frequency ω in (8.7), giving, with (8.7), (8.9), and (8.11) an expression for R:

$$E_n = -\frac{Rhc}{n^2} = -\frac{1}{2}\,\frac{1}{(4\pi\varepsilon_0)^{2/3}}\,(e^4 m_0\,\omega^2)^{1/3}, \qquad \text{with}$$

$$\omega = 2\pi\left(\frac{2Rc}{n^3}\right) \quad \text{and}$$

$$R = \frac{m_0 e^4}{8\,\varepsilon_0^2 h^3 c}. \tag{8.12}$$

From (8.12), we find for the Rydberg constant R (which we denote by R_∞ for reasons which will become apparent below) the numerical value

$$R_\infty = (109\,737.318 \pm 0.012)\ \mathrm{cm}^{-1}. \tag{8.13}$$

This may be compared with the empirical value in (8.2). In *Bohr's* model, R is just the ionisation energy of the ground state of the atom, $n = 1$.

From (8.12), with (8.7) and (8.9), we find the radius r_n of the nth orbital to be

$$r_n = \frac{n^2 \hbar^2 4\pi\varepsilon_0}{e^2 m_0}. \tag{8.14}$$

The quantum number n which occurs in these expressions is called the *principal quantum number.*

In addition, we may calculate the *orbital angular momentum* $\boldsymbol{l} = \boldsymbol{r} \times \boldsymbol{p}$ of an electron having velocity v_n and orbital frequency ω_n in the orbit with radius r_n and find, using (8.11) and (8.14), the quantisation rule

$$|\boldsymbol{l}| = m_0 v_n r_n = m_0 r_n^2 \omega_n = n\hbar \quad \text{with} \quad n = 1, 2, 3, \ldots. \tag{8.15}$$

This quantisation rule is often (but incorrectly) taken to be one of *Bohr's* postulates.

The essential common feature of the Bohr postulates is that they make no statements about processes, but only about states. The classical orbital concept is abandoned. The electron's behaviour as a function of time is not investigated, but only its stationary initial and final states. Figure 8.5 illustrates the model.

Whether spectral lines are observable, either in emission or in absorption, depends on the occupation of the energy terms (also referred to as energy states). Absorption from a state presupposes that this state is occupied by an electron. In emission transitions, an electron falls from a higher state into an unoccupied lower one; the electron must be previously raised to the higher state by an excitation process, i.e. by an input of energy. At normal temperatures only the Lyman series in hydrogen is observable in absorption, since then only the lowest energy term ($n = 1$ in Fig. 8.4) is occupied. When the Balmer lines are observed in the spectra of stars as Fraunhofer lines (that is, these

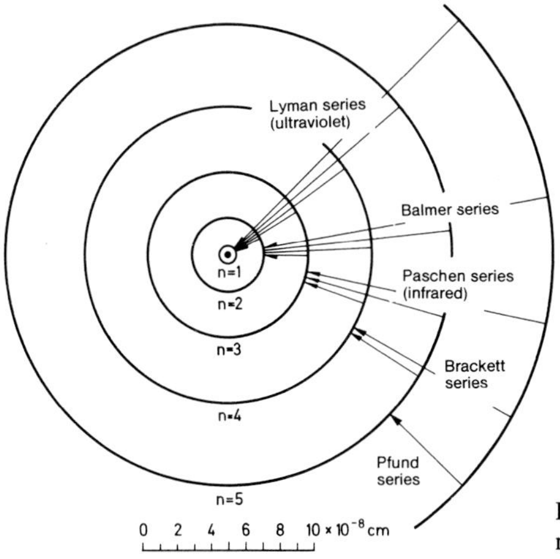

Fig. 8.5. Schematic representation of the Bohr atomic model, showing the first five spectral series

lines are missing in the continuous spectrum because of absorption of light on the way through the stellar atmosphere), then it can be concluded that the temperature of the atmosphere is so high, that the first excited state of the H atom ($n = 2$) is also occupied. This is the basis of spectroscopic temperature determination utilising the Boltzmann distribution (2.8). For example, in the sun, with a surface temperature of 6000 K, only 10^{-8} of the hydrogen atoms in the solar atmosphere are in the $n = 2$ state.

8.4 Some Quantitative Conclusions

We will now treat the Bohr model with arbitrary nuclear charge for *hydrogen-like* systems such as He^+, Li^{2+}, etc. quantitatively. The nucleus with charge Z is orbited by an electron in a circular orbit n at a distance r_n and with the velocity v_n. There is an equilibrium between the Coulomb force and the centrifugal force:

$$\frac{Ze^2}{4\pi\varepsilon_0 r_n^2} = \frac{m_0 v^2}{r_n} = m_0 r_n \omega_n^2 , \tag{8.16}$$

where $\omega_n = v_n/r_n$ is the circular frequency of the electron in its orbit n and m_0 is its mass. For the possible orbital radii follows, see (8.14)

$$r_n = \frac{n^2 \hbar^2 4\pi\varepsilon_0}{Ze^2 m_0} . \tag{8.17}$$

With $Z = 1$, $n = 1$ we find for the smallest orbital radius r_1 in the hydrogen atom $r_1(\mathrm{H}) = 0.529$ Å, the right order of magnitude for the spatial extension of the neutral hydrogen atom. $r_1(\mathrm{H})$ is referred to as the *Bohr radius* of the hydrogen atom in the ground state, abbreviated a_0.

For the possible circular frequencies of the electronic motion we obtain

$$\omega_n = \frac{1}{(4\pi\varepsilon_0)^2} \frac{Z^2 e^4 m_0}{n^3 \hbar^3} . \tag{8.18}$$

For $Z = 1$, $n = 1$, the largest possible circular frequency is seen to be

$$\omega_1(\mathrm{H}) \cong 10^{16}\,\mathrm{Hz} ;$$

ω_n would be the "classical" frequency of the emitted light if the electron behaved like a classical dipole in the atom. This is, however, not the case, see Sect. 8.3. The emitted frequency corresponds to the *difference* of the energy states of two orbits n and n' according to (8.9). The total energy is according to (8.6)

$$E_n = m_0 v_n^2/2 - \frac{Ze^2}{4\pi\varepsilon_0 r_n} . \tag{8.19}$$

Substituting for r_n from (8.17) and v_n, which can be obtained from (8.15), yields the possible energy states:

$$E_n = -\frac{Z^2 e^4 m_0}{32\pi^2 \varepsilon_0^2 \hbar^2} \cdot \frac{1}{n^2} . \tag{8.20}$$

For $Z = 1$, $n = 1$ we find the lowest energy state of the hydrogen atom:

$$E_1(\mathrm{H}) = -13.59 \text{ eV} .$$

This is the ionisation energy of the H atom.

For arbitrary Z, $n = 1$, one obtains

$$E_1(Z) = -Z^2 \cdot 13.59 \text{ eV} .$$

For the wavenumbers of the spectral lines we find, according to (8.3) and (8.9)

$$\bar{\nu} = \frac{1}{hc}(E_n - E_{n'}) = \frac{e^4 m_0 Z^2}{64\pi^3 \varepsilon_0^2 \hbar^3 c}\left(\frac{1}{n'^2} - \frac{1}{n^2}\right) . \tag{8.21}$$

Comparison of this result with the empirically found Balmer formula (see Sect. 8.2) shows complete agreement with respect to the quantum numbers n and n'. The quantum number n which was introduced by *Bohr* is thus identical with the index n of the Balmer formula.

8.5 Motion of the Nucleus

The spectroscopically measured quantity R_H (Sect. 8.2) does not agree exactly with the theoretical quantity R_∞ (8.13). The difference is about 60 cm^{-1}. The reason for this is the motion of the nucleus during the revolution of the electron, which was neglected in the above model calculation. This calculation was made on the basis of an infinitely massive nucleus; we must now take the finite mass of the nucleus into account.

In mechanics it can be shown that the motion of two particles, of masses m_1 and m_2 and at distance r from one another, takes place around the common centre of gravity. If the centre of gravity is at rest, the total energy of both particles is that of a fictitious particle which orbits about the centre of gravity at a distance r and has the mass

$$\mu = \frac{m_1 m_2}{m_1 + m_2} , \tag{8.22}$$

referred to as the *reduced mass*. In all calculations of Sect. 8.4 we must therefore replace the mass of the orbiting electron, m_0, by μ and obtain, in agreement with experiment,

$$R_\mathrm{H} = R_\infty \frac{1}{1 + m_0/M} . \tag{8.23}$$

Here $m_0 \equiv m_1$, the mass of the orbiting electron, and $M \equiv m_2$, the mass of the nucleus. The energy corrections due to motion of the nucleus decrease rapidly with increasing nuclear mass (Table 8.4).

Table 8.4. Energy correction for motion of the nucleus for the Rydberg numbers of several one-electron atoms

Atom	$H(^1H)$	$D(^2H)$	$T(^3H)$	He^+	Li^{2+}
A	1	2	3	4	7
$-\frac{\Delta E}{E}\cdot 10^4$	5.45	2.75	1.82	1.36	0.78
$-\frac{\Delta E}{E}$ %	0.0545	0.0275	0.0182	0.0136	0.0078

This observation makes possible a spectroscopic determination of the mass ratio M/m_0, e.g.

$$M_{\text{proton}}/m_{\text{electron}} = 1836.15\,.$$

Due to the motion of the nucleus, different isotopes of the same element have slightly different spectral lines. This so-called isotope displacement led to the discovery of heavy hydrogen with the mass number $A = 2$ (deuterium). It was found that each line in the spectrum of hydrogen was actually double. The intensity of the second line of each pair was proportional to the content of deuterium. Figure 8.6 shows the H_β line with the accompanying D_β at a distance of about 1 Å in a 1 : 1 mixture of the two gases. The nucleus of deuterium contains a neutron in addition to the proton. There are easily measurable differences in the corresponding lines of the H and D Lyman series, namely

$$R_H = R_\infty \cdot \frac{1}{1 + m_0/M_H} = 109677.584\ \text{cm}^{-1}\,, \tag{8.24}$$

$$R_D = R_\infty \cdot \frac{1}{1 + m_0/M_D} = 109707.419\ \text{cm}^{-1}\,. \tag{8.25}$$

The difference in wavelengths $\Delta\lambda$ for corresponding lines in the spectra of light and heavy hydrogen is:

$$\Delta\lambda = \lambda_H - \lambda_D = \lambda_H\left(1 - \frac{\lambda_D}{\lambda_H}\right) = \lambda_H\left(1 - \frac{R_H}{R_D}\right)\,. \tag{8.26}$$

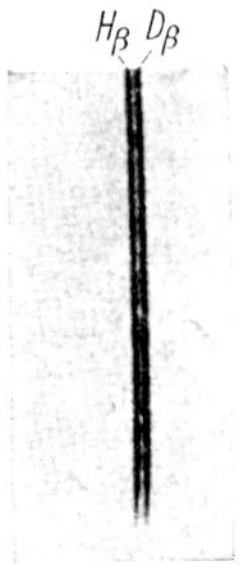

Fig. 8.6. β lines of the Balmer series in a mixture of equal parts hydrogen (^1H) and deuterium (^2H). One sees the isotope effect, which is explained by motion of the nucleus. The lines are about 1 Å apart and have the same intensity here, because the two isotopes are present in equal amounts [from K. H. Hellwege: *Einführung in die Physik der Atome*, Heidelberger Taschenbücher, Vol. 2, 4th ed. (Springer, Berlin, Heidelberg, New York 1974) Fig. 40 a]

Table 8.5 gives the measured values. The agreement between the calculated and measured values is excellent.

Historical remark: a difference of about 0.02% had been found between the values of the molecular weight of hydrogen determined chemically and by mass spectroscopy, because D is present in the natural isotopic mixture of hydrogen. Its mass was included in the results obtained by chemical means, but not by mass spectroscopy. In 1931, however, *Urey* discovered spectral lines which, according to their Rydberg number, belonged to D by observing a gas discharge through the vapour of 3 litres of liquid hydrogen evaporated into a 1 cm^3 volume (Fig. 8.6).

Table 8.5. Comparison of the wavelengths of corresponding spectral lines in hydrogen and deuterium. The lines belong to the Lyman series

λ_D/Å	λ_H/Å
1215.31	1215.66
1025.42	1025.72
972.25	972.53

8.6 Spectra of Hydrogen-like Atoms

According to *Bohr*, the spectra of all atoms or ions with only one electron (one-electron systems) should be the same except for the factor Z^2 and the Rydberg number. The spectrum of hydrogen should thus explain those of the ions He^+, Li^{2+}, Be^{3+} or any other ions which have only one electron. This has been completely verified experimentally (see Table 8.6 and the energy diagram in Fig. 8.7).

For He^+, astronomers found the Fowler series

$$\bar{\nu}_F = 4R_{He}\left(\frac{1}{3^2} - \frac{1}{n^2}\right) \tag{8.27}$$

Table 8.6. Wavelengths λ_{12} of the first Lyman lines, i.e. the spectral lines with $n' = 1$, $n = 2$, of hydrogen and hydrogen-like atomic ions. The mass correction (*first column*) is used to calculate the Rydberg number (*second column*) and thus λ_{12} (*third column*). The calculated values are in good agreement with the measured values (*fourth column*)

	$1 + \frac{m_0}{m_{nucl}}$	R_{nucl} [cm^{-1}]	λ_{12} (calc) [Å]	λ_{12} (meas) [Å]
^1H	1.00054447	109677.6	1215.66	1215.66
^2H	1.00027148	109707.4	1215.33	1215.33
$^4He^+$	1.00013704	109722.3	303.8	303.6
$^7Li^{2+}$	1.00007817	109728.7	135.0	135.0
$^9Be^{3+}$	1.00006086	109730.6	75.9	75.9
$^{10}B^{4+}$	1.00005477	109731.3	48.6 (both B isotopes)	48.6 (both B isotopes)
$^{11}B^{4+}$	1.00004982	109731.8		
$^{12}C^{5+}$	1.00004571	109732.3	33.7	33.7

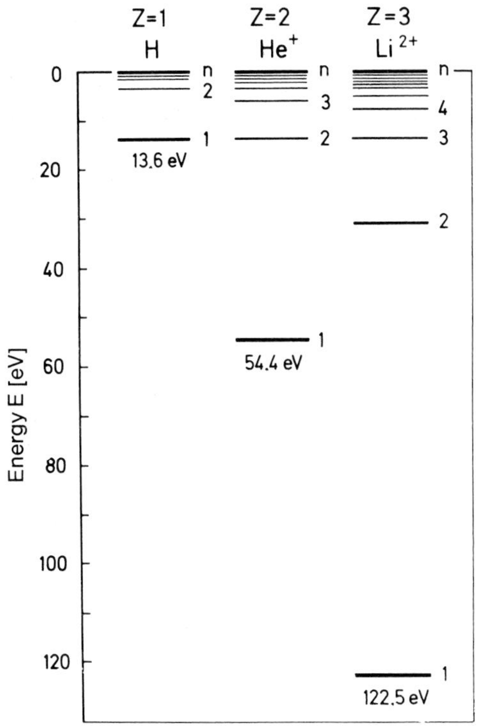

Fig. 8.7. Some energy levels of the atoms H, He^+ and Li^{2+}

Table 8.7. Comparison of the spectral lines of the Balmer series in hydrogen and the Pickering series in the helium ion, in Å

He^+	H
6560.1	6562.8 (H_α)
5411.6	
4859.3	4861.3 (H_β)
4561.6	
4338.7	4340.5 (H_γ)
4199.9	
4100.0	4101.7 (H_δ)

and the Pickering series

$$\bar{\nu}_P = 4R_{He}\left(\frac{1}{4^2} - \frac{1}{n^2}\right), \tag{8.28}$$

which can also be represented as

$$\bar{\nu}_P = R_{He}\left(\frac{1}{2^2} - \frac{1}{(n/2)^2}\right), \quad n = 5, 6 \dots . \tag{8.29}$$

Every other line of the Pickering series thus almost corresponds to one of the Balmer lines of H. This is shown in Table 8.7.

Later other He^+ series were found, such as the

1st Lyman series $\bar{\nu}_{L1} = 4R_{He}\left(\frac{1}{1^2} - \frac{1}{n^2}\right)$, (8.30)

2nd Lyman series $\bar{\nu}_{L2} = 4R_{He}\left(\frac{1}{2^2} - \frac{1}{n^2}\right)$. (8.31)

For Li^{2+}, Be^{3+} and still heavier highly ionised atoms, spectral lines have been observed which can be calculated by multiplying the frequencies of the lines of the

H atom by Z^2 and insertion of the corresponding Rydberg constant. With increasing nuclear charge Z, we quickly reach the region of x-ray wavelengths.

In 1916, the collected spectroscopic experience concerning the hydrogen-similarity of these spectra was generalised in the *displacement theorem* of *Sommerfeld* and *Kossel*, which states:

The spectrum of any atom is very similar to the spectrum of the singly charged positive ion which follows it in the periodic table.

8.7 Muonic Atoms

With the simple Bohr model, the muonic atoms, first observed in 1952, can be described. They contain, instead of an electron, the 207-times heavier μ meson or muon and are, in contrast to the Rydberg atoms, extremely small, in extreme cases hardly larger than the typical diameter of an atomic nucleus.

To produce them, matter is bombarded with energetic protons (about 440 MeV), giving rise to other elementary particles, the pions, according to the following reaction schemes:

$$p+n \rightarrow n+n+\pi^+ \quad \text{or} \quad p+n \rightarrow p+p+\pi^- \,.$$

Here p denotes the proton, n the neutron, and π the pion.

Pions have a charge $+e$ or $-e$ and mass $m_\pi = 273\, m_0$. They decay into other particles, the muons, according to the reactions

$$\pi^+ \rightarrow \mu^+ + \nu_\mu \quad \text{or} \quad \pi^- \rightarrow \mu^- + \bar{\nu}_\mu \,.$$

Here, the symbols ν or $\bar{\nu}$ mean a neutrino or an antineutrino, the index μ means muon neutrino (neutretto), and electron neutrinos carry the index e to distinguish them. The neutrinos are only shown for completeness.

The half-life for this decay is $T_{1/2} = 2.5 \cdot 10^{-8}$ s. Muons may be characterised as heavy electrons; they have a charge e, a mass equal to $206.8\, m_0$, and a half-life $T_{1/2} = 2.2 \cdot 10^{-6}$ s.

Muons decay into electrons (e^-) or into positrons (e^+) according to the reactions

$$\mu^+ \rightarrow e^+ + \nu_e + \bar{\nu}_\mu \,, \qquad \mu^- \rightarrow e^- + \bar{\nu}_e + \nu_u \,.$$

Before they decay, they can be captured into outer atomic orbits by atomic nuclei and can occupy these orbits in the place of electrons. In making transitions from the outer to inner orbits, the muons radiate light of the corresponding atomic transition frequency; this is light in the x-ray region of the spectrum. Since muons behave like heavy electrons, we can simply apply the results of the Bohr model. For the orbital radii we have, see (8.17)

$$r_n = \frac{4\pi\varepsilon_0 \hbar^2}{Ze^2 m_\mu} n^2 \,. \tag{8.32}$$

r_n is thus smaller than the radius of the corresponding orbit which is occupied by an electron by the ratio of the electron to the muon mass.

A numerical example: for the magnesium atom ^{12}Mg we find

$$\text{Electron:} \quad r_1(e^-) = \frac{0.53}{12}\,\text{Å} = 4.5 \cdot 10^{-12}\,\text{m}\,,$$

$$\text{Muon:} \quad r_1(\mu^-) = \frac{r_1(e^-)}{207} = 2.2 \cdot 10^{-14}\,\text{m}\,.$$

The muon is thus much closer to the nucleus than the electron. For the radiation from a transition between the levels with principal quantum numbers 1 and 2 the following expression holds:

$$h\nu = \frac{Z^2 e^4 m_\mu}{32\pi^2\varepsilon_0^2\hbar^2}\left(\frac{1}{1^2} - \frac{1}{2^2}\right), \tag{8.33}$$

that is, the quantum energy is larger by the ratio of the masses than the energy of the corresponding transition in an electronic atom. Finally, the muon decays as described above, or else it is captured by the nucleus, which then may itself decay.

Muonic atoms are observed for the most part by means of the x-radiation which they emit; this radiation decays in intensity with the half-life characteristic of muons. Muonic atoms are interesting objects of nuclear physics research. Since the muons approach the nucleus very closely, much more so than the electrons in an electronic atom, they can be used to study details of the nuclear charge density distribution, the distribution of the nuclear magnetic moment within the nuclear volume and of nuclear quadrupole deformation.

Figure 8.8 shows the spatial distribution of a muon in several orbits of a lead atom. It can be seen that the muons in these orbits spend a considerable amount of time in the nucleus or in its immediate neighbourhood. Since the muons approach the nuclear charge Ze very closely, the binding and excitation energies become extremely large.

Figure 8.9 shows a term diagram of the muonic-atom levels for a nuclear charge number $Z = 60$. The analogy with the hydrogen atom is evident; however, the transitions here are in the energy region of MeV, i.e. in the region of hard x-rays and of

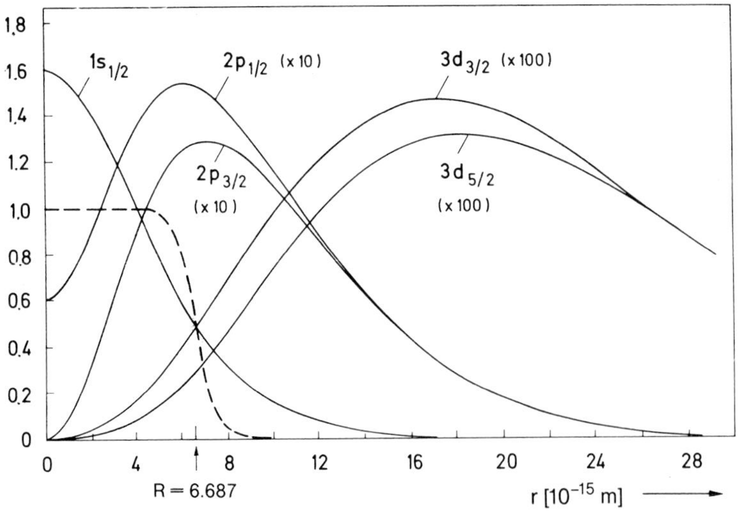

Fig. 8.8. The probability of finding a muon at a distance r from the centre of the nucleus of a lead atom, $Z = 82$, nuclear radius $R = 6.687 \cdot 10^{-15}$ m. The probability distributions are shown for several orbits (——) and the nuclear charge distribution is indicated (– –). The symbols used to denote the various orbits will be explained later

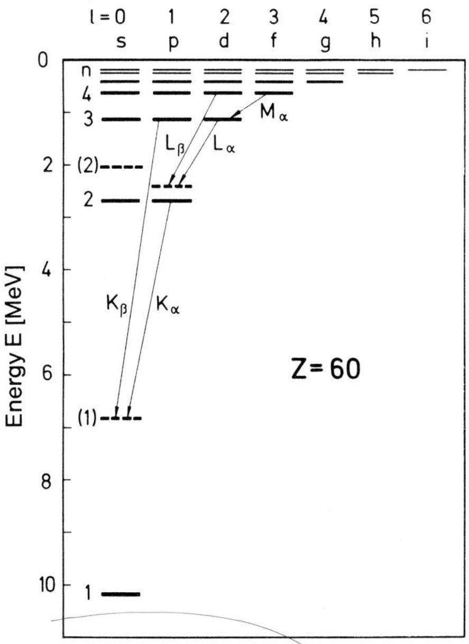

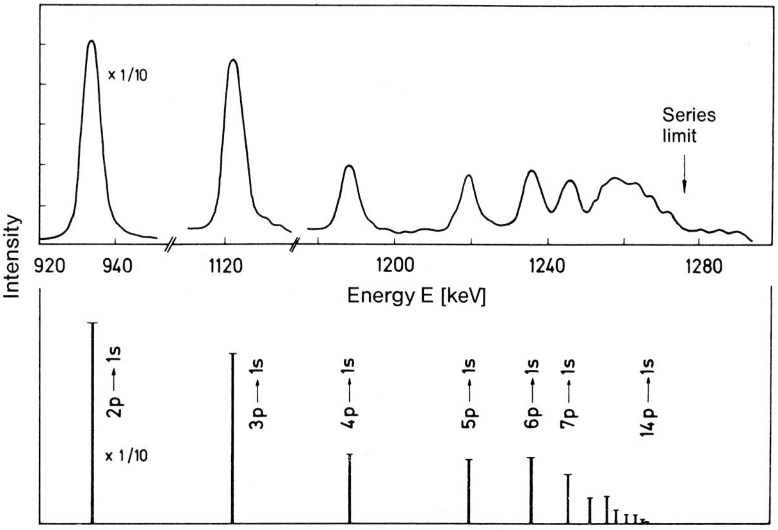

▲
Fig. 8.10. Lyman series ($np \to 1s$) of the muonic transitions in a titanium atom. Note the energy scale

▲
Fig. 8.9. Muonic terms for an atom with $Z = 60$. The fully drawn levels correspond to the assumption of a point nucleus; the dashed levels take account of the finite nuclear size. The notation used for the transitions corresponds to that used for x-ray lines (Chap. 18). Note the energy scale

gamma rays. For the investigation of such muonic atoms, one therefore requires the tools of nuclear physics. Detection of the radiation is carried out with scintillator or semiconductor detectors.

Finally, Fig. 8.10 shows an example of the measurement of radiation from a muonic atom, the Lyman series in the muonic spectrum of titanium. The notations s, p, d, etc. in Figs. 8.8–10 refer to the orbital angular momentum of the electrons (muons). They will be further described in Sect. 8.9.

8.8 Excitation of Quantum Jumps by Collisions

Lenard investigated the ionisation of atoms as early as 1902 using electron collisions. For his measurements, he used an arrangement following the principle of the experimental scheme shown in Fig. 8.11. The free electrons produced by thermionic emission are accelerated by the positive grid voltage V_G and pass through the open-meshed grid into the experimental region. Between the grid and the plate A at the right of the drawing, which serves as the third electrode, a plate voltage V_A is applied. The plate is negatively charged relative to the grid. The voltages are chosen so that the electrons cannot reach the plate; they pass through the grid and are repelled back to it. When an electron has ionised an atom of the gas in the experimental region, however, the ion is accelerated towards the plate A. Ionisation events are thus detected as a current to the plate.

The current is plotted as a function of the grid voltage V_G in the lower part of Fig. 8.11. Only when the electrons have a certain minimum energy eV_i does the current appear. The corresponding accelerating potential V_i is the ionisation potential of the atoms.

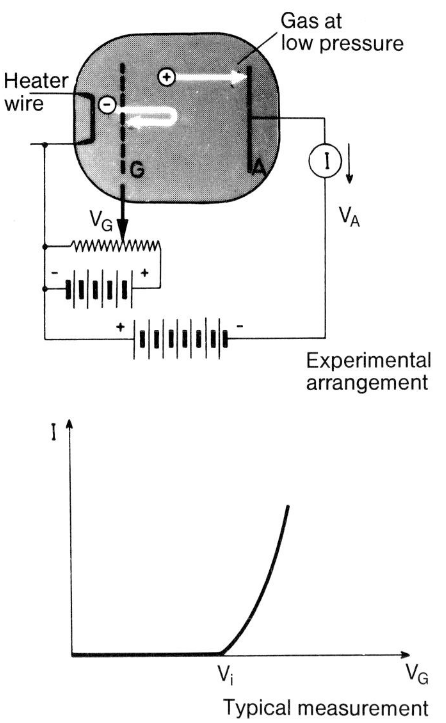

Fig. 8.11. Experimental arrangement for detecting ionisation processes in gases. Only positive ions, which are formed by collisions with electrons, can reach the plate A. In the lower part of the figure, the plate current is plotted as a function of the grid voltage V_G. V_i is the voltage with which the electrons must be accelerated in order to be able to ionise the atoms

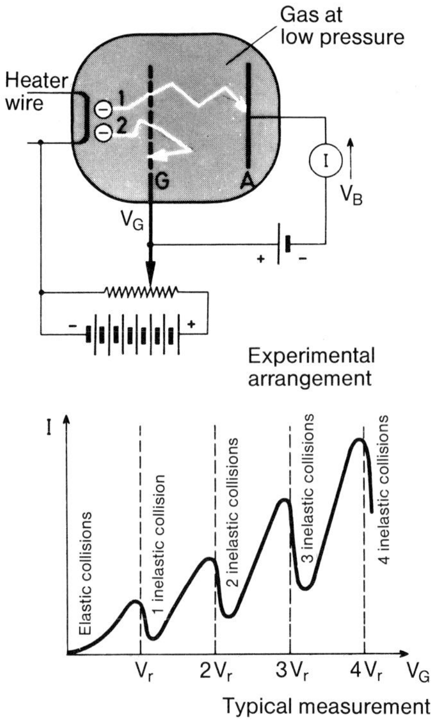

Fig. 8.12. Experimental arrangement of *Franck* and *Hertz* for investigating inelastic collisions between electrons and atoms. Electrons on the way from the grid to the anode can transfer their kinetic energies partially (particle 1) or completely (particle 2) to the gas atoms. The anode current as a function of the grid voltage is plotted in the lower part of the figure. At high grid voltages, several energy-transfer processes can occur one after the other

Franck and *Hertz* showed for the first time in 1913 that the existence of discrete energy levels in atoms can be demonstrated with the help of electron collision processes independently of optical-spectroscopic results. Inelastic collisions of electrons with atoms can result in the transfer of amounts of energy to the atoms which are smaller than the ionisation energy and serve to excite the atoms without ionising them.

The experimental setup is shown schematically in Fig. 8.12. Electrons from a heated cathode are accelerated by a variable voltage V_G applied to a grid. They pass through the grid and are carried by their momenta across a space filled with Hg vapour to an anode A. Between the anode and the grid is a braking voltage of about 0.5 V. Electrons which have lost most of their kinetic energy in inelastic collisions in the gas-filled space can no longer move against this braking potential and fall back to the grid. The anode current is then measured as a function of the grid voltage V_G at a constant braking potential V_B.

The result is shown in the lower part of Fig. 8.12. As soon as V_G is greater than V_B, the current increases with increasing voltage (space-charge conduction law). At a value of $V_G \cong 5$ V (in mercury vapour) the current I is strongly reduced; it then increases again up to $V_G \cong 10$ V, where the oscillation is repeated. The explanation of these results is found by making the following assumptions: when the electrons have reached an energy of about 5 eV, they can give up their energy to a discrete level of the mercury atoms. They have then lost their energy and can no longer move against the braking potential. If their energy is 10 eV, this energy transfer can occur twice, etc. Indeed, one finds an intense line in emission and absorption at $E = 4.85$ eV in the optical spectrum of atomic mercury, corresponding to a wavelength of 2537 Å. This line was also observed by *Franck* and *Hertz* in the optical emission spectrum of Hg vapour after excitation by electron collisions. The excitation or resonance voltages are denoted in Figs. 8.12, 13 as V_r.

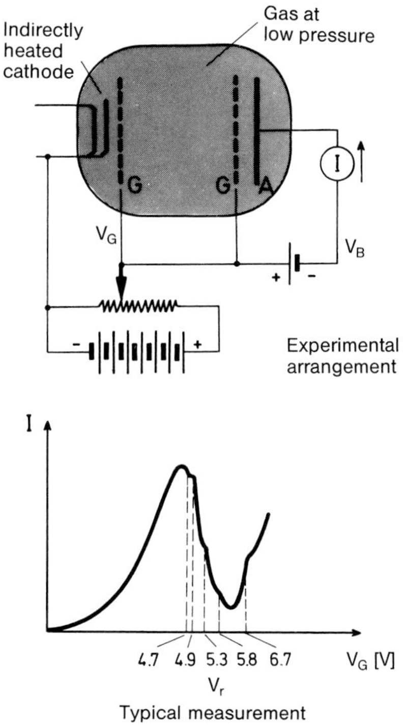

Fig. 8.13. Improved experimental setup for determining atomic excitation energies by electron collisions. The collisions take place in the field-free space between the two grids G. In this way, a high resolution is reached. In the lower part of the figure, an experimental result obtained with Hg vapour is shown in part

The resolving power for the energy loss of the electrons may be improved by using an indirectly heated cathode and a field-free collision region. In this way, one obtains a better uniformity of the energies of the electrons. With an improved experimental arrangement (Fig. 8.13), a number of structures can be seen in the current-voltage curve; these correspond to further excitations of the atoms. The step at 6.73 eV, for example, corresponds to a further intense line in the Hg spectrum; 6.73 eV ≙ 1850 Å.

Not all the maxima in the current-voltage curve can be correlated with observed spectral lines. To explain this fact, we have to assume that optically "forbidden" transitions can, in some cases, be excited by collisions. We shall see later that there are selection rules for optical transitions between energy terms of atoms, according to which not all combinations of terms are possible – one says "allowed". The selection rules for collision excitation of atoms are clearly not identical with those for optical excitation (or de-excitation).

In this connection, the following experiment is interesting: Na vapour at low pressure can be excited to fluorescence by illumination with the yellow Na line (quantum energy 2.11 eV). The excitation occurs only when the light used for illumination has exactly the quantum energy 2.11 eV. Both smaller and larger quantum energies are ineffective in producing an excitation.

Excitation by means of collisions with electrons are in this respect quite different: in this type of excitation, the yellow line is emitted whenever the energy of the electrons is equal to or *larger than* 2.11 eV. This can be explained as follows: the kinetic energy of free electrons is not quantised. After excitation of a discrete atomic energy level by electron collision, the exciting electron can retain an arbitrary amount of energy, depending on its initial value. This remaining energy can, if it is sufficiently large, serve to excite still other atoms in the gas volume.

All in all, these electron collision experiments prove the existence of discrete excitation states in atoms and thus offer an excellent confirmation of the basic assumptions of the Bohr theory. In modern atomic and solid state physics, energy-loss spectra of electrons represent an important aid to the investigation of possible excitation stages of atoms and of the structure of the surfaces of solids.

8.9 Sommerfeld's Extension of the Bohr Model and the Experimental Justification of a Second Quantum Number

The finished picture of the Bohr model still contained some fuzzy details: exact spectral measurements at high resolution showed that the lines of the Balmer series in hydrogen are, in fact, *not* single lines. Each of them consists rather of several components; how many one can distinguish depends on the resolution of the spectrometer employed.

The H_α line of hydrogen with $\bar{\nu} = 15233\ \mathrm{cm}^{-1}$ consists, for example of a multiplet with a wavenumber splitting of $\Delta\bar{\nu} = 0.33\ \mathrm{cm}^{-1}$ between the strongest components (Fig. 8.14). In order to observe this structure, a spectral resolution of nearly $\bar{\nu}/\Delta\bar{\nu} = 100\,000$ is needed. In the spectrum of the one-electron ion He^+, these multiplet lines are more strongly separated, and the splitting is therefore easier to observe. We shall see in Chap. 12 that the splitting increases as the 4th power of the nuclear charge number Z.

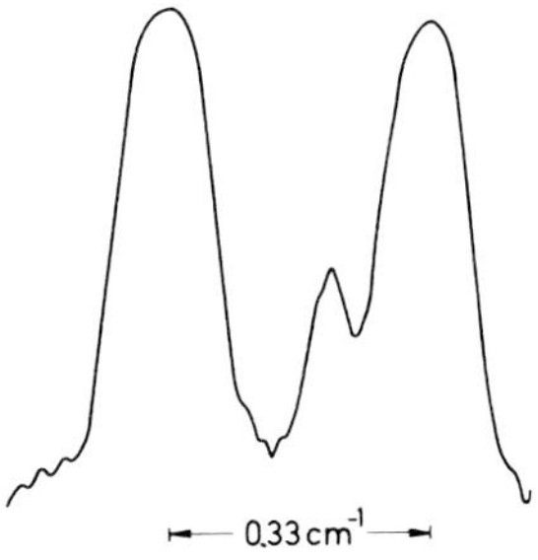

Fig. 8.14. H_α line of the Balmer series at a high spectral resolution. Three components can be distinguished. A still higher resolution is shown in Fig. 12.24. The resolution reached in this spectrum is limited by Doppler broadening

From observations of this type, *Sommerfeld* derived an extension of the Bohr model. It is well known from classical mechanics that, according to Kepler's Laws, not only circular orbits, but also elliptical orbits are possible, having the same energies.

From this, *Sommerfeld* drew the conclusion that the same is true in atoms also. In order to distinguish the elliptical orbits from the circular ones, a new, second quantum number is required. Since *Sommerfeld*'s chain of reasoning was on the one hand of great historical importance in introducing a second quantum number, but has, on the other hand, been made obsolete by the later quantum mechanical treatment, we will only give a brief summary here.

The principal quantum number n remains valid; it continues to determine the total energy of a term according to (8.20), i.e.

$$E_n = -\frac{RhcZ^2}{n^2}\,. \tag{8.34}$$

According to *Sommerfeld*, n also determines the major semiaxis of the ellipse. The minor semiaxis is determined by the second quantum number in such a fashion that the absolute value of the angular momentum remains a whole multiple k of $\hbar$, with $k \leqq n$. The length of the minor semiaxis, that is the eccentricity of the ellipse, has in this model no influence on the total energy. Each principal quantum number n corresponds to one major semiaxis a_n, but to various orbital shapes, characterised by the minor semiaxis $b_{n,k}$. We say that the energy term is n-fold degenerate, by which is meant that different orbits with two quantum numbers n and k belong to one and the same energy value.

We should mention at this point that in quantum theory, the Sommerfeld second quantum number k became the orbital angular momentum quantum number $l(l = k-1)$. The orbital angular momentum of the electron is given by (as we shall show in Chap. 10)

$$|\boldsymbol{l}| = \sqrt{l(l+1)}\,\hbar \quad \text{with} \quad l = 0, 1, 2, \ldots n-1\,. \tag{8.35}$$

In order to distinguish the orbital angular momentum itself, $\boldsymbol{l}$, from its quantum number l, we shall henceforth use the symbol $|\boldsymbol{l}|$ for the absolute value of the angular momentum vector $\boldsymbol{l}$.

For the various numerical values of the angular momentum quantum number, letter symbols s, p, d, f, g, h, etc. have become firmly established; these are listed in the following table:

Quantum number	$l = 0$	1	2	3	4	5
Angular momentum	$\|\boldsymbol{l}\| = 0$	$\sqrt{2}\,\hbar$	$\sqrt{6}\,\hbar$	$\sqrt{12}\,\hbar$	$\sqrt{20}\,\hbar$	$\sqrt{30}\,\hbar$
Name (Symbol)	s	p	d	f	g	h-electron or state.

What this means in terms of the spatial form of the electron orbitals will be explained later, together with the solution of the Schrödinger equation (Chap. 10).

8.10 Lifting of Orbital Degeneracy by the Relativistic Mass Change

We still have no explanation for the doublet or multiplet structure of the spectral lines mentioned at the beginning of the last section. However, we now know that each level is n-fold degenerate; by this we mean the fact that each energy level has various possibilities for the spatial distribution of the electrons occupying it. The number of levels with differing energies, and thus the number of observable spectral lines, however still remains the same.

The lifting of this degeneracy occurs, according to *Sommerfeld* (1916), through the effect of the relativistic mass change, $m = m(v)$, which we have neglected up to now. We can understand this qualitatively as follows: exactly as in planetary motion according to Kepler's Laws, the electrons are accelerated when they come near to the nucleus. This is a result of Kepler's Law of Areas, which requires that the moving electron sweep out equal areas between its orbit and the nucleus in equal times. In the neighbourhood of the nucleus, the electrons are thus faster and, from special relativity, more massive. This leads, in turn, to a decrease in energy: increased mass means, according to *Bohr*, a smaller radius, and this leads to a larger (negative) binding energy, i.e. to a decrease in total energy. The smaller the minor semiaxis of an ellipse, the more significant these relativistic corrections must become.

We will not repeat *Sommerfeld*'s calculation here; we just give the result. The relativistic mass change leads to a rotation of the perihelion point of the orbits; in an intuitive picture, the electron then has a "rosette motion" about the nucleus (Fig. 8.15).

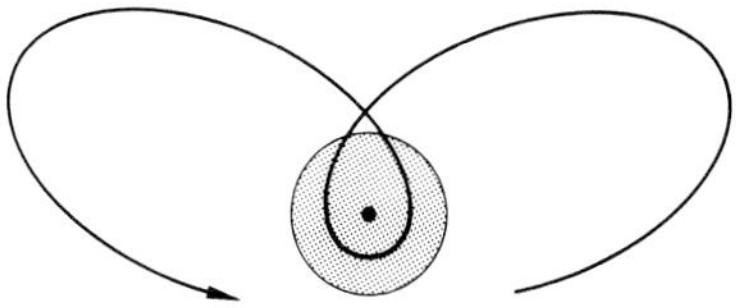

Fig. 8.15. Rotation of the perihelion point in the motion of an electron around the nucleus in a many-electron atom according to the Sommerfeld theory. The shaded region is the electronic shell of the atom. The outer electron follows a so-called "diving orbit" in its motion, i.e., it dives into the atomic shell

In *Sommerfeld*'s calculation, the "fine structure constant" plays a rôle:

$$\alpha = \frac{\text{Velocity of the electron in the 1st Bohr orbit}}{\text{Velocity of light}}$$

$$= \frac{e^2}{2\varepsilon_0 hc} = \frac{1}{137} \quad \text{(dimensionless)}\,.$$

For an electron orbit with the quantum numbers n and k, the result of *Sommerfeld*'s calculation of the relativistic mass effect is

$$E_{n,k} = -Rhc\frac{Z^2}{n^2}\left[1 + \frac{\alpha^2 Z^2}{n^2}\left(\frac{n}{k} - \frac{3}{4}\right) + \text{higher-order corrections}\right]. \qquad (8.36)$$

The relativistic energy change is thus of the order of $\alpha^2 \cong 10^{-5}$, i.e. small, but observable (see Fig. 8.14).

8.11 Limits of the Bohr-Sommerfeld Theory. The Correspondence Principle

The Bohr-Sommerfeld model is theoretically unsatisfying: on the one hand, classical mechanics is set aside, and only certain particular orbits are allowed; on the other hand, classical physics is used to calculate the orbits, see Sect. 8.3. It is as though, "On Mondays, Wednesdays and Fridays one uses the classical laws, on Tuesdays, Thursdays, and Saturdays the laws of quantum physics" (*Bragg*). Furthermore, the model predicts only the frequencies but not the intensities or the time dependence of emitted or absorbed light.

The gap which had opened between classical physics and the (early) quantum theory was bridged by *Bohr* with his *Correspondence Principle.*

According to this principle, for large quantum numbers, the classical and quantum theories approach one another; or, the behaviour of an atom approaches that expected from classical, macroscopic physics, the larger its energy relative to the energy change which occurs in the process considered, i.e. all the more, the higher the level and the smaller the level difference.

Starting from considerations such as the above, one arrives at the following general formulation of the Correspondence Principle:

Every non-classical theory must, in the limit of high energies and small energy changes, yield the results of classical theory.

The intensities, polarisations, and selection rules for spectral lines may be calculated from the laws of classical physics. The Correspondence Principle allows us, within limits, to translate these results, by using a prescription for quantisation, into the quantum theory.

In spite of a series of successes, the application of the Bohr-Sommerfeld theory led to fundamental difficulties. The results were wrong even for atoms with two electrons. The magnetic properties of atoms were not correctly described. The removal of these difficulties was accomplished with the development of modern quantum mechanics. In Chap. 10, we will treat the hydrogen atom problem exactly with the help of quantum theory; we shall find there that some of the results of the Bohr-Sommerfeld theory remain valid, while others must be modified.

8.12 Rydberg Atoms

Atoms in which an electron has been excited to an unusually high energy level illustrate well the logical continuity between the world of classical physics and quantum mechanics.

Such atoms, called Rydberg atoms, have extraordinary properties. They are gigantic: Rydberg atoms are known with diameters reaching 10^{-2} mm, corresponding to a 100000-fold increase over the diameters of atoms in the ground state. Furthermore, these excited states have *extremely long lifetimes.* While typical lifetimes of lower excited states of atoms are about 10^{-8} s, there are Rydberg atoms which have lifetimes of 1 s. The difference in energy between two neighboring states n and n' becomes very small when n is large. The long lifetimes of such states are in part a result of the fact that the probability of a spontaneous transition between two states n and n' is, according to Einstein (Sect. 5.2.3), proportional to ν^3. In addition, Rydberg atoms may be *strongly polarised* by relatively weak electric fields, or even completely ionised.

When the outer electron of an atom is excited into a very high energy level, it enters a spatially extended orbit – an orbital – which is far outside the orbitals of all the other electrons. The excited electron then "sees" an atomic core, consisting of the nucleus and all the inner electrons, which has a charge $+e$, just the same as the charge of the hydrogen nucleus. As long as the excited electron does not approach the core too closely, it behaves as though it belonged to a *hydrogen atom.* Rydberg atoms behave therefore in many respects like highly excited hydrogen atoms.

In interstellar space, there are atoms whose outer electrons are in states with principal quantum numbers n up to 350; this has been observed by radio astronomical methods. In the laboratory, Rydberg atoms with principal quantum numbers between 10 and 290 have been studied.

The orbital radius of an electron in an atom is proportional to n^2 (8.17). The spacing between neighbouring energy levels decreases as n^{-3}. It is because these higher powers of n have especially large effects for large n-values that Rydberg atoms have their unusual properties.

Rydberg atoms are produced by exciting an atomic beam with laser light. To detect the highly excited atoms, an electric field is applied between the plates of a condenser through which the atomic beam passes. Through field ionisation, the atoms can be converted to ions with the aid of small electric fields of the order of a few hundred V cm^{-1}. The ions can be detected by means of their charge, for example with the aid of an electron multiplier or channeltron. An example of an experimental setup is shown in Fig. 8.16; Fig. 8.17 shows some experimental results. In Fig. 8.17, the result of exciting a beam of lithium atoms with three laser beams is shown. The first two excite the atoms into intermediate excited states (e.g. here $n = 3$, $l = 0$), while the third is continuously variable within a small energy range and adds the last necessary energy contribution to put the atoms into a Rydberg state. By continuously changing the frequency of this last

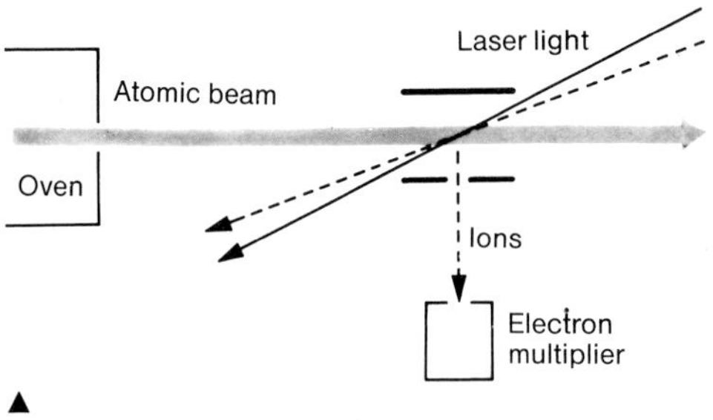

Fig. 8.16. Apparatus for the detection of Rydberg atoms. An atomic beam is crossed by several (here 2) laser beams. They cause the excitation of the atoms into Rydberg states when the sum of the quantum energies of the laser beams corresponds to the excitation energy of a Rydberg state. The Rydberg atoms are ionised in the electric field of a condenser, and the ions are then detected

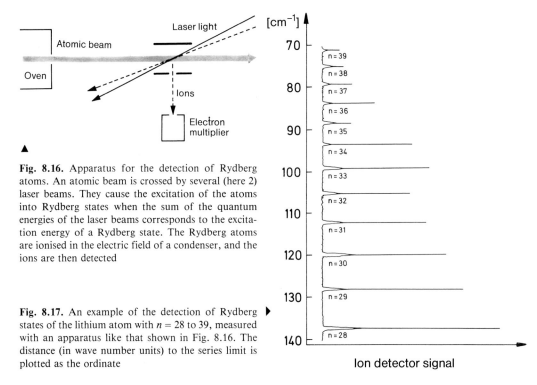

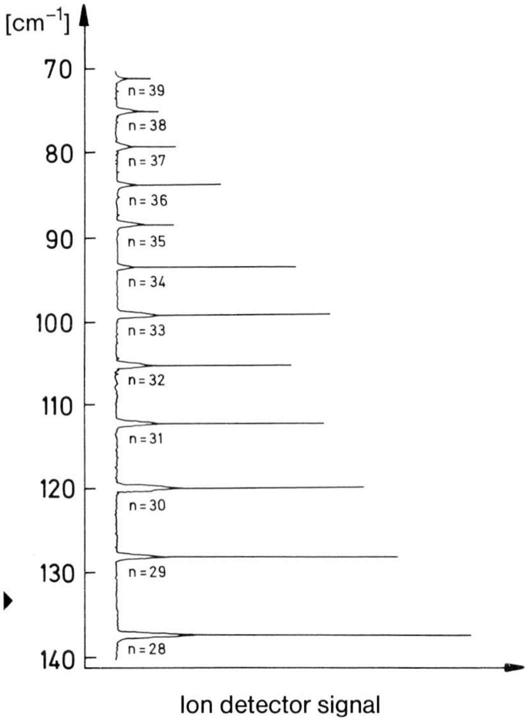

Fig. 8.17. An example of the detection of Rydberg states of the lithium atom with $n = 28$ to 39, measured with an apparatus like that shown in Fig. 8.16. The distance (in wave number units) to the series limit is plotted as the ordinate

laser, the experimenter can excite a series of Rydberg states of the atoms one after another – in the figure, the states with $n = 28$ to 39. Thus, a particular Rydberg state can be chosen and selectively excited in order to investigate its physical properties.

When a Rydberg atom reduces its principal quantum number by 1 in emitting a light quantum, the light is in the far infrared or microwave region of the electromagnetic spectrum. With this radiation, isolated Rydberg atoms were first discovered in 1965 in interstellar space. The density of atoms is so low there that collisions are extremely rare.

It has been possible to investigate Rydberg atoms in the laboratory since narrow-band, tunable lasers have been available (especially dye lasers, see Chap. 21). Since then, the energy levels, lifetimes, spatial extension of the wavefunctions, and the influence of electric and magnetic fields have been studied for quantum numbers which were previously only theoretical. The predictions of theory have been fully confirmed. Table 8.8 contains an overview of the properties of Rydberg atoms.

Table 8.8. Some properties of Rydberg atoms, valid for unperturbed electronic states

Property	General	Rydberg atoms, $n = 30$
Size	$d = a_0 n^2$	10^3 Å
Binding energy	$E_n = -R/n^2$	10^{-2} eV
Transition energy $\Delta n = 1$	$\Delta E = 2R/n^3$	10^{-3} eV ≙ 10 cm^{-1}
Lifetime	$\tau \propto n^3$	$30 \cdot 10^{-6}$ s

Problems

8.1 Calculate the recoil energy and velocity of a hydrogen atom in a transition from the state $n = 4$ to the state $n = 1$, in which a photon is emitted.

8.2 Five of the Balmer series lines of hydrogen have the wavelengths 3669.42 Å, 3770.06 Å, 3835.40 Å, 3970.07 Å and 4340.47 Å. Plot ν as a function of n for the Balmer series. From this, determine the value of n for the upper level of each of the five wavelengths above.

8.3 The absorption spectrum of hydrogen can be obtained by allowing white light to pass through hydrogen gas which is in the ground state and contains atomic hydrogen (not just H_2). Which photon energies are observed in the hydrogen absorption spectrum? Give the wavelengths of these "Fraunhofer lines".

8.4 a) The emission spectrum of the hydrogen atom is taken with a diffraction grating (line spacing $d = 2$ µm). A line of the Balmer series is observed in the second order at an angle $\theta = 29°5'$. What is the quantum number of the excited state from which the transition starts?

b) What is the minimum number of lines necessary in a diffraction grating if the first 30 spectral lines of the Balmer series of the hydrogen atom are to be resolved in the first-order diffraction spectrum?

Hint: In this case, the number of lines corresponds to the required resolution $\lambda/\Delta\lambda$.

8.5 Is it true that in a circular Bohr orbit, the potential energy is equal to the kinetic energy? If not, where does the energy difference go which arises if we assume that the electron and the nucleus are initially infinitely far apart and at rest? How large is E_{pot} compared to E_{kin} for the various Bohr orbits?

8.6 The attractive force between a neutron (mass M) and an electron (mass m) is given by $F = GMm/r^2$. Let us now consider the smallest orbit which the electron can have around the neutron, according to Bohr's theory.

a) Write a formula for the centrifugal force which contains m, r and v; r is the radius of the Bohr orbit, and v is the velocity of the electron in this orbit.
b) Express the kinetic energy in terms of G, M, m and r.
c) Express the potential energy in terms of G, M, m and r.
d) Express the total energy in terms of G, M, m and r.
e) Set up an equation which corresponds to the Bohr postulate for the quantisation of the orbits.
f) How large is the radius r of the orbit with $n = 1$? Express r in terms of $\hbar$, G, M and m; give the numerical value of r.

8.7 For the Bohr model of the atom, calculate the electric current and the magnetic dipole moment of the electron in the first three orbits ($n = 1, 2, 3$).

Hint: Use (12.1 – 7) to calculate the magnetic dipole moment.

8.8 "Positronium" is a bound electron-positron pair. The positron is the anti-particle corresponding to the electron. It has a charge $+e$ and the same rest mass as the electron. On the assumption that e^- and e^+ – in analogy to the H atom – circle the common centre of gravity, calculate the rotational frequency ω, the radius r and the binding energy of the system in the ground state.

8.9 A muonic atom consists of an atomic nucleus with nuclear charge Z and a captured muon, which is in the ground state. The muon is a particle with a mass 207 times that of the electron; its charge is the same as that of the electron.

a) What is the binding energy of a muon which has been captured by a proton?
b) What is the radius of the corresponding Bohr orbit with $n = 1$?
c) Give the energy of the photon which is emitted when the muon goes from the state $n = 2$ to the ground state.

8.10 Estimate the number of revolutions N an electron makes around the nucleus in an excited hydrogen atom during the average lifetime of the excited state – 10^{-8} s – if a) it is in the state with $n = 2$, and b) in the state with $n = 15$, before it returns to the $n = 1$ state. c) Compare these numbers with the number of revolutions the earth has made around the sun in the 4.5×10^9 years of its existence.

8.11 In addition to the isotope ^{4}He, natural helium contains a small amount of the isotope ^{3}He. Calculate the differences in the wavenumbers and energies of the first and third lines of the Pickering series which result from these mass differences. The relative isotopic masses are:

^{3}He: 3.01603 u and ^{4}He: 4.00260 u .

8.12 Which lines of the hydrogen spectrum lie in the visible region of the spectrum (between 4000 Å and 7000 Å)? Which helium lines fall in the same region? How could one tell whether a helium sample has been contaminated with hydrogen?

8.13 Estimate the relative relativistic correction $\Delta E_{n,k}/E_n$ for the $n = 2$ levels in the hydrogen atom.

Hint: Compare (8.29).

8.14 To excite the hydrogen atom into its Rydberg states, one uses the additive absorption of the light from two lasers. Let the first of these have a fixed emission wavelength λ, which corresponds to 11.5 eV. What wavelengths must the second laser have in order to pump atoms into the state with $n = 20$, 30, 40 or 50? How large are the radii and binding energies for these states? What is the maximum possible linewidth for both lasers if only a single n state is to be populated?

8.15 a) Calculate the frequency of the orbital motion of an electron in a hydrogen atom for a level with the quantum number n.

b) Calculate the frequency of the radiation emitted in the transition from the state n to the state $n-1$.

c) Show that the results of a) and b) agree if n is very large.

8.16 Estimate the magnitude of the correction terms which must be applied to the energies of the stationary states of the lightest atoms, i.e. ^{1}H, ^{2}H, ^{3}H, He^+ and Li^{2+}, to account for the motion of the nucleus.

8.17 If one did the Franck-Hertz experiment on atomic hydrogen vapour, which lines in the hydrogen spectrum would one see if the maximum energy of the electrons were 12.5 eV?

9. The Mathematical Framework of Quantum Theory

As we saw in the previous chapter, classical physics is unable to offer a satisfactory explanation of the structure of even the simplest atom, that of hydrogen. This was first achieved by quantum theory. We shall therefore go into the theory in more depth, beginning where Chap. 7 left off. We shall be particularly, but not exclusively, concerned with bound states, of which the simpest example is

9.1 The Particle in a Box

In order to become more familiar with the formalism of quantum theory, which will then lead to quantitative predictions, we first consider the one-dimensional motion of an enclosed particle. "Enclosed" means that it can only move in a "box" of length a. The probability of finding the particle outside the box is zero (Fig. 9.1). We shall now attempt to construct the appropriate wavefunction. We require that

$$\begin{aligned} \psi &= 0 \quad \text{for} \quad x < 0 , \\ \psi &= 0 \quad \text{for} \quad x > a , \end{aligned} \tag{9.1}$$

because the particle cannot be outside the box. We further postulate that the wavefunction $\psi(x)$ inside the box is continuous with the function outside, i.e. that

$$\psi(0) = 0 , \quad \psi(a) = 0 . \tag{9.2}$$

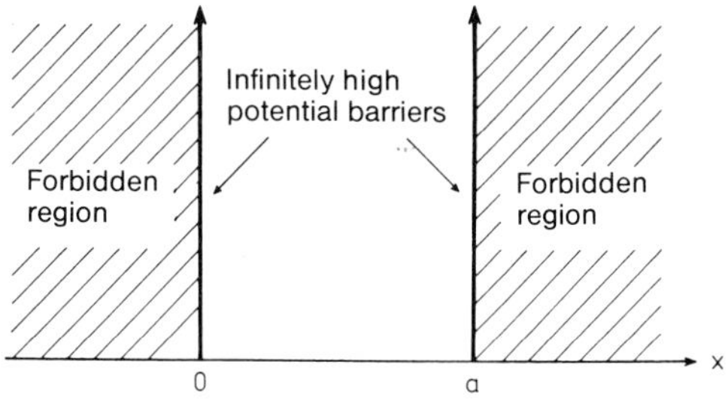

Fig. 9.1. Potential barriers for a particle in a box

We are seeking wavefunctions which describe a particle in this box and simultaneously guarantee that the particle always has a certain definite energy. We recall the de Broglie waves

$$A \exp[\mathrm{i}(kx - \omega t)] . \tag{9.3}$$

According to the basic laws of quantum theory, the wavenumber k and the frequency ω are related to the particle's energy and momentum by the relations

$$E = \hbar\omega \tag{9.4}$$

and

$$p = \hbar k\,. \tag{9.5}$$

From the experiments described above, we know that we can use the relationship from classical physics,

$$E = \frac{p^2}{2m_0}\,. \tag{9.6}$$

If we express p in terms of k, and solve for k, we obtain two possible values for k,

$$k_{1,2} = \pm\frac{1}{\hbar}\sqrt{2m_0E}\,, \tag{9.7}$$

for the given value of total energy E.

In addition to the wavefunction (9.3), the wavefunction

$$A\exp(-\mathrm{i}kx - \mathrm{i}\omega t) \tag{9.8}$$

yields the same energy. This will help us out of a difficulty. As one can see by substituting $x = 0$ and $x = a$ in (9.3), the wavefunction (9.3) does not satisfy the boundary conditions (9.2). One way out is the following: since electron waves display diffraction and interference, we may infer that we can superpose waves in quantum mechanics, as we did in fact with wave packets in Sect. 7.1. We therefore generate a new wavefunction by superposing (9.3) and (9.8):

$$\psi(x,t) = (C_1\mathrm{e}^{\mathrm{i}kx} + C_2\mathrm{e}^{-\mathrm{i}kx})\mathrm{e}^{-\mathrm{i}\omega t}\,, \tag{9.9}$$

where the constants C_1 and C_2 are still unknown.

To abbreviate, we write (9.9) in the form

$$\psi(x,t) = \phi(x)\mathrm{e}^{-\mathrm{i}\omega t} \tag{9.9a}$$

where

$$\phi(x) = C_1\mathrm{e}^{\mathrm{i}kx} + C_2\mathrm{e}^{-\mathrm{i}kx}\,. \tag{9.9b}$$

In order to determine the constants C_1 and C_2, we substitute (9.9) in the first equation (9.2) and obtain

$$\phi(0) = 0\colon\quad C_1 + C_2 = 0\,. \tag{9.10}$$

Thus C_2 can be expressed in terms of C_1. (9.9) then takes the form

$$\phi(x) = C_1(e^{ikx} - e^{-ikx}) = 2iC_1 \sin kx ; \quad (9.11)$$

here we have made use of the definition of the sine function. To fulfil the second condition of (9.2), we substitute (9.11) in (9.2) and obtain:

$$\text{because} \quad \phi(a) = 0 ; \quad \text{the condition} \quad \sin ka = 0 . \quad (9.12)$$

Since the sine can only be zero if its argument is a whole multiple of π, we can only satisfy (9.12) by the choice of

$$k = \frac{n\pi}{a}, \quad n = 1, 2, 3, 4 \dots . \quad (9.13)$$

This result means that the only waves which will fit into the box have a half-wavelength equal to a whole fraction of the length of the box, a (Fig. 9.2). If we substitute (9.13) in the expression for kinetic energy (9.6), we obtain

$$E = \frac{\hbar^2}{2m_0}\left(\frac{n\pi}{a}\right)^2 \quad (9.14)$$

for the energy of the particle, with the condition that $n \geqq 1$ must be an integer. The parameter n cannot be equal to zero, because otherwise the wavefunction would be identically equal to zero. In other words, there would be no particle.

The result (9.14) is typical for quantum theory. The energies are no longer continuous as in classical physics, but are quantised. In order to determine C_1 in (9.11), which is still open, we remember that the wavefunction must be normalised. We thus have the condition $\int \psi^* \psi dx = 1$ to fulfil. If we substitute (9.11) in this, we first obtain

$$\int_0^a |\phi(x)|^2 dx = |C_1|^2 \int_0^a (2 - e^{i\frac{2\pi n}{a}x} - e^{-i\frac{2\pi n}{a}x}) dx . \quad (9.15)$$

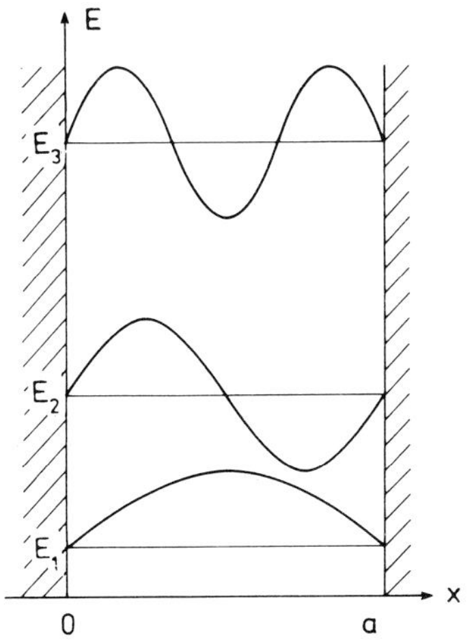

Fig. 9.2. Potential barriers, energies and wavefunctions of the particle in the box. *Two different* parameters are plotted in the same figure. 1) The energies E_1, E_2, E_3 of the first 3 states are plotted along the E (= energy) axis. (There is an infinite series of higher energies above these.) 2) The x axis is drawn to the right of each of the E values, and the wavefunction appropriate to each energy is shown on it. One should notice that the number of times the wavefunction crosses the x axis inside the box increases by 1 for each higher energy state

This integral is easy to evaluate and yields

$$\int_0^a |\phi(c)|^2 dx = |C_1|^2 \cdot 2a \,. \tag{9.16}$$

Because the integral (9.15) has to be equal to 1 to fulfil the normalisation condition, the normalisation constant C_1 has the form

$$C_1 = \frac{1}{\sqrt{2a}} \,. \tag{9.17}$$

It should be remembered that C_1 can only be determined up to the constant phase factor $\exp(\mathrm{i}\alpha)$. As we shall see later, this type of phase factor has no physical meaning, because it disappears during the calculation of expectation values (see below). Our final result thus has the form

$$\phi(x) = \frac{1}{\sqrt{2a}} \exp(\mathrm{i}xn\pi/a) - \frac{1}{\sqrt{2a}} \exp(-\mathrm{i}xn\pi/a) \,, \tag{9.18}$$

or in another notation,

$$\phi(x) = \sqrt{\frac{2}{a}} \cdot \mathrm{i} \sin(x \cdot n\pi/a) \,. \tag{9.19}$$

As we have seen, the wavefunction (9.18) is associated with a *definite energy*. Does this also hold for the momentum? This is clearly not the case, because it describes both a wave with $k = n\pi/a$ and a wave with $k = -n\pi/a$. If we should measure the momentum, we would thus find values $p = \hbar k$ and $p = -\hbar k$ with equal frequencies. In order to derive the probability of occurrence of a given momentum from the wavefunction, let us first consider the wavefunction

$$\frac{1}{\sqrt{a}} \exp(\mathrm{i}xn\pi/a) \tag{9.20}$$

which is obviously normalised in the region from 0 to a:

$$\int_0^a \left| \frac{1}{\sqrt{a}} \exp\left(\mathrm{i}\frac{n\pi}{a}x\right)\right|^2 dx = \frac{1}{a}\int_0^a dx = 1 \,. \tag{9.21}$$

When we measure the momentum, it means that we determine a particular value of k, i.e., we select one of the components of (9.18). This component is a factor of $1/\sqrt{2}$ smaller than the corresponding component of (9.20). On the other hand, we expect for symmetry reasons that both components occur with equal probability $= 1/2$. To go from $1/\sqrt{2}$ to $1/2$, of course, we square $1/\sqrt{2}$. This observation can be generalised: The probability of measuring a given momentum k can be obtained by taking the square of the absolute value of the coefficient in front of the normalised plane wave.

We leave it to the reader as an exercise to explain the relationship between the wavefunction (9.18) and the momentum (9.5) using the Heisenberg uncertainty relation.

9.2 The Schrödinger Equation

As we saw in the preceding example, there are for a given problem, in this case the particle in a box, infinitely many solutions, each with a corresponding energy level (9.14). In this case it was relatively easy to find these solutions, which is decidedly not the case for other quantum mechanical problems. In such cases it is often useful first to look for an equation which determines ψ. In the case of the electron which is not subjected to any forces, we find it as follows: we ask if there is an equation for ψ such that its solutions automatically fulfil the relation

$$\hbar\omega = \frac{\hbar^2 k^2}{2m_0}. \tag{9.22}$$

Since the parameters k and ω are found in the de Broglie wave $\exp(\mathrm{i}kx - \mathrm{i}\omega t)$, we can formulate the question thus: what must be done to obtain $\hbar^2 k^2/2m_0$ from $\exp(\mathrm{i}kx)$ and $\hbar\omega$ from $\exp(-\mathrm{i}\omega t)$, so that the relation

$$\frac{\hbar^2 k^2}{2m_0} = \hbar\omega \tag{9.23}$$

will be fulfilled? If we differentiate $\exp(\mathrm{i}kx)$ twice with respect to x and multiply by $-\hbar^2/2m_0$, we obtain the left side of (9.23) as a factor. Correspondingly, the right side of (9.23) is obtained if we differentiate $\exp(-\mathrm{i}\omega t)$ with respect to time and multiply by $\mathrm{i}\hbar$. In this way we obtain the basic Schrödinger equation for the force-free particle:

$$-\frac{\hbar^2}{2m_0}\frac{d^2}{dx^2}\psi = \mathrm{i}\hbar\dot{\psi}. \tag{9.24}$$

It must be said, however, that it is generally not possible to derive the basic equations of physics from still more fundamental principles. Instead, one must try to comprehend the physics by heuristic thought processes, to arrive at an equation, and then to compare the possible solutions with experimentally testable facts. In this way it has been found that the Schrödinger equation is completely valid in *nonrelativistic* quantum mechanics. We generalise (9.24) to three dimensions by writing the kinetic energy in the form

$$E = \frac{1}{2m_0}(p_x^2 + p_y^2 + p_z^2). \tag{9.25}$$

It seems reasonable to generalise the wavefunction to

$$\exp(\mathrm{i}k_x x + \mathrm{i}k_y y + \mathrm{i}k_z z)\exp(-\mathrm{i}\omega t). \tag{9.26}$$

Instead of (9.23) we have the relation

$$\frac{1}{2m_0}\hbar^2(k_x^2 + k_y^2 + k_z^2) = \hbar\omega. \tag{9.27}$$

The left side of (9.27) is obtained from (9.26) by taking the second derivatives of (9.26) with respect to the position coordinates x, y and z, adding these and multiplying the result by $-\hbar^2/2m_0$. The right side of (9.27) results by differentiation of (9.26) with respect to time and multiplication by $\mathrm{i}\hbar$. We thus obtain the equation

$$-\frac{\hbar^2}{2m_0}\left(\frac{\partial^2}{\partial x^2}+\frac{\partial^2}{\partial y^2}+\frac{\partial^2}{\partial z^2}\right)\psi=\mathrm{i}\hbar\frac{\partial}{\partial t}\psi\,. \tag{9.28}$$

The left side can be abbreviated by introducing the Laplace operator

$$\nabla^2=\frac{\partial^2}{\partial x^2}+\frac{\partial^2}{\partial y^2}+\frac{\partial^2}{\partial z^2}\,, \tag{9.29}$$

which yields the usual form of the Schrödinger equation for the force-free particle in three dimensions,

$$-\frac{\hbar^2}{2m_0}\nabla^2\psi=\mathrm{i}\hbar\frac{\partial}{\partial t}\psi\,. \tag{9.30}$$

Now we are naturally not so interested in the force-free motion of the particle as in its motion in a force field. However, (9.30) gives us a hold on the subject. We see that the left side was derived from the expression $p^2/2m_0$ for the kinetic energy by replacing it by a differentiation rule $-(\hbar^2/2m_0)\nabla^2$. This rule acts on ψ and is called the kinetic energy operator. In the presence of a potential field, the total energy according to classical mechanics is the sum of the kinetic and the potential energy:

$$\frac{1}{2m_0}p^2+V(\boldsymbol{r})=E\,. \tag{9.31}$$

We can arrive heuristically at the *total energy operator* of the quantum treatment by simply adding V to the kinetic energy operator. We thus obtain the *time-dependent Schrödinger equation in the presence of a potential field*:

$$\left(-\frac{\hbar^2}{2m_0}\nabla^2+V(\boldsymbol{r})\right)\psi(\boldsymbol{r},t)=\mathrm{i}\hbar\frac{\partial}{\partial t}\psi(\boldsymbol{r},t)\,. \tag{9.32}$$

The expression

$$\mathscr{H}=-\frac{\hbar^2}{2m_0}\nabla^2+V(\boldsymbol{r}) \tag{9.33}$$

is called the *Hamiltonian* (operator).

The beginner may not be used to working with operators. One can quickly become accustomed to them, if one remembers that they are only convenient abbreviations. One must also remember that such operators are always to be applied to functions.

If the potential field on the left side of (9.32) does not depend on time, we can proceed from the time-dependent to the time-independent Schrödinger equation. In doing so, just as in (9.9a), we separate a time factor $\exp(-\mathrm{i}\omega t)$ from $\psi(\boldsymbol{r}, t)$. In quantum mechanics it is customary to write $E/\hbar$ instead of ω, so that we write

$$\psi(\boldsymbol{r}, t) = \mathrm{e}^{-\mathrm{i}Et/\hbar}\phi(\boldsymbol{r}) . \tag{9.34}$$

Since the time differentiation only applies to ψ on the right side of (9.32), we need here only to differentiate the exponential function with respect to time, which yields the factor E. If we then divide both sides of the corresponding equation by the exponential function, we obtain as the result the *time-independent Schrödinger equation*

$$\left(-\frac{\hbar^2}{2m_0}\nabla^2 + V(\boldsymbol{r})\right)\phi(\boldsymbol{r}) = E\phi(\boldsymbol{r}) . \tag{9.35}$$

As we saw in the preceding example, the wavefunction must ordinarily be subject to boundary conditions (9.2). If these are not specified, we apply the so-called natural boundary conditions, which require that ψ vanishes at infinity, so that the wavefunction can be normalised, i.e.

$$\int |\psi|^2 dV = 1 . \tag{9.36}$$

Before we proceed to the solution of the Schrödinger equation, we shall again take up the question of observations, measured values and operators.

9.3 The Conceptual Basis of Quantum Theory

9.3.1 Observations, Values of Measurements and Operators

Determination and Probability of Position

In the preceding sections, we saw that the explanation of microcosmic processes required new ways of thinking which are fundamentally different from the ideas of classical physics. In classical mechanics, the motion of a body, such as the fall of a stone or the flight of a rocket, can be precisely predicted by the laws of motion. According to these laws, the position and momentum of a body can be determined to as great a precision as is desired.

The wavefunction is the new concept which is central to quantum physics. As the solution of the time-dependent Schrödinger equation, it describes the time evolution of physical processes in the microcosm. In this section we shall explore the physical implications of the wavefunction, or in other words, which experimental results the theoretical physicist can predict for the experimental physicist. The (conceptually) simplest experiment would be to determine the position of a particle. As we already know, the wave function ψ can only make a probabilistic prediction. The expression

$$|\psi(x, y, z)|^2 dx\, dy\, dz \tag{9.37}$$

gives the probability that the particle will be found in a volume element $dx\, dy\, dz$ about the point x, y, z. We now ask whether the wavefunction can also predict the results of observations of momentum.

9.3.2 Momentum Measurement and Momentum Probability

Let us first consider as an example the wavefunction of the particle in a box (Sect. 9.1),

$$\phi(x) = \frac{1}{\sqrt{2}} \underbrace{\frac{1}{\sqrt{a}} \exp(\mathrm{i}kx)}_{u_1(x)} - \frac{1}{\sqrt{2}} \underbrace{\frac{1}{\sqrt{a}} \exp(-\mathrm{i}kx)}_{u_2(x)} . \tag{9.38}$$

The two underlined wavefunctions each satisfy the normalisation conditions (9.36). According to the basic rules of quantum mechanics, the momentum associated with the wavefunction $u_1(x)$ is given by $\hbar k$, while the momentum of the second wavefunction $u_2(x)$ is $\hbar(-k) = -\hbar k$.

Both of these momenta are thus represented by the wavefunction (9.38). If we determine the momentum of the particle in the box described by the wavefunction (9.38), we expect to observe either $+\hbar k$ or $-\hbar k$. However, we cannot predict which of the two momenta we will observe. If we imagine that the particle flies back and forth in the box, it is intuitively clear that we will observe the momenta $\hbar k$ and $-\hbar k$ with a probability of 1/2 each. As we saw in Sect. 9.1, the squares of the absolute values of the coefficients C_1 and C_2 give the probability of finding the corresponding momentum. We generalise this insight to the determination of the probability distribution of the momenta of a generalised wave packet. Here the particle is no longer confined in a box. This type of wave packet has the general form

$$\psi(x) = \int_{-\infty}^{+\infty} a_k \mathrm{e}^{\mathrm{i}kx} dk . \tag{9.39}$$

In order to connect the coefficients a_k with a probabilistic interpretation, we must be sure that the wavefunctions $\exp(\mathrm{i}kx)$ are normalised in infinite space. This is somewhat difficult, and will not be demonstrated here (see Appendix A). We shall simply state the result. If we introduce the momentum variable p in the place of the integration variable k, and at the same time use the correct normalisation of the wavefunction in one dimension, we obtain

$$\psi(x) = \int_{-\infty}^{+\infty} c(p) \underbrace{\frac{1}{\sqrt{2\pi\hbar}} \mathrm{e}^{\mathrm{i}px/\hbar}}\, dp . \tag{9.40}$$

The underlined wavefunction is normalised. As a generalisation of our considerations above, we see $|c(p)|^2 dp$ as the probability of observing momentum p in the interval $p, \ldots, p+dp$. This result can be immediately expanded to three dimensions: if we represent a wavefunction $\psi(x,y,z)$ as a superposition of normalised plane waves,

$$\psi(x,y,z) = \iiint_{-\infty}^{+\infty} c(p_x,p_y,p_z)(2\pi\hbar)^{-3/2} \exp(\mathrm{i}\boldsymbol{p}\cdot\boldsymbol{r}/\hbar)\, d^3p , \quad \text{with} \tag{9.41}$$

$\boldsymbol{p}\cdot\boldsymbol{r} = p_x x + p_y y + p_z z$, then

$$|c(p_x,p_y,p_z)|^2 dp_x\, dp_y\, dp_z$$

is the probability that the components of the observed momentum of the particle p will lie in the intervals $p_x \dots p_x + dp_x$, $p_y \dots p_y + dp_y$, $p_z \dots p_z + dp_z$.

9.3.3 Average Values and Expectation Values

To explain these concepts, we think again about the example of the die. The individual possible "observed values" are the numbers of spots, $1, 2, \dots, 6$. For a single throw we cannot predict which of these numbers we will obtain. We can only make predictions if we throw many times and keep track of the frequency F_n with which we obtain the number n $(n = 1, 2, \dots, 6)$. The average number $\bar{n}$ is then given by

$$\bar{n} = \frac{\sum_{n=1}^{6} nF_n}{\sum_{n=1}^{6} F_n} . \tag{9.42}$$

This average value can be predicted statistically (in the limiting case of an infinite number of throws) through the use of the concept of probability. This is the ratio of the number of times the desired result is obtained divided by the total number of attempts. The probability of obtaining n spots ("desired result" is n) is denoted by P_n. Since each number of spots is equally probable, $P_1 = P_2 \dots = P_6$. Further, since $\sum_{n=1}^{6} P_n = 1$ must hold, we use the equality of the individual probabilities to obtain immediately

$$P_n = 1/6 , \qquad n = 1, 2, \dots 6 . \tag{9.43}$$

(We exclude loaded dice.) According to probability theory, $\bar{n}$ may be expressed in terms of P_n as follows:

$$\bar{n} = \sum_{n=1}^{6} n \cdot P_n = 1 \cdot \tfrac{1}{6} + 2 \cdot \tfrac{1}{6} + \dots 6 \cdot \tfrac{1}{6} . \tag{9.44}$$

These relatively simple concepts may be applied directly to the definition of the mean value of position and of momentum in quantum mechanics. In general, we can make no definite predictions as to which position or which momentum will be measured; we can only give probabilities. If we repeat the measurement of position or of momentum many times and calculate the mean value, the latter may be defined exactly as for the dice. The theoretician can, as we saw in the dice game, predict this mean value for the experimentalist. This mean value is therefore called the *expectation value*; it is defined as follows: *Expectation value = Sum over the individual values measured, times the probability that that value would be found.*

Let us apply this definition to some examples.

a) Mean Value of the Position (one-dimensional example), Fig. 9.3

A single measurement yields the result that the particle is to be found in the interval $x \dots x + dx$. The corresponding probability is $|\psi(x)|^2 dx$. Since the position x is con-

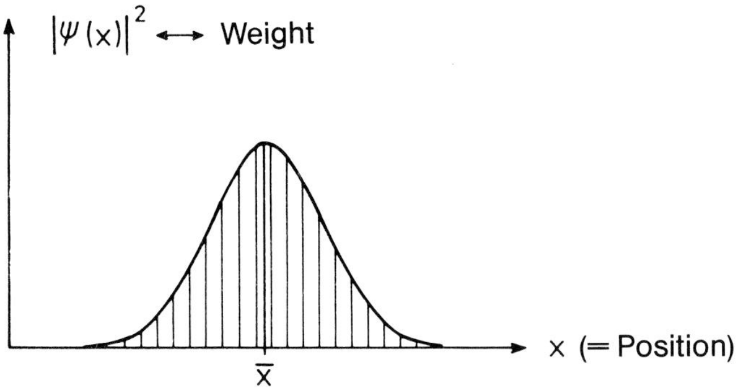

Fig. 9.3. Explanation of the mean value of the position. The location of a vertical line indicates the value of a measurement of the position coordinate x and the length of the line is proportional to the frequency with which that value is found (probability density). If we interpret the latter as a weight, the calculation of $\bar{x}$ corresponds to the calculation of the centre of gravity $\bar{x}$ of an object

tinuously variable, while the number of spots on the die was discrete, we use an integral instead of the sum (9.44). The mean value of the position is thus defined as

$$\bar{x} = \int_{-\infty}^{+\infty} x \, |\psi(x)|^2 dx \, . \tag{9.45}$$

In the calculation of (9.45) and in the following, the normalisation of the wavefunction was assumed, i.e.

$$\int_{-\infty}^{+\infty} |\psi(x)|^2 dx = 1 \, . \tag{9.46}$$

Correspondingly, we can take the nth power of x, x^n, and then generalise the definition (9.45) to obtain the mean value of the nth power:

$$\bar{x}^n = \int_{-\infty}^{+\infty} x^n |\psi(x)|^2 dx \, . \tag{9.47}$$

If we replace the function x^n quite generally by the potential energy function $V(x)$, we obtain the definition of the mean value of the potential energy,

$$\bar{V} = \int_{-\infty}^{+\infty} V(x) \, |\psi(x)|^2 dx \, . \tag{9.48}$$

b) Mean Value of the Momentum (one-dimensional example), Fig. 9.4.

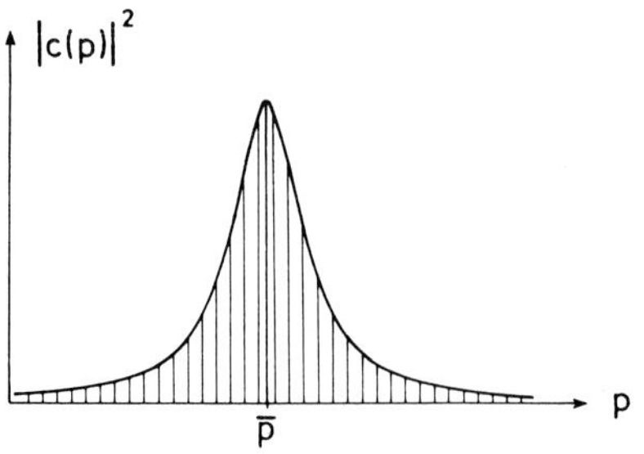

Fig. 9.4. Explanation of the mean value of the momentum. See caption of Fig. 9.3

In this case, we first take the wavefunction to be given by a superposition of plane waves:

$$\psi(x)=\int\limits_{-\infty}^{+\infty} c(p)\frac{1}{\sqrt{h}}\,e^{ipx/\hbar}dp\,. \tag{9.49}$$

If we now measure the momentum, the probability of finding its value in the interval $p\ldots p+dp$ is given by $|c(p)|^2dp$. In complete analogy to the mean value of the position, we find the definition of the mean value of the momentum to be

$$\bar{p}=\int\limits_{-\infty}^{+\infty} p\,|c(p)|^2dp\,, \tag{9.50}$$

or for the nth power

$$\overline{p^n}=\int\limits_{-\infty}^{+\infty} p^n|c(p)|^2dp\,. \tag{9.51}$$

As we shall see later, wavefunctions are normally expressed as functions of position, in the form $\psi(x)$. It is therefore difficult to calculate the expansion (9.49) in detail in order to determine the mean of the momentum, because the coefficients $c(p)$ would first have to be calculated. We shall now show that there is a very simple computational rule which allows us to calculate the mean value of the momentum without following the indirect route via (9.49).

We assert that the mean value of the momentum is given by the basic formula

$$\bar{p}=\int\limits_{-\infty}^{+\infty}\psi^*(x)\left(\frac{\hbar}{i}\frac{d}{dx}\right)\psi(x)\,dx\,. \tag{9.52}$$

The notation $(\hbar/i)(d/dx)\,\psi(x)$ may seem unfamiliar to the reader; it is a common form in quantum mechanics. It means that we are to differentiate $\psi(x)$ by x, that is, we calculate

$$\frac{\hbar}{i}\frac{d\psi}{dx}\,. \tag{9.53}$$

This notation (9.52) is also referred to as applying the "momentum operator" $(\hbar/i)(d/dx)$ to the wavefunction $\psi(x)$. The proof that (9.52) is the same as (9.50) is relatively simple, but requires some basic mathematical knowledge. We begin by substituting (9.49) in (9.52). After differentiation with respect to x and exchanging the order of integration with respect to x and p, we obtain

$$\bar{p}=\int\limits_{-\infty}^{+\infty}dp\int\limits_{-\infty}^{+\infty}dp'\,p'\,c^*(p)\,c(p')\underline{\frac{1}{h}\int\limits_{-\infty}^{+\infty}\exp(-ipx/\hbar)\exp(ip'x/\hbar)\,dx}\,. \tag{9.54}$$

The underlined part, however, is merely the Dirac δ function, $\delta(p-p')$ (Appendix A). The definition of the δ function eliminates the integration over p', and leads to $p'=p$, so that p' is replaced by p. We then obtain directly

$$\bar{p} = \int_{-\infty}^{+\infty} dp\, p\, |c(p)|^2 . \tag{9.55}$$

If we go through the calculation again in detail, we recognise that we have replaced the factor p in (9.50) by the differential operator $(\hbar/\mathrm{i})\, d/dx$. In order to arrive at (9.51), we would have had to apply this operator n times to the wavefunction $\psi(x)$.

c) Average Values of Energy

Our results to this point enable us to calculate average energy values. The kinetic energy of a particle is $p^2/2m_0$. The probability of observing the momentum p in the interval $p \dots p+dp$ is given by $|c(p)|^2 dp$.

Thus the average kinetic energy is given by

$$\bar{E}_{\text{kin}} = \int_{-\infty}^{+\infty} |c(p)|^2 \frac{p^2}{2m_0}\, dp . \tag{9.56}$$

If we use the computational rule discussed above, we immediately obtain

$$\bar{E}_{\text{kin}} = \iiint_{-\infty}^{\infty} \psi^* \left(-\frac{\hbar^2}{2m_0} \nabla^2 \psi \right) dx\, dy\, dz , \tag{9.57}$$

where we have used the abbreviation

$$\nabla^2 = \frac{\partial^2}{\partial x^2} + \frac{\partial^2}{\partial y^2} + \frac{\partial^2}{\partial z^2} \tag{9.58}$$

and generalised the result to three dimensions. Equation (9.48) can be extended in the same way, which yields the expectation value for the potential energy:

$$\bar{E}_{\text{pot}} = \iiint_{-\infty}^{\infty} \psi^* V(\boldsymbol{r})\, \psi\, dx\, dy\, dz . \tag{9.59}$$

Since the total energy is equal to the sum of the kinetic and the potential energy, the expectation value for the total energy is, finally,

$$\bar{E}_{\text{tot}} = \iiint_{-\infty}^{\infty} \psi^* \left[-\frac{\hbar^2}{2m_0} \nabla^2 + V(\boldsymbol{r}) \right] \psi\, dx\, dy\, dz . \tag{9.60}$$

9.3.4 Operators and Expectation Values

With the help of the above results, we can now discuss the conceptual framework and the computational rules of quantum theory. In classical physics, we have certain mechanical parameters, like the position $x(t)$, momentum $p(t)$, energy, etc. In quantum theory, these classical parameters are assigned certain expectation values [compare (9.45, 52, 60)]. These quantum mechanical expectation values can be obtained from classical physics by means of a very simple translation process according to the following "recipe": The classical parameters are assigned operators, which are nothing but multiplication or differentiation rules, which act on the wavefunctions following them. The position operator x is assigned to $x(t)$, which simply says that one multiplies the wavefunction $\psi(x)$ by x. It may seem strange at first that a time-independent operator x can be assigned to a time-dependent parameter $x(t)$. As we shall see below, however, the time-dependence is reintroduced in the process of finding the average, if the wavefunction itself is time-dependent. The momentum is assigned the operator $-\mathrm{i}\hbar(d/dx)$ which differentiates the wavefunction. After carrying out the appropriate operator multiplication or differentiation, one multiplies the result by ψ^* and integrates over all space to obtain the quantum mechanical expectation value.

Using these rules, we can define still other operators which we have not yet considered. One important parameter is the angular momentum $\boldsymbol{l}$, which has the components l_x, l_y and l_z. In classical physics, l_z, for example, is defined as $xp_y - yp_x$. In quantum theory we obtain the corresponding operator by replacing p_x and p_y by $(\hbar/\mathrm{i})\,\partial/\partial x$ and $(\hbar/\mathrm{i})\,\partial/\partial y$, respectively. The z component of the angular momentum *operator* is thus

$$\hat{l}_z = \frac{\hbar}{\mathrm{i}}(x\partial/\partial y - y\partial/\partial x)\,. \tag{9.61}$$

In order to prevent confusion between the classical angular momentum and the angular momentum operator, we use here and in the following text the symbol ^ (read "hat") over the angular momentum operator.

The following table summarises what has been said above.

Classical variable	Operator	Quantum theoretical Expectation value
Position $x(t)$	x	$\bar{x} = \int \psi^*(x,t)\,x\,\psi(x,t)\,dx$
Momentum $p(t)$	$\frac{\hbar}{\mathrm{i}}\frac{d}{dx}$ (Jordan's rule)	$\bar{p} = \int \psi^*(x,t)\frac{\hbar}{\mathrm{i}}\frac{d}{dx}\psi(x,t)\,dx$
Energy $E = \mathscr{H}(x(t),p(t))$	$-\frac{\hbar^2}{2m_0}\frac{d^2}{dx^2} + V(x)$	$\bar{E} = \int \psi^*(x,t)\left[-\frac{\hbar^2}{2m_0}\frac{d^2}{dx^2} + V(x)\right]\psi(x,t)\,dx$
Angular momentum $\boldsymbol{l} = [\boldsymbol{r} \times \boldsymbol{p}]$	$\left[\boldsymbol{r} \times \frac{\hbar}{\mathrm{i}}\nabla\right]$	$\bar{\boldsymbol{l}} = \int \psi^*\left[\boldsymbol{r} \times \frac{\hbar}{\mathrm{i}}\nabla\right]\psi\,dx\,dy\,dz$

In the preceding discussion, we have given no consideration to the wavefunction ψ, which has, so to speak, fallen from heaven. We must still consider the principles by which we can determine the wavefunction, in case it is not determined by the Schrödinger equation.

9.3.5 Equations for Determining the Wavefunction

We have already presented equations which were explicitly or implicitly applicable to the determination of ψ. As the simplest example, let us take the plane wave $\psi \sim \exp(\mathrm{i}kx)$. As we already know, this wave determines the propagation of a particle with momentum $\hbar k$. Can we regard this plane wave as a solution of an equation which relates directly to momentum? This is in fact the case, because if we differentiate the plane wave with respect to x and multiply by $\hbar/\mathrm{i}$, we obtain the relation

$$\frac{\hbar}{\mathrm{i}}\frac{d}{dx}\mathrm{e}^{\mathrm{i}kx} = \hbar k\,\mathrm{e}^{\mathrm{i}kx} \equiv p\,e^{ikx}. \tag{9.62}$$

The plane wave thus satisfies an equation of the following form: The momentum operator $(\hbar/\mathrm{i})\,d/dx$ applied to the plane wave yields $p \equiv \hbar k$ times the plane wave.

As a second example, let us consider the time-independent Schrödinger equation. The application of the Hamiltonian operator to the wavefunction gives an energy $\bar{E}$ times the wavefunction. A glance at the above table shows that the Hamiltonian is precisely that quantum mechanical operator associated with the classical energy expression $E_{\mathrm{kin}} + E_{\mathrm{pot}}$.

When we extract what is common to these examples, we see that these functions are so-called *eigenfunctions* which satisfy the following equation:

$$\text{Operator} \cdot \text{Eigenfunction} = \text{Eigenvalue} \cdot \text{Eigenfunction}\,.$$

If we denote the operator by Ω, the eigenfunction by ϕ and the eigenvalue by ω, this relationship is

$$\Omega\phi = \omega\phi\,. \tag{9.63}$$

The eigenvalue indicated here and in Sect. 9.3.6 following should not be confused with a frequency. It can have quite different physical meanings, e.g. momentum. In the example (9.62), we had

$$\Omega = \frac{\hbar}{\mathrm{i}}\frac{d}{dx}\,, \qquad \phi = \mathrm{e}^{\mathrm{i}kx}\,, \qquad \omega = \hbar k\,.$$

We must now make use of a few basic facts of the mathematical treatment of such eigenvalue equations without being able to derive them here. As can be shown mathematically, eigenfunctions and eigenvalues are determined by (9.63), if appropriate boundary conditions for the wavefunction (eigenfunction) are given. One example for a set of boundary conditions is the particle in a box. If no explicit boundary conditions are given, we must require that the wavefunction be normalisable, which implies that the wavefunction must go to zero rapidly enough as infinity is approached.

When the operator Ω in (9.63) and the boundary conditions are given, there is a particular sequence of eigenvalues, e.g. discrete energy values as in the particle in the box, etc. The calculation of these eigenvalues and the associated eigenfunctions is thus the task of mathematicians or theoretical physicists. In order to make them agree with experimental observations, one makes use of the basic postulate of the quantum theory: *the eigenvalues are identical with the observed values.* This basic postulate has enormous significance, and we can accept it, because it has been repeatedly confirmed in innumerable experiments. If we measure the energy of the electron in a hydrogen atom, for example, this must agree with the quantum mechanically calculated eigenvalues E_n. If there is a discrepancy, one does not impute this to a failure of quantum theory, but rather looks for interactions which have not yet been taken into account. In this way, an excellent agreement has so far been attained.

As we can see from our example (9.62), the Schrödinger equation is only one of many possible ways to determine the wavefunction. We are always concerned here with the physical problem. Thus whenever we use the Schrödinger equation, we would always assume that we have access to observations which measure the energy exactly. When we have then measured the energy, we have identified the associated eigenfunctions as solutions of the Schrödinger equation. We might also wish to measure the momentum. Since the wavefunction is known and, as one can easily demonstrate by Fourier analysis, this function contains several momentum eigenfunctions, we are no longer able to predict exactly the momentum of the particle, but can only calculate the expectation value. The simplest example for this is again the particle in the box.

9.3.6 Simultaneous Observability and Commutation Relations

As we saw above, there is a very close relationship between wavefunctions and eigenvalues on the one hand and individual observations on the other. If a wavefunction is an eigenfunction for a particular operator – that is, if it satisfies an equation like (9.63) – then we know that the eigenvalue will be found by measurement. If we repeat this measurement, we shall find exactly the same eigenvalue. If follows from this that:

If ψ_λ is an eigenfunction of a specific operator Ω, the eigenvalue ω_λ agrees with the expectation value $\bar{\Omega}$. In fact, if we know the operator Ω and the associated eigenvalue ω_λ, then

$$\Omega\psi_\lambda = \omega_\lambda\psi_\lambda\colon\ \bar{\Omega} = \int\psi_\lambda^*\Omega\psi_\lambda dx = \int\psi_\lambda^*\omega_\lambda\psi_\lambda dx = \omega_\lambda\int\psi_\lambda^*\psi_\lambda dx = \omega_\lambda\,.$$

What happens, though, when we want to determine another parameter with the second measurement? One example for this was examined in more detail in Sect. 7.3, where we wanted to measure first the momentum and then the position of the particle. In that case, the measurement of position destroyed the results of the previous momentum determination. On the other hand, we can measure first the momentum and then the kinetic energy of a particle. In the first measurement, we obtain a certain value p. We have then "prepared" the particle into a particular state which is an eigenfunction of the momentum operator; the wavefunction after the measurement is thus (aside from a normalisation factor) given by $\exp(\mathrm{i}px/\hbar)$. If we now measure the kinetic energy, this measurement corresponds to the mathematical operation of applying the kinetic energy operator, $-(\hbar^2/2m_0)d^2/dx^2$. In the process, the "prepared" plane wave yields the eigenvalue $E=p^2/2m_0$, and the plane wave remains as wavefunction. In this case, the second measurement does *not* destroy the result of the first measurement. There exist, apparently, measurements which do not disturb each other, or, in other words, which can be simultaneously carried out with arbitrary accuracy.

We will now derive a necessary criterion for simultaneous measurability. For this purpose, we consider the operators $\Omega^{(1)}$ and $\Omega^{(2)}$, which could, for example, be operators for the momentum and the kinetic energy. We now require that the wavefunction ψ be simultaneously an eigenfunction of both characteristic equations

$$\Omega^{(1)}\psi = \omega^{(1)}\psi \tag{9.64}$$

and

$$\Omega^{(2)}\psi = \omega^{(2)}\psi\,. \tag{9.65}$$

If we apply operator $\Omega^{(2)}$ to the left side of the first equation and operator $\Omega^{(1)}$ to the second equation, then subtract one equation from the other, rearrange, and finally apply (9.64) and (9.65) again, we obtain

$$(\Omega^{(1)}\Omega^{(2)}-\Omega^{(2)}\Omega^{(1)})\,\psi=(\omega^{(1)}\omega^{(2)}-\omega^{(2)}\omega^{(1)})\,\psi=0\,. \tag{9.66}$$

The simultaneous measurability of *all* wavefunctions which simultaneously fulfil (9.64) and (9.65), not merely special cases, should be guaranteed. Therefore the ψ in (9.66) is omitted in quantum theory, and one writes

$$\Omega^{(1)}\Omega^{(2)}-\Omega^{(2)}\Omega^{(1)}=0\,. \tag{9.67}$$

This, however, should be understood to be an abbreviation. When one sees such an equation, one should always remember that any desired wavefunction ψ stands to the right of the operators, i.e., (9.66) applies. It can be mathematically shown that the converse of the above is also true: if two operators $\Omega^{(1)}$ and $\Omega^{(2)}$ fulfil the commutation relation (9.67), then eigenfunctions of $\Omega^{(1)}$ can always be determined to be eigenfunctions of $\Omega^{(2)}$ as well; they fulfil (9.64) and (9.65). If there is only a single eigenfunction belonging to the eigenvalue $\omega^{(1)}$ of $\Omega^{(1)}$, this is itself an eigenfunction of $\Omega^{(2)}$. However, if there are several eigenfunctions of $\Omega^{(1)}$ associated with $\omega^{(1)}$, then it will always be possible to find linear combinations of these which are also eigenfunctions of $\Omega^{(2)}$.

Let us consider a few examples. If we choose as $\Omega^{(1)}$ the momentum operator $(\hbar/\mathrm{i})\,d/dx$, and the kinetic energy operator $(-\hbar^2/2m_0)\,d^2/dx^2$ as $\Omega^{(2)}$, these operators commute. The result of differentiating a wavefunction twice and then once with respect to x is naturally the same as that of differentiating first once and then twice with respect to x:

$$\left(-\frac{\hbar^2}{2m_0}\right)\cdot\frac{\hbar}{\mathrm{i}}\left(\frac{d}{dx}\cdot\frac{d^2}{dx^2}-\frac{d^2}{dx^2}\cdot\frac{d}{dx}\right)=0\,. \tag{9.68}$$

It can be shown in the same way that the x components of the momentum and the y components of the position mutually commute.

Let us take as a second example the x component of the momentum and the coordinate x itself. Thus $\Omega^{(1)}=(\hbar/\mathrm{i})\,d/dx$ and $\Omega^{(2)}=x$:

$$(\Omega^{(1)}\Omega^{(2)}-\Omega^{(2)}\Omega^{(1)})\,\psi=\left(\frac{\hbar}{\mathrm{i}}\,\frac{d}{dx}x-x\frac{\hbar}{\mathrm{i}}\,\frac{d}{dx}\right)\psi\,. \tag{9.69}$$

We now evaluate this expression. First we remove the parentheses:

$$=\frac{\hbar}{\mathrm{i}}\,\frac{d}{dx}x\,\psi-x\frac{\hbar}{\mathrm{i}}\,\frac{d\psi}{dx}\,. \tag{9.70}$$

d/dx means, of course, that everything to the right of the operator is to be differentiated, and

$$\frac{d}{dx}(x\psi) = \frac{dx}{dx}\psi + x\frac{d\psi}{dx}. \tag{9.71}$$

If we substitute this in (9.70), we obtain

$$\frac{\hbar}{\mathrm{i}}\psi. \tag{9.72}$$

If we again write out the right side of (9.69), we obtain the relation

$$\left(\frac{\hbar}{\mathrm{i}}\frac{d}{dx}x - x\frac{\hbar}{\mathrm{i}}\frac{d}{dx}\right)\psi = \frac{\hbar}{\mathrm{i}}\psi. \tag{9.73}$$

Since this relationship holds for any function ψ, one can also write in abbreviated form

$$\frac{\hbar}{\mathrm{i}}\frac{d}{dx}x - x\frac{\hbar}{\mathrm{i}}\frac{d}{dx} = \frac{\hbar}{\mathrm{i}}. \tag{9.74}$$

This is the famous Heisenberg commutation relation between the momentum operator and the position operator. It says that the momentum and the position operators do not commute, which means that the position and momentum cannot be simultaneously determined to any desired degree of precision (see Sect. 7.3).

The following formulation is often used to express the commutation relation between the two operators $\Omega^{(1)}$ and $\Omega^{(2)}$:

$$[\Omega^{(1)}, \Omega^{(2)}] \equiv \Omega^{(1)}\Omega^{(2)} - \Omega^{(2)}\Omega^{(1)}. \tag{9.75}$$

In this form, the Heisenberg commutation relation is

$$\left[\frac{\hbar}{\mathrm{i}}\frac{d}{dx}, x\right] = \frac{\hbar}{\mathrm{i}}. \tag{9.76}$$

We leave it to the reader to derive the following relations:

$$\left[\frac{\hbar}{\mathrm{i}}\frac{d}{dx}, V\right] = \frac{\hbar}{\mathrm{i}}\frac{dV}{dx}.$$

For the components of the angular momentum [compare the definition in (9.61)].

$$[\hat{l}_x, \hat{l}_y] = \mathrm{i}\hbar\hat{l}_z, \tag{9.77}$$

$$[\hat{l}_y, \hat{l}_z] = \mathrm{i}\hbar\hat{l}_x, \tag{9.78}$$

$$[\hat{l}_z, \hat{l}_x] = \mathrm{i}\hbar\hat{l}_y, \tag{9.79}$$

$$[\hat{l}^2, \hat{l}_j] = 0, \quad j = x, y, z. \tag{9.80}$$

These equations say that the components of the angular momentum are not simultaneously measurable, although one component and the square of the angular momentum can be simultaneously measured.

9.4 The Quantum Mechanical Oscillator

Aside from the particle in the box, the harmonic oscillator is one of the simplest examples of quantum theory. Although this example does not apply to the motion of an electron in an atom, because a different force law applies there, the harmonic oscillator has innumerable applications in all areas of quantum physics. We shall return to it repeatedly. In classical physics, the equation of motion of the harmonic oscillator is given by $m_0 \ddot{x} = -kx$ (Fig. 9.5). The associated kinetic energy is $(m_0/2)\dot{x}^2$ and the potential energy $(k/2)x^2$. To convert this to quantum mechanics, we express the velocity $\dot{x}$ in terms of the momentum: $m_0 \dot{x} = p$. We also make use of the classical relation between oscillation frequency ω, mass and force constant, $\omega^2 = k/m_0$. In this way we obtain the following expression for the total energy (or mathematically expressed, for the Hamiltonian function):

$$\mathscr{H} = \frac{p^2}{2m_0} + \frac{m_0}{2}\omega^2 x^2 . \tag{9.81}$$

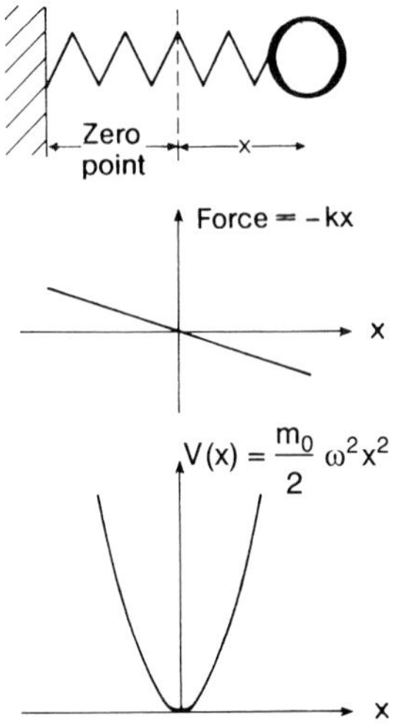

Fig. 9.5. The harmonic oscillator. *Above*, as an example, a point mass on a spring. *Middle*, force as a function of the displacement x. *Below*, potential energy as function of the displacement x

The corresponding Schrödinger equation is

$$\left(-\frac{\hbar^2}{2m_0}\frac{d^2}{dx^2} + \frac{m_0}{2}\omega^2 x^2\right)\psi(x) = E\psi(x) . \tag{9.82}$$

As one can easily convince oneself, the energy E can only have positive values. We ensure this by multiplying both sides of (9.82) by $\psi^*(x)$ and integrating from $x = -\infty$ to $x = +\infty$. The integral on the right side of (9.82), $\int_{-\infty}^{+\infty} \psi^* \psi \, dx$, is positive, because

$\psi^* \psi = |\psi|^2 \geqq 0$. The same applies to the term containing x^2, $\int_{-\infty}^{+\infty} (m_0/2)\,\omega^2 x^2 |\psi|^2 dx$ on the left side of (9.82). The remaining term, $\int_{-\infty}^{+\infty} [-(\hbar^2/2m_0)\,\psi^* \cdot d^2\psi/dx^2]\,dx$, is rearranged by partial integration to yield

$$-\frac{\hbar^2}{2m_0}\,\psi^* \frac{d\psi}{dx}\Bigg|_{x=-\infty}^{x=+\infty} + \frac{\hbar^2}{2m_0}\int_{-\infty}^{+\infty} \frac{d\psi}{dx}\cdot\frac{d\psi^*}{dx}\,dx\,. \tag{9.82a}$$

If we substitute the limits of integration in the first term, it disappears, since we require that $\psi \to 0$ as $x \to \pm\infty$. (Otherwise the normalisation integral $\int_{-\infty}^{+\infty} |\psi|^2 dx = 1$ would not exist!) The integral in (9.82a) is positive, and thus, so is the entire expression corresponding to the left side of (9.82). It now immediately follows that $E \geqq 0$.

Since the Schrödinger equation includes several constants, we first carry out a transformation to a new, dimensionless coordinate ξ and a new energy, by setting

$$x = \sqrt{\frac{\hbar}{m_0\omega}}\,\xi\,; \qquad \varepsilon = \frac{E}{\hbar\omega}\,, \tag{9.83}$$

$$\psi(x) = \phi(\xi)\,. \tag{9.84}$$

Then (9.82) becomes

$$\tilde{\mathscr{H}}\phi \equiv \frac{1}{2}\left(-\frac{d^2}{d\xi^2} + \xi^2\right)\phi(\xi) = \varepsilon\phi(\xi)\,. \tag{9.85}$$

If the differentiation operator $d/d\xi$ were an ordinary number, we could use the rule $-a^2 + b^2 = (-a+b)(a+b)$. Although this is naturally not possible with operators, we shall use it as an heuristic aid and write, so to speak, experimentally,

$$\underbrace{\frac{1}{\sqrt{2}}\left(-\frac{d}{d\xi} + \xi\right)}_{b^+}\underbrace{\frac{1}{\sqrt{2}}\left(\frac{d}{d\xi} + \xi\right)}_{b}\phi(\xi)\,. \tag{9.86}$$

The order of the differentiation steps must be strictly observed here, that is, operators on the right must be applied before operators to the left of them. Let us now "multiply" out the parentheses, strictly observing the order of operations:

$$\frac{1}{2}\left(-\frac{d^2}{d\xi^2} + \xi^2\right)\phi(\xi) + \frac{1}{2}\left(-\frac{d}{d\xi}\,\xi + \xi\frac{d}{d\xi}\right)\phi(\xi)\,. \tag{9.87}$$

This is the left side of (9.85), with an extra term. Just as we did with the Heisenberg commutation relation (9.69), we can apply the differentiation in the extra term to the wavefunction, and we obtain $-\phi(\xi)/2$ for the second expression in (9.87). Equation

(9.86) thus differs from the middle expression in (9.85) only by the term $-\phi/2$. If we observe this and introduce, as shown in (9.86), the abbreviations b and b^+, the original Schrödinger equation (9.82) can be given in the form

$$b^+ b\phi \equiv (\tilde{\mathscr{H}} - \tfrac{1}{2})\phi = (\varepsilon - \tfrac{1}{2})\phi \,. \tag{9.88}$$

In the following it is important to remember that b and b^+ are only certain abbreviations for operators, which are defined in (9.86). If we also substitute $\varepsilon - \frac{1}{2} = n$ and provide the wavefunction ϕ and this n with an index λ, the justification for which will be given below, we finally obtain the Schrödinger equation in the form

$$b^+ b\phi_\lambda = n_\lambda \phi_\lambda \,. \tag{9.89}$$

The operators b and b^+ satisfy the commutation relation

$$bb^+ - b^+ b = 1 \,. \tag{9.90}$$

We shall leave the proof of (9.90) to the reader as an exercise. One needs only to substitute the definition of b^+ and b and then proceed as above with the Heisenberg commutation rule.

Let us first consider (9.89) generally and multiply it from the left by the operator b, i.e., we apply the operator b to the left and right sides of (9.89). We then obtain

$$bb^+ b\phi_\lambda = n_\lambda b\phi_\lambda \,. \tag{9.91}$$

According to the commutation relation (9.90), we can substitute $1 + b^+ b$ for bb^+. When we do this with the first two factors on the left side of (9.91), we obtain

$$b^+ b(b\phi_\lambda) + b\phi_\lambda = n_\lambda b\phi_\lambda \,, \tag{9.92}$$

or, if we combine the terms containing $b\phi_\lambda$ on the right,

$$b^+ b(b\phi_\lambda) = (n_\lambda - 1)(b\phi_\lambda) \,. \tag{9.93}$$

As we see, application of b to the wavefunction ϕ_λ produces a new wavefunction $\phi = (b\phi_\lambda)$ which satisfies (9.89), although its eigenvalue is 1 less: $n_\lambda \to n_\lambda - 1$. The operator b thus reduces the number n by 1. We refer to it as an *annihilation operator*. Since, as we observed earlier, the energy E must be positive, n must have a lower limit. There must therefore be a lowest number n_0 and a corresponding wavefunction ϕ_0 for (9.89). If we were to repeat this formalism on the lowest eigenstate with $\lambda = 0$, we would introduce a contradiction. We would have found a wavefunction with a still smaller eigenvalue, contrary to the assumption that ϕ_0 is already the lowest eigenstate. The contradiction is only resolved if $b\phi_0$ is identically equal to zero. Then (9.89) is fulfilled trivially for each value of n; zero is, however, not a genuine eigenvalue. For the lowest state, we then have the condition

$$b\phi_0 = 0 \,. \tag{9.94}$$

If we replace b with the operator which it symbolises (9.86), then (9.24) is equivalent to

$$\left(\frac{d}{d\xi}+\xi\right)\phi_0=0\,. \tag{9.95}$$

This first-order differential equation can also be written in the form

$$\frac{d\phi_0}{\phi_0}=-\xi\,d\xi\,, \tag{9.96}$$

from which we obtain on integration

$$\ln\phi_0=-\tfrac{1}{2}\xi^2+C'\,, \tag{9.97}$$

or, taking the antilogarithm,

$$\phi_0=C\exp(-\tfrac{1}{2}\xi^2)\,. \tag{9.98}$$

The constant C must be determined by the normalisation condition.

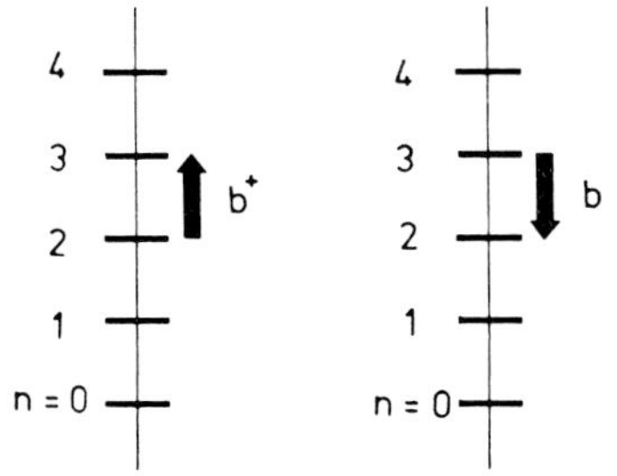

Fig. 9.6. Illustration of the effect of the creation and annihilation operators. *Left:* Application of b^+ means climbing up the "ladder" of states $n=0,1,\dots$ by one rung. *Right:* Application of b corresponds to climbing down by one rung

We will now investigate what happens if we apply not the annihilation operator b but rather the operator b^+ on both sides of (9.89). By analogy to the steps (9.91 – 93), we obtain using (9.90) the relation

$$b^+b(b^+\phi_\lambda)=(n_\lambda+1)(b^+\phi_\lambda)\,, \tag{9.99}$$

i.e. by application of b^+ to ϕ_λ, we increase the eigenvalue by one unit. Therefore, b^+ is called a *creation operator* (Fig. 9.6). If we choose the ground state ϕ_0 for ϕ_λ, we obtain a proportionality

$$\phi_1\propto b^+\phi_0\,,$$

and a second application of b^+ gives

$$\phi_2\propto b^+\phi_1\propto(b^+)^2\phi_0\,,\qquad \text{etc.}$$

Here we have used a proportionality sign and not an equals sign, since we do not yet know whether the functions $b^+\phi_0$, $(b^+)^2\phi_0$, etc. are normalised. In general, we obtain

$$\phi_\lambda = C_\lambda (b^+)^\lambda \phi_0 , \tag{9.100}$$

where the constant factor C_λ serves as normalisation factor.

Since n always increases by an integer on application of b^+, but the lowest eigenvalue is zero ($n_0 = 0$), we may identify the index λ with n. Including the normalisation factor (which we will not derive here), $C_n = 1/\sqrt{n!}$, we find the normalised wavefunctions:

$$\phi_n = \frac{1}{\sqrt{n!}} (b^+)^n \phi_0 . \tag{9.101}$$

Relation (9.101) still looks terribly abstract. We shall therefore show by means of several examples how the explicit wavefunctions may be derived; for this purpose, we shall leave the normalisation factor out of consideration. For $n = 0$ we have already obtained $\phi_0 \propto \exp(-\xi^2/2)$. Using (9.88, 83), we find for the lowest energy value $E_0 = \hbar\omega/2$, the same zero-point energy which we have already discussed in Sect. 7.5. For $n = 1$ we obtain

$$\phi_1 \propto b^+ \phi_0 ,$$

or, using the explicit expressions for b^+ and ϕ_0,

$$\phi_1 \propto \left(-\frac{d}{d\xi} + \xi \right) \exp(-\tfrac{1}{2}\xi^2) .$$

After carrying out the differentiation we have

$$\phi_1 \propto \xi \exp(-\tfrac{1}{2}\xi^2) .$$

The corresponding energy is

$$E = (3/2)\hbar\omega .$$

For $n = 2$ we obtain

$$\phi_2 \propto b^+ \phi_1 \propto \left(-\frac{d}{d\xi} + \xi \right) \cdot \xi \exp(-\tfrac{1}{2}\xi^2) ,$$

or, after differentiating,

$$\phi_2 \propto (2\xi^2 - 1) \exp(-\tfrac{1}{2}\xi^2) .$$

For the energy we find

$$E = (5/2)\hbar\omega .$$

If we continue this procedure, we obtain polynomials through multiplication by ξ or differentiation with respect to ξ. In general, for the nth wavefunction we obtain an expression of the type

$$\phi_n = \mathrm{e}^{-1/2\xi^2}\mathrm{H}_n(\xi)\,, \tag{9.102}$$

where H_n is a polynomial which is known in the mathematical literature as a Hermite polynomial. The corresponding energy is given by

$$E_n = (n+\tfrac{1}{2})\hbar\omega\,, \qquad n = 0, 1, 2 \ldots \tag{9.103}$$

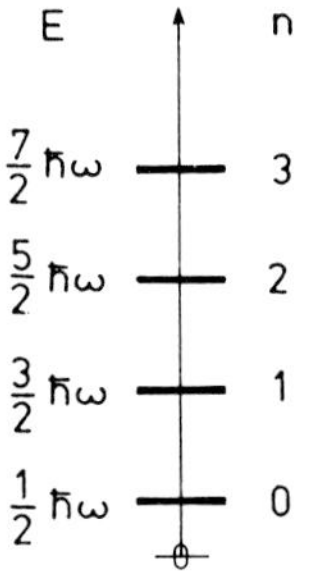

Fig. 9.7. The energy levels of the harmonic oscillator

(Fig. 9.7). For completeness, we shall give the formula for finding the Hermite polynomials. It states

$$\mathrm{H}_n(\xi) = \frac{(-1)^n}{\sqrt{2^n}}\,\mathrm{e}^{\xi^2}\frac{d^n\mathrm{e}^{-\xi^2}}{d\xi^n}\,\frac{1}{\sqrt{n!\sqrt{\pi}}}\,. \tag{9.104}$$

If we return from the coordinate ξ to the original coordinate x, the correctly normalised eigenfunctions of the Schrödinger equation of the harmonic oscillator are given by

$$\psi_n(x) = \sqrt[4]{\frac{m_0\omega}{\hbar}}\,\exp(-\tfrac{1}{2}x^2 m_0\omega/\hbar)\cdot\mathrm{H}_n(x\sqrt{m_0\omega/\hbar})\,. \tag{9.105}$$

In Fig. 9.8, we have plotted the potential $V(x)$. Furthermore, the energy levels $(n+1/2)\hbar\omega$ are given along the ordinate, as are, finally, the wavefunctions themselves. The first four wavefunctions in the energy scale are shown in more detail in Figs. 9.9a, b. Although we will for the most part use the configuration representation $\psi(x)$ for the wavefunctions in this book, the creation and annihilation operators b^+ and b are indispensable in many areas of modern quantum theory.

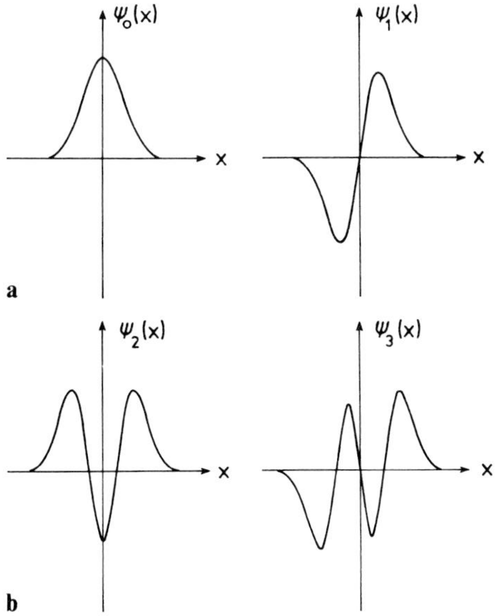

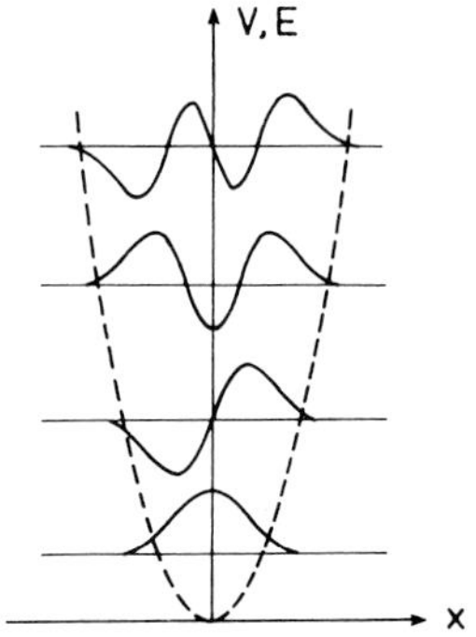

Fig. 9.8. A representation of the quantum mechanical harmonic oscillator which is often found in books. This figure contains three drawings in one: 1) The ordinate means total energy E. The horizontal lines (above the x axis) give the quantised energy levels. 2) The ordinate gives the potential $V(x)$. The dashed curve shows the shape of the potential curve as a function of position x. 3) Each of the horizontal lines serves as an x axis, on which the shape of the wavefunction of the corresponding energy is plotted

Fig. 9.9. a) The wavefunctions of the harmonic oscillator for $n = 0, 1$. **b)** The wavefunctions of the harmonic oscillator for $n = 2, 3$

Problems

9.1 By substituting the wave packet with $\omega = \hbar k^2/(2m_0)$ from Problem 7.1 into the Schrödinger equation, convince yourself that it is a solution for a particle which experiences no forces.

9.2 Let the wavefunctions ϕ_1 and ϕ_2 be solutions of the Schrödinger equation (9.35) with the eigenenergies E_1 and E_2. Show that

$$\psi(r,t) = c_1 \exp(-\mathrm{i}E_1 t/\hbar)\,\phi_1(r) + c_2 \exp(-\mathrm{i}E_2 t/\hbar)\,\phi_2(r)$$

is a solution of the Schrödinger equation (9.32). What conditions must c_1 and c_2 satisfy, in order to normalise $\psi(r,t)$? Generalise this exercise to the wavepacket

$$\psi(r,t) = \sum_j c_j \exp(-\mathrm{i}E_j t/\hbar)\,\phi_j(r)\,.$$

Hint:

$$\int \phi_j^*(r)\,\phi_k(r)\,dV = \delta_{jk} \begin{cases} = 0 & \text{for} \quad j \neq k \\ = 1 & \text{for} \quad j = k\,. \end{cases}$$

9.3 The potential $V(r)$ is represented in one dimension by $-\beta\delta(x)$, where $\delta(x)$ is the Dirac δ function (see the Mathematical Appendix). Solve the Schrödinger equation for bound states, i.e. for $E < 0$.

Hint: Solve the Schrödinger equation for $x < 0$ and $x > 0$, in other words where $\delta(x) = 0$.

Where $x = 0$, the solutions found for ψ_- and ψ_+ must join in a continuous fashion.

Also, derive a second boundary condition ("jump condition") for ψ'_- and ψ'_+ by integrating the Schrödinger equation over $-\varepsilon < x < \varepsilon$, $\varepsilon \to 0$. Write the wavefunction so that it can be normalised, and find the normalisation constants and the energy.

9.4 Find the bound states of a particle in a one-dimensional box, for which the potential is

$$\begin{aligned} V(x) &= 0 && \text{for} \quad x < -L \\ V(x) &= -V_0 < 0 && \text{for} \quad -L \leqslant x \leqslant L \\ V(x) &= 0 && \text{for} \quad x > L\,. \end{aligned}$$

Hint: Solve the Schrödinger equation in the three subregions. Require $\psi(x) \to 0$ for $x \to \pm\infty$; $\psi(x)$ and $\psi'(x)$ are continuous at $x = \pm L$. Display the eigenvalue spectrum for $E < 0$, and discuss its dependence on L and V_0.

9.5 Calculate the "scattering states", in which $E \geq 0$, for a particle moving in the δ potential of Problem 9.3.

Hint: Use the trial solution $\psi(x) = \exp(\mathrm{i}kx) + a\exp(-\mathrm{i}kx)$ for $x \leq 0$ and $\psi = b \exp(\mathrm{i}kx)$ for $x \geq 0$, and determine a and b. What is the physical interpretation of this trial solution in the field of wave optics? It does not need to be normalised. How do a and b change when the sign of β is changed, i.e., when the potential is repulsive?

9.6 Let an otherwise free particle collide with an infinitely high potential barrier. What is its wavefunction (without normalisation)?

9.7 For the one-dimensional wave packet of Problem 7.1, calculate the expectation values of the position x, momentum p, kinetic energy, and x^2. Why is the expectation value of x^2 more informative than that of x?

9.8 Express the energy expectation value for the wave packet of a free particle in Problem 7.1 in terms of the energy eigenvalues of the kinetic energy operator.

9.9 Prove the commutation relations (9.77 – 80) for angular momentum.

Hint: Use the quantum mechanical definition of the angular momentum operator and the commutation relations between position and momentum in three dimensions.

9.10 Demonstrate the commutation relations between $\hat{l}_x$ and x, and between $\hat{l}_x$ and the central potential $V(r)$ which depends only on $r = |\boldsymbol{r}|$.

9.11 The two functions ψ_1 and ψ_2 are to vanish at infinity.

Show that

$$\int_{-\infty}^{+\infty} \psi_1^* x \psi_2 dx = \left(\int_{-\infty}^{+\infty} \psi_2^* x \psi_1 dx \right)^*$$

$$\int_{-\infty}^{+\infty} \psi_1^* \frac{\hbar}{\mathrm{i}} \frac{d}{dx} \psi_2 dx = \left(\int_{-\infty}^{+\infty} \psi_2^* \frac{\hbar}{\mathrm{i}} \frac{d}{dx} \psi_1 dx \right)^*$$

$$\int_{-\infty}^{+\infty} \psi_1^* \mathscr{H} \psi_2 dx = \left(\int_{-\infty}^{+\infty} \psi_2^* \mathscr{H} \psi_1 dx \right)^* .$$

The properties of the operators x, $p = \frac{\hbar}{\mathrm{i}} \frac{d}{dx}$, $\mathscr{H} = \frac{\hbar^2}{2m_0} \frac{d^2}{dx^2} + V(x)$, which are to be proved here, indicate that these operators are Hermitian.

Hint: Carry out partial integrations over d/dx and d^2/dx^2.

9.12 Prove the Ehrenfest theorem

$$m_0 \frac{d}{dt} x = p , \quad \frac{d}{dt} p = - \left(\frac{dV}{dx} \right)$$

for the one-dimensional quantum mechanical motion of a particle.

Hint: Use the definition of the operators x, p and dV/dx, and the fact that ψ (and ψ^*) satisfy a Schrödinger equation with the potential $V(x)$. Make use also of the result of Problem 9.11.

What is the expression for this theorem in three dimensions?

9.13 Calculate the wavefunctions and energy values of a particle which is subjected to a force $F = -kx + k_0$, $(k = m_0\omega^2)$.

Hint: Set up $V(x)$ and derive the new Schrödinger equation from the "old" one for the harmonic oscillator by means of a coordinate transformation.

9.14 Prove the commutation relation (9.90)

$$bb^+ - b^+b = 1$$

for the operators b and b^+ of the harmonic oscillator.

Hint: Use the definitions of b^+ and b (9.86) and the commutation relation between x and $\frac{\hbar}{\mathrm{i}}\frac{d}{dx}$ (9.74).

9.15 Construct the wavepacket

$$\psi = \psi_0 \exp\left(-\mathrm{i}\frac{\omega}{2}t\right) + \psi_1 \exp\left(-\mathrm{i}\frac{3\omega}{2}t\right)$$

from the first two states of the harmonic oscillator and examine the change in $|\psi|^2$ with time by means of a graphical representation.

9.16 Let the Schrödinger equation of the harmonic oscillator be

$$b^+ b\phi_n = n\phi_n \qquad (n = 0, 1, 2, \ldots),$$

where $b^+ = (1/\sqrt{2})\left(-\frac{d}{d\xi} + \xi\right)$, $b = (1/\sqrt{2})\left(\frac{d}{d\xi} + \xi\right)$, $\phi = \phi(\xi)$. For b, b^+, the commutation relation $[b, b^+] = 1$ holds.
Prove the following relations. The integrals extend from $-\infty$ to $+\infty$.

a) $\int [b^+\phi(\xi)]^* \psi(\xi)\, d\xi = \int \phi^*(\xi)\, b\, \psi(\xi)\, d\xi$
$\int [b\phi(\xi)]^* \psi(\xi)\, d\xi = \int \phi(\xi)^* b^+ \psi(\xi)\, d\xi$

b) $\int (b^+\phi_n)^*(b^+\phi_n)\, d\xi = (n+1)\int \phi_n^* \phi_n\, d\xi$

c) If ϕ_n is normalised, then $\phi_{n+1} = 1/\sqrt{n+1}\, b^+\phi_n$ is also normalised.

d) The normalised functions ϕ_n can be expressed as

$$\phi_n = 1/\sqrt{n!}\,(b^+)^n \phi_0 \,, \qquad b\phi_0 = 0 \,.$$

e) $b^+ \phi_n = \sqrt{n+1}\,\phi_{n+1}\,, \qquad b\phi_n = \sqrt{n}\,\phi_{n-1}\,.$

f) $b(b^+)^n - (b^+)^n b = n(b^+)^{n-1}\,, \qquad b^+(b)^n - (b)^n b^+ = -n(b)^{n-1} = -\dfrac{\partial b^n}{\partial b}\,.$

Hints: a) Use the explicit expressions for b^+ and b in terms of ξ, $\dfrac{d}{d\xi}$ and partial integration.
b) Use a), the exchange relation and the Schrödinger equation.
c) Follows from a).
d) Mathematical induction method.
e) Follows from d) and the commutation relations.
f) Solve by the induction method (write $b(b^+)^n - (b^+)^n b$ as $b(b^+)^n - (b^+)^{n-1} b^+ b$).

9.17 Calculate the expectation value of the momentum, the kinetic energy and the potential energy for the nth excited state of the harmonic oscillator.

Hint: According to (9.83 and 84), change from x to ξ, transform ξ and $d/d\xi$ into b^+ and b, and use

$$\int \phi_n^*(\xi)\,\phi_m(\xi)\,d\xi = \delta_{mn} = \begin{cases} 0 & \text{for} \quad m \neq n \\ 1 & \text{for} \quad m = n \,, \end{cases}$$

$n, m = 0, 1, 2, \ldots$.

9.18 Prove for the wavefunctions of the harmonic oscillator, $\phi_n(\xi)$:

$$\int \phi_m^*(\xi)\,\phi_n(\xi)\,d\xi = \delta_{mn}\,.$$

Hint: Use the fact that

$$\phi_n = \frac{1}{\sqrt{n!}}(b^+)^n \phi_0 \,,$$

$$b\phi_0 = 0 \,,$$

and the result a) of Problem 9.16. Proceed by induction.

9.19 *The bra and ket notation*

The English physicist Dirac introduced a very concise notation, especially for expectation values and wavefunctions, which we shall demonstrate here for the case of the harmonic oscillator.

Instead of ϕ_n, one writes $|n\rangle$. The integral $\int \phi_n^*(\xi)\,\phi_n(\xi)\,d\xi$ is presented as $\langle n|n\rangle$, and the expectation value $\int \phi_n^*(\xi)\,b\,\phi_n(\xi)\,d\xi$ as $\langle n|b|n\rangle$. Since $\langle\ \rangle$ is a "bracket", $\langle n|$ is called "bra", and $|n\rangle$ is called "ket". Using the results of Problems 9.16 and 9.18, show that

a) $b^+|n\rangle = \sqrt{n+1}\,|n+1\rangle$
   $b\;|n\rangle = \sqrt{n}\,|n-1\rangle$

b) $\langle n|m\rangle = \delta_{n,m}$

c) $\langle n|b|n\rangle = 0$
   $\langle n|b^+|n\rangle = 0$

d) Calculate $\langle n|(b^+ + b)^2|n\rangle$ and $\langle n|(b^+ - b)^2|n\rangle$.

What is the physical significance of these expectation values?

10. Quantum Mechanics of the Hydrogen Atom

10.1 Motion in a Central Field

In this chapter, we shall solve the Schrödinger equation of the hydrogen atom. For our calculations, we will not initially restrict ourselves to the Coulomb potential of the electron in the field of the nucleus of charge Z, $V(r) = -Ze^2/(4\pi\varepsilon_0 r)$, but rather will use a general potential $V(r)$, which is symmetric with respect to a centre. As the reader may know from the study of classical mechanics, the angular momentum of a particle in a spherically symmetric potential field is conserved; this fact is expressed, for example, in Kepler's law of areas for the motion of the planets in the solar system. In other words, we know that in classical physics, the angular momentum of a motion in a central potential is a constant as a function of time. This tempts us to ask whether in quantum mechanics the angular momentum is simultaneously measurable with the energy. As a criterion for simultaneous measurability, we know that the angular momentum operators must commute with the Hamiltonian. As we have already noted, the components l_x, l_y, and l_z of the angular momentum $\boldsymbol{l}$ are not simultaneously measurable; on the other hand, l_z and $\boldsymbol{l}^2$, for example, *are* simultaneously measurable. A long but straightforward calculation reveals that these two operators also commute with the Hamiltonian for a central-potential problem. Since the details of this calculation do not provide any new physical insights, we shall not repeat it here.

In quantum mechanics as well as in classical mechanics, we may thus measure the total energy, the z component of the angular momentum, and the square of the angular momentum simultaneously to any desired accuracy. In the following, we shall therefore seek the simultaneous eigenfunctions of $\hat{l}^2$, $\hat{l}_z$, and $\mathscr{H}$. We remind the reader that we denote the angular momentum *operators* by a ^ (hat), in order to distinguish them from the classical quantities $\boldsymbol{l}$. Since we are here dealing with a spherically symmetric problem it is reasonable not to use Cartesian coordinates, but to change to another coordinate system which better reflects the symmetry of the problem. This is naturally the spherical polar coordinate system. If we choose a particular point x, y, z in Cartesian coordinates, we shall describe its position by means of the follwing coordinates (Fig. 10.1):

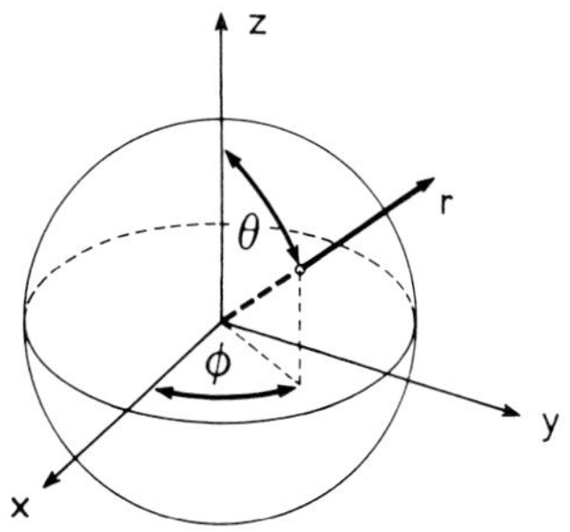

Fig. 10.1. Illustration of spherical polar coordinates

1) its distance from the origin, r,
2) the angle θ between the z axis and the vector $\boldsymbol{r}$,
3) the angle ϕ between the x axis and the projection of $\boldsymbol{r}$ on the x-y plane.

Recalculating the Laplace operator $\nabla^2 = \partial^2/\partial x^2 + \partial^2/\partial y^2 + \partial^2/\partial z^2$ in terms of spherical polar coordinates is a lengthy mathematical procedure, which however contributes nothing to the understanding of quantum mechanics. We shall therefore simply write down the kinetic energy operator in polar coordinates; it is

$$-\frac{\hbar^2}{2m_0}\nabla^2 = -\frac{\hbar^2}{2m_0}\frac{1}{r^2}\frac{\partial}{\partial r}\left(r^2\frac{\partial}{\partial r}\right) + \frac{1}{2m_0 r^2}\hat{l}^2, \tag{10.1}$$

with

$$\hat{l}^2 = -\hbar^2\left[\frac{1}{\sin\theta}\frac{\partial}{\partial\theta}\left(\sin\theta\frac{\partial}{\partial\theta}\right) + \frac{1}{\sin^2\theta}\frac{\partial^2}{\partial\phi^2}\right]. \tag{10.2}$$

We note that the operator $\hat{l}^2$ is nothing other than the square of the angular momentum operator, and contains only derivatives with respect to angles. Since the potential of our problem depends only on the radius coordinate r, it is reasonable to separate the radial and angular functions in a trial wavefunction as follows:

$$\psi(r,\theta,\phi) = R(r)F(\theta,\phi), \tag{10.3}$$

i.e., we write the wavefunction as the product of a function which depends only on r with a second function which depends only on the angles θ and ϕ. If we insert (10.3) into the Schrödinger equation

$$\left[-\frac{\hbar^2}{2m_0}\nabla^2 + V(r)\right]\psi = E\psi, \tag{10.4}$$

we obtain

$$\mathscr{H}\psi = F(\theta,\phi)\left[-\frac{\hbar^2}{2m_0}\frac{1}{r^2}\frac{\partial}{\partial r}\left(r^2\frac{\partial}{\partial r}\right) + V(r)\right]R(r) + \frac{R(r)}{2m_0 r^2}\hat{l}^2 F(\theta,\phi) = ERF. \tag{10.5}$$

We now make use of our recognition of the fact that the wavefunction (10.3) can be chosen to be an eigenfunction of $\hat{l}^2$ and of $\hat{l}_z$ as well as of $\mathscr{H}$. We write the corresponding eigenvalues in the (arbitrary) form $\hbar^2\omega$ and $\hbar m$. These new, additional equations are then[1]

$$\hat{l}^2 F(\theta,\phi) = \hbar^2\omega F(\theta,\phi) \quad \text{and} \tag{10.6}$$

$$\hat{l}_z F(\theta,\phi) = \hbar m F(\theta,\phi). \tag{10.7}$$

Note that m in (10.7) is the "magnetic quantum number" and must not be confused with the mass.

---

[1] In literature one often uses Y instead of F.

By assuming that (10.6) is already solved, we can express the term $R(r)/(2m_0r^2)\hat{l}^2F(\theta,\phi)$ in (10.5) in a simple form through the eigenvalue $\hbar^2\omega$. We then have eliminated all derivatives with respect to θ or ϕ on the left side of (10.5), and we may divide both sides of (10.5) by $F(\theta,\phi)$. We thus obtain an equation for the radial part $R(r)$ alone:

$$\left[-\frac{\hbar^2}{2m_0}\frac{1}{r^2}\frac{\partial}{\partial r}\left(r^2\frac{\partial}{\partial r}\right)+V(r)+\frac{\hbar^2\omega}{2m_0r^2}\right]R(r)=ER(r)\,. \tag{10.8}$$

We have reduced the task of solving the three-dimensional Schrödinger equation (10.4), to that of solving the (as we shall see) simpler equations (10.6, 7 and 8).

Since the quantity $\hbar^2\omega$ in (10.8) is still an unknown parameter, which occurs as an eigenvalue in (10.6), our first problem is to determine this eigenvalue. We thus begin with the task of solving (10.6) and (10.7).

10.2 Angular Momentum Eigenfunctions

The first part of this section is somewhat more abstract. For the reader who would like to see the results first we give them here in compact form:
The eigenvalues of the square of the angular momentum $\hat{l}^2$ are

$$\hbar^2 l(l+1)\,, \tag{10.9}$$

where l is an integer,

$$l=0,1,2\ldots\,.$$

According to (10.7), the eigenvalues of the z component of the angular momentum are

$$\hbar m\,.$$

The integer m is called the magnetic quantum number, and takes on the values

$$-l\leqq m\leqq l\,.$$

The wavefunctions $F(\theta,\phi)$ naturally depend on the quantum numbers l and m and have the form

$$F_{l,m}(\theta,\phi)=\mathrm{e}^{\mathrm{i}m\phi}\mathrm{P}_l^m(\cos\theta)\,. \tag{10.10}$$

These functions are drawn in Fig. 10.2. $\mathrm{P}_l^{(0)}$ is called a Legendre polynomial, and P_l^m with $m\neq 0$ is called an associated Legendre function. The entire function (10.10) is called a spherical harmonic function.

We first address ourselves to the task of finding the eigenfunctions F as the solutions to (10.6, 7). We write (10.6) again, giving the components of $\hat{l}$ explicitly:

$$(\hat{l}_x^2+\hat{l}_y^2+\hat{l}_z^2)F_{l,m}=\hbar^2\omega_lF_{l,m}\,. \tag{10.11}$$

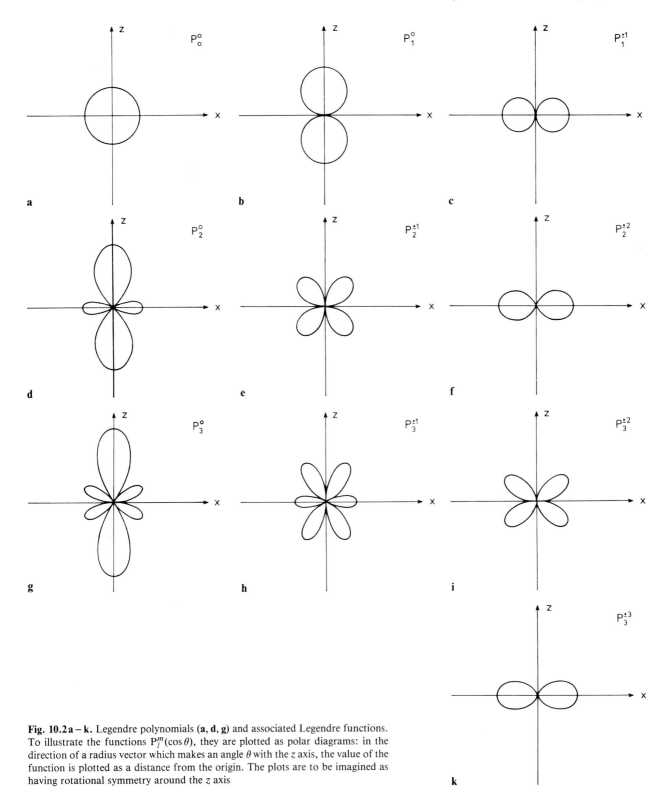

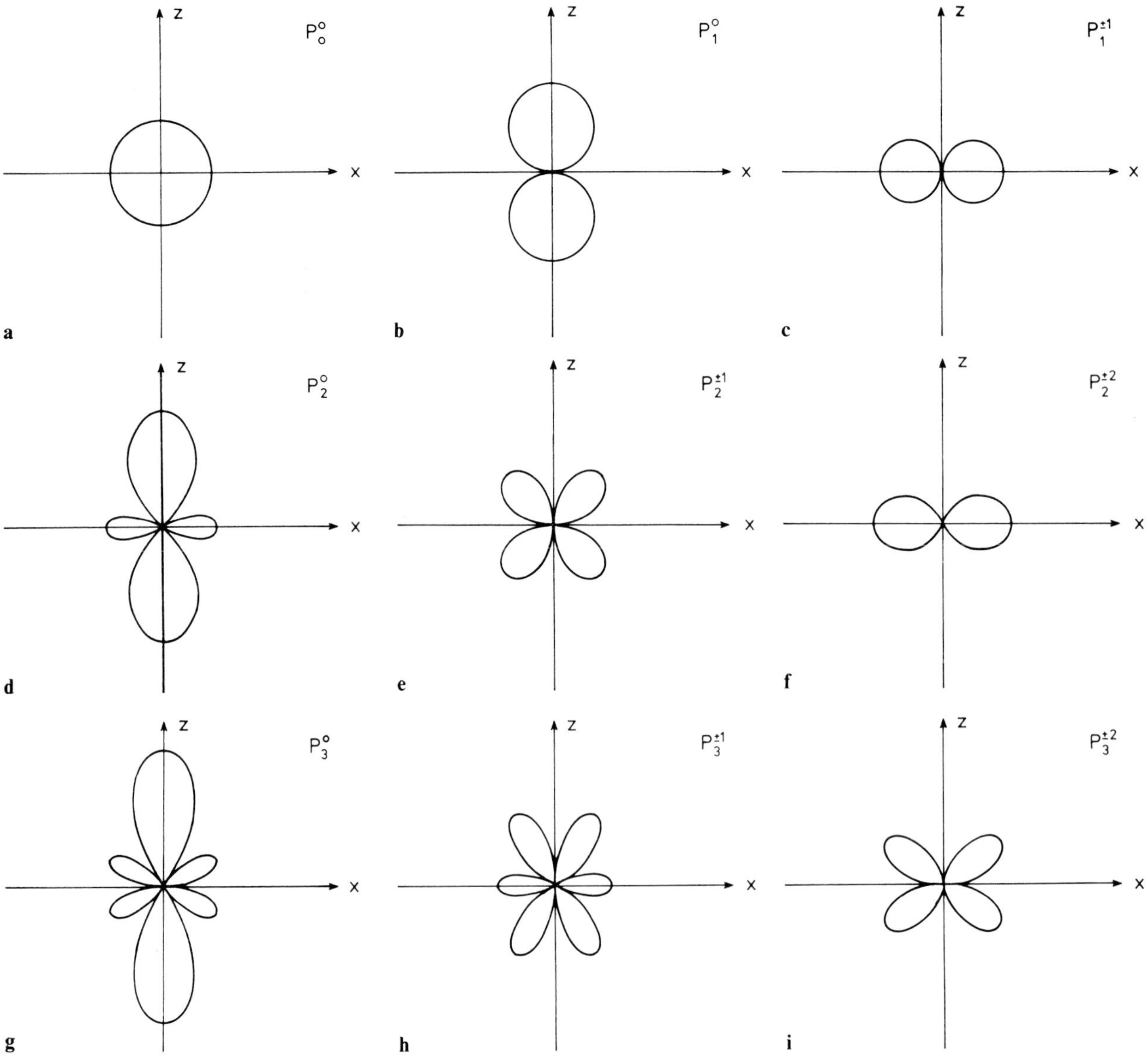

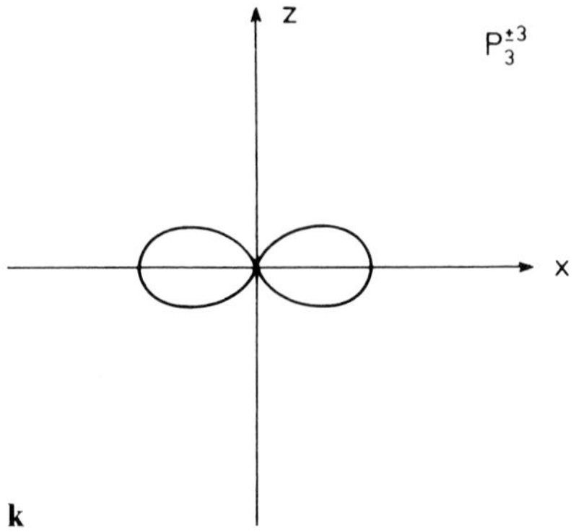

Fig. 10.2a – k. Legendre polynomials (**a, d, g**) and associated Legendre functions. To illustrate the functions $P_l^m(\cos\theta)$, they are plotted as polar diagrams: in the direction of a radius vector which makes an angle θ with the z axis, the value of the function is plotted as a distance from the origin. The plots are to be imagined as having rotational symmetry around the z axis

Furthermore, we derive a new equation from (10.7) by applying the operator $\hat{l}_z$ to both sides and then using (10.7) once more. This yields

$$\hat{l}_z^2 F_{l,m} = \hbar^2 m^2 F_{l,m} . \tag{10.12}$$

If we subtract (10.12) from (10.11), we obtain

$$(\hat{l}_x^2 + \hat{l}_y^2) F_{l,m} = \hbar^2(\omega_l - m^2) F_{l,m} . \tag{10.13}$$

If we multiply both sides of this equation from the left by $F^*_{l,m}$ and integrate over the coordinates θ and ϕ, we can show, in a way similar to that used with the harmonic oscillator, that

$$\omega_l - m^2 \geqq 0 . \tag{10.13 a}$$

In analogy to the harmonic oscillator, it seems reasonable to write $\hat{l}_x^2 + \hat{l}_y^2$ as the product of two factors $\hat{l}_+ = \hat{l}_x + \mathrm{i}\hat{l}_y$ and $\hat{l}_- = \hat{l}_x - \mathrm{i}\hat{l}_y$. We might suspect that these new linear combinations, like the operators b^+ and b with the harmonic oscillator, are a kind of creation and annihilation operators. As already stated in (9.77 – 80), the following commutation relations hold between the angular momentum operators:

$$\begin{aligned}
[\hat{l}^2, \hat{l}_j] &= 0 , \quad j = x, y, z , \\
[\hat{l}_x, \hat{l}_y] &= \mathrm{i}\hbar \hat{l}_z , \\
[\hat{l}_y, \hat{l}_z] &= \mathrm{i}\hbar \hat{l}_x , \\
[\hat{l}_z, \hat{l}_x] &= \mathrm{i}\hbar \hat{l}_y .
\end{aligned} \tag{10.14}$$

Further commutation relations can be derived from these by simple algebraic transformations:

$$[\hat{l}^2, \hat{l}_\pm] = 0 , \tag{10.15^2}$$

$$[\hat{l}_z, \hat{l}_\pm] = \pm \hbar \hat{l}_\pm , \tag{10.16}$$

$$[\hat{l}_\pm, \hat{l}_z] = \mp \hbar \hat{l}_\pm . \tag{10.17}$$

In order to demonstrate that $\hat{l}_+$ is a kind of creation operator and $\hat{l}_-$ a corresponding annihilation operator, we consider

$$\hat{l}_\pm F_{l,m} . \tag{10.18}$$

To find an equation for this quantity, we apply $\hat{l}_\pm$ to the left of each side of (10.6) and then obtain, due to the commutativity with $\hat{l}^2$, the equation

$$\hat{l}^2(\hat{l}_\pm F_{l,m}) = \hbar^2 \omega_l (\hat{l}_\pm F_{l,m}) . \tag{10.19}$$

This means that if $F_{l,m}$ is an eigenfunction of (10.6), so is the function (10.18). We now apply $\hat{l}_\pm$ to the left of both sides of (10.7) and then, because of the commutation relation (10.17), we obtain after rearranging

$$\hat{l}_z(\hat{l}_\pm F_{l,m}) = \hbar(m \pm 1)(\hat{l}_\pm F_{l,m}) . \tag{10.20}$$

[1] $\hat{l}_\pm$ means that (10.15) holds for both $\hat{l}_+$ and $\hat{l}_-$. (10.16) and (10.17) are to be understood in the same way. In each case, the two upper signs belong together, as do the two lower signs

$\hat{l}_{\pm}$ thus increases (or decreases) the eigenvalue m by 1. Leaving off the normalisation factor, we can therefore write

$$\hat{l}_{\pm} F_{l,m} = F_{l,m\pm 1} \cdot (\text{Numerical normalisation factor})\,. \tag{10.21}$$

Equation (10.13a), which requires that m^2 cannot be larger than ω_l, applies here. Therefore the series of new eigenfunctions $\hat{l}_{\pm} F_{l,m}$ must terminate at a maximum $m = m_{\max}$ and at a negative, minimum $m = m_{\min}$. Thus, just as in the case of the harmonic oscillator, we must require that

$$\hat{l}_{+} F_{l,m_{\max}} = 0 \tag{10.22}$$

and

$$\hat{l}_{-} F_{l,m_{\min}} = 0\,. \tag{10.23}$$

If we apply $\hat{l}_{-}$ to the left of (10.22), and make use of the relations

$$\hat{l}_{\mp}\hat{l}_{\pm} = \hat{l}_x^2 + \hat{l}_y^2 \mp \hat{l}_z \hbar = \hat{l}^2 - \hat{l}_z(\hat{l}_z \pm \hbar) \tag{10.24}$$

and the fact that $F_{l,m}$ is an eigenfunction of $\hat{l}^2$ and $\hat{l}_z$, we obtain the basic equation

$$\hat{l}_{-}\hat{l}_{+} F_{l,m_{\max}} = \hbar^2(\omega_l - m_{\max}^2 - m_{\max}) F_{l,m_{\max}} = 0\,. \tag{10.25}$$

In analogous fashion, by applying $\hat{l}_{+}$ to (10.23), we obtain

$$\hat{l}_{+}\hat{l}_{-} F_{l,m_{\min}} = \hbar^2(\omega_l - m_{\min}^2 + m_{\min}) F_{l,m_{\min}} = 0\,. \tag{10.26}$$

Since the eigenfunctions $F_{l,m}$ are not zero, the factors by which they are multiplied must vanish. It must therefore hold that

$$m_{\max}(m_{\max}+1) = m_{\min}(m_{\min}-1) = \omega_l\,. \tag{10.27}$$

This can be rearranged to

$$(m_{\max}+m_{\min})(m_{\max}-m_{\min}+1) = 0\,. \tag{10.28}$$

Since $m_{\max} \geqq m_{\min}$, it follows that the second factor in (10.28) must be different from zero, and therefore that the first factor is equal to zero. From this,

$$m_{\max} = -m_{\min}\,. \tag{10.29}$$

As we have seen, each application of $\hat{l}_{+}$ to $F_{l,m}$ increases the eigenvalue m by 1. Therefore the difference $m_{\max} - m_{\min}$ must be an integer. If follows from (10.29) that

$$m_{\max} = \frac{\text{integer}}{2} \geqq 0\,. \tag{10.30}$$

So far we have only made use of the fact that $F_{l,m}$ satisfies the Eqs. (10.6) and (10.7), and that the commutation relations (10.14) apply. As we shall see later, we must require for the *orbital* motion of the electron that all values of m, and thus in particular $m_{\max}$, must be integers. Interestingly, the electron and also a few other elementary particles have their own angular momentum, which is independent of the orbital angular momentum, for which $m_{\max} = \frac{1}{2}$. This independent angular momentum is called "spin". We shall return to it in Sect. 14.2.1.

If we set $m_{max} = l$, there are $2l+1$ integers m between $+l$ and $-l$ which satisfy the condition

$$-l \leqq m \leqq l . \tag{10.31}$$

From (10.27) we know that the parameter ω_l, which appears in (10.6), is

$$\omega_l = l(l+1) . \tag{10.32}$$

The eigenvalue of the operator "angular momentum squared" is thus

$$m_{max}(m_{max}+1)\hbar^2 = l(l+1)\hbar^2 . \tag{10.33}$$

With these results, we can give the original equations (10.6) and (10.7) with their exact eigenvalues

$$\hat{l}^2 F_{l,m} = \hbar^2 l(l+1) F_{l,m} , \tag{10.34}$$

$$\hat{l}_z F_{l,m} = \hbar m F_{l,m} . \tag{10.35}$$

The application of $\hat{l}_+$ to $F_{l,m}$ leads to a new function $F_{l,m+1}$ for which the normalisation factor N remains undetermined:

$$F_{l,m+1} = N \hat{l}_+ F_{l,m} . \tag{10.36}$$

It can be shown that

$$N = \frac{1}{\hbar} \frac{1}{\sqrt{(l-m)(l+m+1)}} . \tag{10.37}$$

We again proceed in analogy to the harmonic oscillator. There we constructed the eigenfunctions in space, in that we applied the operators b and b^+ successively to the ground state. Here we do exactly the same. First, one can express the angular momentum operators, which were given in Cartesian coordinates according to (9.61), in polar coordinates. As can be shown mathematically, the result is

$$\hat{l}_z = \frac{\hbar}{\mathrm{i}} \frac{\partial}{\partial \phi} , \tag{10.38}$$

$$\hat{l}_x = -\frac{\hbar}{\mathrm{i}} \left(\sin\phi \frac{\partial}{\partial \theta} + \cot\theta \cos\phi \frac{\partial}{\partial \phi} \right) , \tag{10.39}$$

$$\hat{l}_y = \frac{\hbar}{\mathrm{i}} \left(\cos\phi \frac{\partial}{\partial \theta} - \cot\theta \sin\phi \frac{\partial}{\partial \phi} \right) . \tag{10.40}$$

Using (10.35) and (10.38), we represent $F_{l,m}$ as the following product:

$$F_{l,m} = \mathrm{e}^{\mathrm{i}m\phi} f_{l,m}(\theta) , \tag{10.41}$$

where we write the second factor on the right in the form

$$\mathrm{P}_l^m(\cos\theta) \tag{10.42}$$

for later use. If we increase ϕ by 2π, we must naturally obtain a single-valued function $F_{l,m}$. This can only be guaranteed if m is an integer. Therefore the odd multiples of 1/2 which would satisfy (10.30) are excluded.

We now calculate $F_{l,m}$ for $m=-l$ from the condition (10.23). If we substitute (10.39) and (10.40) into this equation, we obtain in simple fashion

$$(\hat{l}_x - \mathrm{i}\hat{l}_y)F_{l,-l} = -\hbar \mathrm{e}^{-\mathrm{i}\phi}\mathrm{e}^{-\mathrm{i}l\phi}\left[\frac{\partial}{\partial\theta} - l\cot\theta\right]f_{l,-l}(\theta) = 0\,. \tag{10.43}$$

The exponential functions can be removed from the second equation to give

$$\frac{\partial f_{l,-l}(\theta)}{\partial\theta} = l\cot\theta f_{l,-l}(\theta)\,. \tag{10.44}$$

The solution of this differential equation is

$$f_{l,-l}(\theta) = C(\sin\theta)^l\,, \tag{10.45}$$

as the reader can be convinced by substitution. Here C must be determined by the normalisation. The condition

$$\int_0^{2\pi}\int_0^{\pi}|F|^2\sin\theta\, d\theta\, d\phi = 1 \tag{10.46}$$

yields the coefficient C after carrying out the integration:

$$C = \frac{1}{\sqrt{4\pi}}\,\frac{\sqrt{(2l+1)!}}{l!\,2^l}\,.$$

If we now apply $\hat{l}_+$ to $F_{l,m}$ consecutively in the form

$$\hat{l}_+ F_{l,m} = \hbar \mathrm{e}^{\mathrm{i}\phi}\left[\frac{\partial}{\partial\theta} - m\cot\theta\right]F_{l,m}\,, \tag{10.47}$$

we can construct all the angular momentum eigenfunctions.

In the following, we give the expressions obtained thus for $l=0, 1$, and 2. The functions $F_{l,m}$ are normalised according to (10.46). They are given both as functions of the angular coordinates θ and ϕ and as functions of the Cartesian coordinates x, y, z (with $r=\sqrt{x^2+y^2+z^2}$); they are denoted in the standard notation by $\mathrm{Y}_{l,m}(\theta,\phi)$.

$l=0$

$$\mathrm{Y}_{0,0} = \frac{1}{\sqrt{4\pi}} \tag{10.48}$$

$$
\left.
\begin{aligned}
&l = 1\\
&\mathrm{Y}_{1,0} = \sqrt{\frac{3}{4\pi}}\cos\theta = \sqrt{\frac{3}{4\pi}}\,\frac{z}{r}\\
&\mathrm{Y}_{1,\pm 1} = \mp\sqrt{\frac{3}{8\pi}}\sin\theta\,\mathrm{e}^{\pm\mathrm{i}\phi} = \mp\sqrt{\frac{3}{8\pi}}\,\frac{x\pm\mathrm{i}y}{r}
\end{aligned}
\right\} \tag{10.49}
$$

$$
\left.
\begin{aligned}
&l = 2\\
&\mathrm{Y}_{2,0} = \sqrt{\frac{5}{4\pi}}\left(\frac{3}{2}\cos^2\theta - \frac{1}{2}\right) = \frac{1}{2}\sqrt{\frac{5}{4\pi}}\,\frac{2z^2-x^2-y^2}{r^2}\\
&\mathrm{Y}_{2,\pm 1} = \mp\frac{1}{2}\sqrt{\frac{15}{2\pi}}\sin\theta\cos\theta\,\mathrm{e}^{\pm\mathrm{i}\phi} = \mp\frac{1}{2}\sqrt{\frac{15}{2\pi}}\,\frac{(x\pm\mathrm{i}y)z}{r^2}\\
&\mathrm{Y}_{2,\pm 2} = \frac{1}{4}\sqrt{\frac{15}{2\pi}}\sin^2\theta\,\mathrm{e}^{\pm 2\mathrm{i}\phi} = \frac{1}{4}\sqrt{\frac{15}{2\pi}}\left(\frac{x\pm\mathrm{i}y}{r}\right)^2 .
\end{aligned}
\right\} \tag{10.50}
$$

10.3 The Radial Wavefunctions in a Central Field*

Before we turn to the problem of hydrogen, let us consider the general case of an electron in a centrally symmetrical potential field $V(r)$, of which we assume only that it vanishes at infinity. The starting point is then (10.8), which we repeat here:

$$
\left[-\frac{\hbar^2}{2m_0}\underline{\frac{1}{r^2}\frac{d}{dr}\left(r^2\frac{d}{dr}\right)} + \frac{\hbar^2 l(l+1)}{2m_0 r^2} + V(r)\right]R(r) = ER(r)\,. \tag{10.51}
$$

Let us rewrite the underlined differential expression:

$$
\frac{d^2}{dr^2} + \frac{2}{r}\frac{d}{dr} \tag{10.52}
$$

and multiply the equation by $-2m_0/\hbar^2$ to obtain

$$
\frac{d^2R}{dr^2} + \frac{2}{r}\frac{dR}{dr} + \left[A - \tilde{V}(r) - \frac{l(l+1)}{r^2}\right]R = 0\,, \tag{10.53}
$$

where we have used the abbreviations

$$
\begin{aligned}
&A = \frac{2m_0}{\hbar^2}\mathrm{E} = \begin{cases} -\kappa^2 & \text{for} \quad E<0\\ k^2 & \text{for} \quad E>0\,, \end{cases}\\
&\tilde{V} = \frac{2m_0}{\hbar^2}\mathrm{V}(r)\,.
\end{aligned} \tag{10.54}
$$

We shall now see what happens to the solution $R(r)$ if we allow r to become very large. We begin with the function

$$R=\frac{u(r)}{r}\,. \tag{10.55}$$

If we substitute this in (10.53), we obtain

$$\frac{d^2}{dr^2}u(r)+\left[A-\tilde{V}(r)-\frac{l(l+1)}{r^2}\right]u(r)=0\,. \tag{10.56}$$

Since both $\tilde{V}$ and $1/r^2$ go to zero at infinity, we neglect these two parameters. The remaining equation has two types of solution:

1) $E>0\,,\quad$ i.e. $\quad A>0\,.$

In this case the general solution of (10.56) is

$$u=c_1\,\mathrm{e}^{\mathrm{i}kr}+c_2\,\mathrm{e}^{-\mathrm{i}kr} \tag{10.57}$$

and thus the original solution $R(r)$, according to (10.55), is

$$R=\frac{1}{r}(c_1\,\mathrm{e}^{\mathrm{i}kr}+c_2\,\mathrm{e}^{-\mathrm{i}kr})\,. \tag{10.58}$$

To illustrate the meaning of this solution, let us imagine it to be multiplied by the time-dependence factor $\exp(-\mathrm{i}\omega t)$ which would occur in the solution of the time-dependent Schrödinger equation. We see then that $r^{-1}\exp(\mathrm{i}kr)\exp(-\mathrm{i}\omega t)$ represents a spherical wave propagating outwards, while $r^{-1}\exp(-\mathrm{i}kr)\exp(-\mathrm{i}\omega t)$ is a spherical wave coming inwards. These spherical waves which come in from infinity and travel outwards again correspond to the hyperbolic orbits in the classical Kepler problem.

Now let us investigate the case

2) $E<0\,,\quad$ i.e. $\quad A<0\,.$

Then the solution of (10.56) is

$$u=c_1\,\mathrm{e}^{\kappa r}+c_2\,\mathrm{e}^{-\kappa r}\,. \tag{10.59}$$

Since the solution naturally must not become infinite at large distances, which the exponential function $\exp(\kappa r)$ would do, we must require that the coefficient $c_1=0$. We then obtain according to (10.55) a solution of the type

$$R=\frac{c}{r}\mathrm{e}^{-\kappa r}\,. \tag{10.60}$$

Since the absolute square of R represents the probability of finding the particle, and this quantity decreases exponentially for increasing r, we see that in (10.60), the electron is localised within a certain area in space. This is the quantum mechanical analogy to the closed elliptical orbits of classical physics (see Sect. 8.9).

10.4 The Radial Wavefunctions of Hydrogen

We will now attack the problem of solving (10.51) for the case of a Coulomb potential

$$V = -\frac{Ze^2}{4\pi\varepsilon_0 r}. \tag{10.61}$$

For this purpose it is convenient to use dimensionless quantities. We thus introduce a new distance variable

$$\varrho = 2\kappa r \tag{10.62}$$

with κ defined by (10.54). Corresponding to this, we introduce a new function $\tilde{R}(\varrho)$, which is related to $R(r)$ by $R(r) = \tilde{R}(2\kappa r) \equiv \tilde{R}(\varrho)$. We then divide (10.53) by $(4\kappa^2)$ and obtain:

$$\tilde{R}'' + \frac{2}{\varrho}\tilde{R}' + \left(-\frac{1}{4} + \frac{B}{\kappa\varrho} - \frac{l(l+1)}{\varrho^2}\right)\tilde{R} = 0 \tag{10.63}$$

in which we have used the abbreviation

$$B = \frac{m_0 Ze^2}{\hbar^2 4\pi\varepsilon_0}. \tag{10.64}$$

The primes on $\tilde{R}$ denote derivatives with respect to ϱ.

Having seen before that the wavefunction decays exponentially at large distances, it would appear reasonable for us to use an exponential function as trial solution. It will later prove useful to adopt the form

$$\tilde{R} = \mathrm{e}^{-\varrho/2} v(\varrho). \tag{10.65}$$

If we insert this trial solution in (10.63) and carry out the differentiation of the exponential function and of the function $v(\varrho)$, we obtain

$$v'' + \left(\frac{2}{\varrho} - 1\right)v' + \left[\left(\frac{B}{\kappa} - 1\right)\frac{1}{\varrho} - \frac{l(l+1)}{\varrho^2}\right]v = 0. \tag{10.66}$$

It is shown in the study of differential equations that (10.66) is satisfied by a trial solution in the form of a power series in ϱ, which we will express in the convenient form

$$v = \varrho^{\mu} \sum_{\nu=0}^{\infty} a_\nu \varrho^{\nu} \equiv \sum_{\nu=0}^{\infty} a_\nu \varrho^{(\nu+\mu)}, \tag{10.67}$$

in which it is assumed that $a_0 \neq 0$.

In this expression, the exponent μ and the coefficients a_ν are still to be determined. We insert the trial solution (10.67) in (10.66), re-order according to powers of ϱ, and require that the coefficient of each power of ϱ should be independently equal to zero. The lowest power which occurs is $\varrho^{\mu-2}$. The corresponding coefficient is

$$a_0\mu(\mu-1)+a_0 2\mu-a_0 l(l+1)=0\,. \qquad (10.68)$$

Since we have assumed that a_0 is nonvanishing, the common factor of a_0 must be zero, i.e.

$$\mu(\mu+1)=l(l+1)\,. \qquad (10.69)$$

Of the two possible solutions $\mu=l$ and $\mu=-l-1$, only the first is usable for us, since the other solution leads to a function v which diverges at the origin (10.67), causing the trial solution for $\tilde{R}$ also to diverge (10.65); however, we require the solutions of the Schrödinger equation to be well-behaved in the entire range.

We now investigate the coefficients of the higher powers of $\varrho(\nu\neq 0)$. For $\varrho^{\nu+l-2}$ we find

$$a_\nu(\nu+l)(\nu+l-1)+a_\nu 2(\nu+l)-a_\nu l(l+1)-a_{\nu-1}(\nu+l-1)+a_{\nu-1}(n-1)=0\,, \qquad (10.70)$$

where the abbreviation

$$n=B/\kappa \qquad (10.71)$$

has been employed. Relation (10.70) connects the coefficient a_ν with the preceding coefficient $a_{\nu-1}$. We thus obtain from (10.70), after an elementary rearrangement, the recursion formula

$$a_\nu=\frac{\nu+l-n}{\nu(\nu+2l+1)}\,a_{\nu-1}\,. \qquad (10.72)$$

This recursion relation permits two quite different types of solutions, depending on whether the chain of the a's is terminated or not. If it is not, the sum in (10.67) contains infinitely many terms, and it may be shown mathematically that then $v(\varrho)$ is practically equal to an exponential function which diverges at infinity. We must therefore restrict ourselves to the case when the chain of the a's *does* terminate; this is in fact possible if n is an integer. We then obtain a cutoff for $\nu=\nu_0$, where

$$\nu_0=n-l\,. \qquad (10.73)$$

Since we must have $\nu_0\geqq 1$, we obtain from this a condition for l:

$$l\leqq n-1\,. \qquad (10.74)$$

In the following, we shall refer to n as the *principal quantum number* and to l as the *angular momentum quantum number*. According to (10.74), the angular momentum quantum number cannot be larger than $n-1$.

We now calculate the energy value, which, as we will see immediately, is already determined in principle by our assumptions. For this purpose, we express E in terms of κ (10.54); κ is, however, already determined by (10.71).

In (10.71), as we have just seen, n is an integer, $n=1,2,\ldots$. Furthermore, B is defined in (10.64). We thus obtain for E:

$$E = -\frac{m_0 Z^2 e^4}{2\hbar^2(4\pi\varepsilon_0)^2}\cdot\frac{1}{n^2}\,. \tag{10.75}$$

If we think back through the whole derivation, we see that the energy values E came about through the requirement that the series (10.72) be terminated, or, in order to find the actual solution, that the wavefunction should vanish at infinity. n is allowed to take on integral values $1,2,3,\ldots$ in (10.75), so that we obtain the energy level diagram of Fig. 8.4. The same energy values have already been derived in Chap. 8, starting with the Bohr postulates.

For $E>0$, i.e. for non-bound states, the energies form a continuous distribution of values. We will not give the corresponding wavefunction here.

Since the series (10.67) has a cutoff, $v(\varrho)$ is a polynomial. If we recall the trial solution for $\tilde{R}(\varrho)$ and the abbreviation for ϱ, (10.62),

$$\tilde{R} = \mathrm{e}^{-\varrho/2}v(\varrho)\,, \tag{10.76}$$

we finally arrive at an expression for the original R of the form

$$R_{n,l}(r) = N_{n,l}\exp(-\kappa_n r)\,r^l\,\mathrm{L}_{n+1}^{2l+1}(2\kappa_n r)\,. \tag{10.77}$$

The various quantities have the following meanings:

$N_{n,l}$ is the normalisation factor, which is determined by the condition

$$\int_0^\infty R_{n,l}^2(r)\,r^2 dr = 1\,. \tag{10.78}$$

(The factor r^2 in the integrand results from the use of spherical polar coordinates.)

κ_n has the dimensions of an inverse radius and is given explicitly by (10.71, 64)

$$\kappa_n = \frac{1}{n}\cdot\frac{m_0 Z e^2}{\hbar^2 4\pi\varepsilon_0}\,. \tag{10.79}$$

L_{n+1}^{2l+1} is the mathematical symbol for the polynomial which occurs in (10.77), whose coefficients are determined by the recursion formula (10.72). It may be shown that L_{n+1}^{2l+1} can be obtained from the so-called Laguerre Polynomials L_{n+l} by $(2l+1)$-fold differentiation:

$$\mathrm{L}_{n+1}^{2l+1} = d^{2l+1}\mathrm{L}_{n+l}/d\varrho^{2l+1}\,. \tag{10.80}$$

The Laguerre Polynomials, in turn, are obtained from the relation

$$\mathrm{L}_{n+l}(\varrho) = \mathrm{e}^{\varrho}d^{n+l}[\exp(-\varrho)\,\varrho^{n+l}]/d\varrho^{n+l}\,. \tag{10.81}$$

A series of examples of (10.77) is given in Fig. 10.3 for various values of the quantum numbers. In Fig. 10.3 a, the radial wavefunction is plotted as a function of the dimensionless radius variable ϱ (10.62). The parentheses (1,0), (2,0) etc. contain the

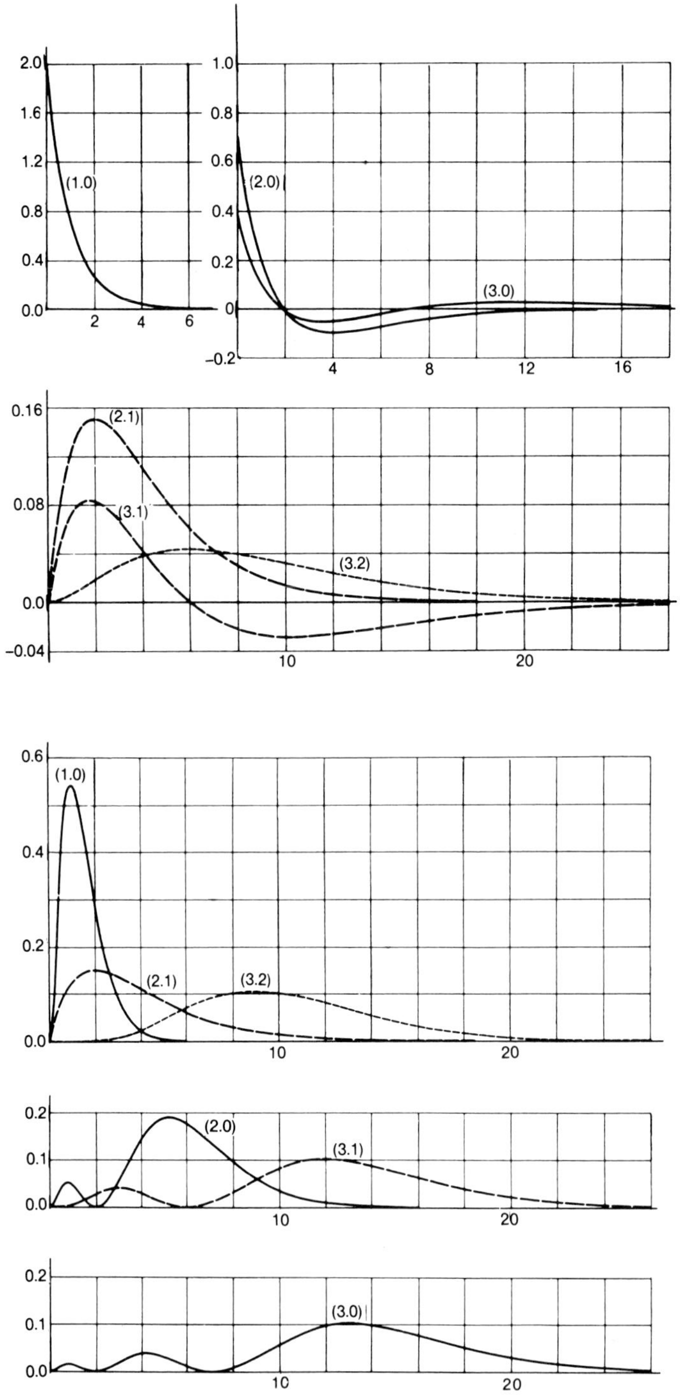

a

b

Fig. 10.3. a) The radial wavefunctions $\tilde{R}(\varrho) = R(r)$ of the hydrogen atom (10.77) are plotted vs the dimensionless coordinate ϱ. The indices on the curves, (1, 0), (2, 0) etc. correspond to (n, l) where n is the principal quantum number and l the angular momentum quantum number. **b)** The corresponding probability densities in the radial coordinate, i.e. $4\pi\varrho^2\tilde{R}^2(\varrho)$ are plotted against the dimensionless coordinate ϱ

values of n and l (n, l). In Fig. 10.3b, $4\pi\varrho^2\tilde{R}^2(\varrho)$ is plotted for various values of n and l. $\tilde{R}^2(\varrho)d\varrho$ gives the probability of finding the particle in a particular direction in space in the interval $\varrho \ldots \varrho+d\varrho$. If, on the other hand, we wish to know the probability of finding the particle at the distance ϱ in the interval $\varrho \ldots \varrho+d\varrho$ *independently* of direction, we must integrate over a spherical shell. Since the volume of a spherical shell is just $4\pi\varrho^2 d\varrho$, we are led to the above quantity, $4\pi\varrho^2\tilde{R}^2(\varrho)$. The maxima of these curves is displaced to regions of greater distance with increasing quantum number n, so that here we see an indication, at least, of the classical orbits.

Let us summarise our results. The wavefunction of the hydrogen atom may be written in the form

$$\psi_{n,l,m}(r,\theta,\phi) = e^{im\phi}P_l^m(\cos\theta)R_{n,l}(r) . \tag{10.82}$$

Here n is the principal quantum number, l the angular momentum quantum number, and m the *magnetic* quantum number or *directional* quantum number. These quantum numbers can assume the following values:

$$\begin{aligned} &n = 1, 2, \ldots , \\ &0 \leqq l \leqq n-1 , \\ &-l \leqq m \leqq +l . \end{aligned} \tag{10.83}$$

Some examples for the density distribution of the electron (= probability density distribution $|\psi_{n,l,m}(r,\theta,\phi)|^2$) are represented in Fig. 10.4.

The density of points shown here was calculated by computer. It represents the probability density of the electron. Since the hydrogen functions are partially complex, combination of functions which belong to $+m$ and $-m$ yields real functions. These linear combinations are also solutions of the Schrödinger equation of the hydrogen problem. They still have the quantum numbers n and l, but they are no longer eigenfunctions for the z component of the angular momentum, so that this quantum number is lost. Figures 10.4a, b and e represent solutions with $l = 0$, which yield spherically symmetrical distributions. The sections c, d, f and g represent $l = 1$. Here one notices the dumbbell shaped distribution along one axis. There is a further linear combination possible in each case, but not shown here, in which the long axis of the dumbbell would lie along the third coordinate. Sections h and i represent $l = 2$, with $m = 0$ in h, and *i* represents a linear combination of $m = \pm 1$. Figure 10.4 does not show the wavefunctions with $l = 2$, $m = \pm 2$.

The energy corresponding to (10.82) is given by (10.75). It clearly depends only on the principal quantum number n. Since each energy level E_n (with the exception of $n = 1$) contains *several* different wavefunctions, these levels are called degenerate. This degeneracy is typical of the hydrogen atom problem with the Coulomb potential.

The degeneracy with respect to l is lifted, i.e. the energy levels become dependent upon l, if the potential no longer has the form $-\text{const}/r$, but is still spherically symmetric (Sect. 11.2). We will be led to consider effective departures from the Coulomb potential for all atoms with more than one electron (see below). The l degeneracy is also lifted even for hydrogen if we treat the problem relativistically, which is necessary for the exact treatment of the spectra (Sect. 12.11). The m degeneracy can only be lifted by superimposing a *non*-spherically symmetric potential on the central potential of the atom, i.e. an electric or a magnetic field (Chaps. 13 and 14).

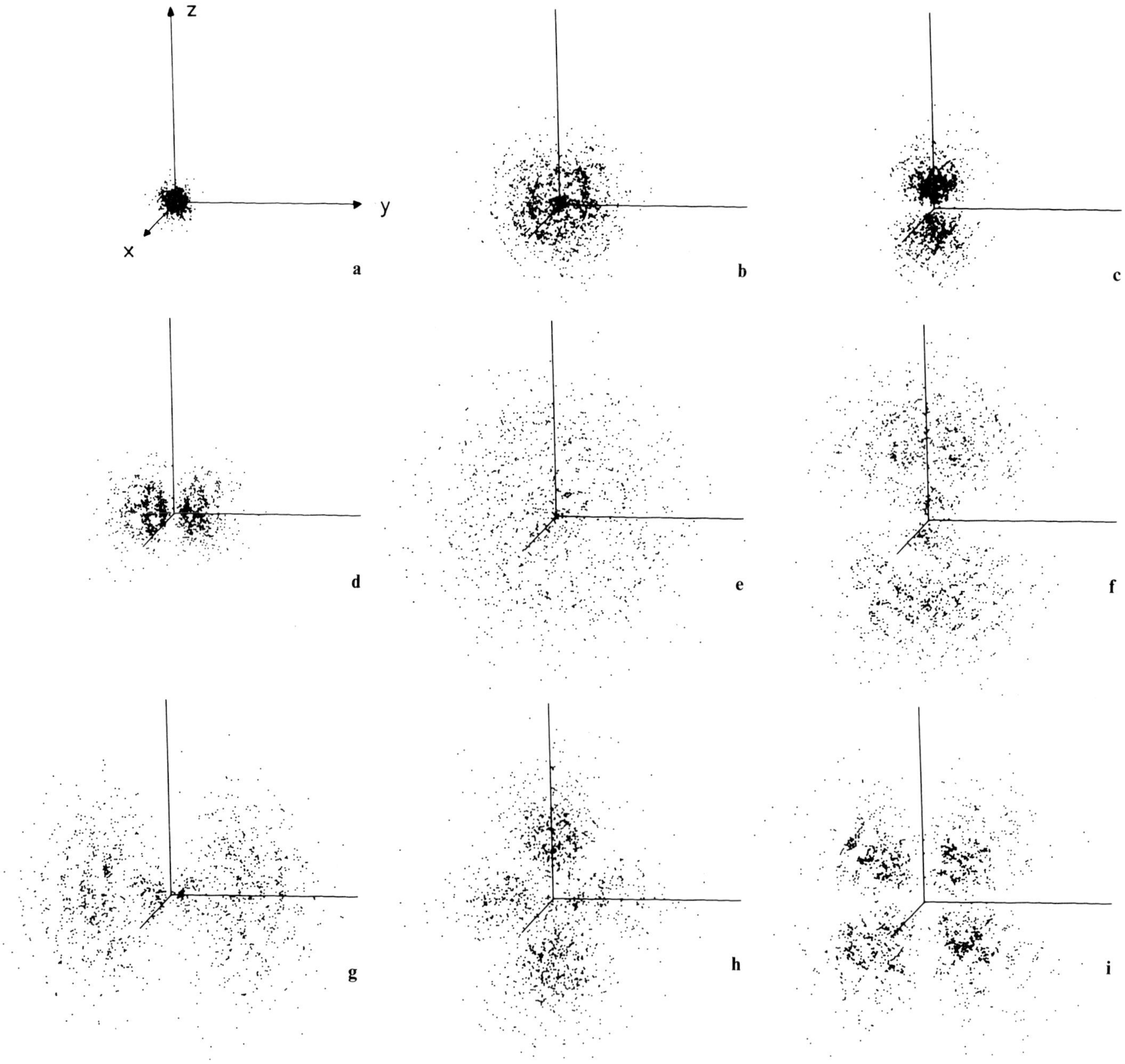

Fig. 10.4a – i. The density distribution (= localisation probability density $|\psi|^2$) of the electron in the H atom represented by the density of dots (according to *H. Ohno*). The wavefunctions represented are:

a) (10.82), $n = 1$, $l = 0$, $m = 0$

b) (10.82), $n = 2$, $l = 0$, $m = 0$

c) (10.82), $n = 2$, $l = 1$, $m = 0$

d) linear combination $(\psi_{2,1,1} + \psi_{2,1,-1})\frac{i}{\sqrt{2}}$

e) (10.82), $n = 3$, $l = 0$, $m = 0$

f) (10.82), $n = 3$, $l = 1$, $m = 0$

g) linear combination $(\psi_{3,1,1} + \psi_{3,1,-1})\frac{i}{\sqrt{2}}$

h) (10.82), $n = 3$, $l = 2$, $m = 0$

i) linear combination $(\psi_{3,2,1} + \psi_{3,2,-1})\frac{i}{\sqrt{2}}$

The linear combinations given are also solutions of the Schrödinger equation of the hydrogen problem with the energies E_n, but they are not eigenfunctions of l_z

Problems

10.1 Calculate the expectation values of the kinetic and potential energies

a) for the ground state of the hydrogen atom, $n=1$, $l=m=0$,
b) for the wavefunctions $n=2$, $l=0$, $m=0$ and $n=2$, $l=1$, $m=\pm 1, 0$.

Hint: Use spherical polar coordinates, so that for the volume element dV, $dV = \sin\theta d\theta d\phi r^2 dr$.

10.2 As will be discussed in Sect. 15.2.3, in quantum mechanics dipole matrix elements between two states with the wavefunctions ψ_1 and ψ_2 are defined by

$$\boldsymbol{D} = \int \psi_1^* e\boldsymbol{r}\, \psi_2 dx\, dy\, dz\,.$$

Why is $\boldsymbol{D}$ a vector? Calculate the components of $\boldsymbol{D}$ when

a) $\psi_1 = \psi_2 = \psi_{1,0,0}$,
b) $\psi_1 = \psi_{1,0,0}$; $\psi_2 = \psi_{2,0,0}$
or $\psi_2 = \psi_{2,1,0}$
or $\psi_2 = \psi_{2,1,\pm 1}$.

Here $\psi_{n,l,m}$ is the wavefunction of the hydrogen atom with the quantum numbers n, l and m.

10.3 Calculate κ (10.79) and E_n (10.75) numerically for the first three values of n for the hydrogen atom.

10.4 Using the ground state of hydrogen as an example, we discuss here the variation principle of quantum mechanics. This says, in general, that the wavefunction ψ of the ground state of a Schrödinger equation $\mathscr{H}\psi = E\psi$ can be found (aside from solving the equation directly) by minimising the expectation value of the energy by a suitable choice of ψ: $\bar{E} = \int \psi^* \mathscr{H} \psi\, dx\, dy\, dz = \min$. ψ must simultaneously satisfy the additional condition that $\int \psi^* \psi\, dx\, dy\, dz = 1$.
This principle can also be used to estimate wavefunctions, and especially energies.

Problem: a) Take the trial solution $\psi = N\exp(-r^2/r_0^2)$. Calculate the normalisation factor N. Then calculate $\bar{E}$ as a function of r_0, and minimise $\bar{E}$ by a suitable choice of r_0. Then compare $\bar{E}_{\min}$ with the exact value of the energy.
b) Repeat the procedure for $\psi = N\exp(-r/r_0)$.

10.5 Solve the one-dimensional Schrödinger equation

$$-\frac{\hbar^2}{2m_0}\frac{d^2\psi(x)}{dx^2} + \left(-\frac{c_1}{x} + \frac{c_2}{x^2}\right)\psi(x) = E\psi$$

for $x \ge 0$, $c_1 > 0$, $c_2 > 0$, $E < 0$.

Hint: First examine the limiting case $x \to \infty$ and determine the asymptotic form of $\psi(x)$. Then try the solution

$$\psi(x) = x^{\sigma} e^{-x\sqrt{\varepsilon}} g(x) ,$$

where $\sigma = \frac{1}{2} + \sqrt{\frac{1}{4} + \tilde{c}_2}$ with $\tilde{c}_2 = 2 m_0 c_2/\hbar^2$ and $\varepsilon = -2 m_0 E/\hbar^2$.

Calculate $g(x)$ with a power series which, however, must be terminated. Why?

11. Lifting of the Orbital Degeneracy in the Spectra of Alkali Atoms

11.1 Shell Structure

After the spectra of atoms with only one electron, the next simplest spectra are those of alkali atoms.

The alkali atoms have a weakly bound outer electron, the so-called valence electron, and all other $(Z-1)$ electrons are in closed shells. What the atomic-physical meaning of a closed shell is, we will discover later. At present we shall only say that even when several electrons are bound to a nucleus, their individual electron states can be characterised by the three quantum numbers n, l and m, but the corresponding energies are strongly modified, with respect to the one-electron problem, by the interactions of the electrons with each other. The Pauli principle (Sect. 17.2) says that a state characterised by specific values of n, l and m can be occupied by at most two electrons. In the ground state of an atom, the states with the lowest energies are naturally the occupied ones. A particular state of occupation of the energy levels or terms of an atom by electrons is called the *electron configuration* of the atom in that state – in this case, the ground state. A closed shell or noble gas configuration occurs whenever the next electron to be added would occupy the s state of the next higher principal quantum number n. It is not necessary that all the states belonging to lower principal quantum numbers be filled; more about this will be said in Chap. 20. The electrons in the closed shells are closer, as a rule, to the nucleus than the valence electron, and are more strongly bound. The total angular momentum of a closed shell vanishes. The closed shell is spherically symmetrical and is especially stable.

How is this known? Firstly, from chemistry: all alkali metal atoms have a valence of one. Each alkali metal is preceded in the periodic system by a noble gas, each of which has one electron fewer and has a particularly stable electron configuration – a closed shell. These gases are chemically inactive. Compared to those of their neighbours in the periodic table, their ionisation potentials are large. The neighbours with one more nuclear charge unit, the alkali metals, have very low ionisation potentials. For example, the ionisation energy of the noble gas helium is 24.46 eV. The next element in the periodic table, the alkali metal lithium, has an ionisation energy of only 5.40 eV. The ionisation energies of the heavier alkali metal atoms are even lower, as can be seen from Table 11.1. The table also shows that the ionisation energy for the removal of the second electron from an alkali metal atom is very large, because the electron configuration of the singly charged positive ion is a closed shell. In Fig. 11.1, the simplified term diagrams of the alkali metals are compared to that of the H atom.

The comparison shows that in the alkali atoms, the l degeneracy is lifted. States with the same principal quantum number n and different orbital angular momentum quantum numbers l have different energies. Relative to the terms of the hydrogen atom, those of the alkalis lie lower – this means a larger (negative) binding energy – and this shift increases, the smaller l is. For larger principal quantum numbers, i.e.

Table 11.1. Work of ionisation for the elements with $Z = 1$ to $Z = 20$. Values are given for the neutral atom, and for singly, doubly and triply charged ions. The ionisation energy is always especially large for a noble gas configuration (closed shell). It is especially low if there is only one electron more than a noble gas configuration which is indicated by bold-face numbers

Element	Work of ionisation [eV] for the transition from the –			
	neutral atom to singly charged	singly to doubly charged	doubly to triply charged	triply to quadruply charged
$_1$H	13.59	–	–	–
$_2$He	**24.5**	54.1	–	–
$_3$Li	5.4	**75**	122	–
$_4$Be	9.3	18.2	**154**	217
$_5$B	8.3	25.1	38	**259**
$_6$C	11.3	24.5	48	64.5
$_7$N	14.6	29.6	47	77.4
$_8$O	13.6	35.2	55	77.4
$_9$F	17.4	34.9	62.7	87.3
$_{10}$Ne	**21.6**	41.0	63.9	96.4
$_{11}$Na	5.14	**47.3**	71.7	98.9
$_{12}$Mg	7.64	15.0	**80.2**	109.3
$_{13}$Al	5.97	18.8	28.5	**120**
$_{14}$Si	8.15	16.4	33.5	44.9
$_{15}$P	10.9	19.7	30.2	51.4
$_{16}$S	10.4	23.4	35.1	47.1
$_{17}$Cl	12.9	23.7	39.9	53.5
$_{18}$Ar	**15.8**	27.5	40.7	ca. 61
$_{19}$K	4.3	**31.7**	45.5	60.6
$_{20}$Ca	6.1	11.9	**51**	67

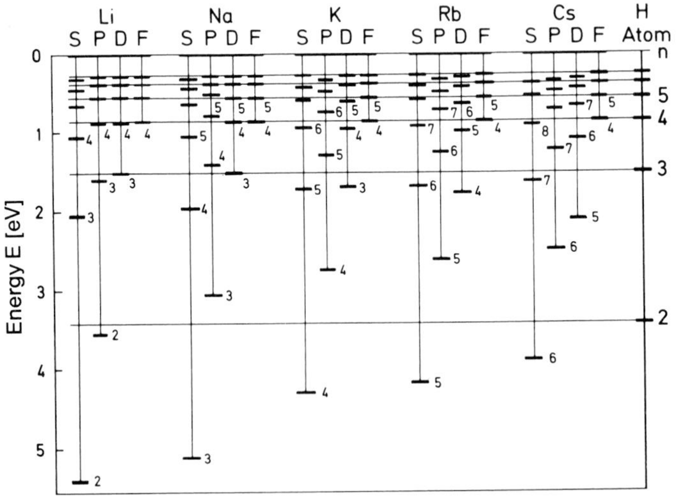

Fig. 11.1. Simplified term diagrams for the alkali metal atoms, showing the empirical positions of the most important energy terms. The principal quantum number n is indicated by numerals, the secondary quantum number l by the letters S, P, D, and F. For comparison, the levels of the H atom are given on the right

greater orbital radii, the terms are only slightly different from those of hydrogen. Here also, however, electrons with small l are more strongly bound and their terms lie lower in the term diagram. This effect becomes stronger with increasing Z. We would like to understand this effect, at least qualitatively.

11.2 Screening

In order to understand the term diagrams of the alkali atoms, we will use the following model (Fig. 11.2):

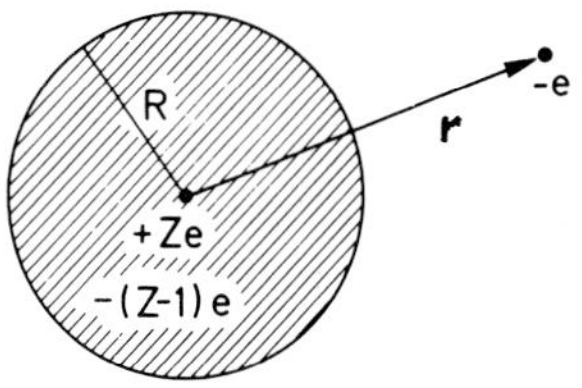

Fig. 11.2. Model of an alkali atom. The valence electron is screened from the nuclear charge $+eZ$ by the $(Z-1)$ inner electrons

A "valence electron" is located at a relatively large distance r from the nucleus. It moves in the electrostatic field of the nuclear charge $+eZ$, which is for the most part screened by the $(Z-1)$ inner electrons. We describe the screening effect of the inner electrons together with the nuclear potential by means of an effective potential $V_{eff}(r)$ for the valence electron. In this way we reduce the original many-body problem to a single-particle system, and we can treat the energy levels of an alkali atom as terms of a single-electron atom.

The shape of the effective potential $V_{eff}(r)$ is shown schematically in Fig. 11.3. If the valence electron moves at a great distance from the nucleus, its potential energy is $-e^2/(4\pi\varepsilon_0 r)$. The nuclear charge which attracts the valence electron is in this case

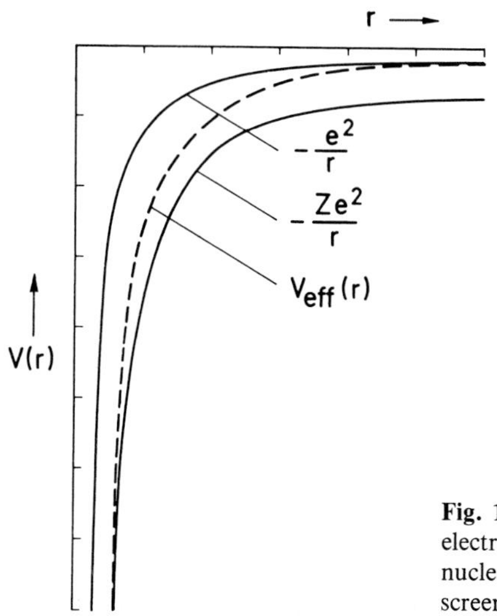

Fig. 11.3. Effective potential $V_{eff}(r)$ for an alkali atom. At small electron-nuclear distances, V_{eff} has the shape of the unscreened nuclear Coulomb potential; at large distances, the nuclear charge is screened to one unit of charge

compensated down to one unit of charge by the inner electrons. However, the nearer the valence electron approaches the nucleus, the more it experiences the unscreened nuclear potential. The potential energy approaches $V = -Ze^2/(4\pi\varepsilon_0 r)$. The effective potential $V_{\rm eff}(r)$ is no longer proportional to r^{-1}. This proportionality was, as we recall (Chap. 10), responsible for the l degeneracy.

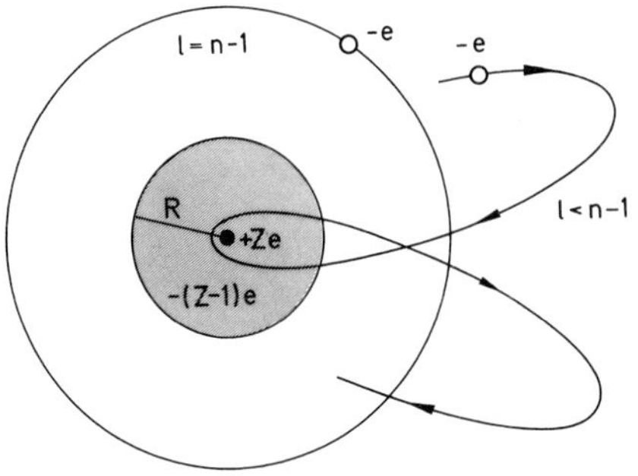

Fig. 11.4. Variation in screening for electrons with different radial probability densities. In the Sommerfeld model it is intuitively clear that electrons with "diving orbits" approach the nucleus closely and are affected at least partially by the unscreened nuclear potential. It has to be remembered, however, that in the modern quantum theory, the electrons are described as charge clouds rather than as orbiting particles

In the Sommerfeld picture, the so-called diving orbits demonstrate especially clearly that electrons with differing orbital angular momenta, i.e. differing orbital shapes, experience different degrees of screening. This is illustrated in Fig. 11.4. Quantum mechanically, this picture remains valid to a large extent. In Chap. 10 it was shown that the probability density of the electrons in the neighbourhood of the nucleus decreases in the order $l = 0, 1, 2, \ldots$. The s electrons are thus most strongly affected by the unscreened field of the nucleus. For a given principal quantum number n, the energy terms of the s electrons are therefore shifted the most strongly to negative values relative to the H atom (Fig. 11.1).

11.3 The Term Diagram

For the alkali atoms, we thus obtain a term diagram like that shown in Fig. 11.5 for lithium. This term diagram permits a classification of the spectral lines to series, if one employs the additional selection rule for optical transitions $\Delta l = \pm 1$, i.e. in an optical transition, the quantum number l must change by 1. Such selection rules will be treated in detail in Chap. 16.

The series in the emission spectra of the neutral alkali atoms can be described by series formulae similar to the Balmer series formula. For the energy terms $E_{n,l}$ which are determined by the quantum numbers n and l, an effective principal quantum number $n_{\rm eff}$ may be defined, so that, e.g. for sodium we have

$$E_{n,l} = -R_{\rm Na} hc \frac{1}{n_{\rm eff}^2} = -R_{\rm Na} hc \left\{ \frac{1}{[n - \Delta(n,l)]^2} \right\} .$$

Here the multiplication by the factor hc is necessary if the Rydberg number $R_{\rm Na}$ is measured in cm^{-1}, as is customary. Here $n_{\rm eff} = n - \Delta(n,l)$ is a principal quantum

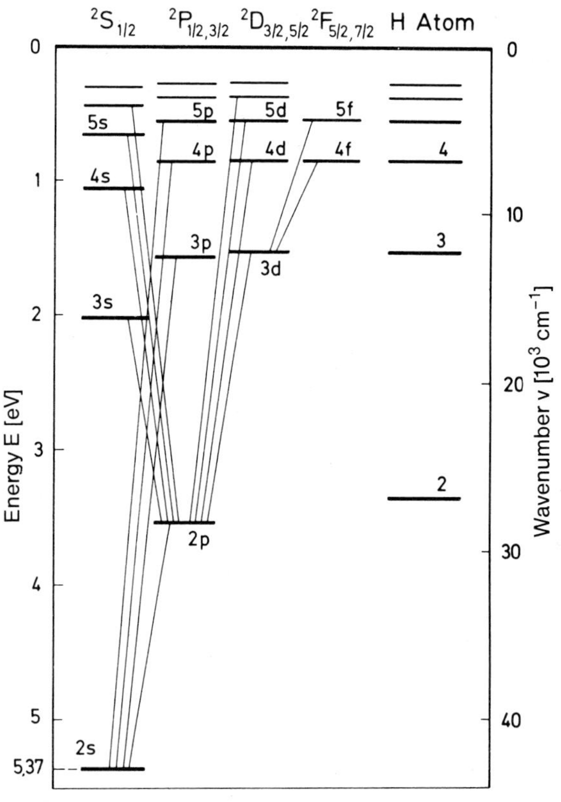

Fig. 11.5. Term diagram of the lithium atom with the most important transitions. This is called a Grotrian diagram. The term symbols given along the top of the figure are explained in Chaps. 12 and 17

number, in general not an integer, and $\Delta(n,l) = n - n_{\text{eff}}$ is the so-called quantum defect associated with the quantum numbers n and l. The empirically determined numerical values for the quantum defects (see Table 11.2) are largest for s electrons, decrease with increasing orbital angular momentum quantum l, and are largely independent of the principal quantum number n. They increase down the column of alkali atoms from lithium to cesium, or with increasing nuclear charge number Z. These quantum defects are empirical expressions of the different degrees of screening of the s, p, d, etc. electrons by the electrons of the inner shells.

For the sodium atom, the decomposition of the total spectrum into series is represented in Fig. 11.6. Figure 11.7 shows the transitions in the form of a Grotrian

Table 11.2. Quantum defects $\Delta(n,l)$ for the spectra of the Na atom [from F. Richtmyer, E. Kennard, J. Cooper: *Introduction to Modern Physics*, 6th ed. (McGraw-Hill, New York 1969)]. These are empirical values

	Term	$n=3$	4	5	6	7	8
$l=0$	s	1.373	1.357	1.352	1.349	1.348	1.351
1	p	0.883	0.867	0.862	0.859	0.858	0.857
2	d	0.010	0.011	0.013	0.011	0.009	0.013
3	f	–	0.000	−0.001	−0.008	−0.012	−0.015

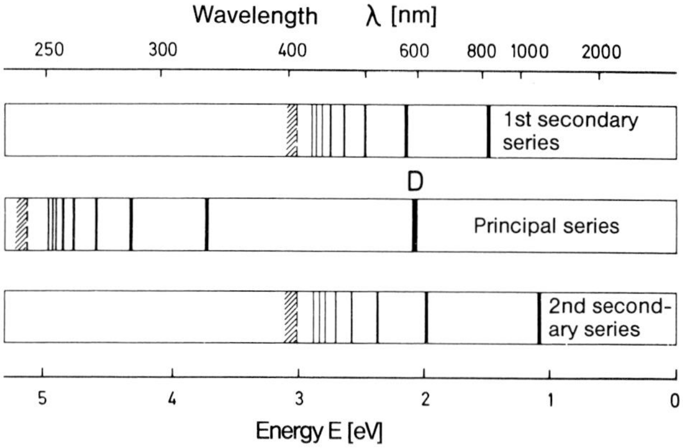

Fig. 11.6. The three shortest-wave spectral series of the sodium atom. The series limits are indicated by shading. The emission spectrum is a composite of these series. In absorption spectra, normally only the principal series is observed, because in the ground state of the Na atom the highest occupied term is the $3s$ term. The yellow colour of sodium lamps is due to the longest-wave resonance line of the main series, the transition $3s \leftrightarrow 3p$. This is the sodium D line, a terminology which has been retained for historical reasons

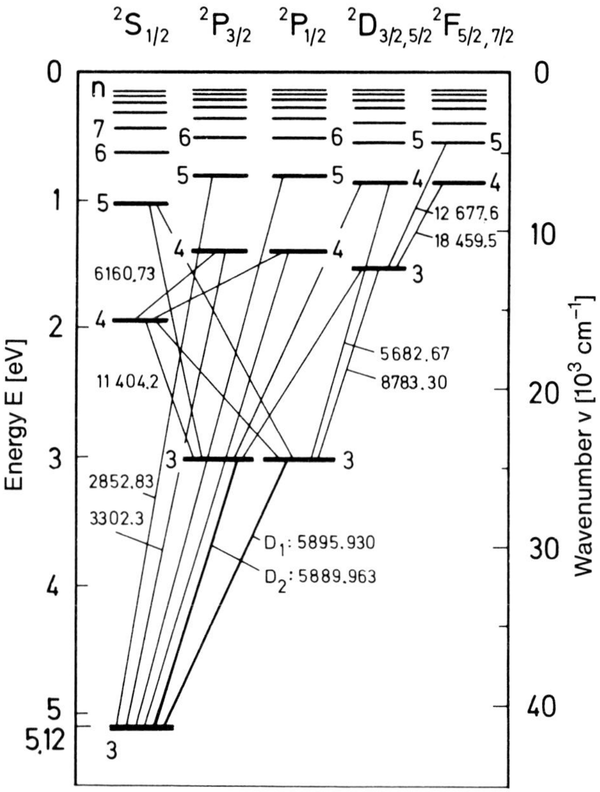

Fig. 11.7. Term scheme (Grotrian diagram) of the sodium atom. Some of the shortest-wave transitions from the principal series, the two secondary series and the Bergmann series have been included. The numbers in the diagram indicate the wavelength of the transition in Ångstrom units. The term symbols indicated on the upper edge of the figure also represent the quantum numbers for the multiplicity and the total angular momentum. These are explained in Chaps. 12 and 17

diagram. The most important series are the *principal* series, with transitions from p to s electron terms:

$$\bar{\nu}_p = R_{\mathrm{Na}}\left[\frac{1}{[n_0 - \Delta(n_0,0)]^2} - \frac{1}{[n - \Delta(n,1)]^2}\right], \qquad n \geqq n_0,\ n_0 = 3\,,$$

the *sharp* or second secondary series with transitions from s to p electron terms:

$$\bar{\nu}_s = R_{\mathrm{Na}}\left[\frac{1}{[n_0 - \Delta(n_0,1)]^2} - \frac{1}{[n - \Delta(n,0)]^2}\right], \qquad n \geqq n_0 + 1\,,$$

the *diffuse* or first secondary series with transitions from d to p electron terms:

$$\bar{\nu}_d = R_{\mathrm{Na}}\left[\frac{1}{[n_0 - \Delta(n_0,1)]^2} - \frac{1}{[n - \Delta(n,2)]^2}\right], \qquad n \geqq n_0\,,$$

and the Bergmann (*fundamental*) series with transitions from f to d electron terms:

$$\bar{\nu}_f = R_{\text{Na}}\left[\frac{1}{[n_0-\Delta(n_0,2)]^2}-\frac{1}{[n-\Delta(n,3)]^2}\right], \qquad n \geqq n_0+1 \, .$$

R_{Na} is again the Rydberg number of the sodium atom and n_0 is the integral principal quantum number of the lowest state. This is 2 for Li, 3 for Na, 4 for K, 5 for Rb and 6 for Cs. We are jumping slightly ahead in saying that the valence electron of the alkali atoms begins a new shell in each element. The principal quantum number of the ground state therefore increases by one in each successive alkali element of the periodic system.

The names for the series and the system of indicating the electrons with orbital angular momentum 0, 1, 2, 3, 4, ... as s, p, d, f, g... are historic. p is for principal, s for sharp, d for diffuse and f for fundamental.

Under normal conditions, only the principal series is observed by absorption spectroscopy, because unless the temperature is extremely high, only the ground state of the atoms is sufficiently populated for transitions into higher states to be observed. The lines of the principal series are thus resonance lines. The best known is the D line of sodium, which is the transition $3s-3p$. The sum of the s terms can also be designated S, and of the p terms, P, so that the sodium series can be written:

Principal series $3S \leftrightarrow nP$

Secondary series $3P \leftrightarrow nS$

$3P \leftrightarrow nD$ with $n \geqq 3$.

Capital letters are used for terms which apply to several electrons in an atom, and lower case letters for the terms for individual electrons. In the alkali atoms, which have only one valence electron, the two notations are equivalent.

The screening effect of the inner electrons can be quantitatively calculated, if one knows their charge distribution with sufficient accuracy. Qualitatively, we wish to demonstrate the effect of the nuclear charge on a single $3d$ or $4s$ electron in the atoms H ($Z = 1$) and K ($Z = 19$).

In the H atom, the charge cloud of a $3d$ electron is, on the average, closer to the nucleus than that of a $4s$ electron (Fig. 11.8). Therefore, the $3d$ electron is more strongly bound to the H atom. It is different, however, in the K atom. The configuration of the atomic core, i.e. the noble gas configuration of Ar, consists of two s electrons with $n = 1$ (symbol $1s^2$), two s electrons with $n = 2$ (symbol $2s^2$), six p electrons

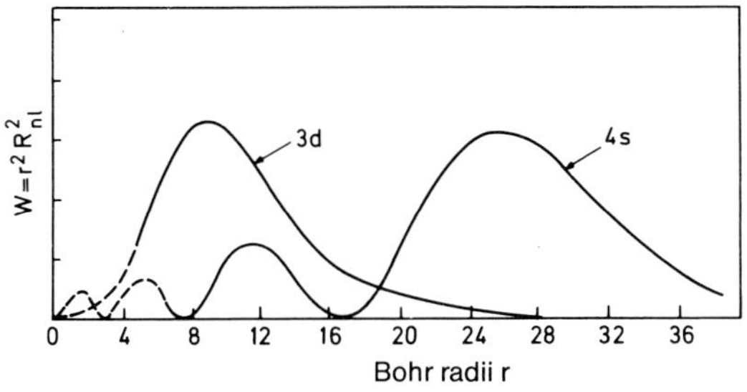

Fig. 11.8. Radial probability densities for a $4s$ and a $3d$ electron in the H atom. The $4s$ electron is, on the average, farther from the nucleus, but the probability of finding it near the nucleus is greater than the probability of finding the $3d$ electron there

with $n = 2$ (symbol $2p^6$), two s electrons with $n = 3$ (symbol $3s^2$) and six p electrons with $n = 3$ (symbol $3p^6$):

$$1s^2\,2s^2\,2p^6\,3s^2\,3p^6 - \text{ or [Ar] for argon .}$$

Now the question is, does K, with one more electron, have the configuration [Ar]$4s^1$, or does [Ar] $3d^1$ have a greater binding energy? Is the 19th electron added as a $4s$ or a $3d$ electron?

From our consideration of the H atom, we would predict that the [Ar] $3d^1$ configuration is more stable. However, it must now be determined what the effects of shielding are on the $3d$ and $4s$ electrons. Because the $4s$ electron has a higher probability of being very close to the nucleus, and thus unscreened, it turns out that the $4s$ energy level is energetically somewhat lower than the $3d$. The 20th electron is also an s electron; see Table 3.1. The element following potassium in the periodic table, calcium, has the configuration [Ar] $4s^2$. It thus becomes clear how decisively the screening affects the binding energies of the outer electrons, in a manner dependent on the orbital angular momentum quantum number l.

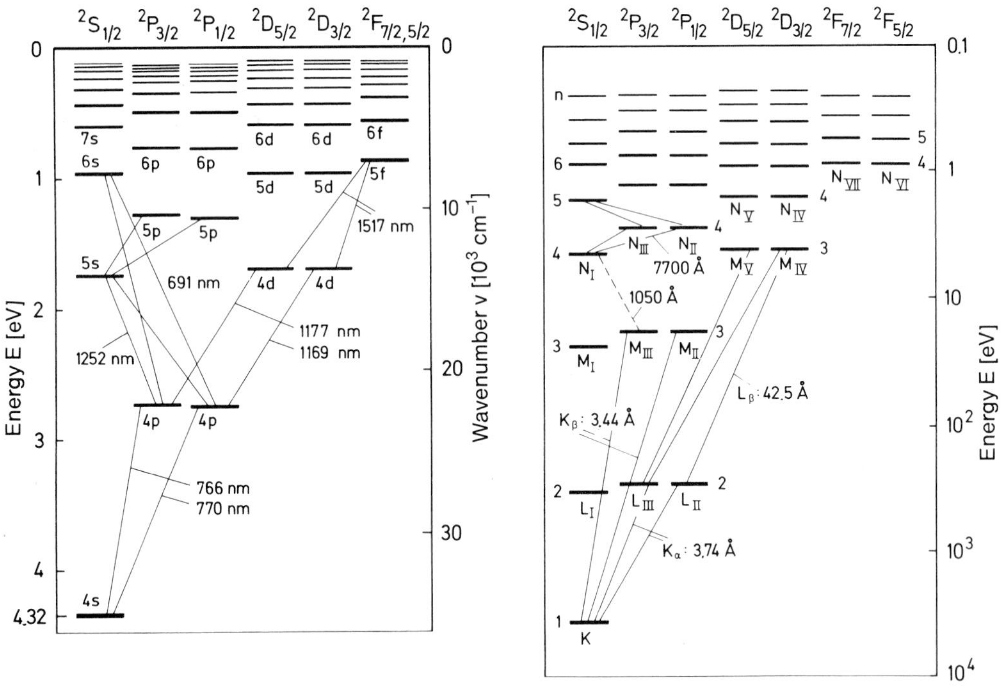

Fig. 11.9. *Left.* Grotrian diagram for the neutral potassium atom in the visible and infrared regions. The wavelengths for a few transitions (in nm) are indicated. The term symbols on the upper edge of the diagram are explained in Sect. 12.8 and Chap. 17. *Right.* Term scheme for the potassium atom in the infrared, visible, ultraviolet and x-ray regions. The term symbols used in this diagram are explained in Chaps. 12 and 17. (One should observe that the energy ranges in the two halves of the figure are different.) The x-ray spectrum also includes terms with lower principal quantum numbers than the visible spectrum. Terms with $n = 1, 2, 3\ldots$ are referred to in the x-ray region as the $K, L, M\ldots$ shells, see Chap. 18

11.4 Inner Shells

So far we have treated only the optical spectra of the alkali atoms. The valence electron could only have the principal quantum number $n \geqq 2$ for Li, $\geqq 3$ for Na, $\geqq 4$ for K, etc. The states with lower principal quantum numbers were completely occupied. Transitions involving these inner electrons were not discussed. However, they are also possible. Since the inner electrons are more strongly bound, such transitions take place at higher energies. We will introduce such transitions later, in the discussion of x-ray spectra (Chap. 18).

Figure 11.9 shows, in addition to the optical term scheme for the valence electron of the K atom, a complete term scheme. This includes the transitions in the x-ray region of the spectrum, in which an electron is removed from a closed inner shell and replaced by an electron from further out.

Problems

11.1 The energy levels of the valence electrons of an alkali atom are given, to a good approximation, by the expression

$$E_n = -Rhc \cdot 1/[n - \Delta(n,l)]^2 .$$

Here $\Delta(n,l)$ is the quantum defect (which depends on the values of n and l of the valence electron in question). For lithium and sodium, $\Delta(n,l)$ have been measured:

	s	p	d
Li ($Z = 3$)	0.40	0.04	0.00
Na ($Z = 11$)	1.37	0.88	0.01

Calculate the energy of the ground state and the first two excited states of the valence electron in lithium and sodium.

11.2 The ionisation energy of the Li atom is 5.3913 eV, and the resonance line ($2s \leftrightarrow 2p$) is observed at 6710 Å. Lithium vapour is selectively excited so that only the $3p$ level is occupied. Which spectral lines are emitted by this vapour, and what are their wavelengths?

Hint: Start from the fact that the quantum defect is independent of n, the principal quantum number.

11.3 Explain the symbols for the $3\,^2D \rightarrow 3\,^2P$ transition in sodium. How many lines can be expected in the spectrum?

12. Orbital and Spin Magnetism. Fine Structure

12.1 Introduction and Overview

We have not yet discussed the magnetic properties of atoms. It turns out that the study of these properties yields a deeper insight into the shell structure of atoms.

The impetus to study the magnetic properties was given by a few fundamental experiments, which we shall discuss in this chapter. The most important are

- Measurements of the macroscopic magnetisation and of the gyromagnetic properties of solids, known as the Einstein-de Haas effect.
- Measurements of directional quantisation and of the magnetic moments of atoms in atomic beams, made by *Stern* and *Gerlach*.
- Observation of the so-called fine structure in the optical spectra of atoms.

We shall begin with the third point. Many of the lines in the spectra of alkali atoms are double, and are called doublets. They occur because all the energy terms $E_{n,l}$ of atoms with single valence electrons, except for the s terms (energy levels with no orbital angular momentum), are split into two terms. This splitting cannot be understood in terms of the theory discussed so far. It is fundamentally different from the lifting of orbital degeneracy discussed in the last chapter. If the orbital degeneracy has already been lifted, there must be a new effect involved, one which has not yet been taken into account. Let us take as an example the D line in the spectrum of the sodium atom, i.e. the transition $3P \leftrightarrow 3S$ (Fig. 11.7 and 12.1). With sufficient spectral resolution, one can see two lines: $D_1 = 589.59$ nm $\triangleq 16956$ cm^{-1} and $D_2 = 588.96$ nm $\triangleq 16973$ cm^{-1}. In the following we shall often use this pair of lines as an example for explanation and experimental demonstration of spectroscopic results. Like the Balmer series of the H atom, the sodium D lines are especially suitable for demonstration of basic concepts in atomic spectroscopy – so much so, that they have become the "guinea pigs" of the field.

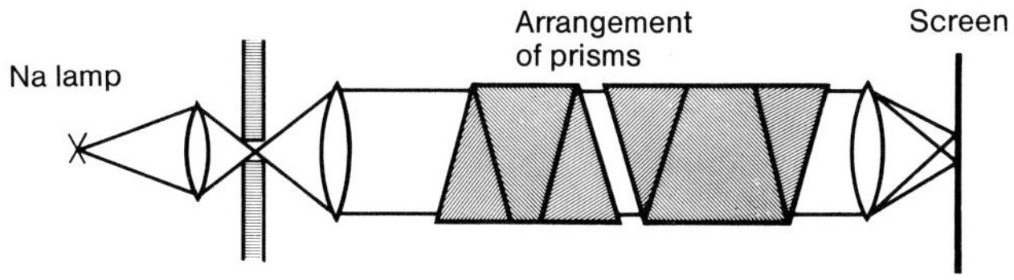

Fig. 12.1. Arrangement for spectral separation of the two components D_1 and D_2 of the sodium D line. With this arrangement, the splitting can easily be demonstrated in the lecture hall by replacing the screen with a television camera. To separate the lines distinctly, one needs two commerically available straight-through prisms

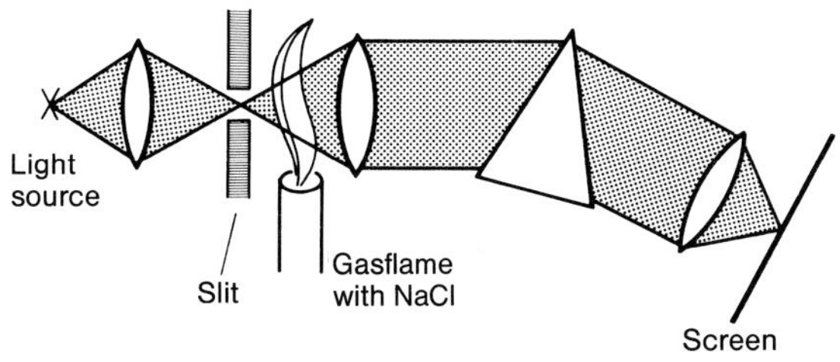

Fig. 12.2. Arrangement for observation of the sodium D lines by absorption (as the so-called Fraunhofer line). The continuous spectrum of an arc lamp or xenon high-pressure lamp is spread out by a prism. A gas flame made yellow by addition of NaCl, or better still, sodium vapour from a heated piece of sodium metal, absorbs the light of the D line from the continuous spectrum. On the screen, therefore, the line is seen as a black band on the continuous spectrum

To explain the doublet structure, one needs three additions to our previous picture:
- A *magnetic moment* μ_l is associated with the orbital angular momentum l.
- The electron also has a *spin* s. It too is associated with a magnetic moment, μ_s.
- The two magnetic moments μ_s and μ_l interact. They can be parallel or antiparallel to each other. The two configurations have slightly different binding energies, which leads to the *fine structure* of the spectrum.

Two demonstrations of the yellow sodium lines are shown in Figs. 12.1 and 12.2; other experiments follow in Chap. 13.

12.2 Magnetic Moment of the Orbital Motion

An electron moving in an orbit is equivalent to a circular electric current. We know from electrodynamics that a circular electric current generates a magnetic dipole field. We expect that the orbiting electron will do the same, and it does in fact have a magnetic dipole moment. This we shall now calculate.

The magnetic dipole moment of a conducting loop is defined as

$$\boldsymbol{\mu} = I \cdot \boldsymbol{A} \qquad [\mathrm{Am}^2] , \tag{12.1}$$

where I is the current, and $\boldsymbol{A}$ is a vector which is perpendicular to the plane of the conducting loop and which has a magnitude equal to the area enclosed by the loop. Thus the vector $\boldsymbol{\mu}$ is also perpendicular to the plane of the loop.

If we bring this magnetic dipole into a homogeneous magnetic field $\boldsymbol{B}$, a torque $\boldsymbol{\tau}$ is applied to the dipole:

$$\boldsymbol{\tau} = \boldsymbol{\mu} \times \boldsymbol{B} . \tag{12.2}$$

The magnetic potential energy of the dipole is (Fig. 12.3)

$$V_{\mathrm{mag}} = -\boldsymbol{\mu} \cdot \boldsymbol{B} = \int_{\pi/2}^{\alpha} \tau \, d\alpha = -\mu B \cos\alpha , \tag{12.3}$$

where α is the angle between $\boldsymbol{\mu}$ and $\boldsymbol{B}$.

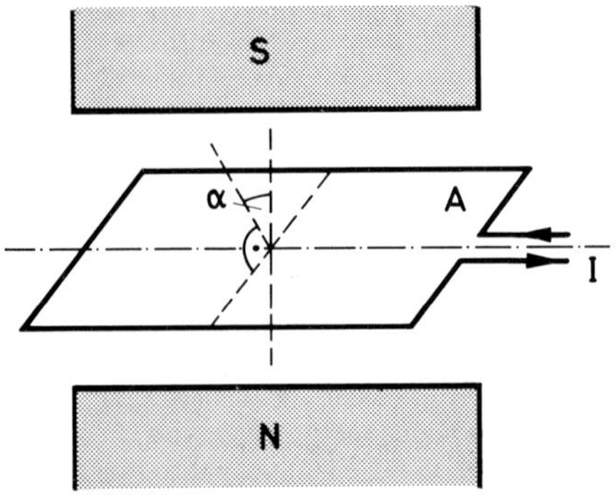

Fig. 12.3. Calculation of the potential energy of a conducting loop in a magnetic field. The magnetic moment is the product of the current I and the area vector $\boldsymbol{A}$. The potential energy depends on the angle α between the normal to the plane of the loop and the direction of the magnetic field

The magnetic moment can be defined either in terms of the torque in the field (12.2) or the potential energy (12.3).

In atomic and nuclear physics, the magnetic moment is often defined as the torque in a uniform field of strength H (not of strength B). Accordingly,

$$\tau = \mu' \times H\,, \qquad \mu' = \mu_0 I A\,, \tag{12.4}$$

if we indicate magnetic moments which are defined w.r.t. H by μ'. Because of the relation $\boldsymbol{B} = \mu_0 \boldsymbol{H}$, the induction constant $\mu_0 = 4\pi \cdot 10^{-7}$ Vs/Am occurs in (12.4).

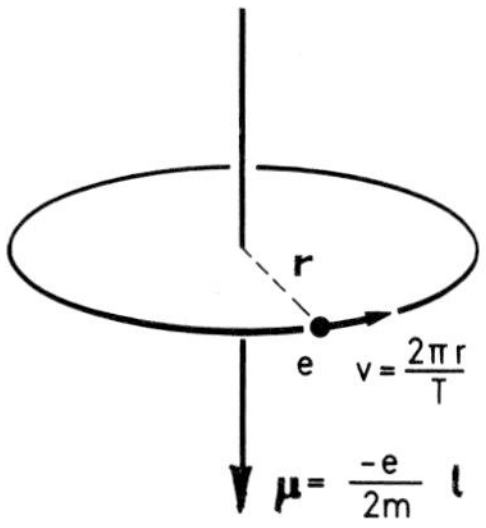

Fig. 12.4. Calculation of the orbital moment. The circulating electron has an angular momentum $\boldsymbol{l}$ and a magnetic dipole moment $\boldsymbol{\mu}_l$. For a negative charge, the vectors $\boldsymbol{l}$ and $\boldsymbol{\mu}_l$ point in opposite directions

We now wish to transfer the definition of the magnetic dipole moment to the atom and calculate the magnetic moment of an electron with the charge $q = -e$ which is in a circular orbit moving with the velocity v (Fig. 12.4). If the time for a single revolution is $T = 2\pi/\omega$, a current

$$I = \frac{q}{T} = -\frac{e\omega}{2\pi} \tag{12.5}$$

is flowing. Here we have used e (without a sign) for the elementary unit of charge. Here and in the following, we use a negative sign for the electron.

The magnetic moment μ of this circular current is then, according to (12.1),

$$\mu = IA = -\tfrac{1}{2} e\omega r^2\,. \tag{12.6}$$

If we introduce[1] the orbital angular momentum $|\boldsymbol{l}| = mvr = m\omega r^2$, we can rewrite (12.6) as a relation between the magnetic moment and the orbital angular momentum

$$\boldsymbol{\mu} = -\frac{e}{2m_0}\boldsymbol{l}\,. \tag{12.7}$$

If the charge q is positive, the vectors $\boldsymbol{\mu}$ and $\boldsymbol{l}$ point in the same direction; if it is negative, as with the electron, they point in opposite directions. Therefore (12.7) holds. We have introduced the symbol m_0 to make it clear that the rest mass is what is meant.

[1] The orbital angular momentum is given by $\boldsymbol{l}$, and its magnitude by $|\boldsymbol{l}|$. This is to prevent confusion with the quantum number l of the orbital angular momentum. See also (8.28)

The proportionality between angular momentum and magnetic moment is also known as the magnetomechanical parallelism. The fact that it is valid in atoms is by no means self-evident, and follows from experimental observations which will be discussed below.

As the unit of magnetic moment in atoms, we use the strength of the moment which corresponds to an electron with the orbital angular momentum $|\boldsymbol{l}| = h/2\pi$. This is the orbital angular momentum in the first Bohr orbit of the hydrogen atom in the old Bohr model. An electron with $|\boldsymbol{l}| = h/2\pi$ or $\hbar$ produces a magnetic moment given by the *Bohr Magneton*:

$$\mu_B = \frac{e}{2m_0}\hbar = 9.274078 \cdot 10^{-24}\,\mathrm{Am}^2\,. \tag{12.8}$$

It is an unfortunate – but, because of its wide usage, unavoidable – inelegance that the same symbol μ is used both for the magnetic moments μ and μ_B and for the induction or permeability constant of vacuum, μ_0.

The magnetic moments of electrons are frequently given in units of μ_B. For the magnitude of the magnetic moment of an orbit with the angular momentum quantum number l, the following expression is valid:

$$\mu_l = \mu_B\sqrt{l(l+1)} = \frac{e}{2m_0}\hbar\sqrt{l(l+1)}\,. \tag{12.9}$$

This expression is also valid for vectors, in the form

$$\boldsymbol{\mu}_l = -g_l\mu_B\frac{\boldsymbol{l}}{\hbar}\,. \tag{12.10}$$

Equation (12.10) thus defines the g factor, which we shall often meet in the following. It is dimensionless and here has the numerical value $g_l = 1$. It is a measure of the ratio of the magnetic moment (in Bohr magnetons) to the angular momentum (in units of $\hbar$). It was introduced by Landé, in the presence of spin-orbit coupling (Sects. 12.7, 8), in order to characterise the ratio of the magnetic moment (in μ_B) and the total angular momentum (in units of $\hbar$).

With "angular momentum", we often denote – briefly but inaccurately – the quantum number l, i.e. the maximum component in the z direction, $l_z/\hbar$. The maximum component of μ in the z direction is then given by $(\mu_z)_{\max} = g_l l \mu_B$. We will treat the g factors for other cases of the angular momentum later. They are always defined as the ratio of the magnetic moment to the corresponding angular momentum, in units of μ_B and $\hbar$, respectively.

12.3 Precession and Orientation in a Magnetic Field

An applied field with the magnetic flux density $\boldsymbol{B}$ acts on the orbital magnetic moment $\boldsymbol{\mu}_l$ by trying to align the vectors $\boldsymbol{\mu}_l$ and $\boldsymbol{B}$ parallel to one another, since the potential energy is a minimum in this orientation (12.3). The electrons, which are moving in their orbits, behave mechanically like gyroscopes and carry out the usual precession about the direction of the field. The precession frequency ω_p of a gyroscope under the action of a torque τ is

$$\omega_p = \frac{|\tau|}{|l|\sin\alpha}, \tag{12.11}$$

where l is the angular momentum of the gyroscope, and α the angle between the directions of l and B_0 (Fig. 12.5).

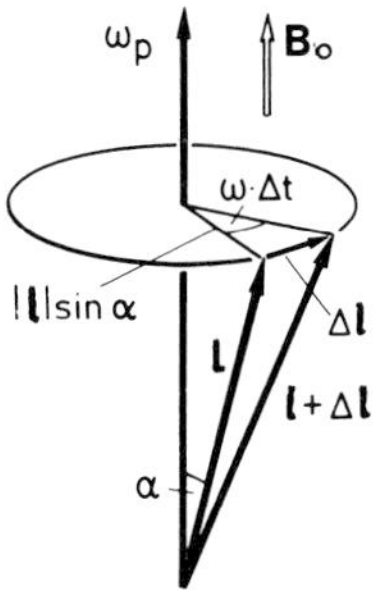

Fig. 12.5. Vector diagram for the calculation of the precession frequency ω_p of a gyroscope with angular momentum l and magnetic moment μ. The angle between the field B_0 and the direction of l (and μ) is denoted by α; the vectors Δl and τ are perpendicular to l and B

These considerations may be directly transferred to the case of the atomic gyroscope. The precession frequency of the electron orbit, the *Larmor frequency*, is found from (12.10) and (12.11) to be

$$\omega_L = \frac{|\tau|}{|l|\sin\alpha} = \frac{\mu_l B \sin\alpha}{|l|\sin\alpha} = \frac{g_l \mu_B}{\hbar} B = \gamma B\,. \tag{12.12}$$

The new quantity γ which we have introduced here is called the *gyromagnetic ratio*. It gives the precession frequency in a field with a magnetic flux density of 1 $Vs/m^2 = 1$ tesla. The sign and direction of the vectors is indicated in Fig. 12.6. As can be seen from (12.12), the Larmor frequency ω_L is independent of the angle α.

We have already seen that the orientation of the vector l in space is not random. The solution of the Schrödinger equation (Sect. 10.2) implies that when one axis is established, a component of the angular momentum is quantised. This axis can be

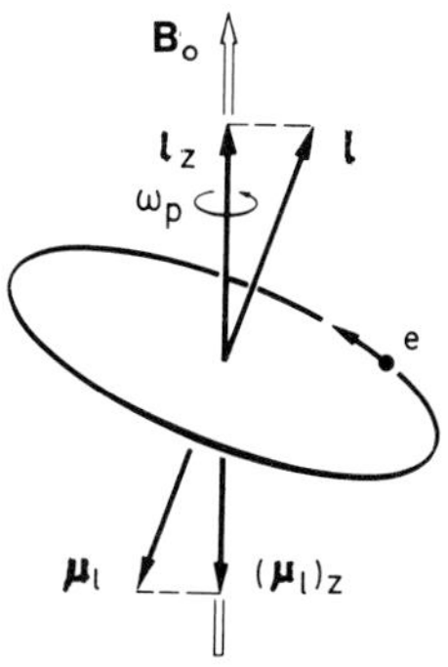

Fig. 12.6. Directional quantisation: Only the projections of the vectors l and μ_l on a chosen axis z can be observed. Here the z direction is the direction of B_0

determined by a magnetic field, for example. Therefore only discrete values of α, the angle between $\boldsymbol{B}$ and $\boldsymbol{l}$ or $\boldsymbol{\mu}_l$, are allowed.

According to Sect. 10.2, the following holds for the components of the angular momentum in the z direction:

$$l_z = m_l \hbar \,, \qquad \text{with} \qquad m_l = 0, \pm 1 \ldots \pm l \,. \tag{12.13}$$

Here m_l is used instead of m in Sect. 10.2. In this way we emphasise that $m(\equiv m_l)$ is associated with the *orbital* angular momentum. m_l is the *magnetic quantum number*. It can have $2l+1$ different values. Here l is again the angular momentum quantum number, $|\boldsymbol{l}| = \sqrt{l(l+1)}\,\hbar$. The largest possible component of $\boldsymbol{l}$ in the z direction thus has the value $l \cdot \hbar$.

The magnetic moment μ_l associated with the orbital angular momentum is correspondingly quantised. For its component in the z direction the quantisation rule is

$$\mu_{l,z} = \frac{-e}{2m_0} l_z = -m_l \mu_B \,. \tag{12.14}$$

The maximum value in the z direction is $l \cdot \mu_B$. As a simplification (but not accurately), it is said that the state has the magnetic moment $l \cdot \mu_B$.

Since $\boldsymbol{\mu}$ precesses around the direction of $\boldsymbol{B}$, it is intuitively clear that in an observation of the energy of interaction between the magnetic moment and the magnetic field, the x and y components of $\boldsymbol{\mu}$ are averaged out over time. However, the z component can be observed.

The experimental demonstration of the existence of a directional quantisation was provided by the Stern and Gerlach experiment (see Sect. 12.6).

12.4 Spin and Magnetic Moment of the Electron

The s states with orbital angular momentum $\boldsymbol{l} = 0$ have no orbital magnetic moment. Therefore, a one-electron atom should be diamagnetic in the ground state, when it has one valence electron in an outer shell and all the others in closed shells. However, these atoms are actually paramagnetic.

The reason is the existence of electron spin and the associated magnetic moment. Electron spin was introduced by *Uhlenbeck* and *Goudsmit* in 1925 to explain spectroscopic observations.

The splitting of many spectral lines in a magnetic field, which will be discussed later (the anomalous Zeeman effect) can only be explained if the electron has a spin angular momentum $\boldsymbol{s}$,

$$|\boldsymbol{s}| = \sqrt{s(s+1)}\,\hbar \tag{12.15}$$

and the associated magnetic moment

$$\boldsymbol{\mu}_s = -g_s \frac{e}{2m_0} \boldsymbol{s} \,, \tag{12.16}$$

where e is again the unit charge of the electron, without the negative sign. $s = 1/2$ is a new quantum number, the *spin quantum number*. The similarity of (12.16) and (12.10) is apparent. The two expressions differ only in that (12.16) contains the new factor g_s, the so-called g factor of the electron. Although the expected value for this proportionality constant on the basis of classical theory would be 1, the value has been empirically determined to be $g_s = 2.0023$. Figure 12.7 represents the spin and magnetic moment of the electron schematically.

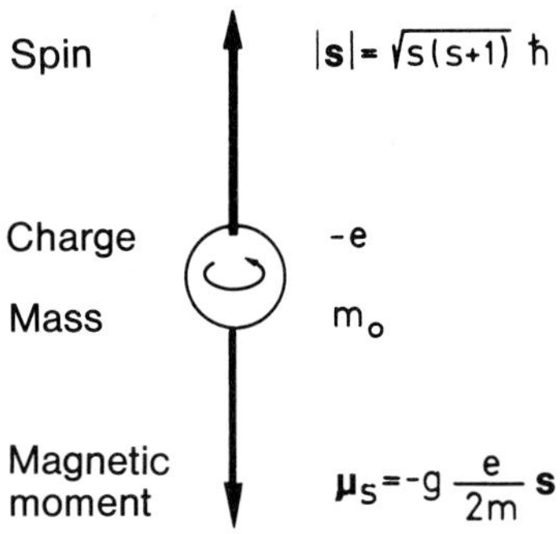

Fig. 12.7. Spin and magnetic moment of the electron

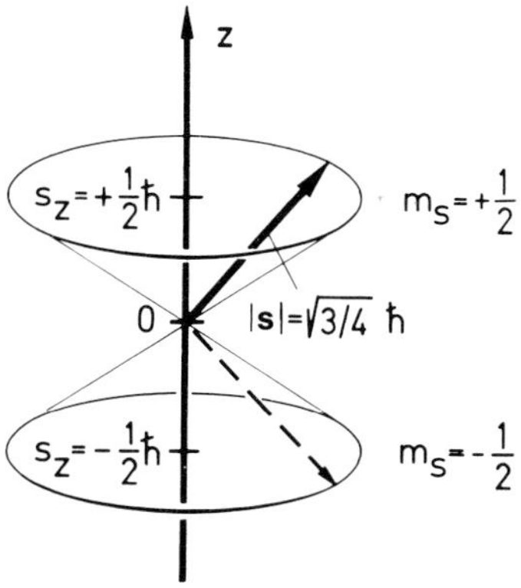

Fig. 12.8. The electron spin has two possible orientations in a magnetic field in the z direction. These are characterised by the quantum number $m_s = \pm 1/2$

Dirac showed in 1928 that the spin of the electron is a necessary consequence of a relativistic quantum theory (the Schrödinger theory is non-relativistic). The g factor $g_s = 2$ could also be thus derived. The slight difference between the predicted value of 2 and the empirical value can only be understood if the interaction of the electron with its own radiation field is taken into account through quantum electrodynamics.

As first shown by the experiment of *Stern* and *Gerlach* (Sect. 12.6), the spin can only have two orientations in an external magnetic field $\boldsymbol{B}$ (or in the presence of a defined z axis): "parallel" and "antiparallel" to the field (Fig. 12.8). Its components in this defined z direction are

$$s_z = m_s \hbar \quad \text{with} \quad m_s = \pm \tfrac{1}{2}; \tag{12.17}$$

m_s is the magnetic quantum number of the spin.

It follows from the orientation of the angular momentum that the magnetic moment is also oriented. The z component is

$$\mu_{s,z} = -g_s m_s \mu_B, \tag{12.18}$$

or numerically,

$$\mu_{s,z} = \pm 1.00116\,\mu_B. \tag{12.19}$$

Intuitively speaking, the spin and the magnetic moment precess around the field axis, leaving the z component constant (compare Sect. 12.3).

The gyromagnetic ratio, which was defined above (12.12) as the ratio between the magnetic moment and the angular momentum,

$$\gamma = \frac{|\mu|}{|l|} \quad \text{or} \quad \gamma = \frac{|\mu|}{|s|}, \tag{12.20}$$

is thus not the same for orbital (12.10) and spin (12.16) magnetism. For pure orbital magnetism,

$$\gamma_l = \frac{1}{2}\frac{e}{m_0},$$

and for pure spin magnetism,

$$\gamma_s = 1.00116\frac{e}{m_0}.$$

The previously mentioned g factor is also used instead of the gyromagnetic ratio γ. g is obtained by multiplying γ by $\hbar$ and is defined for pure orbital magnetism as

$$\gamma_l \hbar = \frac{1}{2}\frac{e}{m_0}\hbar = g_l \mu_B \tag{12.21}$$

and for pure spin magnetism by

$$\gamma_s \hbar = 1.00116\frac{e}{m_0}\hbar = g_s \mu_B = 2.0023\,\mu_B. \tag{12.22}$$

In the following, the reader will see that the easiest and most definitive way to calculate the magnetic properties of atoms is often to make use of measurements of the ratio γ or g.

12.5 Determination of the Gyromagnetic Ratio by the Einstein-de Haas Method

The gyromagnetic ratios of macroscopic samples can be measured as shown in Fig. 12.9. An iron needle is magnetised by a coil. If one changes the magnetisation of the sample – and this means changing the direction of the atomic magnetic moments in the sample – one will also change the direction of the atomic angular momenta, and this must be observable as a change in the angular momentum of the whole sample, according to the law of conservation of angular momentum. If the magnetisation is changed by 180° by reversing the poles of the coil, the angular momentum vector must also be rotated through 180°. Quantitatively, the change $\Delta\mu_N$ in the magnetisation of

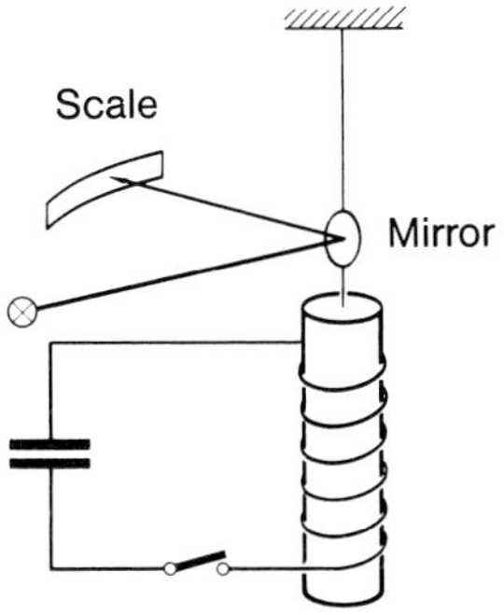

Fig. 12.9. *Einstein-de Haas* experiment. When the current in the coil is reversed, the magnetisable bar hanging in it turns. The torsion of the fibre on which the bar is suspended is measured with a mirror and light beam

the needle, measured with a detection coil and a ballistic galvanometer, can be represented as the sum of the changes for the individual electrons,

$$\sum_1^n \Delta\mu_z = n \cdot 2\mu_z ,$$

if n electrons have reversed their directions.

Likewise, the macroscopic change in the angular momentum of the needle, ΔL_N, measured by means of the torsion fibre, is the sum of the changes of the atomic angular momenta:

$$\sum_1^n \Delta l_z = n \cdot 2 l_z .$$

For macroscopic samples, the measured ratio

$$\frac{\Delta\mu_N}{\Delta L_N} = \frac{\mu_z}{l_z} = \frac{e}{m_0} .$$

Thus according to the definition of (12.20),

$$\gamma = \frac{e}{m_0} \quad \text{or} \quad g = 2 .$$

From this experiment it can be seen that there is an angular momentum associated with the magnetism of atoms, and that it can be calculated as derived above.

In general, gyromagnetic ratio measurements, first described in 1915 by *Einstein* and *de Haas*, can indicate how much of the magnetism in a given sample is due to spin and how much to orbital angular momentum. However, a quantitative understanding of this type of measurement requires a deeper knowledge of solid state physics.

12.6 Detection of Directional Quantisation by Stern and Gerlach

In 1921, the deflection of atomic beams in inhomogeneous magnetic fields made possible

- the experimental demonstration of *directional quantisation* and
- the direct *measurement of the magnetic moments* of atoms.

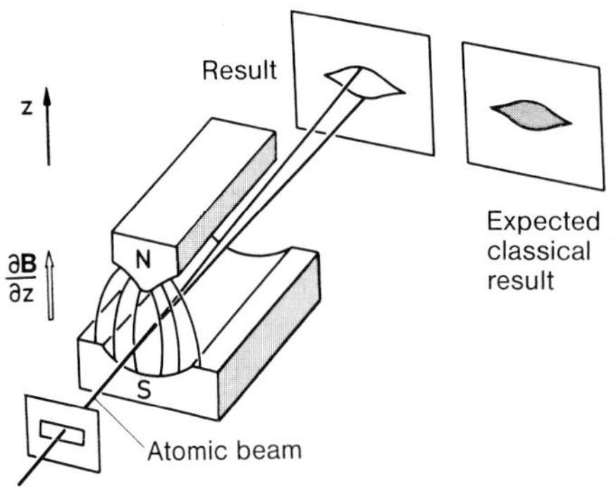

Fig. 12.10. *Stern-Gerlach* experiment. The atomic beam passes through an inhomogeneous magnetic field. One observes the splitting of the beam into two components

In the experiment (Fig. 12.10), one first generates a beam of atoms. In the first experiments of *Stern* and *Gerlach*, this was a beam of silver atoms which was generated in an atomic beam furnace and collimated by a series of slits. Later, hydrogen atoms from a gas discharge were also used. The collimated beam passes through a highly inhomogeneous magnetic field, with the direction of the beam perpendicular to the direction of the field and of the gradient. The directions of the field and gradient are the same. Without the field, the vectors of the magnetic moments and angular momenta of the atoms are randomly oriented in space. In a homogeneous field, these vectors precess around the field direction z.

An inhomogeneous field exerts an additional force on the magnetic moments. The direction and magnitude of this force depends on the relative orientation between the magnetic field and the magnetic dipole. A magnetic dipole which is oriented parallel to the magnetic field moves in the direction of increasing field strength, while an antiparallel dipole moves towards lower field strength. A dipole which is perpendicular to the field does not move.

The deflecting force can be derived from the potential energy in the magnetic field $V_{\mathrm{mag}} = -\boldsymbol{\mu} \cdot \boldsymbol{B}$:

$$F_z = \mu_z \frac{dB}{dz} = \mu \frac{dB}{dz} \cos\alpha \,, \tag{12.23}$$

where α is the angle between the magnetic moment and the direction of the field gradient.

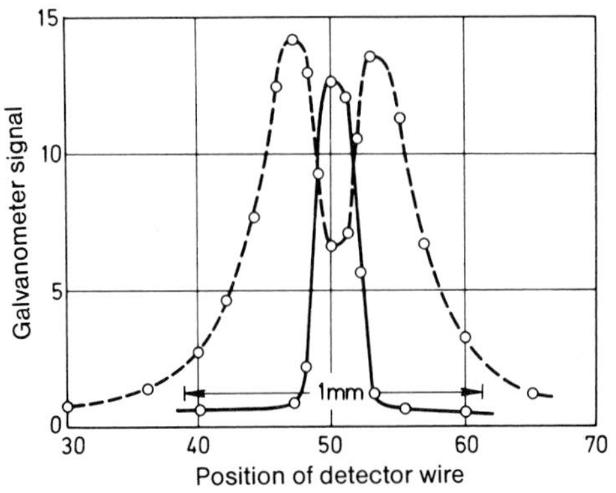

Fig. 12.11. *Stern-Gerlach* experiment. Observed intensity distribution of an atomic beam with and without an applied magnetic field, $^2S_{1/2}$ state [from H. Kopfermann: *Kernmomente*, 2nd ed. (Akademische Verlagsgesellschaft, Frankfurt 1956)]

In classical mechanics, any orientation α of the atomic magnet with respect to the field is allowed. Atoms with magnetic moments perpendicular to the field gradient are not deflected. Those in which the vectors are parallel are deflected the most, and all possible intermediate values can occur. In the classical picture one thus expects a continuum of possible deflections. With H and Ag atoms, however, two rather sharp peaks separated by 2δ were observed on the detector (Fig. 12.11).

This experiment and similar measurements on other atoms permit the following conclusions:

- There is a *directional quantisation.* There are only discrete possibilities for the orientation relative to a field $\boldsymbol{B}_0$, in this case two, parallel and antiparallel.
- From a quantitative evaluation of the observed deflection δ in the above example, one obtains the value $\mu_z = \pm\mu_B$. In general this method provides observed values for *atomic magnetic moments* if the magnitude of the field gradient is known.
- For all atoms which have an s electron in the outermost position, one obtains the same value for the deflecting force, from which it follows that the *angular momenta and magnetic moments of all inner electrons cancel each other* and one measures only the effect of the outermost s electron.
- The s electron has an *orbital angular momentum* $l = 0$ and one observes only spin magnetism.
- Like gyroscopes, atoms maintain the magnitude and direction of their angular momenta in the course of their motion in space.

This experiment provides the basis for the knowledge of the angular momenta and magnetic moments of atoms which was summarised in Sects. 12.2 and 12.3.

12.7 Fine Structure and Spin-Orbit Coupling: Overview

In the introductory section to this chapter we mentioned that all energy terms – with the exception of the s states of one-electron atoms – are split into two substates. This produces a doublet or multiplet structure of the spectral lines, which is denoted by the generic name *fine structure.*

The fine structure cannot be explained with the Coulomb interaction between the nucleus and the electrons. Instead, it results from a magnetic interaction between the orbital magnetic moment and the intrinsic moment of the electron, the *spin-orbit coupling.* Depending on whether the two moments are parallel or antiparallel, the energy term is shifted somewhat.

The magnetic coupling energy between the orbital moment and the spin moment will be calculated in Sect. 12.8. The coupling of the magnetic moments leads to an addition of the two angular momenta to yield a total angular momentum.

The following conclusions are then valid (Fig. 12.12):

- $\boldsymbol{l}$ and $\boldsymbol{s}$ add to give a total angular momentum $\boldsymbol{j}$;
- $\boldsymbol{j}$ has the magnitude $\sqrt{j(j+1)}\,\hbar$ with $j = |l \pm s|$, i.e. $j = |l \pm \frac{1}{2}|$ for the case treated here of a single-electron system with $s = \frac{1}{2}$. The quantum number j is a new quantity: the quantum number of the total angular momentum. We shall show with a quantum mechanical calculation in Sect. 14.3 that $\boldsymbol{j}$ has the magnitude given above.
- For a p electron with $l = 1$, $s = \frac{1}{2}$, we find the following possibilities:

$$j = 3/2\,, \quad |\boldsymbol{j}| = \sqrt{15}/2\;\hbar\,, \quad \text{and}$$
$$j = 1/2\,, \quad |\boldsymbol{j}| = \sqrt{3}/2\;\hbar\,;$$

- when $l = 0$, $j = s$ and there is no doublet splitting;
- for $\boldsymbol{j}$, just as for $\boldsymbol{l}$, there is a directional quantisation. The z components must obey the condition

$$j_z = m_j \hbar\,, \quad m_j = j, j-1, \ldots -j \quad (2j+1 \text{ possibilities})\,.$$

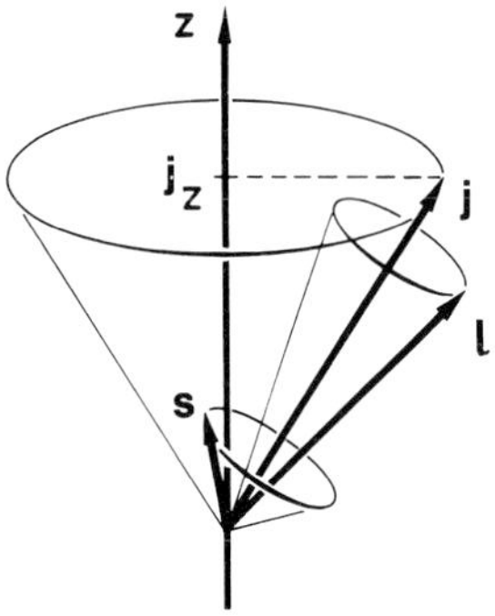

Fig. 12.12. Coupling of the vectors of spin s and orbital angular momentum $\boldsymbol{l}$ to give the total (resultant) angular momentum $\boldsymbol{j}$ in the vector model. The vectors s and $\boldsymbol{l}$ precess about the resultant $\boldsymbol{j}$. In a magnetic field applied in the z direction, $\boldsymbol{j}$ precesses about z. The opening angle of the cone of precession is determined by the magnetic quantum number m_j. The figure shows the case $s = 1/2$, $l = 2$, $j = 5/2$

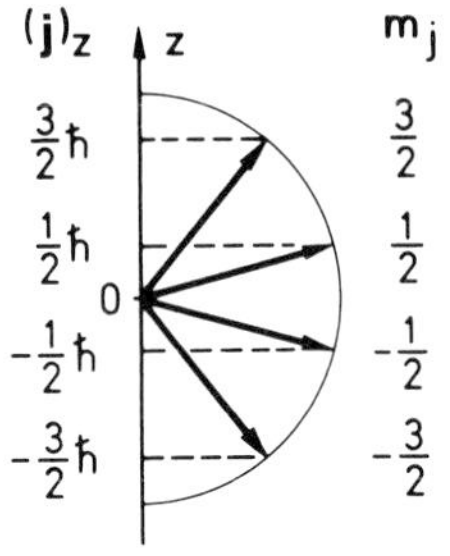

Fig. 12.13. Directional quantisation: for the z component of the angular momentum $\boldsymbol{j}$, only certain discrete values are allowed. They are denoted by the magnetic quantum number m_j. For the case illustrated, $j = 3/2$, the magnitude of the vector is $|\boldsymbol{j}| = \sqrt{(3/2)(5/2)}\,\hbar$. Four orientations are allowed: $m_j = 3/2,\ 1/2,\ -1/2,\ -3/2$

For example, a state with $j = 3/2$ is fourfold degenerate (Fig. 12.13).
- A magnetic moment $\boldsymbol{\mu}_j$ is associated with j; this will be calculated in Sect. 13.3.5.
- For optical transitions, a selection rule $\Delta j = 0$ or ± 1 is valid; however, a transition from $j = 0$ to $j = 0$ is always forbidden. This selection rule may be considered to be an empirical result, derived from the observed spectra. The reasons for it will become clear later (Chap. 16).

12.8 Calculation of Spin-Orbit Splitting in the Bohr Model

In this section, we shall calculate the energy difference between the parallel and the antiparallel orientations of the orbital angular momentum and the spin. For this purpose, the simple Bohr model will be used as starting point; the quantum mechanical treatment will be discussed in Sect. 14.3.

The motion of the electron around the nucleus generates a magnetic field $\boldsymbol{B}_l$ at the site of the electron. This field interacts with the magnetic moment of the electron. To determine the magnitude of this magnetic field, we borrow from relativity theory and assume that the electron is stationary and that the nucleus moves instead (Fig. 12.14). We replace the position vector for the orbiting electron, $\boldsymbol{r}$, by the vector $-\boldsymbol{r}$.

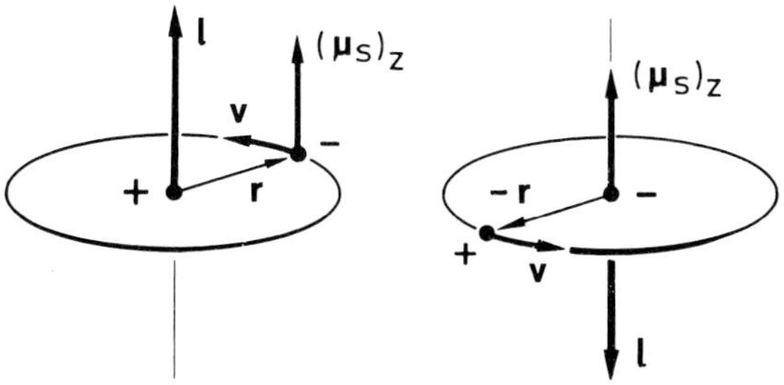

Fig. 12.14. For the calculation of spin-orbit coupling, the system with the nucleus at rest (*left*) is transformed to the system with the electron at rest (*right*). The vector $\boldsymbol{r}$ is replaced by the reversed vector $-\boldsymbol{r}$

The magnetic field of the moving charge $+Ze$ is found from the Biot-Savart law to be

$$\boldsymbol{B}_l = +\frac{Ze\mu_0}{4\pi r^3}[\boldsymbol{v}\times(-\boldsymbol{r})] \tag{12.24}$$

or

$$\boldsymbol{B}_l = -\frac{Ze\mu_0}{4\pi r^3}[\boldsymbol{v}\times\boldsymbol{r}]\,. \tag{12.25}$$

Angular momentum is defined as $\boldsymbol{l} = \boldsymbol{r}\times m_0\boldsymbol{v}$ or $-\boldsymbol{l} = m_0\boldsymbol{v}\times\boldsymbol{r}$. Then

$$\boldsymbol{B}_l = \frac{Ze\mu_0}{4\pi r^3 m_0}\boldsymbol{l}\,, \tag{12.26}$$

where m_0 is the rest mass of the electron.

The magnetic field which is generated by the relative motion of the nucleus and the electron is thus proportional and parallel to the orbital angular momentum of the electron. We still require the back transformation to the centre-of-mass system of the atom, in which the nucleus is essentially at rest and the electron orbits around it. A factor 1/2 occurs in this back transformation, the so-called Thomas factor, which can only be justified by a complete relativistic calculation. The particle in its orbit is accelerated, and from the viewpoint of the proton, the rest system of the electron rotates one additional time about its axis during each revolution around the orbit. The back transformation is therefore complicated and will not be calculated in detail here.

The magnetic moment of the electron, and with it, its coupled spin vector, precess about the magnetic field $\boldsymbol{B}_l$ produced by the orbital motion (cf. Fig. 12.15).

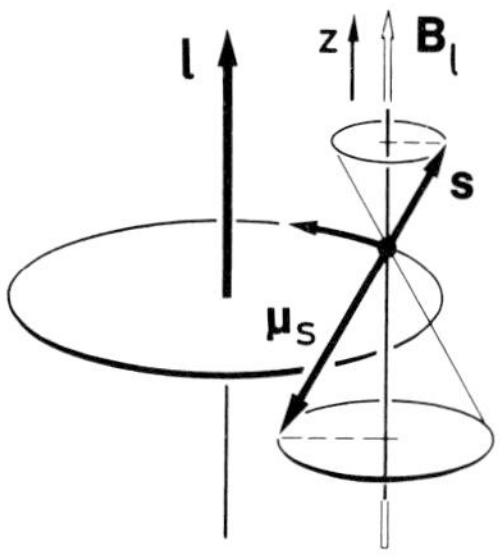

Fig. 12.15. Precession of the spin about the magnetic field $\boldsymbol{B}_l$ associated with the orbital angular momentum, with the components s_z and $\mu_{s,z}$

The interaction energy between the spin and the orbital field is thus

$$V_{l,s} = -\mu_s \cdot B_l .$$

Substituting – see (12.16), $g_s = 2$ – we find

$$V_{l,s} = 2\frac{e}{2m_0}(s \cdot B_l) ,$$

and with (12.26)

$$= \frac{Ze^2\mu_0}{8\pi m_0^2 r^3}(s \cdot l) . \quad (12.27)$$

Here we have included the (underived) Thomas correction; this gives the factor 8 in the denominator (instead of 4).

In order to get a feeling for the order of magnitude, we set $Z = 1$ and $r = 1$ Å and obtain $V_{l,s} \cong 10^{-4}$ eV. The field produced by the orbital motion is found to be about 1 tesla = 10^4 gauss. The fields associated with the orbital angular momentum are thus – for small values of Z – of the same order of magnitude as may be readily produced in the laboratory.

Equation (12.27) may also be written in the form

$$V_{l,s} = \frac{a}{\hbar^2} l \cdot s = \frac{a}{\hbar^2} |l||s| \cos(l,s) \quad (12.28)$$

where $a = Ze^2\mu_0\hbar^2/(8\pi m_0^2 r^3)$. The scalar product $l \cdot s$ may be expressed in terms of the vectors l and s by using the law of cosines according to Fig. 12.16, where we recall that l^2 must be replaced by its quantum value $l(l+1)\hbar^2$, etc. We thus obtain for the spin-orbit coupling energy

$$V_{l,s} = \frac{a}{2\hbar^2}(|l+s|^2 - |l|^2 - |s|^2)$$

$$= \frac{a}{2\hbar^2}(|j|^2 - |l|^2 - |s|^2)$$

$$= \frac{a}{2}[j(j+1) - l(l+1) - s(s+1)] . \quad (12.29)$$

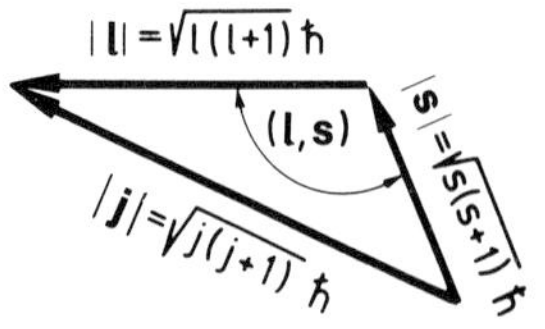

Fig. 12.16. Vector addition of the angular momentum vectors to the total angular momentum j, explanation of (12.29)

The spin-orbit coupling energy is thus expressed in terms of the quantum numbers j, l and s, as well as a constant a, known as the spin-orbit coupling constant. The latter is directly measurable by determination of the doublet structure in the optical spectra.

Comparison with (12.27) shows that the orbital radius r is included in this coupling constant a. We must remember, however, that there are no fixed orbits in the quantum theoretical description of the atom. Therefore it is necessary to replace r^{-3} by the corresponding quantum theoretical average $\overline{1/r^3} = \int(|\psi|^2/r^3)\,dV$, where ψ is the wavefunction of the electron and dV the volume.

If we use the radius r_n of the nth Bohr radius as a rough approximation for r,

$$r_n = \frac{4\pi\varepsilon_0\hbar^2 n^2}{Ze^2 m_0}\,, \tag{12.30}$$

we obtain

$$a \sim \frac{Z^4}{n^6}\,.$$

If instead we use the above-defined average value $\overline{r^{-3}}$, we obtain for atoms similar to H

$$a \sim \frac{Z^4}{n^3 l(l+\frac{1}{2})(l+1)}\,, \tag{12.31}$$

which will not be derived here.

Let us again summarise what we know about the fine structure of one-electron states:

- Interaction of the electron with the orbital angular momentum or the orbital moment splits each level into two. The result is doublet levels; for example in the upper state of the sodium D lines, the $3P$ state is split into the $3P_{1/2}$ and the $3P_{3/2}$ states (Fig. 12.17).

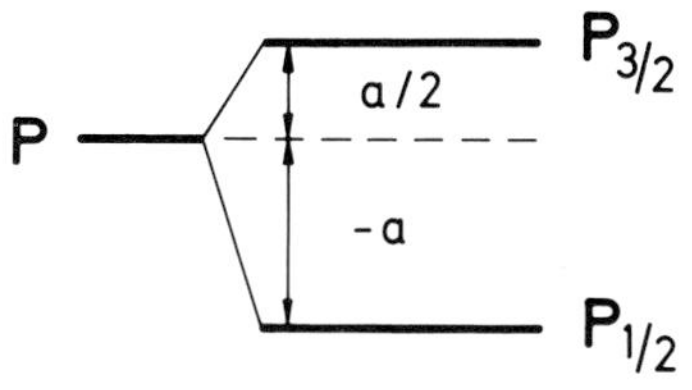

Fig. 12.17. Fine structure splitting of the P state in the one electron system into the two states $P_{3/2}$ and $P_{1/2}$. The magnitude of the splitting is given by (12.29). Since only one electron is involved, one can also use lower case "p"

- For s terms there is no splitting, because there is no magnetic field with which the spin can align itself.
- Levels with higher values of the quantum number j have higher energies (12.29).
- The fine structure splitting $V_{l,s}$ is proportional to the fourth power of the nuclear charge.

  The fine structure is therefore difficult to observe in the H atom. For the H_α, H_β, and H_γ lines of the Balmer series (6562.79, 4861.33 and 4340.46 Å), the splitting is 0.14, 0.08 and 0.07 Å, respectively. This corresponds to a wavenumber of 0.33 cm^{-1} for the H_α line, which is in the microwave range – if one wished to observe it directly. A direct observation of the splitting of optical spectral lines into two very close components is not possible by conventional spectroscopy because of Doppler broadening of the lines. By contrast, the observed values for a line pair of the 1st primary series of cesium are $\lambda = 8943$ Å and 8521 Å. The splitting is thus $\Delta\lambda = 422$ Å or $\Delta\bar{\nu} = 554\ \mathrm{cm}^{-1}$. It is so large, in fact, that the two lines are difficult to recognise as components of a pair. The sodium atom lies between these extremes: the yellow D lines D_1 and D_2 are separated by $\Delta\lambda = 6$ Å, which corresponds to 17.2 cm^{-1}.

– The splitting is greatest for the smallest principal quantum number n (12.31).

We can now expand upon the symbolism needed to identify the energy terms of atoms. The terms for orbital angular momentum are generally indicated by upper case letters S, P, D, F, etc. The principal quantum number n is written as an integer in front of the letter, and the total angular momentum quantum number j as a subscript. The multiplicity $2s+1$ is indicated by a superscript to the left of the orbital angular momentum letter. For single-electron systems, the terms are doublet terms, because the spin of the single electron can have two orientations with respect to the orbital angular momentum.

The S terms are not split. Nevertheless, one writes the multiplicity 2 even for S terms in one-electron systems.

One thus has the following symbols:

$2^2S_{1/2}$ for a state in which the valence electron has the quantum numbers $n = 2$, $l = 0$, $j = 1/2$.

$\left.\begin{matrix} 2^2P_{1/2} \\ 2^2P_{3/2} \end{matrix}\right\}$ for states in which the valence electron has the quantum numbers $n = 2$, $l = 1$, $j = 1/2$ or $3/2$, respectively.

In general, the symbolism is $n^{2S+1}L_J$. The upper case letters S (spin quantum number), L (orbital angular momentum quantum number) and J (total angular momentum quantum number) apply to several-electron atoms, while the corresponding lower case letters apply to single electrons.

12.9 Level Scheme of the Alkali Atoms

For an atom with one electron in the incomplete outer shell, the results of Sect. 12.7 can be summarised in the term scheme of Fig. 12.18. This figure should make it clear that both the lifting of orbital degeneracy (i.e. the energy difference between terms with the

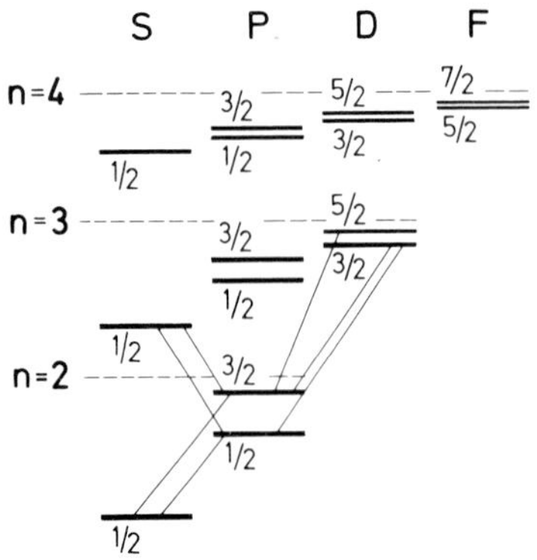

Fig. 12.18. Term scheme for alkali atoms, i.e. one-electron states, including the spin-orbit splitting. The levels are not shown to scale. A few allowed transitions are indicated. The terms are displaced with respect to those of the H atom ($n = 2, 3, 4$, left side, dashed lines), the s terms most. The fine structure splitting decreases with increasing values of n and l

same n but different l quantum numbers) and spin-orbit splitting become smaller as the quantum numbers n and l increase.

The optical transitions in the term scheme obey the rules $\Delta l = \pm 1$, $\Delta j = \pm 1$ or 0. Optical transitions are thus allowed only if the angular momentum changes. The total angular momentum j, however, can remain the same. This would happen if the orbital angular momentum and the spin changed in opposite directions.

The first principal series of the alkali atom arises from transitions between the lowest $^2S_{1/2}$ term (i.e. $n = 2, 3, 4, 5, 6$ for Li, Na, K, Rb, Cs) and the P terms $^2P_{1/2}$ and $^2P_{3/2}$. Since the S terms are single-valued, one sees pairs of lines. The same holds for the sharp secondary series, which consists of transitions between the two lowest P terms $n^2P_{3/2}$ ($n = 2, 3, 4, 5, 6$ for Li, Na, K, Rb, Cs) and all higher $^2S_{1/2}$ terms. The lines of the diffuse secondary series, however, are triple (Fig. 12.19), because both the P and the D terms are double.

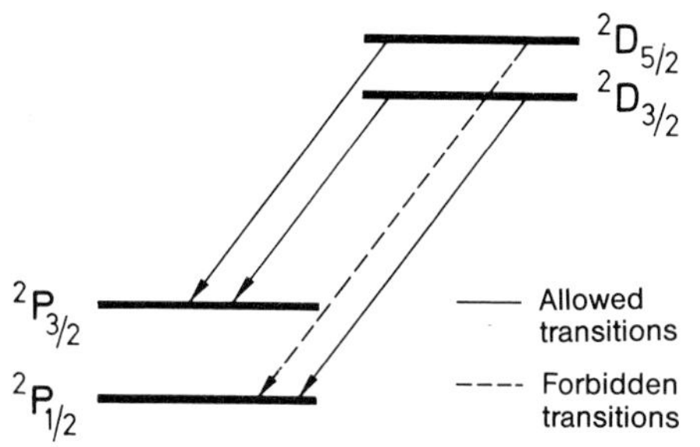

Fig. 12.19. Allowed and forbidden transitions between P and D states of the alkali atom, here for a triplet of the diffuse secondary series. This is a section from Fig. 12.18

12.10 Fine Structure in the Hydrogen Atom

Since the wavefunctions of the H atom are known explicitly (Chap. 10), its fine structure can be exactly calculated. The starting point is the expression derived above (12.27) for the spin-orbit splitting energy:

$$V_{l,s} = \frac{e^2 \mu_0}{8 \pi m_0^2} \cdot \frac{1}{r^3} (s \cdot l) . \tag{12.32}$$

We use the solution of the non-relativistic Schrödinger equation for the H atom, which provides the energy states $E_{n,l}$ (Sect. 10.4). For the H atom, both the relativity correction (cf. Sect. 8.10) and the fine structure interaction are small compared to the energies $E_{n,l}$, but the two are of comparable magnitude. One can therefore calculate the two corrections separately and write

$$E_{n,l,j} = E_{n,l} + E_{\text{rel}} + E_{l,s} .$$

The two correction terms, the one for the relativistic mass change E_{rel} and the other for the spin-orbit coupling $E_{l,s}$, together give the fine structure correction E_{FS}. These terms will not be calculated in detail here. The complete calculation was carried out by *Dirac*.

As a result, one obtains

$$E_{FS} = -\frac{E_n \alpha^2}{n}\left(\frac{1}{j+1/2} - \frac{3}{4n}\right), \tag{12.33}$$

where

$$\alpha = \frac{e^2}{4\pi\varepsilon_0 \hbar c}\left(\text{or } \frac{\mu_0 c}{4\pi\hbar}e^2\right),$$

which is the Sommerfeld fine structure constant which was introduced in Sect. 8.10.

By including the spin-orbit coupling, one thus obtains the same result as earlier (Sect. 8.10) in the calculation of the relativistic correction, the only change being that l has been replaced by j. The energy shift with respect to the previously calculated energy terms $E_{n,l}$ is of the order of α^2, i.e. $(1/137)^2$, and is thus difficult to measure.

The most important result of (12.33) is the fact that the fine structure energy of the H atom depends only on j, not on l (Fig. 12.20).

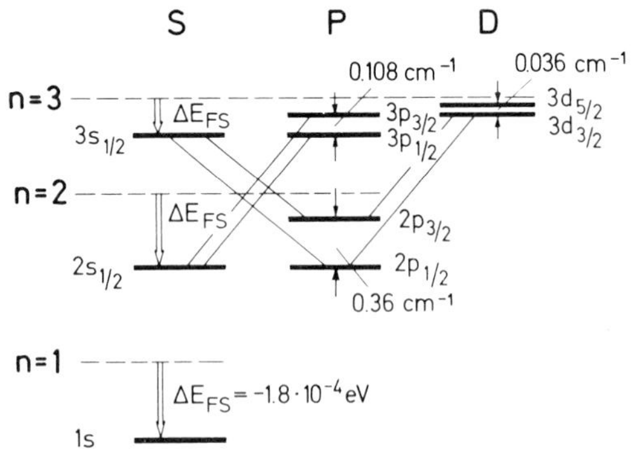

Fig. 12.20. Fine structure splitting of the states with $n = 1, 2$ and 3 (far left, dashed lines, the levels without fine structure), according to *Dirac* (not to scale). The fine structure shifts are indicated by open arrows. States with the same l are degenerate without fine structure interactions. States with the same j have the same energy if fine structure is taken into account

The fine structure of the hydrogen lines is thus quantitatively accounted for. The fine structure energies of heavier atoms are larger and are thus easier to observe. Their calculations, however, are far more difficult, because the exact calculation of the wavefunctions of atoms with more than one electron is far more complex.

12.11 The Lamb Shift

In the years 1947 – 1952, *Lamb* and *Retherford* showed that even the relativistic Dirac theory did not completely describe the H atom. They used the methods of high-frequency and microwave spectroscopy to observe very small energy shifts and splitting in the spectrum of atomic hydrogen. In other words, they used the absorption by H atoms of electromagnetic radiation from high-frequency transmitters or klystron tubes. They could, in this way, observe energy differences between terms with the same j, namely differences of 0.03 cm^{-1} – this corresponds to a difference of 900 MHz – between the terms $2^2S_{1/2}$ and $2^2P_{1/2}$.

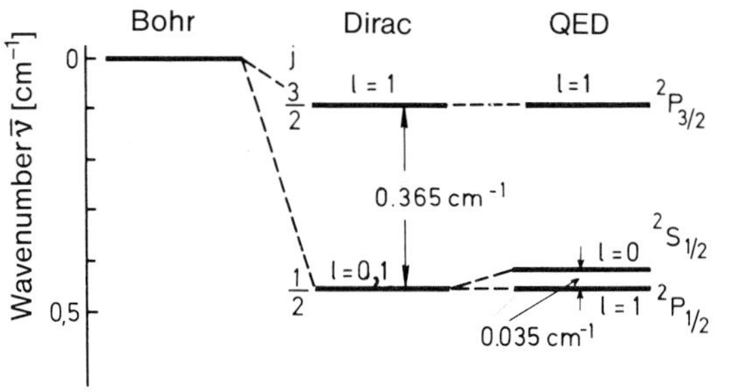

Fig. 12.21. The Lamb shift: fine structure of the $n = 2$ level in the H atom according to *Bohr*, *Dirac* and quantum electrodynamics taking into account the Lamb shift. The j degeneracy is lifted

They achieved a precision of 0.2 MHz. Figure 12.21 shows the corresponding energy diagram.

Like the fine structure, this small energy shift was not observable by means of optical spectroscopy as a splitting of the H_α line of hydrogen, because the Doppler broadening of the spectral lines due to the motion of the atoms exceeded the magnitude of the splitting.

The Lamb-Retherford result can be generalised: levels with the same quantum numbers n and j, but different l, are not exactly the same. Rather, all $S_{1/2}$ terms are higher than the corresponding $P_{1/2}$ terms by an amount equal to about 10% of the energy difference $(P_{3/2} - P_{1/2})$, and the $P_{3/2}$ terms are higher than the $D_{3/2}$ terms by about 2‰ of $(D_{5/2} - D_{3/2})$.

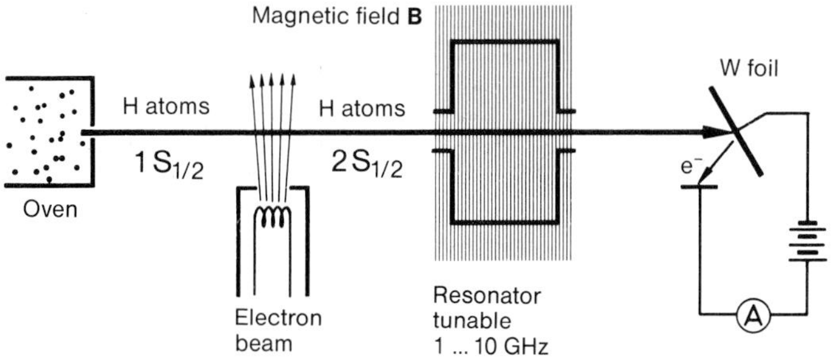

Fig. 12.22. Arrangement for measurement of the Lamb shift. A beam of H atoms is excited to the metastable $2S_{1/2}$ state by bombardment with electrons. The beam passes through a resonator. If electromagnetic transitions are induced there, the number of excited atoms reaching the tungsten foil receiver is lower, and the measured electron current correspondingly drops

The Lamb and Retherford experiment is shown in Fig. 12.22. A beam of hydrogen atoms is generated from H_2 molecules by thermal dissociation at 2500 °C. A small number of these atoms is excited to the metastable state $2^2S_{1/2}$ by bombardment by electrons. Optical transitions between this state and the ground state $1^2S_{1/2}$ are forbidden. The atoms then pass through a tunable resonator for high-frequency or microwave radiation, to a tungsten foil. There the metastable atoms can give up their excitation energy, thereby releasing electrons from the surface of the metal. The electron current is measured and serves as an indicator of the rate at which atoms in the $2^2S_{1/2}$ state arrive at the detector. Those atoms which are excited to the $2^2P_{3/2}$ state by absorbing microwaves in the range of 10000 MHz in the resonator (compare term scheme in Fig. 12.21) can emit light at the wavelength of the H_α line (or more exactly, of one

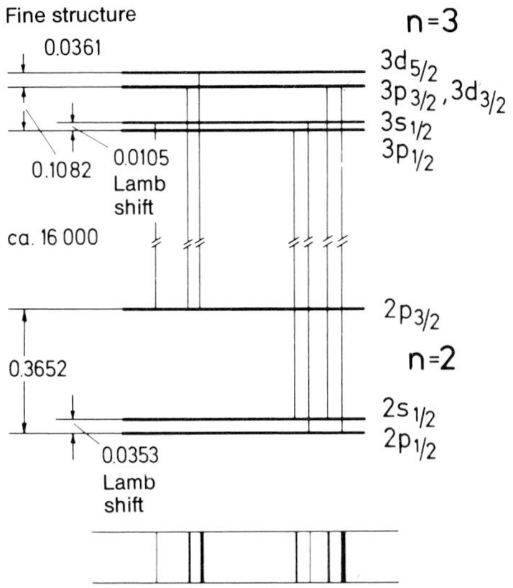

Fig. 12.23. Structure of the H_α line of hydrogen and term scheme including the fine structure. The expected optical spectrum is shown below (ignoring the line widths). Darker lines indicate higher intensity. The wavenumbers are in cm^{-1}

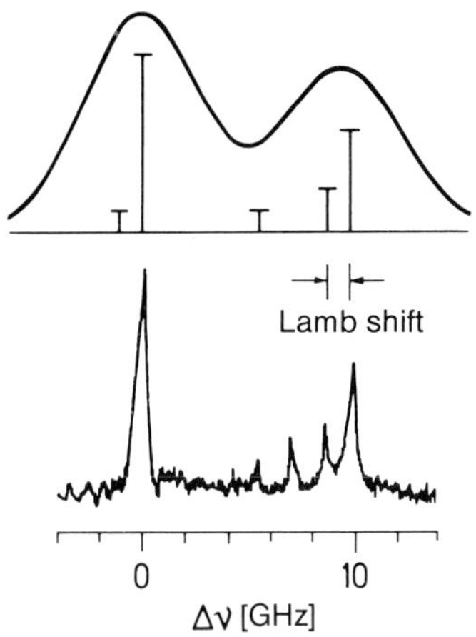

Fig. 12.24. *Above:* Structure of the H_α line of the hydrogen atom at room temperature. The linewidth and thus the spectral resolution is determined by the Doppler width. *Below:* The new method of Doppler-free spectroscopy (saturated absorption using a dye laser, Sect. 22.3) allows resolution of the individual components of the H_α line (after *Hänsch* et al.). The two additional very weak lines shown in Fig. 12.23 are omitted in Fig. 12.24

component of this line) and return to the ground state. When absorption of this type occurs, the electron current in the tungsten foil decreases. *Lamb* and *Retherford* found in 1947 that the same effect, a decrease in the electron current, occurred on absorption of radiation at a frequency of about 1000 MHz. This was due to the transition from the $2^2S_{1/2}$ to the $2^2P_{1/2}$ state. From the latter state, radiative transitions to the ground state are also allowed. It was thus shown that even states with the same total angular momentum j are energetically different.

The term scheme of an atom can be refined for optical transitions as well. Figure 12.23 shows the complete term scheme for the H_α line of the hydrogen atom. This line

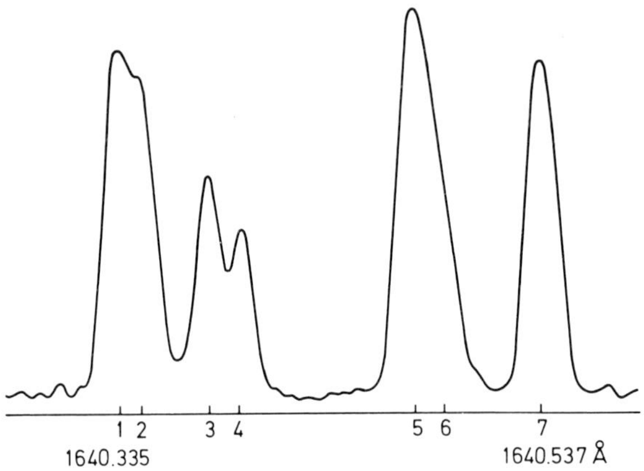

Fig. 12.25. Lamb shift and fine structure of the helium atom: fine structure at 1640 Å. Seven components are observed. The lines 1, 2 and 3, 4 would be unsplit without the Lamb shift. [From G. Herzberg: Trans. Roy. Soc. Can. **5**, (1967) Fig. 5]

consists of 7 components of different intensities in an energy range of about 0.4 cm^{-1}. The upper part of Fig. 12.24 shows the structure of the H_α line, as it can be observed in the presence of Doppler broadening. The lower part shows a curve obtained by the modern method of Doppler-free spectroscopy (Sect. 22.3). With this method, the Lamb shift can also be seen in optical spectra. The fine structure splitting and the Lamb shift are larger in the heavier helium atom, so that direct optical observation of the splitting is easier than with the H atom. Even without removing the Doppler broadening, the fine structure can be resolved, as Fig. 12.25 shows.

The Lamb shift was of utmost importance for the development of quantum electrodynamics. Previously, this theory treated only the emission and absorption of light quanta in atomic transitions. To explain the Lamb shift, it was necessary to go one step further. It had to be assumed that the electrons in an atom were continually emitting and reabsorbing light quanta, in which process energy conservation can apparently be violated.

This "violation of energy conservation" must not, to be sure, be taken too literally. According to the uncertainty relation between energy and time (7.29), the energy is only sharply defined when a measurement is performed over a sufficiently long period of time. It is thus completely consistent with energy conservation that an electron can emit a quantum even without having the necessary energy, as long as the quantum is reabsorbed quickly enough. Much more decisive for the theoreticians was, however, the recognition that the energy shifts in the atomic levels (on a negative energy scale) produced by these "virtual" processes were infinitely large. A free electron can also continually emit and absorb virtual quanta; its energy decreases infinitely in the process. Energy shifts caused by virtual processes are termed *self energy*. Experimentally, a free electron, like a bound electron, has a well-defined, *finite* energy. The basic idea for solving the "infinity problem" of the energy shift was the recognition that only the difference between the energies of bound and free electrons is physically interesting. Or, in other words: to calculate the energy shift of bound electrons, one must subtract the self-energy of a free electron from that of a bound electron in a particular atomic state (cum grano salis). This process is termed "renormalisation". Since the masses also become infinite due to virtual processes, they must also be "renormalised". Naturally, at first glance it seems very adventurous to subtract two infinite quantities from one another in order to obtain a well-defined finite result. In the framework of quantum electrodynamics, however, it was found possible to set up well-defined rules for the renormalisation procedure, and the Lamb shift can be calculated today with great precision. The important result is that the validity of quantum electrodynamics can therefore be tested – and has been verified – in an excellent manner.

A summary of the theoretical treatment is given in Sect. 15.5.2. In preparation for this treatment, in Sect. 15.5.1 we introduce the quantisation of the electromagnetic field, which follows immediately from the quantisation of the harmonic oscillator. As is shown in one of the problems for Sect. 15.5.1, the theory of the Lamb shift has a surprisingly simple physical explanation: the quantum-mechanical zero-point fluctuations of the electromagnetic field act statistically on the electrons and thus cause a shift of their potential energy.

Problems

12.1 Calculate the precessional frequency of electrons and of protons [$I = 1/2$, magnetic moment $= (2.79/1836) \cdot \mu_B$] in the magnetic field of the earth ($\simeq 2 \cdot 10^{-5}$ tesla).

12.2 In the Stern-Gerlach experiment, a beam of silver atoms in the ground state ($5^2S_{1/2}$) is directed perpendicularly to a strong inhomogeneous magnetic field. The field gradient is $dB/dz = 10^3$ tesla/m. In the direction of the atomic beam, the magnetic field extends a distance of $l_1 = 4$ cm, and the catcher screen is a distance $l_2 = 10$ cm from the magnet. Calculate the components of the magnetic moment in the direction of the magnetic field, if the splitting of the beam at the screen is observed to be $d = 2$ mm, and the velocity of the atoms is $v = 500$ m/s. The average mass of silver atoms is $M = 1.79 \cdot 10^{-25}$ kg. Why doesn't the nuclear spin affect the experiment?

12.3 How large is the magnetic field generated by the electron in the ground state of a hydrogen atom, at the position of the proton if it would circulate according to Bohrs model on the shell $n = 1$?

12.4 How large is the magnetic moment of the orbital motion in a muonium atom, in which the electron of a ground-state hydrogen atom has been replaced by a muon? How large is the moment in positronium (an electron and a positron, i.e. particles with the mass of the electron and opposite charges, moving around the common centre of mass)?

12.5 Calculate the spin-orbit splitting of the states of the hydrogen atom with $n = 2$ and $n = 3$ using the relations

$$V_{l,s} = \frac{Ze^2\mu_0}{8\pi m_0^2 r^3}(\boldsymbol{s} \cdot \boldsymbol{l}) ,$$

and

$$r^{-3} = \frac{Z^3}{a_0^3 n^3 l(l+\frac{1}{2})(l+1)} .$$

What are the values for a Rydberg state with $n = 30$ for the largest ($l = 1$) and the smallest ($l = 29$) splitting?
a_0 is the radius of the innermost Bohr orbit.

12.6 In the cesium atom, spin-orbit splitting between the states $6P_{1/2}$ and $6P_{3/2}$ leads to a wavelength difference of $\Delta\lambda = 422$ Å for the first line pair of the primary series, with $\lambda = 8521$ Å for the line with the shorter wavelength. Calculate from this the fine structure constant a and the field at the nucleus $\boldsymbol{B}_1$. Use (12.27).

12.7 Sketch the energy levels of the hydrogen atom, including the fine structure, up to $n = 3$. Show the possible transitions. How many different lines are there?

12.8 The fine structure in hydrogen-like ions (ions with only one electron) is described by (12.33).

a) Show that the correction term does not disappear for any possible combination of the quantum numbers n and j, but that it always reduces the value of the uncorrected energy.
b) Into how many energy levels are the terms of singly charged helium with the principal quantum numbers $n = 3$ and $n = 4$ split by the fine structure interaction?
c) Sketch the positions of these levels relative to the non-shifted terms and give the amount of the shift.
d) Determine which transitions are allowed, using the selection rules $\Delta l = \pm 1$, $\Delta j = 0$ or ± 1.

12.9 Give the relative splitting of the various levels of an LSJ multiplet due to spin-orbit interaction for the ^3F and ^3D multiplets. Sketch the energy levels of these multiplets and indicate with arrows the allowed $^3F \rightarrow {}^3D$ transitions. Repeat the above process for the $^4D \rightarrow {}^4P$ and $^4P \rightarrow {}^4S$ transitions.

12.10 The interaction energy E between two magnetic moments $\boldsymbol{\mu}_1$ and $\boldsymbol{\mu}_2$ is ($\boldsymbol{r}$ = the radius vector of $\boldsymbol{\mu}_1$ and $\boldsymbol{\mu}_2$):

$$E = -\frac{\mu_0}{4\pi}\left\{\frac{\boldsymbol{\mu}_1 \cdot \boldsymbol{\mu}_2}{r^3} - 3\frac{(\boldsymbol{\mu}_1 \cdot \boldsymbol{r})(\boldsymbol{\mu}_2 \cdot \boldsymbol{r})}{r^5}\right\}.$$

a) Under which conditions is $E = 0$ for a given $|\boldsymbol{r}|$?
b) For parallel moments, which arrangement yields an extreme value for E?
c) For case b) with $|\boldsymbol{r}| = 2$ Å, calculate the energy for the electron-electron and proton-proton interactions. In each case, how large is the magnetic field at $\boldsymbol{\mu}_2$ due to $\boldsymbol{\mu}_1$ ($\mu_{\text{proton}} = 1.4 \cdot 10^{-26}\,\text{A m}^2$)?

13. Atoms in a Magnetic Field: Experiments and Their Semiclassical Description

13.1 Directional Quantisation in a Magnetic Field

In the previous chapters, we have already seen that a directional quantisation exists. The angular momentum vectors in an atom can only orient themselves in certain discrete directions relative to a particular axis (the quantisation axis). The directional quantisation is described by the magnetic quantum number m. In an applied magnetic field $\boldsymbol{B}_0$, the interaction energy between the field and the magnetic moment of the electrons in an atom, which we have already calculated, leads to a splitting of the energy terms, which is described by the different possible values of the magnetic quantum number. We shall concern ourselves in this chapter with the measurement of this energy splitting.

A first application of the splitting of atomic states in a magnetic field to the determination of the magnetic moments of the atoms was already discussed in the treatment of the Stern-Gerlach experiment. In the following, we shall consider some other types of experiments.

13.2 Electron Spin Resonance

The method of electron spin resonance (abbreviated ESR, sometimes EPR for electron paramagnetic resonance) involves the production of transitions between energy states of the electrons which are characterised by different values of the magnetic quantum number m. In general, the degeneracy is lifted by the application of an external magnetic field; the transition frequencies, which are usually in the range of microwave frequencies, depend on the strength of the applied field. With this technique, one can observe transitions between states of different magnetic quantum number directly. In Zeeman spectroscopy, to be described later, the transitions observed are in the optical region, and their response to magnetic fields is studied; in this case, the transitions cause changes in not only the magnetic quantum number, but also in the other quantum numbers.

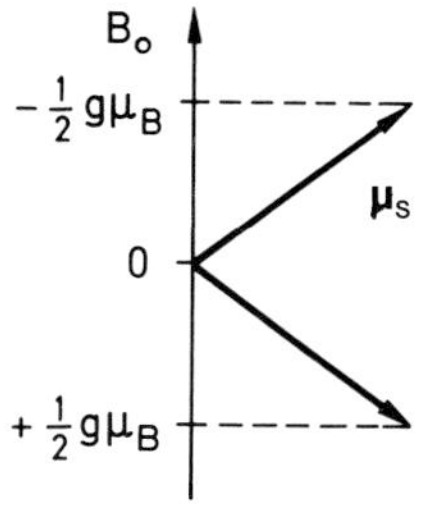

Fig. 13.1. The magnetic moment and the spin of an electron have two possible orientations in an applied magnetic field. They correspond to two values of the potential energy

The principle of ESR may be easily understood by considering the magnetic moment produced by the spin of a free electron in a magnetic field $\boldsymbol{B}_0$ (Fig. 13.1).

An electron has the magnetic moment

$$\mu_s = \sqrt{s(s+1)}\,\mu_B g_s \tag{13.1}$$

with the possible components along the quantisation axis z of the field $\boldsymbol{B}_0$

$$(\mu_s)_z = \pm\tfrac{1}{2} g_s \mu_B\,. \tag{13.2}$$

The potential energy of these two orientations differs by the amount

$$\Delta E = g_s \mu_B B_0\,. \tag{13.3}$$

If a sinusoidally varying magnetic field $\boldsymbol{B}_1 = \tilde{\boldsymbol{B}}_1 \sin \omega t$ is now applied in a direction perpendicular to $\boldsymbol{B}_0$, transitions between the two states are induced if the frequency $\nu = \omega/2\pi$ fulfils the condition

$$\Delta E = h\nu = g_s \mu_B B_0\,, \tag{13.4}$$

or, in numbers,

$$\nu = 2.8026 \cdot 10^{10} \cdot \boldsymbol{B}_0\,\text{Hz (tesla)}^{-1}\,. \tag{13.5}$$

The transitions with $\Delta m = \pm 1$ are allowed magnetic dipole transitions. A quantum mechanical treatment of ESR will follow in Chap. 14. The frequency which must be used depends, according to (13.5), on the choice of the applied magnetic field $\boldsymbol{B}_0$. For reasons of sensitivity, usually the highest possible frequencies are used, corresponding to the highest possible magnetic fields. The fields and frequencies used in practice are, of course, limited by questions of technical feasibility; usually, fields in the range 0.1 to 1 T are chosen (T = tesla). This leads to frequencies in the GHz region (centimetre waves).

What we have here described for a free electron is also valid for a paramagnetic atom. In this case, the total resultant magnetic moment produced by the spin and orbital angular momenta of the atom, $\boldsymbol{\mu}_j$, must be used in (13.3 – 5).

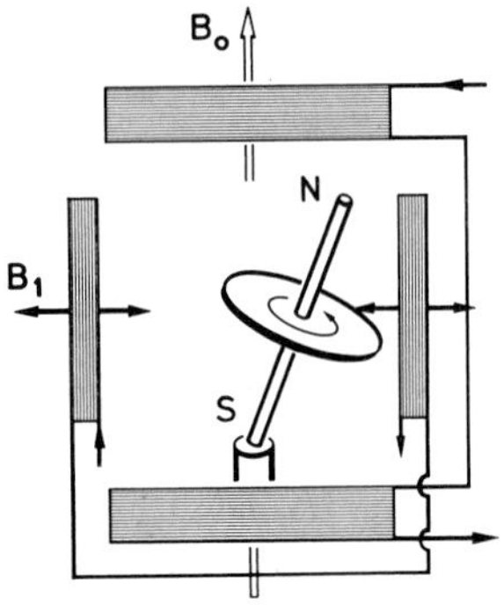

Fig. 13.2. Demonstration experiment for electron spin resonance: a gyroscope whose axle is a bar magnet is processing in a magnetic field $\boldsymbol{B}_0$ (as well as in the gravitational field of the earth). The inclination of the axis of the gyroscope relative to $\boldsymbol{B}_0$ may be changed by means of an oscillating field $\boldsymbol{B}_1$ if the frequency of $\boldsymbol{B}_1$ is equal to the precession frequency of the gyroscope. For a lecture demonstration, it is expedient to construct the gyroscope in such a way that it is driven from the support pedestal S, for example using compressed air and following the principle of a water turbine

The fundamental idea of ESR may be illustrated by a mechanical model (Fig. 13.2): a gyroscope containing a bar magnet in its axis is precessing in a magnetic field. The precession frequency is (neglecting gravitational force)

$$\omega_L = \frac{|\boldsymbol{\mu}| \cdot |\boldsymbol{B}_0|}{|\boldsymbol{L}|}, \tag{13.6}$$

where $\boldsymbol{\mu}$ is the magnetic moment of the bar magnet and $\boldsymbol{L}$ is the angular momentum of the gyroscope.

The precession frequency ω_L of a magnetic gyroscope in a magnetic field is independent of the angle α between $\boldsymbol{\mu}$ and $\boldsymbol{B}_0$, since the torque produced by the field and the rate of change of the angular momentum vector both depend in the same way on the sine of the angle α (12.12). When gravitational force is neglected, the frequency ω_L is determined only by the magnetic moment $\boldsymbol{\mu}$ and the angular momentum $\boldsymbol{L}$ of the gyroscope, as well as by the torque produced by the field $\boldsymbol{B}_0$.

When we now let an additional oscillating field $\boldsymbol{B}_1$ with the frequency ω act perpendicular to $\boldsymbol{B}_0$, we observe a continuous increase or decrease in the angle of inclination α, depending on whether the field is in phase or out of phase with the motion of the gyroscope, provided that the frequency ω is equal to ω_L.

This model may be immediately transferred to the atom. We replace the magnetic moment of the bar magnet by the moment of the atom and obtain for the circular frequency of the electron spin resonance the following condition:

$$\omega_L = \frac{|\boldsymbol{\mu}| \cdot |\boldsymbol{B}_0|}{|\boldsymbol{l}|} = \gamma B_0 \,. \tag{13.7}$$

This is the Larmor frequency, which was already introduced in Sect. 12.3.

In the classical gyroscope model, the tip of the gyroscope axle moves on a spiral orbit from one stable position to another. This picture may be applied with considerable accuracy to the motion of the spin or the orbital angular momentum in an atom. There is an additional possibility for picturing the resonant transitions, which makes use of the fact that the spin or the angular momentum of an atom has only certain discrete allowed stationary orientations in a constant magnetic field $\boldsymbol{B}_0$. In this picture, the spin makes transitions between these discrete energy levels under the influence of the oscillating field $\boldsymbol{B}_1$. In particular, this means in the case of spin 1/2 that the spin flips from the one possible orientation to the other when the resonance condition (13.7) is fulfilled.

Electron spin resonance was observed for the first time in 1945 by the Russian physicist *Zavoisky*. The analogous spin resonance of paramagnetic atomic *nuclei* is seen under otherwise identical conditions at a frequency which is 3 orders of magnitude smaller, due to the fact that nuclear moments are about a factor of 1000 smaller than atomic magnetic moments; the corresponding frequencies are in the radio frequency region. This *nuclear magnetic resonance* (NMR) was observed in the solid state for the first time in 1946 by *Bloch* and *Purcell*, nearly 10 years after it had first been used by *Rabi* to measure the gyromagnetic ratio of nuclei in gas atoms (cf. Sect. 20.6).

A schematic of an ESR apparatus is shown in Fig. 13.3. Today, ESR spectrometers count as standard spectroscopic accessories in many physical and chemical laboratories. For technical reasons, usually a fixed frequency is used in the spectrometers;

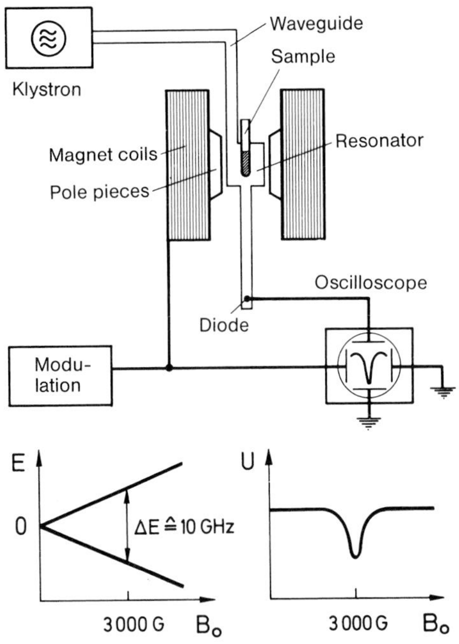

Fig. 13.3. Electron spin resonance. *Above:* Schematic representation of the experimental setup. The sample is located in a resonant cavity between the pole pieces of an electromagnet. The microwaves are generated by a klystron and detected by a diode. To increase the sensitivity of detection, the field $\boldsymbol{B}_0$ is modulated. *Below, left:* Energy states of a free electron as functions of the applied magnetic field. *Below, right:* Signal U from the diode as a function of B_0 for resonance

the magnetic field is varied to fulfil the resonance condition and obtain ESR transitions in absorption and sometimes in emission. The sample is usually placed in a microwave resonator; a frequently used wavelength is 3 cm (the so-called X-band). The microwave radiation is generated by a klystron and detected by a high frequency diode or a bolometer.

ESR is utilised for

- precision determinations of the gyromagnetic ratio and the g factor of the electron;
- measurement of the g factor of atoms in the ground state and in excited states for the purpose of analysing the term diagram;
- the study of various kinds of paramagnetic states and centres in solid state physics and in chemistry: molecular radicals, conduction electrons, paramagnetic ions in ionic and metallic crystals, colour centres.

The full importance of ESR will only become clear after we have treated the topic of hyperfine structure, i.e. when we discuss the interaction of the electronic spin with the spins of the neighbouring nuclei. Using this interaction, termed hyperfine splitting (Chap. 20), one can determine the spatial distribution of the electrons in molecules, in liquids, and in solids.

13.3 The Zeeman Effect

13.3.1 Experiments

The splitting of the energy terms of atoms in a magnetic field can also be observed as a splitting of the frequencies of transitions in the optical spectra (or as a shift). A splitting of this type of spectral lines in a magnetic field was observed for the first time in 1896

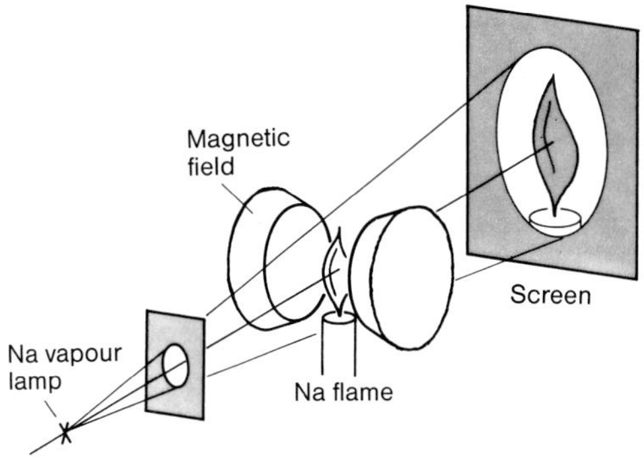

Fig. 13.4. Demonstration experiment for the Zeeman effect. A flame coloured with sodium or NaCl appears dark when projected using light from a Na vapour lamp. Upon switching on a magnetic field, it brightens, since the resonance between the light from the lamp and the light of the sodium flame is destroyed by the Zeeman effect. The wavelength of the light from the flame is shifted slightly by the magnetic field; this suffices to remove the resonance

by *Zeeman*. The effect is small; for its observation, spectral apparatus of very high resolution is required. These are either diffraction grating spectrometers with long focal lengths and a large number of lines per cm in the grating, or else interference spectrometers, mainly Fabry-Perot interferometers. We shall discuss this topic in more detail in Chap. 22.

There is, however, a simple lecture demonstration (Fig. 13.4) which shows the shift of the spectral lines in a magnetic field in a drastic manner: a flame, coloured yellow with sodium, is opaque to the yellow light of a sodium vapour lamp, because the latter represents resonance light, i.e. light whose wavelength matches the absorption and emission wavelength in the flame. If, however, a magnetic field is applied to the flame, the resonance between the light source (Na lamp) and the absorber (Na flame) is destroyed. On the observation screen, the previously "dark" flame brightens, because it has now become transparent to the light from the Na vapour lamp.

With a Fabry-Perot interferometer or with a grating spectrometer of sufficient resolution, the splitting in magnetic fields may be quantitatively measured. The splitting behaviour observed in moderate magnetic fields is illustrated in Figs. 13.5 and 13.6. The splitting of the cadmium line in Fig. 13.5 is called the "ordinary" Zeeman effect; using transverse observation (i.e. observation perpendicular to the direction of the applied magnetic field, Fig. 13.7), one sees the unshifted line as well as two

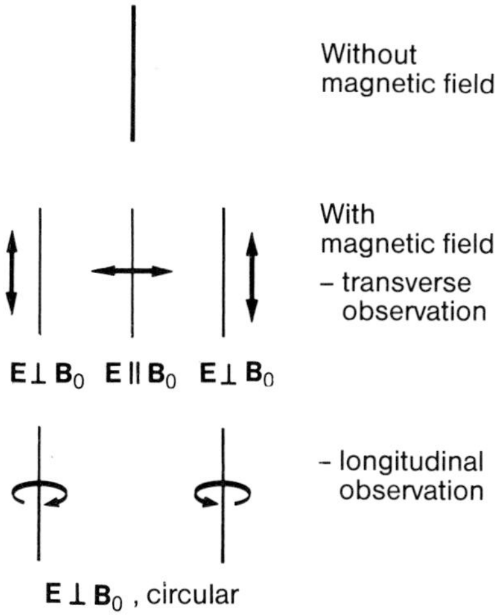

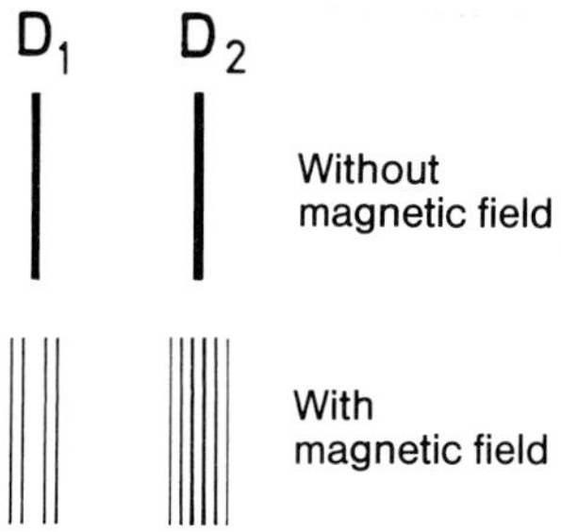

Fig. 13.6. Anomalous Zeeman effect, here using the sodium D lines. The D_1 line splits into four components, the D_2 line into six in a magnetic field. The wavelengths of the D_1 and D_2 lines are 5896 and 5889 Å; the quantum energy increases to the right in the diagram

◀ **Fig. 13.5.** Ordinary Zeeman effect, e.g. for the atomic Cd line at $\lambda = 6438$ Å. With transverse observation the original line and two symmetrically shifted components are seen. Under longitudinal observation, only the split components are seen. The polarisation (E vector) is indicated

symmetrically split components, each linearly polarised. With longitudinal observation (parallel to the field lines), only the two shifted components are seen; they are circularly polarised in this case.

The splitting behaviour of the D lines of the sodium atom shown in Fig. 13.6 is typical of the *anomalous* Zeeman effect. The number of components into which the spectral lines are split is greater than in the normal Zeeman effect. Both the ordinary and the anomalous Zeeman effects merge to the so-called Paschen-Back effect in sufficiently large magnetic fields $\boldsymbol{B}_0$. We shall now discuss these three effects of the influence of magnetic fields on the spectral lines and the energy terms of atoms.

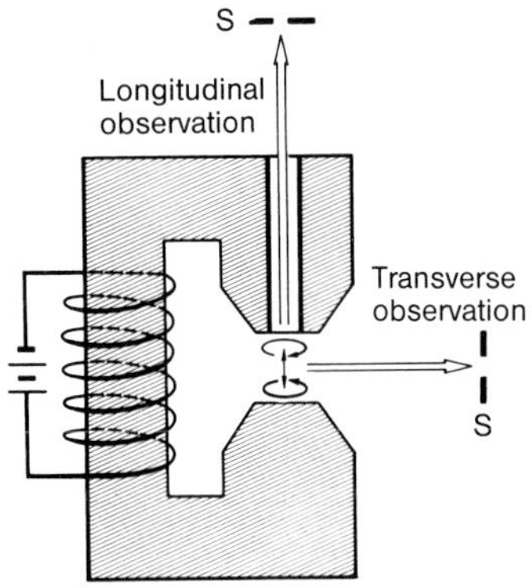

Fig. 13.7. Transverse and longitudinal observation of spectral lines in a magnetic field. The three component electrons used in the classical description of the Zeeman effect are indicated (orbits with arrows in the pole gap of the magnet). The emission of a light source in the magnetic field is observed either transversely or longitudinally (through a hole drilled in the magnet pole piece). S is the entrance slit of a spectrometer

13.3.2 Explanation of the Zeeman Effect from the Standpoint of Classical Electron Theory

The Zeeman effect may be understood to a large extent using classical electron theory, as was shown by *Lorentz* shortly after its discovery. We shall restrict ourselves to the ordinary Zeeman effect – the splitting of states with pure orbital angular momentum. If the resultant angular momentum is composed of both spin and orbital contributions, one speaks of the anomalous Zeeman effect.

We discuss the emission of light by an electron whose motion about the nucleus is interpreted as an oscillation, for example by considering the projection of the motion on a certain direction. We ask the question, "What force does a magnetic field exert on a radiating electron?" The radiating electron is treated as a linear oscillator with a random orientation with respect to the magnetic lines of force (Fig. 13.8).

In the model, we replace the electron by three component oscillators according to the rules of vector addition: component oscillator 1 oscillates linearly, parallel to the direction of $\boldsymbol{B}_0$. Oscillators 2 and 3 oscillate circularly in opposite senses and in a plane

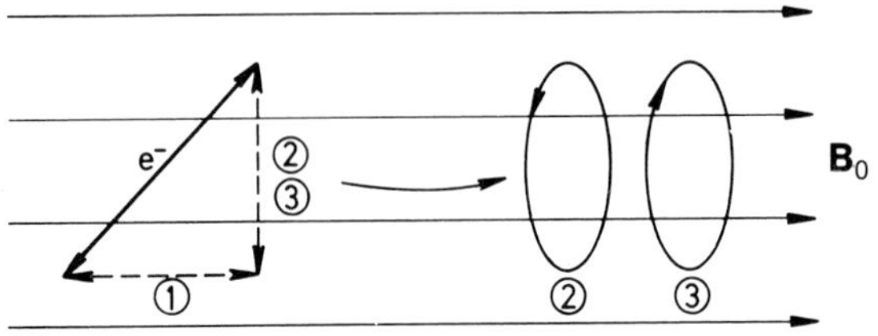

Fig. 13.8. Classical explanation of the Zeeman effect. An oscillating electron is resolved into three component oscillators. Further details in the text

perpendicular to the direction of $\boldsymbol{B}_0$. This resolution into components is allowed, since any linear oscillation may be represented by the addition of two counterrotating circular ones. Without the field $\boldsymbol{B}_0$, the frequency of all the component oscillators is equal to that of the original electron, namely ω_0.

We now inquire as to the forces which the magnetic field exerts on our three component electron-oscillators:

- Component 1, parallel to $\boldsymbol{B}_0$, experiences no force. Its frequency remains unchanged; it emits light which is linearly polarised with its $\boldsymbol{E}$ vector parallel to the vector $\boldsymbol{B}_0$.
- The circularly oscillating components 2 and 3 are accelerated or slowed down by the effect of magnetic induction on turning on the field $\boldsymbol{B}_0$, depending on their directions of motion. Their circular frequencies are increased or decreased by an amount

$$\delta\omega = \tfrac{1}{2}(e/m_0)B_0 = (\mu_B/\hbar)B_0 . \tag{13.8}$$

This is almost the same expression as that which we have already come to know as the Larmor frequency. It differs from the Larmor frequency only by a factor 2, because we are here dealing with an orbital moment ($g = 1$) instead of a spin moment ($g = 2$) as in the case of the Larmor frequency, which applies to electron spin resonance.

Classically, one can calculate the frequency shift $\delta\omega$ for the component oscillators as follows: without the applied magnetic field, the circular frequency of the component electrons is ω_0. The Coulomb force and the centrifugal force are in balance, i.e.

$$m\omega_0^2 \boldsymbol{r} = \frac{Ze^2}{4\pi\varepsilon_0 r^3}\boldsymbol{r} .$$

In a homogeneous magnetic field $\boldsymbol{B}_0$ applied in the z direction, the Lorentz force acts in addition; in Cartesian coordinates, the following equations of motion are then valid:

$$m\ddot{x} + m\omega_0^2 x - e\dot{y}B_0 = 0 , \tag{13.9a}$$

$$m\ddot{y} + m\omega_0^2 y + e\dot{x}B_0 = 0 , \tag{13.9b}$$

$$m\ddot{z} + m\omega_0^2 z = 0 . \tag{13.9c}$$

From (13.9c), we immediately find the solution for component oscillator 1, $z = z_0 \exp(\mathrm{i}\omega_0 t)$, i.e. the frequency of the electron which is oscillating in the z direction remains unchanged.

To solve (13.9a) and (13.9b), we substitute $u = x + \mathrm{i}y$ and $v = x - \mathrm{i}y$. It is easy to show that the equations have the following solutions (with the condition $eB_0/2m \ll \omega_0$):

$$u = u_0 \exp[\mathrm{i}(\omega_0 - eB_0/2m)t] \quad \text{and} \quad v = v_0 \exp[\mathrm{i}(\omega_0 + eB_0/2m)t] .$$

These are the equations of motion for a left-hand and a right-hand circular motion with the frequencies $\omega_0 \pm \delta\omega$, with $\delta\omega = eB_0/2m$. The component electron oscillators 2 and 3 thus emit or absorb circularly polarised light with the frequency $\omega_0 \pm \delta\omega$.

The splitting observed in the ordinary Zeeman effect is therefore correctly predicted in a classical model.

The frequency change has the magnitude:

$$\delta\nu = \delta\omega/2\pi = \frac{1}{4\pi}\frac{e}{m_0}B_0 . \qquad (13.10)$$

For a magnetic field strength $B_0 = 1$ T, this yields the value

$$\delta\nu = 1.4 \cdot 10^{10}\,\mathrm{s}^{-1} \triangleq 0.465\,\mathrm{cm}^{-1} . \qquad (13.11)$$

Independently of the frequency ν, we obtain the same frequency shift $\delta\nu$ for each spectral line with a given magnetic field B_0. Theory and experiment agree completely here. For the polarisation of the Zeeman components, we find the following predictions: component electron oscillator 1 has the radiation characteristics of a Hertzian dipole oscillator, oscillating in a direction parallel to $\boldsymbol{B}_0$. In particular, the $\boldsymbol{E}$ vector of the emitted radiation oscillates parallel to $\boldsymbol{B}_0$, and the intensity of the radiation is zero in the direction of $\boldsymbol{B}_0$. This corresponds exactly to the experimental results for the unshifted Zeeman component; it is also called the π component (π for parallel). If the radiation from the component electron oscillators 2 and 3 is observed in the direction of $\boldsymbol{B}_0$, it is found to be circularly polarised; observed in the direction perpendicular to $\boldsymbol{B}_0$, it is linearly polarised. This is also in agreement with the results of experiment. This radiation is called σ^+ and σ^- light, were σ stands for perpendicular (German "senkrecht") and the + and − signs for an increase or a decrease of the frequency. The σ^+ light is right-circular polarised, the σ^- light is left-circular polarised. The direction is defined relative to the lines of force of the $\boldsymbol{B}_0$ field, *not* relative to the propagation direction of the light.

The differing polarisations of the Zeeman components are used in optical pumping. In this technique, the exciting light can be polarised so as to populate individual Zeeman levels selectively, and thus to produce a spin orientation. More about this in Sect. 13.3.7.

13.3.3 Description of the Ordinary Zeeman Effect by the Vector Model

In the preceding section, we gave a purely classical treatment of the ordinary Zeeman effect; we now take the first step towards a quantum mechanical description. For this purpose, we employ the vector model which has been already introduced in Sect. 12.2 (see also Fig. 13.9). A complete quantum mechanical treatment will be given in Chap. 14. The angular momentum vector $\boldsymbol{j}$ and the magnetic moment $\boldsymbol{\mu}_j$, which is coupled to $\boldsymbol{j}$, precess together around the field axis $\boldsymbol{B}_0$. The additional energy of the atom due to the magnetic field is then (Chap. 12)

$$V_{m_j} = -(\mu_j)_z \cdot B_0 = -m_j g_j \mu_B B_0 \quad \text{with} \quad m_j = j,\ j-1, \ldots -j . \qquad (13.12)$$

Here the factor g_l in (12.6) was replaced by g_j, because the total angular momentum is being considered.

The $(2j+1)$-fold directional degeneracy is thus lifted, and the term is split into $2j+1$ components. These are energetically equidistant. The distance between two components with $\Delta m_j = 1$ is

$$\Delta E = g_j \mu_B B_0 .$$

If we ignore the spin and consider only orbital magnetism (i.e. the ordinary Zeeman effect), g_j has a numerical value of 1 and we obtain

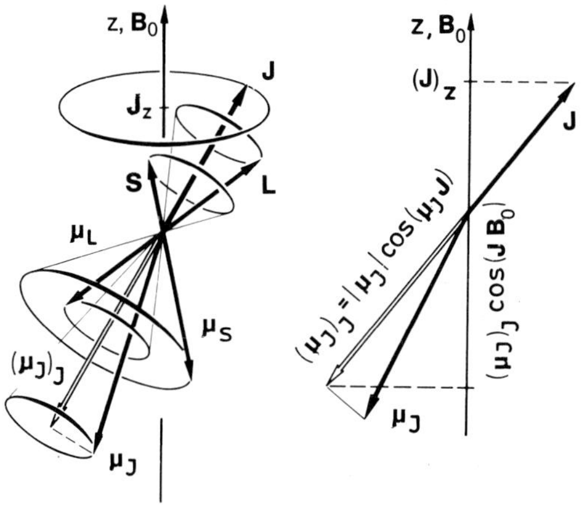

Fig. 13.9. *Left:* The relation between the angular momentum J, the magnetic moment μ_J and their orientation with respect to the magnetic field $\boldsymbol{B}_0$ for strong spin-orbit coupling. The angular momentum vectors $\boldsymbol{S}$ and $\boldsymbol{L}$ combine to form $\boldsymbol{J}$. Likewise, the associated magnetic moments μ_L and μ_S combine to μ_J. Because spin and orbital magnetism have different gyromagnetic ratios, the directions of the vectors $\boldsymbol{J}$ and μ_J do not coincide. What can be observed is the projection of μ_J on $\boldsymbol{J}$, as the time average of many precession cycles. That is, one observes the component $(\mu_J)_J$, which is therefore represented as $\bar{\mu}_J$ or $\bar{\mu}_S$, see the right-hand diagram. In the one-electron system, lower case letters can be used instead of S, L and J, as is done in the text. *Right:* The projection of μ_J on the vector $\boldsymbol{J}$ is $(\mu_J)_J$. The projection of $(\mu_J)_J$ on $\boldsymbol{B}_0$ is calculated using the Landé factor. Because the angular momenta $\boldsymbol{S}$ and $\boldsymbol{L}$ are strongly coupled, the vector μ_J precesses rapidly around the negative extension of the vector $\boldsymbol{J}$. Only the time average $(\mu_J)_J$ in the $\boldsymbol{J}$ direction can be observed. This precesses slowly, because of weak coupling, around the axis of $\boldsymbol{B}_0$. The magnetic energy is the product of the field strength B_0 and the component of $(\mu_J)_J$ in the direction of $\boldsymbol{B}_0$, i.e. $(\mu_J)_{J,z}$ or $(\bar{\mu}_J)_z$. Lower case letters can be used instead of S, L, J in the one electron system. Figure 13.9 illustrates the anomalous Zeeman effect (Sect. 13.3.4). The ordinary Zeeman effect (Sect. 13.3.3) is more simple. From $S=0$ follows $\mu_J=\mu_L$, and the directions of the vectors $-\mu_J$ and $J=L$ coincide

$$\delta\nu = \frac{1}{4\pi}\frac{e}{m_0}B_0\,. \tag{13.13}$$

The magnitude of the splitting is thus the same as in classical theory. For optical transitions, one must also make use of the selection rule

$$\Delta m_j = 0, \pm 1\,.$$

One thus obtains from quantum theory, too, the result that the number of lines is always three: the ordinary Zeeman triplet.

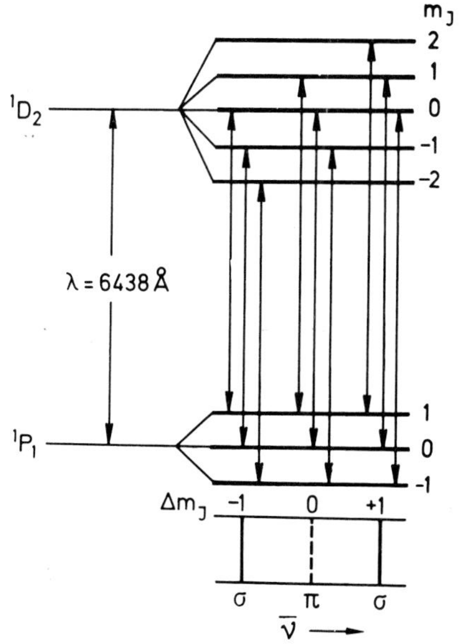

Fig. 13.10. Ordinary Zeeman effect. Splitting of the λ = 6438 Å line of the neutral Cd atom, transition ${}^1P_1 - {}^1D_2$, into three components. The transitions with $\Delta m_J = 0$ are called π transitions; those with $\Delta m_J = \pm 1$ are σ transitions. The quantum number J is written as a capital letter because the atom has several electrons (see Chap. 17)

As an example, Fig. 13.10 shows the splitting diagram for a cadmium line. We must point out that the orbital angular momentum for the states of the Cd atom comprises the orbital angular momenta of two electrons, and is therefore indicated by a capital letter L. The spins of the two electrons are antiparallel and thus compensate each other, giving a total spin $S = 0$. Transitions between the components of different terms (e.g. ${}^1P_1$ or ${}^1D_2$ in Fig.13.10) with the same Δm_j are energetically the same. The splitting is equal in each case because only orbital magnetism is involved. [See the discussion of the Landé g factor in Sect. 13.3.5, especially (13.18).] The undisplaced line corresponds to transitions with $\Delta m = 0$, while the displaced lines are the transitions with $\Delta m = \pm 1$. They are circularly polarised.

Polarisation and ordinary Zeeman splitting are a good example of the correspondence principle (Sect. 8.11). Based on the conservation of angular momentum for the system of electrons and light quanta, the polarisation behaviour of the Zeeman effect implies that light quanta have the angular momentum $1 \cdot \hbar$.

13.3.4 The Anomalous Zeeman Effect

One speaks of the anomalous Zeeman effect when the angular momentum and magnetic moment of the two terms between which an optical transition occurs cannot be described by just one of the two quantum numbers s or l (or S or L), but are determined by both. This is the general case, in which atomic magnetism is due to the superposition of spin and orbital magnetism. The term "anomalous" Zeeman effect is historical, and is actually contradictory, because this is the normal case.

In cases of the anomalous Zeeman effect, the two terms involved in the optical transition have different g factors, because the relative contributions of spin and orbital magnetism to the two states are different. The g factors are determined by the total angular momentum j and are therefore called g_j factors. The splitting of the terms in the ground and excited states is therefore different, in contrast to the situation in the normal Zeeman effect. This produces a larger number of spectral lines. The calculation of the g_j factors follows in Sect. 13.3.5.

We will use the Na D lines (Fig. 13.11) as an example for a discussion of the anomalous Zeeman effect.

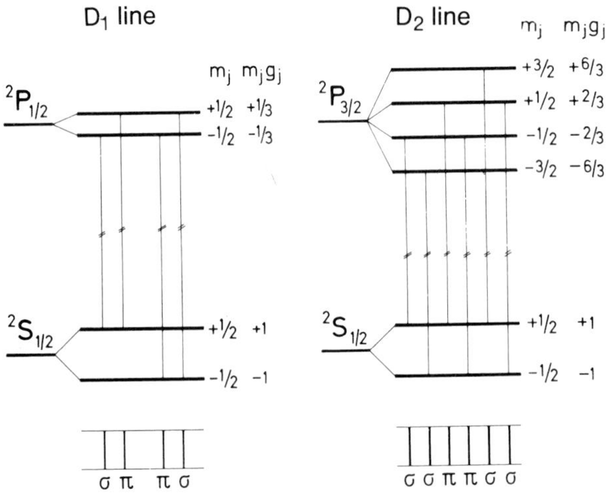

Fig. 13.11. Anomalous Zeeman effect. Splitting of the D_1 and D_2 lines of the neutral Na atom, transitions ${}^2S_{1/2}-{}^2P_{1/2}$ and ${}^2S_{1/2}-{}^2P_{3/2}$, into 4 and 6 components, respectively, in a magnetic field

For the three terms involved in the transitions which produce the Na D line, namely the $^2S_{1/2}$, the $^2P_{1/2}$ and the $^2P_{3/2}$, the magnetic moments in the direction of the field are

$$(\mu_j)_{j,z} = m_j g_j \mu_B \,, \tag{13.14}$$

and the magnetic energy is

$$V_{m_j} = -(\mu_j)_{j,z} B_0 \,. \tag{13.15}$$

The number of splitting components in the field is given by m_j and is again $2j+1$. The distance between the components with different values of m_j – the so-called Zeeman components – is no longer the same for all terms, but depends on the quantum numbers l, s, and j:

$$\Delta E_{m_j, m_{j-1}} = g_j \mu_B B_0 \,. \tag{13.16}$$

Experimentally, it is found that $g_j = 2$ for the ground state $^2S_{1/2}$, 2/3 for the state $^2P_{1/2}$ and 4/3 for the state $^2P_{3/2}$. We shall explain these g_j factors in the next section. For optical transitions, the selection rule is again $\Delta m_j = 0, \pm 1$. It yields the 10 lines shown in Fig. 13.11.

The significance of the Zeeman effect is primarily its contribution to empirical term analysis. Term splitting depends unequivocally on the quantum numbers l, s and j or, in many-electron atoms, L, S and J (Chap. 17). The quantum numbers can therefore be determined empirically from measurements of the Zeeman effect.

13.3.5 Magnetic Moments with Spin-Orbit Coupling

In anomalous Zeeman splitting, other values of g_j than 1 (orbital magnetism) or 2 (spin magnetism) are found. We can understand these quantitatively through the vector model.

The g_j factor links the magnitude of the magnetic moment of an atom to its total angular momentum. The magnetic moment is the vector sum of the orbital and spin magnetic moments,

$$\boldsymbol{\mu}_j = \boldsymbol{\mu}_s + \boldsymbol{\mu}_l \,.$$

The directions of the vectors $\boldsymbol{\mu}_l$ and $\boldsymbol{l}$ are antiparallel, as are those of the vectors $\boldsymbol{\mu}_s$ and $\boldsymbol{s}$. In contrast, the directions of $\boldsymbol{j}$ and $-\boldsymbol{\mu}_j$ do not in general coincide. This is a result of the difference in the g factors for spin and orbital magnetism. This is demonstrated in Figs. 13.12 and 13.9.

The magnetic moment $\boldsymbol{\mu}_j$ resulting from vector addition of $\boldsymbol{\mu}_l$ and $\boldsymbol{\mu}_s$ precesses around the total angular momentum vector $\boldsymbol{j}$, the direction of which is fixed in space. Due to the strong coupling of the angular momenta, the precession is rapid. Therefore only the time average of its projection on $\boldsymbol{j}$ can be observed, since the other components cancel each other in time. This projection $(\boldsymbol{\mu}_j)_j$ precesses in turn around the $\boldsymbol{B}_0$ axis of the applied magnetic field B_0. In the calculation of the magnetic contribution to the energy V_{m_j}, the projection of $\boldsymbol{\mu}_j$ on the $\boldsymbol{j}$ axis $(\boldsymbol{\mu}_j)_j$ must therefore be inserted in (13.15). Its magnitude can be calculated from the vector model: from Figs. 13.9, 12, the j component of $\boldsymbol{\mu}_j$ is

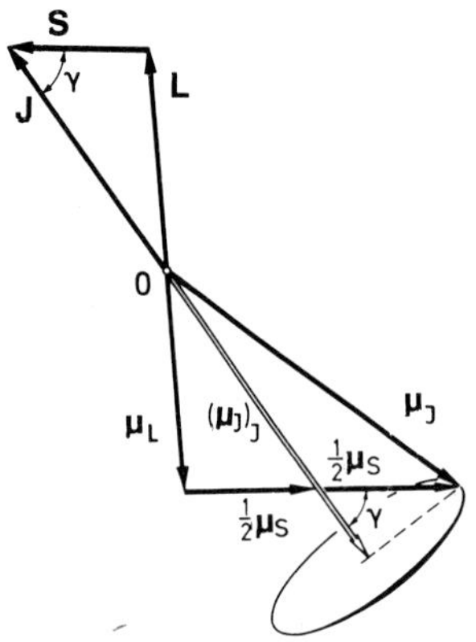

Fig. 13.12. Calculation of the J components of $\boldsymbol{\mu}_J$ and interpretation of the differing g factors of orbital and spin magnetism. Again, lower case letters s, l and j apply to single-electron systems, upper case S, L and J to many-electron systems

$$|(\boldsymbol{\mu}_j)_j| = |\boldsymbol{\mu}_l|\cos(\boldsymbol{l},\boldsymbol{j}) + |\boldsymbol{\mu}_s|\cos(\boldsymbol{s},\boldsymbol{j})$$
$$= \mu_B[\sqrt{l(l+1)}\cos(\boldsymbol{l},\boldsymbol{j}) + 2\sqrt{s(s+1)}\cos(\boldsymbol{s},\boldsymbol{j})] .$$

The expressions for $\cos(\boldsymbol{l},\boldsymbol{j})$ and $\cos(\boldsymbol{s},\boldsymbol{j})$ are derived from Figs. 13.12 and 13.9 using the law of cosines. The length of the vectors is again $\sqrt{l(l+1)}\,\hbar$ or $\sqrt{s(s+1)}\,\hbar$, respectively. We shall present a deeper quantum theoretical justification for this in Sect. 14.3.

We then have for the magnitude of $(\boldsymbol{\mu}_j)_j$,

$$|(\boldsymbol{\mu}_j)_j| = \frac{3j(j+1)+s(s+1)-l(l+1)}{2\sqrt{j(j+1)}}\,\mu_B = g_j\sqrt{j(j+1)}\,\mu_B\,, \tag{13.17}$$

and for the moment

$$(\boldsymbol{\mu}_j)_j = -g_j\mu_B\boldsymbol{j}/\hbar$$

with

$$g_j = 1 + \frac{j(j+1)+s(s+1)-l(l+1)}{2j(j+1)}\,, \tag{13.18}$$

and for the component in the z direction,

$$(\mu_j)_{j,z} = -m_j g_j \mu_B\,. \tag{13.19}$$

The Landé factor g_j defined in this way has a numerical value of 1 for pure orbital magnetism ($s=0$) and 2 (more exactly, 2.0023) for pure spin magnetism ($l=0$). For mixed magnetism, one observes values which differ from these two cases. By making the appropriate substitutions, one can easily see that the g factors given in the preceding section for the terms of the sodium atom are obtained from (13.18). In many-electron atoms, the quantum numbers s, l and j are replaced by S, L and J, as already mentioned (but see Sect. 17.3.3). This has been done in Figs. 13.9 and 13.12.

13.4 The Paschen-Back Effect

The preceding considerations on the splitting of spectral lines in a magnetic field hold for "weak" magnetic fields. "Weak" means that the splitting of energy levels in the magnetic field is small compared to fine structure splitting; or, in other words, the coupling between the orbital and spin moments, the so-called spin-orbit coupling, is stronger than the coupling of either the spin or the orbital moment alone to the external magnetic field. Since spin-orbit coupling increases rapidly with increasing nuclear charge Z (Sect. 12.8), the conditions for a "strong" field are met at a much lower field with light atoms than with heavy atoms. For example, the spin-orbit splitting of the sodium D lines is 17.2 cm^{-1}, while the splitting for the corresponding lines of the lithium atom is 0.3 cm^{-1}. The Zeeman splitting in an external field B_0 of 30 kG (3 T) is the same in both cases, about 1 cm^{-1}. Thus this field is a "strong" magnetic field for lithium, but a "weak" field for sodium.

When the magnetic field $\boldsymbol{B}_0$ is strong enough so that the above condition is no longer fulfilled, the splitting picture is simplified. The magnetic field dissolves the fine structure coupling. $\boldsymbol{l}$ and $\boldsymbol{s}$ are, to a first approximation, uncoupled, and precess independently around $\boldsymbol{B}_0$. The quantum number for the total angular momentum, j, thus loses its meaning. This limiting case is called the Paschen-Back effect.

The components of the orbital and spin moments $(\mu_l)_z$ and $(\mu_s)_z$ in the field direction are now individually quantised. The corresponding magnetic energy is

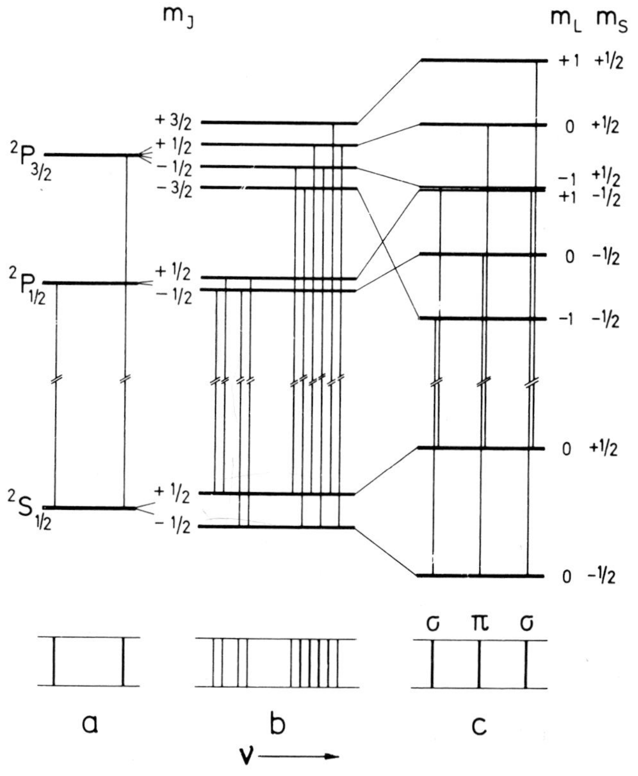

Fig. 13.13a–c. Paschen-Back effect (**c**) and Zeeman effect (**b**) with the D_1 and D_2 lines of the neutral sodium atom (**a**). In the limiting case of strong magnetic fields, one observes one unshifted and two symmetrically split lines, as in the ordinary Zeeman effect

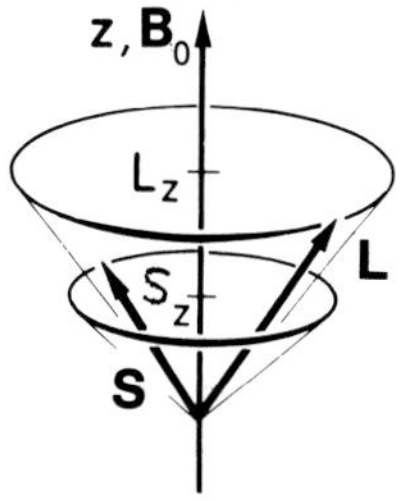

Fig. 13.14. Paschen-Back effect. In the limiting case of a strong magnetic field B_0, the spin $\boldsymbol{S}$ and orbital $\boldsymbol{L}$ angular momenta align independently with the field $\boldsymbol{B}_0$. A total angular momentum $\boldsymbol{J}$ is not defined

$$V_{m_s, m_l} = (m_l + 2\,m_s)\,\mu_B B_0 \tag{13.20}$$

and the splitting of the spectral lines is

$$\Delta E = (\Delta m_l + 2\,\Delta m_s)\,\mu_B B_0\,. \tag{13.21}$$

For optical transitions, there are again selection rules, and as before, $\Delta m_l = 0$ or ± 1 for π or σ transitions. Since electric dipole radiation cannot, to a first approximation, effect a spin flip, it also holds that $\Delta m_s = 0$. With these rules, (13.21) yields a triplet of spectral lines like those of the ordinary Zeeman effect.

Figure 13.13 shows the splitting scheme of the Na D lines. A vector model is shown in Fig. 13.14, which makes it clear that a total angular momentum vector $\boldsymbol{j}$ cannot even be defined here. Like the Zeeman effect, the Paschen-Back effect is chiefly used in empirical term analysis. In many-electron atoms, where the single-electron quantum numbers j, l, and s are replaced by the many-electron quantum numbers J, L and S, this method is especially important (Chap. 17).

The area between the limiting cases of weak fields (Zeeman effect) and strong fields (Paschen-Back effect) is difficult to analyse, both theoretically and experimentally.

13.5 Double Resonance and Optical Pumping

One can make use of the difference in polarisation of the various Zeeman components in order to populate selectively individual Zeeman levels, even when the spectral resolution is insufficient or the linewidth is too great to obtain the excited state otherwise. This is the simplest case of optical pumping.

The first experiment of this type is represented in Fig. 13.15 (*Brossel*, *Bitter* and *Kastler* 1949 – 1952). Mercury atoms in an external magnetic field $\boldsymbol{B}_0$ are excited by irradiation with linearly polarised light in a π transition to the $m_J = 0$ level of the $^3P_1$ excited state. The emission from these atoms is also linearly polarised π light. Now one can induce transitions $\Delta m = \pm 1$ with a high-frequency coil perpendicular to $\boldsymbol{B}_0$, as shown in Fig. 13.15, and thus populate the Zeeman substates $m = 1$ and $m = -1$. The light emitted from these levels, however, is circularly polarised σ light. The emission of circularly polarised light in a direction perpendicular to that of the π emission can thus be used for the detection and measurement of $\Delta m = \pm 1$ transitions between Zeeman substates.

Here, the same transitions as in electron spin resonance are observed, but they are detected optically. By means of this double resonance technique (double excitation with light and with high-frequency radiation), an extremely high detection sensitivity can be reached, because the small high-frequency quanta are detected via the much larger light quanta. In this way, the detection of spin resonance in a short-lived excited state becomes possible. Double resonance methods of this type have attained considerable importance in spectroscopy in the past 25 years.

The principle of *optical pumping* may be explained conveniently using the example of the sodium D lines, e.g. the transition from the $^2S_{1/2}$ ground state to the $^2P_{1/2}$ excited state. In an applied magnetic field, both terms are split into the Zeeman terms $m_j = \pm 1/2$ (Figs. 13.11 and 13.16). If the "pumping" light is now circularly polarised,

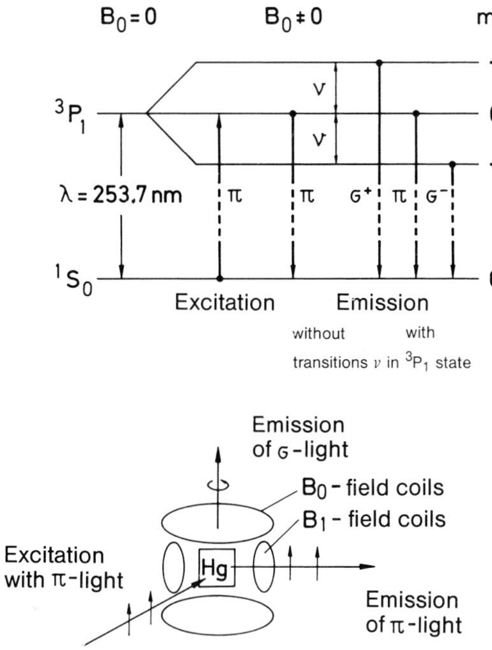

Fig. 13.15. Double resonance, after the method of *Brossel*, *Bitter* and *Kastler*. In the upper part of the figure, the three Zeeman levels of the excited state $^3P_1$ are shown. The lower part shows the experimental arrangement. The mercury atoms are contained in a cuvette between two pairs of coils, which produce the constant field $\boldsymbol{B}_0$ and the high-frequency field $\boldsymbol{B}_1$. The transition occurs between the ground state of the Hg atom $6s^2(^1S_0)$ and the excited state $6s6p(^3P_1)$

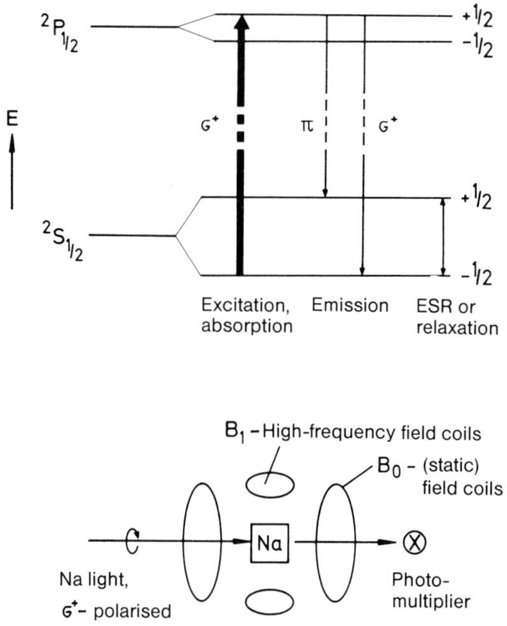

Fig. 13.16. Optical pumping of the transition $^2S_{1/2}-{}^2P_{1/2}$ of the sodium atom. In the field $\boldsymbol{B}_0$, the terms split up into Zeeman terms with $m_j = \pm 1/2$. Only atoms in the ground state $m_j = -1/2$ absorb the σ^+ light with which the sample is irradiated. π transitions occurring in emission from the excited state lead to an increase in the population of atoms in the ground state with $m_j = +1/2$. With the high-frequency field, transitions from $m_j = +1/2$ to $m_j = -1/2$ are induced, increasing the number of atoms which are able to absorb the pumping light

for example as σ^+ light, only transitions from $m_j = -1/2$ in the ground state to the $m_j = +1/2$ excited state can take place, populating only the latter state. Emission from this state can occur either as σ^+ light, leading to the $^2S_{1/2}$, $m_j = -1/2$ initial state, or as π light, leading to the ground state term with $m_j = +1/2$. Overall, this *pumping cycle* increases the population of the terms with $m_j = +1/2$ in the ground state at the cost of those with $m_j = -1/2$. An equalisation of the populations can occur through *relaxation* processes, for example by means of collisions of the Na atoms with one another or with the walls of the container. If these processes are not sufficiently rapid, one can induce transitions in the ground state by irradiation with microwaves. These electron spin resonance transitions change the populations of the Zeeman terms. The *detection* of this ESR can be accomplished *optically*, namely through the change in the intensity of the absorption from $^2S_{1/2}$, $m_j = -1/2$ to $^2P_{1/2}$, $m_j = +1/2$, provided that the population of the ground state terms was altered by optical pumping. Figure 13.16 shows the experimental arrangement schematically. This is thus also a double resonance method.

Problems

13.1 What frequency is required to induce electron spin transitions from the parallel to the antiparallel configuration, or vice versa, if the magnetic field is 10^{-1} tesla?

13.2 Why is the $^4D_{1/2}$ term not split in a magnetic field? Explain this in terms of the vector model.

13.3 Calculate the angle between the total and the orbital angular momenta in a $^4D_{3/2}$ state.

13.4 The spectral lines corresponding to the $3p \leftrightarrow 3s$ transition in sodium have the wavelengths $\lambda_2 = 5895.9$ Å and $\lambda_1 = 5890.0$ Å.

a) Calculate the magnetic field strength at which the lowest Zeeman level of the $^2P_{3/2}$ term would coincide with the highest level of the $^2P_{1/2}$ term, if the conditions for the anomalous Zeeman effect were still fulfilled.
b) How large are the frequency differences between the outer two components of the D_1 line and of the D_2 line in a magnetic field of 1 tesla?

13.5 Discuss the splitting of the lines in the $3d \leftrightarrow 2p$ transition in the presence of a magnetic field which is weak compared to the spin-orbit interaction.

13.6 Sketch the Zeeman splitting in the lines of the hydrogen atom Balmer series.

Calculate the magnetic moments of the states $P_{1/2}$, $P_{3/2}$, $D_{3/2}$ and $D_{5/2}$.

Also sketch the splitting in the Paschen-Back effect. At what magnetic field does the transition from the Zeeman to the Paschen-Back effect occur?

13.7 a) Consider hydrogen atoms in a magnetic field $B_0 = 4.5$ tesla. At this field strength, is the splitting of the H_α line ($n = 3 \rightarrow n = 2$) due to the anomalous Zeeman effect or the Paschen-Back effect? Support your answer. (The spin-orbit coupling between the $3^2P_{1/2}$ and $3^2P_{3/2}$ terms of the hydrogen atom is $0.108\ \mathrm{cm}^{-1}$.)

b) Sketch the splitting of the energy levels in the given magnetic field and show the transitions which contribute to the H_α line. Into how many components is the H_α line split?

c) Determine the specific charge e/m of the electron, given that the frequency splitting between two neighbouring components is $6.29 \cdot 10^{10}$ Hz. The fine structure can be ignored here.

d) Is the wavelength splitting in the first line of the Lyman series ($n = 2 \rightarrow n = 1$) smaller, larger or the same as that of the H_α line?

14. Atoms in a Magnetic Field: Quantum Mechanical Treatment

14.1 Quantum Theory of the Ordinary Zeeman Effect

The ordinary Zeeman effect is a beautiful example of the fact that even with classical physics, one can obtain results similar to those of strict quantum theory. In order to set our earlier results on a firm basis, however, we shall now go through the strict quantum theoretical treatment.

This chapter is somewhat more demanding, because we shall have to make use of some of the basic theory of electromagnetism. As is shown in this theory, a magnetic field $\boldsymbol{B}$ can be expressed as the curl of the vector potential $\boldsymbol{A}$:

$$\boldsymbol{B} = \operatorname{curl} \boldsymbol{A} . \tag{14.1}$$

The electric field strength $\boldsymbol{F}$[1] can be obtained in a similar way from the electric potential $\tilde{V}$ and the vector potential $\boldsymbol{A}$ according to the rule

$$\boldsymbol{F} = -\operatorname{grad} \tilde{V} - \frac{d\boldsymbol{A}}{dt} . \tag{14.2}$$

Furthermore, we remember that the equation of motion of a particle with charge $-e$ (we are thinking specifically of electrons here) and mass m_0 is

$$m_0 \ddot{\boldsymbol{r}} = (-e)(\boldsymbol{F}) + (-e)(\boldsymbol{v} \times \boldsymbol{B}) . \tag{14.3}$$

The second term on the right is the so-called Lorentz force, $\boldsymbol{v}$ is the particle velocity. It can be shown that this equation of motion can be obtained, using the Hamilton equations

$$\dot{\boldsymbol{p}} = -\operatorname{grad}_r H(\boldsymbol{p},\boldsymbol{r}) \quad \text{and} \tag{14.4}$$

$$\dot{\boldsymbol{r}} = \operatorname{grad}_p H(\boldsymbol{p},\boldsymbol{r}) , \tag{14.5}$$

from the Hamiltonian function

$$H = \frac{1}{2m_0}(\boldsymbol{p} + e\boldsymbol{A})^2 + V . \tag{14.6}$$

The potential energy V of the electron is related to the electric potential $\tilde{V}$: $V = -e\tilde{V}$.

---

[1] In order to avoid confusion between the energy E and the electric field strength, we denote the latter by F

At this point, it is only important to remember that in quantum theory, we always start from a Hamiltonian function. As we saw in Sect. 9.3.4, the Hamiltonian function is converted to an *operator* in quantum mechanics by using the Jordan rule to replace the momentum, according to

$$p \to \frac{\hbar}{\mathrm{i}}\,\mathrm{grad}\,. \tag{14.7}$$

By applying this technique here, we arrive at the Hamiltonian operator

$$\mathcal{H} = [(1/2m_0)\,[\hbar/\mathrm{i})\,\mathrm{grad} + e\boldsymbol{A}]^2 + V\,. \tag{14.8}$$

When we multiply out the squared term, taking care to maintain the order of the factors, we obtain

$$\mathcal{H} = -\frac{\hbar^2}{2m_0}\nabla^2 + \frac{\hbar e}{2m_0\mathrm{i}}\boldsymbol{A}\,\mathrm{grad} + \frac{\hbar e}{2m_0\mathrm{i}}\,\mathrm{grad}\boldsymbol{A} + \frac{e^2\boldsymbol{A}^2}{2m_0} + V\,. \tag{14.9}$$

In applying the various differential operators, however, we must be careful, since we know that $\mathcal{H}$ is to operate on the wavefunction ψ. Thus we must interpret

$$\mathrm{grad}\boldsymbol{A} \tag{14.10}$$

exactly as

$$\mathrm{grad}(\boldsymbol{A}\,\psi)\,. \tag{14.11}$$

On differentiating the product in (14.11) and then again applying (14.7), we obtain for the Hamiltonian

$$\mathcal{H} = -\frac{\hbar^2}{2m_0}\nabla^2 + \frac{e}{m_0}\boldsymbol{A}\boldsymbol{p} + \frac{e\hbar}{2m_0\mathrm{i}}\,\mathrm{div}\boldsymbol{A} + \frac{e^2\boldsymbol{A}^2}{2m_0} + V\,. \tag{14.12}$$

(The operators *grad*ient, *div*ergence, and curl used here are vector differential operators which are often abbreviated using the Nabla symbol ∇, with $\nabla f \equiv \mathrm{grad} f$, $\nabla\cdot\boldsymbol{F} \equiv \mathrm{div}\boldsymbol{F}$, $\nabla\times\boldsymbol{F} \equiv \mathrm{curl}\boldsymbol{F}$, and $\nabla\cdot\nabla f = \nabla^2 f = \text{Laplacian } f$, where f is a scalar function and $\boldsymbol{F}$ a vector function.)

We now choose, as always in this book, the constant magnetic field $\boldsymbol{B}$ in the z direction:

$$\boldsymbol{B} = (0, 0, B_z)\,. \tag{14.13}$$

It can be demonstrated that the vector potential $\boldsymbol{A}$ in (14.1) cannot be uniquely determined. One possible representation, which is convenient for the present calculation, is

$$A_x = -\frac{B_z}{2}y\,, \quad A_y = \frac{B_z}{2}x\,, \quad A_z = 0\,. \tag{14.14}$$

With this, the Schrödinger equation with the Hamiltonian (14.12) becomes

$$\left[-\frac{\hbar^2}{2m_0}\nabla^2 + B_z\frac{e}{2m_0}\frac{\hbar}{\mathrm{i}}\left(x\frac{\partial}{\partial y} - y\frac{\partial}{\partial x}\right) + \frac{e^2B_z^2}{8m_0}(x^2+y^2) + V(r)\right]\psi = E\psi\,. \tag{14.15}$$

In the following, we shall assume a spherically symmetrical potential for V.

We recall the following relation from Sect. 10.2:

$$\frac{\hbar}{\mathrm{i}}\left(x\frac{\partial}{\partial y} - y\frac{\partial}{\partial x}\right) = \hat{l}_z = \frac{\hbar}{\mathrm{i}}\frac{\partial}{\partial \phi}\,, \tag{14.16}$$

where $\hat{l}_z$ is the angular momentum operator in the z direction. In general, the term in (14.15) containing (x^2+y^2) can be neglected in comparison to the preceding term with $\hat{l}_z$, if the magnetic field is not too large, and as long as the magnetic quantum number $m \neq 0$. Leaving out the term with x^2+y^2, and using the usual formula for the wavefunction,

$$\psi(r) = R_{n,l}(r)\,\mathrm{e}^{\mathrm{i}m\phi}\mathrm{P}_l^m(\cos\theta)\,, \tag{14.17}$$

we recognise that (14.15) is identically satisfied. The energy is now

$$E = E_n^0 + B_z\frac{e\hbar}{2m_0}\cdot m\,, \qquad -l \leqq m \leqq l\,. \tag{14.18}$$

The energy E is thus shifted with respect to the unperturbed energy E_n^0 by an amount which depends on the magnetic quantum number m, and the energy level is split. The factor $\mu_\mathrm{B} = e\hbar/(2m_0)$ is the Bohr magneton which was introduced earlier. With the addition of the selection rules for optical transitions,

$$\Delta m = 0 \quad \text{or} \quad \pm 1\,,$$

the above derivation leads to the splitting of spectral lines known as the ordinary Zeeman effect (Sect. 13.3).

14.2 Quantum Theoretical Treatment of the Electron and Proton Spins

14.2.1 Spin as Angular Momentum

As we saw in Sect. 12.4, the electron has three degrees of freedom in its translational motion, and a fourth in its spin. As we know, a number of other elementary particles, including the proton, have spins too. Our quantum mechanical calculations to this point, especially our derivation of the Schrödinger equation and its application to the hydrogen atom, have not included spin. In the following, we shall show how spin is included in the quantum theoretical treatment of atomic states. This is necessary, for example in spin-orbit coupling, in the anomalous Zeeman effect, in spin resonance, and in an adequate formulation of the Pauli principle, which will be discussed later. Like every angular momentum, the spin of the electron is a vector with three spatial components s_x, s_y and s_z:

$$s = (s_x, s_y, s_z) . \tag{14.19}$$

In the following development of the spin formalism, we must account for the experimental observation that the spin has only two possible orientations such that the spin component in a chosen direction, e.g. the z direction, can only have the value $+\hbar/2$ or $-\hbar/2$. In this sense, it is a genuine two-level system.

14.2.2 Spin Operators, Spin Matrices and Spin Wavefunctions

Since it is intuitive to think of one of the states of spin as "spin up" and the other as "spin down", we shall first introduce in a purely formal way two "wave" functions which correspond to these spin directions, i.e. $\phi_\uparrow$ and $\phi_\downarrow$. If we proceed strictly according to quantum formalism, measurement of the z component of the spin corresponds to applying the operator $\hat{s}_z$ to a wavefunction. (As with the angular momentum $\boldsymbol{l}$, we distinguish the spin operator from the corresponding classical parameter by using the "hat" sign.) We can choose the wavefunctions in such a way that the application of the operator gives the observed values of the wavefunction. Because we have only two observed values, namely $\hbar/2$ and $-\hbar/2$, we expect that

$$\hat{s}_z \phi_\uparrow = \frac{\hbar}{2} \phi_\uparrow , \quad \text{and} \tag{14.20a}$$

$$\hat{s}_z \phi_\downarrow = -\frac{\hbar}{2} \phi_\downarrow . \tag{14.20b}$$

These can be summarised as

$$\hat{s}_z \phi_{m_s} = \hbar m_s \phi_{m_s} , \tag{14.21}$$

where $m_s = +1/2$ (corresponding to $\uparrow$) or
$m_s = -1/2$ (corresponding to $\downarrow$).

m_s is thus the quantum number of the z component of the spin.

We are now looking for a formalism which will more or less automatically give us the relations (14.20a, b). It has been found that this is most easily done by using matrices. A matrix, in mathematics, is a square array, for example

$$M = \begin{pmatrix} a & b \\ c & d \end{pmatrix} . \tag{14.22}$$

There is a multiplication rule for this array. As an example, let us imagine a vector $\boldsymbol{v}$ with the components x and y in a plane, or $\boldsymbol{v} = \begin{pmatrix} x \\ y \end{pmatrix}$. We can produce a new vector x', y' by multiplying $\begin{pmatrix} x \\ y \end{pmatrix}$ by M. This is done according to the rule

$$\begin{pmatrix} x' \\ y' \end{pmatrix} = M \begin{pmatrix} x \\ y \end{pmatrix} \equiv \begin{pmatrix} a & b \\ c & d \end{pmatrix} \begin{pmatrix} x \\ y \end{pmatrix} = \begin{pmatrix} ax+by \\ cx+dy \end{pmatrix} . \tag{14.23}$$

We thus are looking for a "vector" ϕ and a matrix M such that $M\phi$ yields exactly either $\frac{\hbar}{2}\phi$ or $-\frac{\hbar}{2}\phi$.

We shall simply give the result, and then verify it. We choose $\hat{s}_z$ in the form

$$\hat{s}_z = \frac{\hbar}{2} \begin{pmatrix} 1 & 0 \\ 0 & -1 \end{pmatrix} \tag{14.24}$$

and the spin functions in the form

$$\phi_\uparrow = \begin{pmatrix} 1 \\ 0 \end{pmatrix}, \quad \phi_\downarrow = \begin{pmatrix} 0 \\ 1 \end{pmatrix} . \tag{14.25}$$

With the help of (14.23), it can be immediately calculated that substitution of (14.24 and 25) in (14.20a and b) actually yields the relations $M\phi_\uparrow = (\hbar/2)\,\phi_\uparrow$, $M\phi_\downarrow = -(\hbar/2)\,\phi_\downarrow$. We obtain the most general spin function by superposition of $\phi_\uparrow$ and $\phi_\downarrow$ with the coefficients a and b, as we have done before with wave packets:

$$\phi = a\phi_\uparrow + b\phi_\downarrow = \begin{pmatrix} a \\ b \end{pmatrix} . \tag{14.26}$$

In order to arrive at a normalisation condition, we must now introduce the "scalar product" for the ϕ's. If we have a general ϕ_1 in the form

$$\phi_1 = \begin{pmatrix} a_1 \\ b_1 \end{pmatrix} \tag{14.27}$$

and another ϕ_2 in the form

$$\phi_2 = \begin{pmatrix} a_2 \\ b_2 \end{pmatrix}, \tag{14.28}$$

we define the scalar product as

$$\bar{\phi}_1 \phi_2 = (a_1^*, b_1^*) \begin{pmatrix} a_2 \\ b_2 \end{pmatrix} = (a_1^* a_2 + b_1^* b_2) . \tag{14.29}$$

These are calculation rules, which should be familiar to the reader from vector calculations. If we substitute in (14.29) $\bar{\phi}_1 = \bar{\phi}_\uparrow$, $\phi_2 = \phi_\uparrow$, we obtain

$$\bar{\phi}_\uparrow \phi_\uparrow = 1 \tag{14.30}$$

and correspondingly,

$$\bar{\phi}_{\downarrow}\phi_{\downarrow} = 1\,. \tag{14.31}$$

Thus the wavefunctions are normalised. With $\bar{\phi}_1 = \bar{\phi}_{\downarrow}$ and $\phi_2 = \phi_{\uparrow}$, we have

$$\bar{\phi}_{\downarrow}\phi_{\uparrow} = (1-1) = 0\,, \tag{14.32}$$

i.e. the wavefunctions are mutually orthogonal.

With (14.24), we have the first part of the solution of the entire problem. The representation of the operators for the x and y directions of the angular momentum is naturally still open. Because we are talking about *angular momenta*, it seems reasonable to require the usual commutation relations for angular momenta (10.14). We do not wish to go into the mathematics of the problem here. For the purposes of this book, it is sufficient simply to choose $\hat{s}_x$ and $\hat{s}_y$ appropriately. It turns out that

$$\hat{s}_x = \frac{\hbar}{2}\begin{pmatrix} 0 & 1 \\ 1 & 0 \end{pmatrix} \tag{14.33 a}$$

and

$$\hat{s}_y = \frac{\hbar}{2}\begin{pmatrix} 0 & -\mathrm{i} \\ \mathrm{i} & 0 \end{pmatrix} \tag{14.33 b}$$

are suitable. If we calculate $\hat{s}^2 = \hat{s}_x^2 + \hat{s}_y^2 + \hat{s}_z^2$ with the matrices (14.24, 33 a and 33 b), we obtain after a short calculation

$$\hat{s}^2 = \frac{\hbar^2}{4}\begin{pmatrix} 3 & 0 \\ 0 & 3 \end{pmatrix} = \hbar^2\frac{3}{4}\begin{pmatrix} 1 & 0 \\ 0 & 1 \end{pmatrix} = \hbar^2\frac{3}{4}\cdot(\text{unit matrix})\,.$$

Therefore, it we apply $\hat{s}^2$ to any spin function ϕ, in particular to ϕ_{m_s}, it will always yield

$$\hat{s}^2\phi_{m_s} = \hbar^2\tfrac{3}{4}\phi_{m_s}\,.$$

The analogy between this equation and the eigenvalue equation for the orbital angular momentum $\boldsymbol{l}^2$ with the eigenvalue $\hbar^2 l(l+1)$ (10.6) is especially clear if we write $\hbar^2\,3/4$ in the form $\hbar^2 s(s+1)$, with $s = 1/2$:

$$\hat{s}^2\phi_{m_s} = \hbar^2 s(s+1)\,\phi_{m_s}\,. \tag{14.34}$$

14.2.3 The Schrödinger Equation of a Spin in a Magnetic Field

We shall now proceed to the formulation of a Schrödinger equation for the spin in a magnetic field. A magnetic moment

$$\mu_{\mathrm{B}} = \frac{e\hbar}{2m_0} \tag{14.35}$$

is associated with the electron spin of $\hbar/2$. Here m_0 is the rest mass of the electron and e is the positive unit charge. This magnetic moment, the "Bohr magneton", was presented in Sect. 12.2. Since the magnetic moment is a vector oriented antiparallel to the electron spin, we can write more generally

$$\mu = -\frac{e}{m_0} s \,, \tag{14.36}$$

where the factor $\hbar/2$ is now naturally included in the angular momentum s. The following calculations can be directly applied to the spin of a proton, if the Bohr magneton μ_B is consistently replaced by the so-called nuclear magneton $-\mu_N$ and $-e/m_0$ by e/m_p. μ_N is defined as $-(m_0/m_p)\mu_B$, and m_p is the mass of the proton. The negative sign comes from the fact that the charge of the proton is the negative of the electron charge.

The energy of a spin in a spatially homogeneous magnetic field $\boldsymbol{B}$ is, as is shown in electrodynamics,

$$V_S = -\mu \cdot \boldsymbol{B} \,. \tag{14.37}$$

We are trying to find an equation analogous to the Schrödinger equation, and we realise from the previous discussion of quantum mechanics that the Schrödinger equation was obtained from energy expressions (Sect. 9.2). There the energy expressions were the Hamilton functions, which were then converted to the Hamiltonian operator. In a similar way, we now make the energy expression (14.37) into an operator and write the equation

$$\frac{e}{m_0} \boldsymbol{B} \cdot \hat{s}\, \phi = E \phi \,. \tag{14.38}$$

If the magnetic field has the components B_x, B_y and B_z, the left side of (14.38) is

$$\frac{e}{m_0} (B_x \hat{s}_x + B_y \hat{s}_y + B_z \hat{s}_z) \phi \,. \tag{14.39}$$

Now $\hat{s}_x$, $\hat{s}_y$ and $\hat{s}_z$ are the matrices (14.33a, b and 24), respectively. Therefore (14.39) is also a matrix. According to the rules for the addition of matrices, it is

$$\frac{e\hbar}{2m_0} \begin{pmatrix} B_z & B_x - \mathrm{i} B_y \\ B_x + \mathrm{i} B_y & -B_z \end{pmatrix} . \tag{14.40}$$

The characteristic of being an operator thus accrues to the left side of (14.38) from $\hat{s}$, which was defined above as the spin operator. If we choose the field $\boldsymbol{B}$ in the z direction, as above,

$$\boldsymbol{B} = (0, 0, B_z) \,, \tag{14.41}$$

the left side of (14.38) is the same, except for the numerical factor eB_z/m_0, as the left side of (14.20a or b) which shows us that the functions introduced above (14.25) are also eigenfunctions of the operator in (14.38) with the corresponding eigenvalues

$$E = \pm \mu_B B_z . \tag{14.42}$$

The spin energy in a constant magnetic field in the z direction is thus just given by the expression which we would expect in classical theory for the interaction of an antiparallel spin moment with a magnetic field. Of course, instead of (14.38), we could have formulated the corresponding time-dependent Schrödinger equation

$$\frac{e}{m_0} \boldsymbol{B} \cdot \hat{\boldsymbol{s}} \phi = \mathrm{i} \hbar \frac{d\phi}{dt} . \tag{14.43}$$

This equation must be used, in particular, if we are dealing with a time-dependent magnetic field.

14.2.4 Description of Spin Precession by Expectation Values

It is, however, also interesting to determine the time-dependent solution of (14.43) for a constant magnetic field. If we choose a magnetic field in the z direction, the Schrödinger equation is given by

$$\mu_B B_z \begin{pmatrix} 1 & 0 \\ 0 & -1 \end{pmatrix} \phi = \mathrm{i} \hbar \frac{d\phi}{dt} . \tag{14.44}$$

The general solution is found as a superposition of $\phi_\uparrow$ and $\phi_\downarrow$ (14.26). Since the Schrödinger equation contains a derivative with respect to time on the right-hand side, we have to include in $\phi_\uparrow$ and $\phi_\downarrow$ the corresponding time functions

$$\exp(-\mathrm{i} E_\uparrow t/\hbar) \quad \text{and} \quad \exp(-\mathrm{i} E_\downarrow t/\hbar) ,$$

where $E_\uparrow$ and $E_\downarrow$ may be written in the form

$$E_\uparrow = (\hbar/2)\,\omega_0 , \quad E_\downarrow = -(\hbar/2)\,\omega_0 , \quad \text{and} \quad \omega_0 = \frac{e}{m_0} B_z . \tag{14.45}$$

Since a linear combination may also contain constant coefficients, we use the more general trial solution for (14.44):

$$\phi(t) = a \exp(-\mathrm{i}\,\omega_0 t/2)\,\phi_\uparrow + b \exp(\mathrm{i}\,\omega_0 t/2)\,\phi_\downarrow . \tag{14.46}$$

We require ϕ to be normalised, as always in quantum mechanics, i.e. that the scalar product $\bar{\phi}\phi$ (14.29) be equal to one. This means

$$|a|^2 + |b|^2 = 1 . \tag{14.47}$$

The physical meaning of (14.46) will become clear when we form the expectation value of the spin operator $\hat{s}$ with this wavefunction. To do this, we must first recall how expectation values are to be calculated, and refer to Sect. 9.3. The "recipe" given there states:

1) Take the wavefunction ψ,
2) allow the "operator for the measurable quantity" Ω of which the expectation value is to be found, to operate on it,
3) then multiply with ψ^* and integrate:

$$\int \psi^*(x)\,\Omega\,\psi(x)\,dx\,.$$

The steps 1 – 3) can easily be transformed into three analogous rules for calculating with the spin formalism:
1) Take the spinfunction ϕ, e.g. (14.46),
2) let the spin operator $\hat{s}_x$, $\hat{s}_y$, or $\hat{s}_z$ operate on (14.46), i.e. form, for example $\hat{s}_z\phi$;
3) multiplication by $\bar{\phi}$ and integration are replaced by the rules for calculating the scalar product:
we multiply $\hat{s}_z\phi$ from the left by $\bar{\phi}$.
As an abbreviation we set

$$\begin{aligned} a\exp(-\mathrm{i}\,\omega_0 t/2) &= \alpha\,,\\ b\exp(\mathrm{i}\,\omega_0 t/2) &= \beta\,. \end{aligned} \tag{14.48}$$

The individual steps 1 – 3) are now as follows:

$$1)\quad \phi = \alpha\phi_\uparrow + \beta\phi_\downarrow = \begin{pmatrix} \alpha \\ \beta \end{pmatrix}, \tag{14.49}$$

$$2)\quad \hat{s}_z\phi = \frac{\hbar}{2}\begin{pmatrix} 1 & 0 \\ 0 & -1 \end{pmatrix}\begin{pmatrix} \alpha \\ \beta \end{pmatrix}. \tag{14.50}$$

Using the rule (14.23), this is equal to

$$\frac{\hbar}{2}\begin{pmatrix} \alpha \\ -\beta \end{pmatrix}. \tag{14.51}$$

$$3)\quad \bar{\phi}\hat{s}_z\phi = \bar{\phi}\frac{\hbar}{2}\begin{pmatrix} \alpha \\ -\beta \end{pmatrix}. \tag{14.52}$$

According to rule (14.29), the right-hand side is equal to

$$\frac{\hbar}{2}(|\alpha|^2 - |\beta|^2)\,.$$

Writing the expectation value of $\hat{s}_z$ as $\langle\hat{s}_z\rangle$, we have found:

$$\langle\hat{s}_z\rangle = \frac{\hbar}{2}(|\alpha|^2 - |\beta|^2)\,. \tag{14.53}$$

We leave it to the reader as an exercise to show that

$$\langle\hat{s}_x\rangle = \frac{\hbar}{2}(\alpha^*\beta + \alpha\beta^*)\,, \tag{14.54}$$

and

$$\langle \hat{s}_y \rangle = \frac{\hbar}{2} \mathrm{i}(\alpha\beta^* - \alpha^*\beta) . \tag{14.55}$$

Since we can see all the essentials by assuming a and b in (14.48) to be real numbers, we shall do so and insert (14.48) into (14.53 – 55). This yields

$$\langle \hat{s}_z \rangle = \frac{\hbar}{2}(a^2 - b^2) = \text{const w.r.t. time} . \tag{14.56}$$

The expectation value of the z component of the spin thus remains constant in time.

$$\langle \hat{s}_x \rangle = ab\hbar \cos\omega_0 t , \tag{14.57}$$

$$\langle \hat{s}_y \rangle = ab\hbar \sin\omega_0 t . \tag{14.58}$$

The component of the spin in the $x-y$ plane rotates with the angular velocity ω_0. The expectation values (14.56 – 58) can be interpreted as a *precessional motion* of the spin (Fig. 14.1). Thus the model used in Chap. 13 is justified by quantum theory.

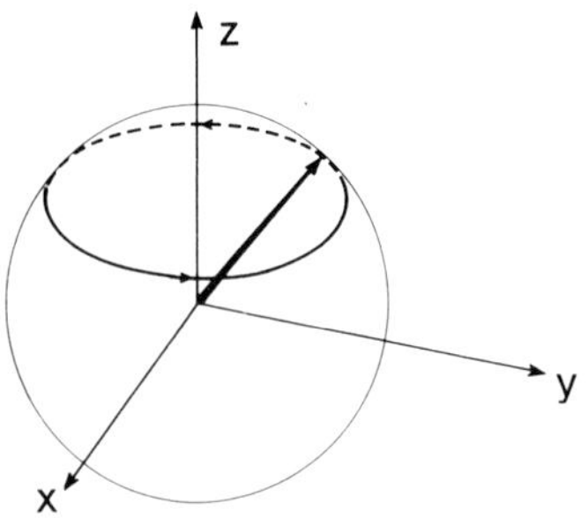

Fig. 14.1. Precessional motion of the spin

14.3 Quantum Mechanical Treatment of the Anomalous Zeeman Effect with Spin-Orbit Coupling*

In this section we shall continue the complete quantum mechanical treatment of spin-orbit coupling. The goal is to give an exact justification for the vector model of spin-orbit coupling introduced in Chap. 12. Specifically, we are concerned with LS coupling and wish to show the justification for the rule that l^2, s^2 and j^2 can be replaced by $l(l+1)$, $s(s+1)$ and $j(j+1)$, respectively. If we ignore spin-orbit coupling for a moment, then the energies of the orbital motion and of the spin (magnetic moment) in a magnetic field are additive. This means that the total Hamiltonian is simply the sum of the Hamiltonians for orbital motion (14.8) and for the spin (14.38). Thus we have the Schrödinger equation

$$\left[\frac{1}{2m_0}\left(\frac{\hbar}{\mathrm{i}}\operatorname{grad} + e\boldsymbol{A}\right)^2 + V + \frac{e}{m_0}\hat{\boldsymbol{s}} \cdot \boldsymbol{B}\right]\psi = \mathrm{i}\hbar\frac{\partial\psi}{\partial t} . \tag{14.59}$$

This is also known in the literature as the Pauli equation.

Because the Hamiltonians (14.8) and (14.38) are additive and apply to entirely different degrees of freedom, the wavefunction ψ can be written as a product of the wavefunction of the orbital motion and that of the spin motion. Finally, we can also treat the spin-orbit coupling introduced in Sect. 12.8 quantum mechanically. For this we need only to introduce the previously derived expression (12.27), which represents an interaction energy, into quantum mechanics. This is done, as usual, by assigning the angular momentum $\boldsymbol{l}$ to the angular momentum *operator* $\hat{\boldsymbol{l}}$ (Sect. 9.3.4.) and the spin $\boldsymbol{s}$ to the spin operator $\hat{\boldsymbol{s}}$ [compare with (9.61)]. The resulting term

$$W(\hat{\boldsymbol{l}},\hat{\boldsymbol{s}}) = \frac{\mu_0 Z e^2}{8\pi m_0^2}\,\frac{1}{r^3}(\hat{\boldsymbol{l}}\cdot\hat{\boldsymbol{s}}) = \frac{\mu_0 Z}{4\pi r^3}(\hat{\boldsymbol{\mu}}_{\text{orbit}}\cdot\hat{\boldsymbol{\mu}}_{\text{spin}}) \tag{14.60}$$

is introduced into the Schrödinger equation to give the Schrödinger equation of an electron with spin in a magnetic field, where spin-orbit coupling is taken into account. The time-independent form of this equation is

$$\left[-\frac{\hbar^2}{2m_0}\nabla^2 + \frac{e\hbar}{m_0 \mathrm{i}}\boldsymbol{A}\,\mathrm{grad} + \frac{e\hbar}{2m_0\mathrm{i}}\,\mathrm{div}\boldsymbol{A} + \frac{e^2\boldsymbol{A}^2}{2m_0} + V \right.$$
$$\left. + \frac{e}{m_0}\hat{\boldsymbol{s}}\cdot\boldsymbol{B} + \frac{\mu_0 Z e^2}{8\pi m_0^2}\,\frac{1}{r^3}(\hat{\boldsymbol{l}}\cdot\hat{\boldsymbol{s}})\right]\psi = E\psi\,. \tag{14.61}$$

As we saw in Sect. 13.3, spin-orbit coupling dominates in low magnetic fields. Therefore we shall first examine the Schrödinger equation in the absence of a magnetic field:

$$\left[-\frac{\hbar^2}{2m_0}\nabla^2 - \frac{Ze^2}{4\pi\varepsilon_0 r} + \frac{\mu_0 Z e^2}{8\pi m_0^2}\,\frac{1}{r^3}(\hat{\boldsymbol{l}}\cdot\hat{\boldsymbol{s}})\right]\psi(\boldsymbol{r}) = E\,\psi(\boldsymbol{r})\,. \tag{14.62}$$

Equation (14.62) includes the spin operator $\hat{\boldsymbol{s}}$ which, as we know, is a matrix. Therefore the wavefunction $\psi(\boldsymbol{r})$ has two components:

$$\psi(\boldsymbol{r}) = \begin{pmatrix}\psi_1(\boldsymbol{r})\\ \psi_2(\boldsymbol{r})\end{pmatrix},$$

where ψ_1 corresponds to spin ↑ and ψ_2 to spin ↓.

Spin-orbit coupling mixes orbital and spin states, and makes it necessary to introduce new quantum numbers. Without spin-orbit coupling, the wavefunction would have the form

$$\psi_{n,l,m,m_s} = \underbrace{R_{n,l}(r)F_{l,m}(\theta,\phi)}_{\text{orbit}}\,\underbrace{\phi_{m_s}}_{\text{spin}}\,. \tag{14.63}$$ [2]

2 The letter ϕ in (14.63) has two entirely different meanings: in $F_{l,m}(\theta,\phi)$, it indicates an angular coordinate, while in ϕ_{m_s} it indicates one of the spin wavefunctions (14.25)

It is characterised by the principal quantum number n, the orbital angular momentum quantum number l, the magnetic quantum number $m(\equiv m_l)$ and the spin quantum number m_s. In order to determine the quantum numbers applicable to spin-orbit coupling, we must expand on the considerations on orbital angular momentum presented in Sect. 10.2, and examine the parameters to decide which can be observed simultaneously. As we know, this can be done with the help of commutation relations (Sect. 9.3). If, as in Sect. 12.7, we introduce the total spin operator $\hat{j} = \hat{l} + \hat{s}$, and its component in the z direction, $\hat{j}_z$, the following parameters can be observed to any desired precision simultaneously:

The square of the orbital angular momentum $\boldsymbol{l}^2$

The square of the spin $\boldsymbol{s}^2$

The square of the total angular momentum $\boldsymbol{j}^2$

Component j_z

$\boldsymbol{l} \cdot \boldsymbol{s}$ and $\boldsymbol{j} \cdot \boldsymbol{s}$.

Because $\boldsymbol{l} \cdot \boldsymbol{s}$ occurs in (14.62), we can characterise the wavefunction by choosing those quantum numbers which are eigenvalues for the operators $\hat{j}^2$, $\hat{l}^2$, $\hat{s}^2$ and $\hat{j}_z$. We therefore obtain the following relations between operators and quantum numbers

$$\begin{array}{ll} \hat{j}^2\text{: quantum number } j & \hat{j}_z\text{: quantum number } m_j \\ \hat{s}^2\text{: quantum number } s & \hat{l}^2 \text{ quantum number } l \,. \end{array} \tag{14.64}$$

Since the spin-orbit coupling is much smaller than the term spacing, the principal quantum number n is still a good quantum number, i.e. it still characterises the eigenfunction to a good approximation. The wavefunction is now characterised by

$$\psi_{n,j,m_j,l,s} = R(r) \cdot (\text{Function of angle and spin}) \,. \tag{14.65}$$

The spin-orbit coupling leads to the relative orientations of the spin and orbital moments, as was discussed in detail in Sect. 12.8.

We now examine the effect of a magnetic field on an electron, taking spin-orbit coupling into account. It can be shown that in the Schrödinger equation (14.59), the $\boldsymbol{A}^2$ term is much smaller than the other terms, if the magnetic field is not too large, and can be ignored. Let us again choose the magnetic field $\boldsymbol{B}$ in the z direction and

$$A_x = -\tfrac{1}{2}B_y\,, \quad A_y = \tfrac{1}{2}B_x \quad \text{and} \quad A_z = 0\,.$$

div$\boldsymbol{A}$ is then zero. The Schrödinger equation is then

$$\Big[\underbrace{-\frac{\hbar^2}{2m_0}\nabla^2-\frac{Ze^2}{4\pi\varepsilon_0 r}}_{\mathscr{H}^0}+\underbrace{\frac{e}{2m_0}B\hat{l}_z+\frac{e}{m_0}\hat{s}_z B}_{W_{\text{magn}}}+\underbrace{\frac{Ze^2\mu_0}{8\pi m_0^2 r^3}(\hat{l}\cdot\hat{s})}_{W_{\text{spin-orbit}}}\Big]\psi=E\psi\,. \tag{14.66}$$

We are treating the case of a weak magnetic field in which the spin-orbit coupling is larger than the interaction with the external magnetic field. We are now in a position to justify quantum mechanically the vector model introduced in Chap. 13. Let us consider the operator occurring in (14.66):

$$W_{\text{magn}}\equiv\frac{eB}{2m_0}(\hat{l}_z+2\hat{s}_z)=\frac{eB}{2m_0}(\hat{j}_z+\hat{s}_z) \tag{14.67}$$

(it leads to an additional magnetic energy, which we called V_{m_j} in Sects. 13.3.4, 5). If we here had $\hat{l}_z+\hat{s}_z$ instead of $\hat{l}_z+2\hat{s}_z$, the solution would be very simple, and analogous to the treatment of an electron without a spin in the magnetic field (Sect. 14.1). In that case, the wavefunction ψ, which is already characterised by the quantum number m_j, would also be an eigenfunction of the operator $\hat{j}_z=\hat{l}_z+\hat{s}_z$. We must therefore see how we can deal with the additional $\hat{s}_z$ in (14.67). Let us consider

$$\hat{s}_z\hat{j}^2=\hat{s}_z(\hat{j}_x^2+\hat{j}_y^2+\hat{j}_z^2)\,, \tag{14.68}$$

which can be rewritten as

$$\hat{j}_z(\hat{s}\cdot\hat{j})+\underbrace{(\hat{s}_z\hat{j}_x-\hat{j}_z\hat{s}_x)\hat{j}_x+(\hat{s}_z\hat{j}_y-\hat{j}_z\hat{s}_y)\hat{j}_y}_{q}\,. \tag{14.69}$$

It can be shown that the matrix elements of the operator q disappear when it is applied to wavefunctions with the same quantum number j, or, in other words, the operator q can only couple wavefunctions with different values of j. If the externally applied field is small, we can also expect that such transitions will make only a small contribution and can therefore be ignored. In the following, we shall therefore leave out the operator q. With this approximation, (14.68) can then be written as

$$\hat{s}_z\hat{j}^2=\hat{j}_z\tfrac{1}{2}(\hat{j}^2-\hat{l}^2+\hat{s}^2)\,, \tag{14.70}$$

where we have replaced $\hat{s}\cdot\hat{j}$ by the corresponding expression on the right side of (14.70). It is important to note that all the parameters in (14.70) are *operators*. We now apply both sides of (14.70) to a wavefunction ψ, which is characterised by the *quantum numbers* j, m_j, l, and s. We obtain

$$\underbrace{\hat{s}_z\hat{j}^2}_{\text{Operators}}\cdot\psi=\underbrace{\hat{s}_z}_{\text{Operator}}\cdot\underbrace{\hbar^2 j(j+1)}_{\text{Numbers}}\psi$$

$$=\hbar^2\cdot\underbrace{\hat{j}_z}_{\text{Operator}}\cdot\underbrace{\tfrac{1}{2}[j(j+1)-l(l+1)+s(s+1)]}_{\text{Numbers}}\psi\,. \tag{14.71}$$

If we divide the right half of the double equation (14.71) by $\hbar^2 j(j+1)$, we obtain

$$\hat{s}_z \psi = \frac{j(j+1) - l(l+1) + s(s+1)}{2j(j+1)} \hat{j}_z \psi , \qquad s = 1/2 . \tag{14.72}$$

If we write W_{magn} (14.67) in the form

$$W_{\text{magn}} = \frac{eB}{2m_0} (\underset{\substack{\uparrow}}{\hat{j}_z} + \underset{\substack{\uparrow}}{\hat{s}_z}) , \tag{14.73}$$

Operators

we finally obtain

$$W_{\text{magn}} \psi = \frac{eB}{2m_0} \cdot \underset{\substack{\uparrow \\ \text{Operator}}}{\hat{j}_z} \cdot \left[1 + \underbrace{\frac{j(j+1) - l(l+1) + s(s+1)}{2j(j+1)}}_{\text{Numbers}} \right] \psi . \tag{14.74}$$

The additional energy due to the orientation of the total moment $\boldsymbol{j}$ in the magnetic field is represented by (14.74).

If we write the energy change of a quantum state n, j, l, m_j in the form

$$\Delta E_{j,l,m_j} = \frac{e\hbar}{2m_0} Bg \cdot m_j , \tag{14.75}$$

we can infer the *Landé factor* by comparison with (14.74) to be

$$g = 1 + \frac{j(j+1) - l(l+1) + s(s+1)}{2j(j+1)} . \tag{14.76}$$

We derived this Landé factor earlier, in an intuitive way with the help of the vector model, but we had to make use of the law of cosines in an ad hoc fashion when we replaced $\boldsymbol{j}^2$ by $j(j+1)\hbar^2$, $\boldsymbol{l}^2$ by $l(l+1)\hbar^2$ and $\boldsymbol{s}^2$ by $s(s+1)\hbar^2$. The quantum mechanical calculation presented here gives the exact basis for this substitution.

14.4 Quantum Theory of a Spin in Mutually Perpendicular Magnetic Fields, One Constant and One Time Dependent

A number of important experiments on spin have been carried out with the following arrangement: both a constant, spatially homogeneous magnetic field in the z direction and an oscillating field in the x-y plane are applied. We shall see that this leads to the interesting phenomenon of spin flipping. These experiments make possible, among other things, the exact measurement of magnetic moments, and permit detailed analysis of the structure of and relaxation processes in liquids and solids.

We shall see that we can easily solve these problems using the spin formalism introduced in Sect. 14.2. We write the magnetic field expressed as a time-dependent and a time-independent part:

$$\boldsymbol{B} = \boldsymbol{B}_0 + \boldsymbol{B}^{\mathrm{s}}(t)\,, \tag{14.77}$$

where the vectors of the magnetic fields are defined as

$$\boldsymbol{B}_0 = (0, 0, B_z^0) \qquad \text{and} \tag{14.78}$$

$$\boldsymbol{B}^{\mathrm{s}}(t) = (B_x^{\mathrm{s}}(t), B_y^{\mathrm{s}}(t), 0)\,. \tag{14.79}$$

Naturally we cannot expect that the spin will always point up or down in a time-dependent magnetic field. Rather, we must expect time-dependent transitions. We take these into account by writing the wave function which is to be a solution of the Schrödinger equation (14.43) in the general form

$$\phi(t) = c_1(t)\,\phi_\uparrow + c_2(t)\,\phi_\downarrow \equiv \begin{pmatrix} c_1(t) \\ c_2(t) \end{pmatrix}. \tag{14.80}$$

To arrive at equations for the still unknown coefficients c_1 and c_2, we substitute (14.80) in (14.43), observing the decomposition (14.77 – 79). If we multiply (14.39) out like a normal scalar product and observe the matrix form of $\hat{s}_x$, $\hat{s}_y$, and $\hat{s}_z$ – see (14.40) – we obtain the Schrödinger equation (14.43) in the form

$$\mu_{\mathrm{B}} \begin{pmatrix} B_z^0 & B_x^{\mathrm{s}} - \mathrm{i} B_y^{\mathrm{s}} \\ B_x^{\mathrm{s}} + \mathrm{i} B_y^{\mathrm{s}} & -B_z^0 \end{pmatrix} \cdot \begin{pmatrix} c_1 \\ c_2 \end{pmatrix} = \mathrm{i}\hbar \begin{pmatrix} \dot{c}_1 \\ \dot{c}_2 \end{pmatrix}. \tag{14.81}$$

If we multiply the matrices according to the rule (14.23), we obtain these equations instead of (14.81):

$$(\tfrac{1}{2}\hbar\omega_0)\,c_1 + \mu_{\mathrm{B}}(B_x^{\mathrm{s}} - \mathrm{i} B_y^{\mathrm{s}})\,c_2 = \mathrm{i}\hbar\dot{c}_1\,, \tag{14.82}$$

$$\mu_{\mathrm{B}}(B_x^{\mathrm{s}} + \mathrm{i} B_y^{\mathrm{s}})\,c_1 - \tfrac{1}{2}\hbar\omega_0 c_2 = \mathrm{i}\hbar\dot{c}_2\,. \tag{14.83}$$

Here we have introduced the frequency

$$\hbar\omega_0 = 2\mu_{\mathrm{B}} B_z^0 \tag{14.84}$$

as an abbreviation. In order to simplify the following calculation, let us think of the transverse magnetic field as rotating with the frequency ω. In other words, the magnetic field has the form

$$\begin{aligned} B_x^{\mathrm{s}} &= F\cos\omega t\,, \\ B_y^{\mathrm{s}} &= F\sin\omega t\,. \end{aligned} \tag{14.85}$$

Since B_x^{s} and B_y^{s} appear in (14.82, 83) in a combined form, let us first consider these expressions. We can express them as an exponential function, due to elementary relationships between sines and cosines:

$$B_x^{\mathrm{s}} \pm \mathrm{i} B_y^{\mathrm{s}} = F(\cos\omega t \pm \mathrm{i}\sin\omega t) = F\exp(\pm\mathrm{i}\omega t)\,. \tag{14.86}$$

Then (14.82, 83) simplify to

$$(\hbar\omega_0/2)c_1 + \mu_B F \exp(-i\omega t)c_2 = i\hbar\dot{c}_1 , \tag{14.87}$$

$$\mu_B F \exp(i\omega t)c_1 - (\hbar\omega_0/2)c_2 = i\hbar\dot{c}_2 . \tag{14.88}$$

We shall solve these two equations in two steps. In the first, we put the coefficients $c_j(t)$ into the form

$$c_1(t) = d_1(t)\exp(-i\omega_0 t/2) ; \qquad c_2(t) = d_2(t)\exp(i\omega_0 t/2) . \tag{14.89}$$

If we differentiate (14.89) with respect to time and rearrange slightly, we obtain

$$i\hbar\dot{c}_1 = (\hbar\omega_0/2)c_1 + i\hbar\dot{d}_1 \exp(-i\omega_0 t/2) . \tag{14.90}$$

If we substitute this in (14.87), we see that the term $(\hbar\omega_0/2)c_1$ on both sides cancels out. The same thing happens with c_2 in (14.88), so that (14.87) and (14.88) simplify to

$$\mu_B F \exp[-i(\omega-\omega_0)t]d_2 = i\hbar\dot{d}_1 , \tag{14.91}$$

$$\mu_B F \exp[i(\omega-\omega_0)t]d_1 = i\hbar\dot{d}_2 . \tag{14.92}$$

These equations become very simple when we set the rotational frequency of the magnetic field ω equal to the spin frequency ω_0:

$$\omega = \omega_0 . \tag{14.93}$$

We then obtain

$$\mu_B F d_2 = i\hbar\dot{d}_1 , \tag{14.94}$$

$$\mu_B F d_1 = i\hbar\dot{d}_2 . \tag{14.95}$$

To solve these equations, we first take the time derivative of (14.94):

$$\mu_B F\dot{d}_2 = i\hbar\,\ddot{d}_1 , \tag{14.96}$$

and then, according to (14.95), we replace $\dot{d}_2$ by $(\mu_B F d_1)/(i\hbar)$, and thus obtain

$$\ddot{d}_1 + \frac{\mu_B^2 F^2}{\hbar^2} d_1 = 0 . \tag{14.97}$$

If we simplify the expression by setting $\mu_B F/\hbar = \Omega$, we recognise (14.97) as a typical oscillator equation with the general solution

$$d_1 = a\sin(\Omega t + \Phi) , \tag{14.98}$$

where the amplitude a and phase Φ are free to vary. Using (14.98) and (14.94) we obtain

$$d_2 = i a\cos(\Omega t + \Phi) . \tag{14.99}$$

With the proper choice of the zero time, we can set $\Phi = 0$. The normalisation conditions for the spin wavefunction require that $a = 1$. If we substitute (14.99) in (14.89) and this in (14.80), and do the same with (14.98), we obtain the desired spin wavefunction

$$\phi(t) = \sin(\Omega t)\exp(-\mathrm{i}\,\omega_0 t/2)\,\phi_\uparrow + \mathrm{i}\cos(\Omega t)\exp(\mathrm{i}\,\omega_0 t/2)\,\phi_\downarrow\,. \tag{14.100}$$

The spin functions and the spin formalism naturally seem very unintuitive. In order to see the meaning of the above equations, let us remember that the immediate predictions of quantum mechanics can be read from the corresponding expectation values (Sect. 9.3). We will first develop the expectation value of the spin operator in the z direction. A comparison of (14.49) with (14.100) shows that we can now express the α and β of (14.49) in the form

$$\begin{aligned}\alpha &= \sin(\Omega t)\exp(-\mathrm{i}\,\omega_0 t/2)\,,\\ \beta &= \mathrm{i}\cos(\Omega t)\exp(\mathrm{i}\,\omega_0 t/2)\,.\end{aligned} \tag{14.101}$$

These can be immediately substituted into the end results (14.53 – 55), however, to give

$$\begin{aligned}\langle \hat{s}_z\rangle &= (\hbar/2)\sin^2(\Omega t) - \cos^2(\Omega t)\\ &= -(\hbar/2)\cos(2\,\Omega t)\,.\end{aligned} \tag{14.102}$$

According to (14.102), the z component of the spin oscillates with the frequency $2\,\Omega$. If the spin is originally down at $t = 0$, it flips up, then down again, and so on.

For the other components,

$$\langle \hat{s}_x\rangle = -\frac{\hbar}{2}\sin(2\,\Omega t)\sin(\omega_0 t)\,, \tag{14.103}$$

$$\langle \hat{s}_y\rangle = \frac{\hbar}{2}\sin(2\,\Omega t)\cos(\omega_0 t)\,. \tag{14.104}$$

These equations indicate that the spin motion in the x-y plane is a superposition of two motions, a rapid rotational motion with the frequency ω_0 and a modulation with the frequency $2\,\Omega$. The entire result (14.102 – 104) can be very easily interpreted if we think of the expectation value of the spin as a vector $\boldsymbol{s}$ with the components $\langle \hat{s}_x\rangle$, $\langle \hat{s}_y\rangle$, and $\langle \hat{s}_z\rangle$. Obviously the projection of the vector on the z axis is $(-\hbar/2)\cos(2\,\Omega t)$, while the projection in the x-y plane is $(\hbar/2)\sin(2\,\Omega t)$. As can be seen from the formulae, the spin gradually tips out of the $-z$ direction toward the horizontal, and then further into the $+z$ direction, while simultaneously precessing. The spin thus behaves exactly like a top under the influence of external forces, as we indicated in previous chapters.

We shall consider this process again, in more detail. At a time $t = 0$,

$$\langle \hat{s}_z\rangle = -\hbar/2\,. \tag{14.105}$$

We now ask when the spin, considered intuitively, is in the horizontal position, i.e. when

$$\langle \hat{s}_z\rangle = 0\,. \tag{14.106}$$

This is clearly the case when the cosine function vanishes, that is when

$$2\Omega t = \pi/2 \tag{14.107}$$

holds, or when the time

$$t = \pi/(4\Omega) = \pi\hbar/(4\mu_B F) \tag{14.108}$$

has passed. If one allows the transverse magnetic field to act upon the spins for this time, they will be pointing in the horizontal position (Fig. 14.2). In other words, they have been rotated by an angle $\pi/2$. We therefore speak of a $\pi/2$ or of a 90° pulse. Naturally, we may allow the magnetic field to act for a longer time, for example until the spins are pointing up, i.e.

$$\langle \hat{s}_z \rangle = \hbar/2 \,. \tag{14.109}$$

This occurs when

$$\cos(2\Omega t) = -1 \tag{14.110}$$

is fulfilled, i.e. after the time

$$t = \pi\hbar/(2\mu_B F) \,. \tag{14.111}$$

In this case, we speak of a π or of a 180° pulse (Fig. 14.2).

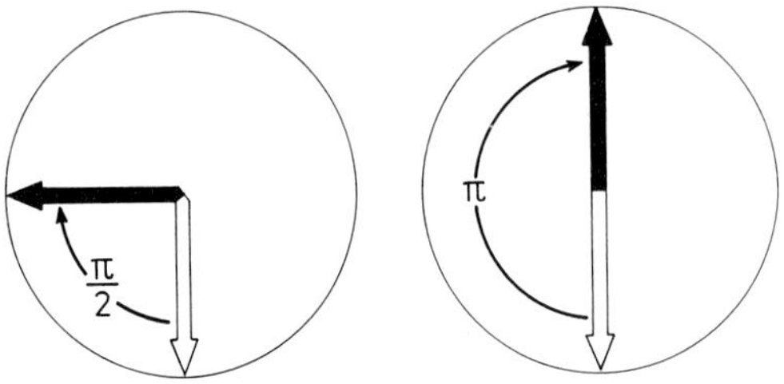

Fig. 14.2. *Left:* Spin flip through $\pi/2$. *Right:* Spin flip through π

With these considerations, we have sketched the most important traits of spin resonance. By applying a rotating magnetic field, we can cause the spin to flip from one direction to another. In practice, of course, one does not apply a magnetic field rotating with the spin frequency, but a linearly oscillating magnetic field. This can be pictured as a superposition of two fields rotating in opposite directions. Then one of the fields rotates with the spin, as before, while the other rotates with twice the frequency, as seen from the point of view of the rotating spin system.

The corresponding equations have practically the same form as those above, except for an additional, rapidly oscillating term, which comes from the "oppositely rotating" magnetic field. To a good approximation, this can be ignored; the result is the "rotating wave approximation".

14.5 The Bloch Equations

As we have just seen, the behaviour of the expectation values of the spin operators can be very simply interpreted. It is thus reasonable to ask whether it would not be possible to derive equations for these expectation values themselves. This is in fact the case. To derive these equations, we use the explicit form which we have just derived for the expectation values of the spin operator. We differentiate $\langle \hat{s}_x \rangle$ with respect to time and make use of (14.103) to obtain

$$\frac{d}{dt}\langle \hat{s}_x \rangle = -(\hbar/2)2\Omega\cos(2\Omega t)\sin(\omega_0 t) - (\hbar/2)\,\omega_0\sin(2\Omega t)\cos(\omega_0 t)\,. \qquad (14.112)$$

The first term on the right side contains $-(\hbar/2)\cos(2\Omega t)$, which, however, is none other than the expectation value of the z component of the spin. We also recognise that the second term on the right side contains the expectation value of the y component of the spin. Equation (14.112) therefore has the form

$$\frac{d}{dt}\langle \hat{s}_x \rangle = \hbar^{-1}2\mu_B F\sin(\omega_0 t)\langle \hat{s}_z \rangle - \omega_0\langle s_y \rangle\,. \qquad (14.113)$$

However, we have seen the factors in front of the expectation values on the right side before. $F\sin(\omega_0 t)$ is just B_y, while ω_0 is proportional to B_z. If we also take into account the relationships (14.84) and (14.85), (14.112) becomes

$$\frac{d}{dt}\langle \hat{s}_x \rangle = \frac{e}{m_0}\langle \hat{s}_z \rangle B_y - \frac{e}{m_0}B_z\langle \hat{s}_y \rangle\,. \qquad (14.114)$$

In a similar way, we find that the time derivative of the y component of the spin is

$$\frac{d}{dt}\langle \hat{s}_y \rangle = -\frac{e}{m_0}\langle \hat{s}_z \rangle B_x + \frac{e}{m_0}B_z\langle \hat{s}_x \rangle\,. \qquad (14.115)$$

If we differentiate the expression (14.102) for $\langle \hat{s}_z \rangle$, we immediately obtain

$$\frac{d}{dt}\langle \hat{s}_z \rangle = \frac{\hbar}{2}2\Omega\sin(2\Omega t)\,. \qquad (14.116)$$

Since we expect that the right side of (14.116) can be expressed in terms of the expectation values of the spin components, like (14.114) and (14.115), we take advantage of the relation

$$\sin^2\omega_0 t + \cos^2\omega_0 t = 1 \qquad (14.117)$$

to write the right side of (14.116) in the form

$$\hbar\Omega\sin(2\Omega t)[\sin(\omega_0 t)\sin(\omega_0 t) + \cos(\omega_0 t)\cos(\omega_0 t)]\,. \qquad (14.118)$$

It is now easy to convince oneself that (14.116) can also be written in the form

$$\frac{d}{dt}\langle\hat{s}_z\rangle = -\frac{e}{m_0}\langle\hat{s}_x\rangle B_y + \frac{e}{m_0}\langle\hat{s}_y\rangle B_x . \tag{14.119}$$

Equations (14.114, 115 and 119) can be written in the form

$$\frac{d}{dt}\langle\hat{s}\rangle = \boldsymbol{\mu}\times\boldsymbol{B} , \tag{14.120}$$

as can be easily seen from the rules of vector multiplication. Here we have assembled the expectation values of the three components of the spin operator into the vector

$$\langle\hat{s}\rangle = \begin{pmatrix} \langle\hat{s}_x\rangle \\ \langle\hat{s}_y\rangle \\ \langle\hat{s}_z\rangle \end{pmatrix} . \tag{14.121}$$

This is strongly reminiscent of the torque equation for a top, if we identify s as the angular momentum and take into account that

$$\boldsymbol{\mu} = -\frac{e}{m_0}\langle\hat{s}\rangle . \tag{14.122}$$

Equation (14.120) is not quite adequate for the interpretation of many experiments, because in many cases, the spin of the particle interacts with its environment. For example, the orbital motion of the spins is continually perturbed by lattice oscillations. This results in continual phase shifts in the precession of the spin. In this case, it is no longer sufficient to regard the equations of a single spin as representative for those of all the spins, as we have implicitly done up to this point. Instead we must consider an "ensemble" of spins, and accordingly, we must in a certain sense subject the pure quantum mechanical expectation values we have used so far to another averaging process. We have to take into account the fact, for instance, that the x component of the spin no longer has a definite value at a definite time, but rather a distribution of values. As time passes, the distribution of values widens out, so that the probability that the value of $\langle\hat{s}_x\rangle$ is positive approaches the same value as the probability that it is negative. This means, however, that in the course of time, the average value of s_x goes to zero. In order to take this decay into account, we add more terms to (14.120) which reflect this incoherent spin motion.

These qualitative considerations are reflected by the phenomenological rule

$$\frac{d}{dt}\langle\hat{s}_x\rangle_{\mathrm{incoh}} = -\frac{1}{T_2}\langle\hat{s}_x\rangle . \tag{14.123}$$

Since $\hat{s}_x$ and $\hat{s}_y$ play the same rôle, we must naturally assume the corresponding rule for $\langle\hat{s}_y\rangle$:

$$\frac{d}{dt}\langle\hat{s}_y\rangle_{\mathrm{incoh}} = -\frac{1}{T_2}\langle\hat{s}_y\rangle . \tag{14.124}$$

Since the spins precess around the z axis, (14.123) and (14.124) indicate how quickly the components transverse to s_z decay. T_2 is therefore often called the transverse relaxation time. It is a measure of the speed with which the individual precessional movements of the spins get out of phase.

Since the z component of the spin is directed along a predetermined, constant field, it must be treated differently from the other two. In this case, too, we would expect a relaxation due to the interaction of the spin with its environment. It will naturally depend upon the orientation of the spin with respect to the external magnetic field – whether the field lies in the positive or negative z direction. The spin can give up energy through its coupling to the environment, and will attempt to reach the lowest state if the environment is at the absolute zero of temperature, $T=0$. On the other hand, if the environment is at a finite temperature, the system of the spins and their environment will attempt to come to thermal equilibrium. At thermal equilibrium, some of the spins will be in the higher state, and others in the lower. If the spin system is displaced from thermal equilibrium, it will naturally attempt to return to it, and in a certain time interval which we call T_1. T_1 is often referred to as the longitudinal relaxation time. What we have just said can be put in mathematical form, if we take

$$\frac{d}{dt}\langle\hat{s}_z\rangle_{\text{incoh}} = \frac{s_0 - \langle\hat{s}_z\rangle}{T_1} \tag{14.125}$$

for the incoherent relaxation of $\langle\hat{s}_z\rangle$. Here s_0 is the value of $\langle\hat{s}_z\rangle$ which the spin component would assume at thermal equilibrium. We arrive at the Bloch equations by adding the "incoherent" terms (14.123 – 125) to the equation (14.120) describing the "coherent" motion of the spin.

The Bloch equations thus have the form

$$\frac{d}{dt}\langle\hat{s}\rangle = -\frac{e}{m_0}\langle\hat{s}\rangle \times \boldsymbol{B} + \begin{pmatrix} -\dfrac{1}{T_2}\langle\hat{s}_x\rangle \\ -\dfrac{1}{T_2}\langle\hat{s}_y\rangle \\ \dfrac{s_0 - \langle\hat{s}_z\rangle}{T_1} \end{pmatrix} . \tag{14.126}$$

The relaxation times T_1 and T_2 are a measure of the strength of the coupling of the electron (or proton) spin to its environment. Measurement of T_1 and T_2 often provides important information about processes in the environment of the spin being investigated, e.g. motion in liquids and solids. We shall discuss a typical and especially elegant experiment in Sect. 15.4.

14.6 The Relativistic Theory of the Electron. The Dirac Equation

In order to correctly describe the interaction of an electron with a magnetic field, we introduced spin operators, which represent the intrinsic degree of freedom of the electron. It was shown by *Dirac* that this intrinsic degree of freedom follows quite automatically from relativistic quantum theory. We therefore wish to treat the Dirac equation in this section. In order to arrive at a relativistic wave equation, it would seem appropriate to attempt to derive it in the same manner as the non-relativistic Schrödinger equation (see Sect. 9.2).

The derivation given there can be summarised in the following "recipe" (cf. also Sect. 9.3): one starts with the classical relation between energy and momentum for a force-free particle

$$E = p^2/2m_0 \tag{14.127}$$

and replaces the energy E and the components of the momentum $\boldsymbol{p}$ by operators according to

$$E \rightarrow \mathrm{i}\hbar \frac{\partial}{\partial t} \tag{14.128}$$

and

$$p_x \rightarrow \frac{\hbar}{\mathrm{i}} \frac{\partial}{\partial x}, \qquad p_y \rightarrow \frac{\hbar}{\mathrm{i}} \frac{\partial}{\partial y}, \qquad p_z \rightarrow \frac{\hbar}{\mathrm{i}} \frac{\partial}{\partial z}. \tag{14.129}$$

The last equivalence can be abbreviated as

$$\boldsymbol{p} \rightarrow \frac{\hbar}{\mathrm{i}} \nabla. \tag{14.130}$$

Following the computational rules of quantum mechanics (cf. Sect. 9.2, 3), these operators act on wavefunctions Ψ, whereby (14.127) thus becomes the well-known Schrödinger equation

$$\mathrm{i}\hbar \frac{\partial}{\partial t} \Psi = -\frac{\hbar^2}{2m_0} \nabla^2 \Psi. \tag{14.131}$$

We shall now attempt to apply this recipe to the relativistic relation between energy and momentum. The latter is

$$E = \sqrt{p^2 c^2 + m_0^2 c^4}. \tag{14.132}$$

If we replace E and $\boldsymbol{p}$ by operators according to (14.128) and (14.130) and allow the resulting expressions on both sides of (14.132) to act upon a wavefunction Ψ, we obtain the equation

$$\mathrm{i}\hbar \frac{\partial}{\partial t} \Psi = \sqrt{-c^2\hbar^2 \nabla^2 + m_0^2 c^4}\, \Psi. \tag{14.133}$$

This equation contains the Laplace operator ∇^2 under a square-root sign, which may at first appear to be only a cosmetic defect. However, this approach failed utterly when the attempt was made to include the effects of electric and magnetic fields on the electron in such a wave equation. The theory had entered a cul-de-sac. Physicists chose two routes to lead it out again:

Route 1: The Klein-Gordon Equation

Since all of the difficulties stem from the square root in (14.133), one has to consider ways to avoid it. To this end, we square both sides of (14.132) and obtain

$$E^2 = p^2c^2 + m_0^2c^4\,, \tag{14.134}$$

which, of course, may be immediately translated into the wave equation

$$-\hbar^2\frac{\partial^2}{\partial t^2}\Psi = (-c^2\hbar^2\nabla^2 + m_0^2c^4)\,\Psi\,, \tag{14.135}$$

called the *Klein-Gordon equation.*

The latter may be arranged in a more elegant (and relativistically more obvious) form by dividing both sides by $c^2\hbar^2$ and introducing the operator

$$\Box^2 = \nabla^2 - \frac{1}{c^2}\,\frac{\partial^2}{\partial t^2}\,. \tag{14.136}$$

The Klein-Gordon equation is then given as

$$\Box^2\Psi = \frac{m_0^2c^2}{\hbar^2}\,\Psi\,. \tag{14.137}$$

Let us examine its solutions. Since, for a force-free particle, we expect the solutions to be de Broglie waves, we use the trial function

$$\Psi = \exp\left[\mathrm{i}(\boldsymbol{k}\cdot\boldsymbol{r} - \omega t)\right] \tag{14.138}$$

in which, as usual,

$$\omega = \frac{E}{\hbar} \quad \text{and} \quad \boldsymbol{p} = \hbar\boldsymbol{k}\,. \tag{14.139}$$

If we insert (14.138) in (14.135), we obtain (14.134) as the immediate result.

In order to determine the energy E itself, we must naturally take the square root. We thus obtain not only a positive energy

$$E = +\sqrt{p^2c^2 + m_0^2c^4} \tag{14.140}$$

but also a negative energy

$$E = -\sqrt{p^2c^2 + m_0^2c^4}\,. \tag{14.141}$$

Since free particles can have only positive energies, we are faced here with a difficulty! Furthermore, the analysis of the solutions reveals that the particle *density* can also become negative, which is also an unphysical result. The Klein-Gordon equation was reinterpreted by Pauli and Weisskopf, who used the charge density instead of the mass density, and thus found it to be applicable in quantum field theory to particles with spin zero. However, further development of that topic lies outside the framework of this book.

Route 2: The Dirac Equation

Dirac considered the question as to whether the root in (14.132) could not be extracted in some simple manner. In the limit $p=0$, we find

$$\sqrt{p^2c^2+m_0^2c^4} \rightarrow m_0c^2 ;$$

and, for $m_0=0$, (14.142)

$$\sqrt{p^2c^2+m_0^2c^4} \rightarrow pc .$$

In order to understand Dirac's approach, let us first consider the one dimensional case and generalise (14.142) to

$$\sqrt{p_x^2c^2+m_0^2c^4} = \alpha c p_x + \beta m_0 c^2 . \qquad (14.143)$$

This relation can clearly not be fulfilled in the general case $p_x \neq 0$, $m_0 \neq 0$ by ordinary numbers α and β; however, it can be, when α and β are *matrices*, as we shall proceed to demonstrate. We square both sides of (14.143), remembering that matrices do not commute, in general, so that we must maintain the order of α and β in multiplying out the right-hand side of (14.143). We then obtain

$$p_x^2c^2+m_0^2c^4 = \alpha^2c^2p_x^2+(\alpha\beta+\beta\alpha)m_0c^3p_x+\beta^2m_0^2c^4 . \qquad (14.144)$$

For the left and right sides of this equation to be equal, we clearly require that

$$\alpha^2=1 ; \quad \alpha\beta+\beta\alpha=0 ; \quad \beta^2=1 . \qquad (14.145)$$

These relations are familiar from the (Pauli) spin matrices! (compare Problem 14.2). Unfortunately, we cannot use the latter directly, since we wish to describe a three-dimensional, not a one-dimensional motion. Thus, we require

$$\sqrt{(p_x^2+p_y^2+p_z^2)c^2+m_0^2c^4} = \alpha_1 c p_x + \alpha_2 c p_y + \alpha_3 c p_z + \beta m_0 c^2 . \qquad (14.146)$$

Squaring (14.146) leads, analogously to the one-dimensional case, to

$$\begin{aligned} &\alpha_j^2=1 ; \quad \beta^2=1 ; \quad \alpha_j\beta+\beta\alpha_j=0 ; \quad \text{and} \\ &\alpha_j\alpha_k+\alpha_k\alpha_j=0 \quad \text{for} \quad j\neq k ; \quad j=1,2,3 \quad \text{and} \quad k=1,2,3 . \end{aligned} \qquad (14.147)$$

In addition, as always in quantum mechanics, the operators (matrices) are Hermitian. These relations may be fulfilled in various (but physically equivalent) ways, for example

$$\alpha_j = \begin{bmatrix} 0 & \sigma_j \\ \sigma_j & 0 \end{bmatrix}, \quad \beta = \begin{bmatrix} 1 & 0 \\ 0 & -1 \end{bmatrix}, \qquad (14.148)$$

where the σ_j are the Pauli spin matrices (cf. 14.24, 33 without $\hbar$). The "1"s in β represent 2×2 identity matrices, so that β may be written in the conventional notation as

$$\beta = \begin{bmatrix} 1 & 0 & 0 & 0 \\ 0 & 1 & 0 & 0 \\ 0 & 0 & -1 & 0 \\ 0 & 0 & 0 & -1 \end{bmatrix} . \tag{14.149}$$

After these intermediate steps, we can again attack the Dirac equation, employing the translation rules (14.128) and (14.130) and applying them to the equation

$$E = \alpha_1 c p_x + \alpha_2 c p_y + \alpha_3 c p_z + \beta m_0 c^2 . \tag{14.150}$$

This leads to

$$\mathrm{i}\hbar \frac{\partial}{\partial t} \Psi = (\alpha_1 c p_x + \alpha_2 c p_y + \alpha_3 c p_z + \beta m_0 c^2)\, \Psi , \tag{14.151}$$

the *Dirac equation*.

Since α_j and β are 4×4 matrices, they must operate on *vectors* with four components, i.e. Ψ must be of the form

$$\Psi = \begin{pmatrix} \Psi_1 \\ \Psi_2 \\ \Psi_3 \\ \Psi_4 \end{pmatrix} . \tag{14.152}$$

In the preceding sections dealing with the electronic spin, we became acquainted with wavefunctions having 2 components; in the Dirac theory, they have four! This is a result of the fact that the Dirac equation allows both positive and negative-energy solutions for free particles.

As the reader may verify in one of the problems to this chapter, the Dirac equation yields the same energy spectrum as the Klein-Gordon equation; it is given in (14.140, 141), and in Fig. 14.3. One can easily convince oneself that the solutions of the Dirac equation for force-free particles are plane waves having the form

$$\Psi(\boldsymbol{r}, t) = \begin{pmatrix} \Psi_1 \\ \Psi_2 \\ \Psi_3 \\ \Psi_4 \end{pmatrix} \exp(\mathrm{i}\boldsymbol{k}\cdot\boldsymbol{r} - \mathrm{i}\omega t) , \tag{14.153}$$

where the constants $\Psi_1, \dots \Psi_4$ are computed in Problem 14.6.

In the Dirac equation in the form (14.151), the time derivative plays a special rôle relative to the spatial-coordinate derivatives. However, in relativity theory, time and space coordinates have a symmetric position as components of space-time four-vectors; thus, in the literature, a symmetrised form of the Dirac equation is often used. It is obtained by multiplying (14.151) on both sides from the left by $\gamma^0 = \beta$, and introducing new matrices

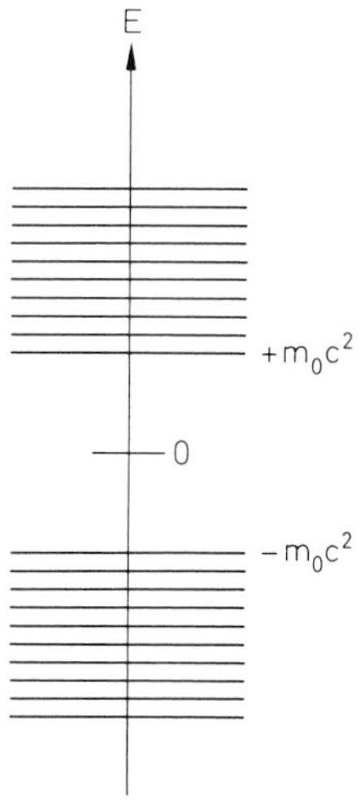

Fig. 14.3. Energy spectrum of the Klein-Gordon equation and the Dirac equation

$$\gamma^j = \beta \alpha_j, \quad \text{with} \quad j = 1, 2, 3 . \tag{14.154}$$

It may be shown that the resulting equation is "Lorenz covariant":

$$\mathrm{i}\hbar \left(\gamma^0 \frac{\partial}{\partial x^0} + \gamma^1 \frac{\partial}{\partial x_1} + \gamma^2 \frac{\partial}{\partial x_2} + \gamma^3 \frac{\partial}{\partial x_3} \right) \Psi = m_0 c \Psi , \tag{14.155}$$

with $x^0 = ct$, $x_1 = x$, $x_2 = y$, $x_3 = z$.

The explicit forms for the matrices γ^0 and γ^j are

$$\gamma^0 = \begin{bmatrix} 1 & 0 \\ 0 & -1 \end{bmatrix} , \quad \gamma^j = \begin{bmatrix} 0 & \sigma^j \\ -\sigma^j & 0 \end{bmatrix} \tag{14.156}$$

where the σ^j are again the Pauli matrices.

Finally, we discuss the inclusion of the action of electric and magnetic fields on the electron in the Dirac equation. For this purpose, we use the procedure of the Schrödinger theory once again:

1) The potential energy $V(\boldsymbol{r}) = -e\tilde{V}$, which results from the electrostatic potential $\tilde{V}$, is added in analogy to (9.32). This can also be expressed by adopting the following expression:

$$\mathrm{i}\hbar \frac{\partial}{\partial t} \rightarrow \mathrm{i}\hbar \frac{\partial}{\partial t} + e\tilde{V} . \tag{14.157}$$

2) The magnetic field is taken into account by replacing the momentum operator (as in Sect. 14.1) by

$$\frac{\hbar}{\mathrm{i}} \nabla \rightarrow \frac{\hbar}{\mathrm{i}} \nabla + e\boldsymbol{A} \tag{14.158}$$

where $\boldsymbol{A}$ is the vector potential.

The resulting Dirac equation has been solved for several cases, in particular for the hydrogen atom. The results are in very good agreement with experiment, apart from the corrections due to quantum electrodynamics (Lamb shift).

In spite of the success of the Dirac theory, the question of the meaning of the negative energy values for free particles remained open. They would permit an electron with a positive energy to emit light and drop down to deeper-lying, i.e. negative energy levels, and thus all particles with positive energies would finally fall into this energy chasm.

Dirac had the ingenious idea of assuming that all the states of negative energies were already occupied with electrons, following the Pauli principle, according to which each state can contain at most two electrons with antiparallel spins. The infinitely large negative charge of this so-called "Dirac sea" can be thought to be compensated by the positive charges of the protons, which likewise obey the Dirac equation and must fill a corresponding positively charged Dirac sea. The vacuum would, in this interpretation, consist of the two filled Dirac seas.

If we now add sufficient energy that an electron from the Dirac sea can cross over the energy gap of $2m_0c^2$, an electron with positive energy would appear, leaving behind a hole in the Dirac sea. Since this hole is a missing negative charge, but the Dirac sea(s) were previously electrically neutral, the hole acts like a positive charge $(+e)$. Furthermore, it has the same properties as a particle, so that it appears as such. The creation of electron-hole pairs can indeed be observed; the positively charged particles are experimentally known as *positrons*.

In modern quantum field theory, the creation of positrons can be described directly by means of a formal trick, without having to invoke the infinite filled Dirac sea. On the other hand, precisely this idea of a Dirac sea provides an intuitive picture for the appearance of positively charged electrons, i.e. the positrons.

Problems

14.1 *The Landau levels*

If an otherwise free electron is moving in a magnetic field, it is forced into a circular path in the plane perpendicular to the magnetic field. It thus has a periodic motion and would be subject to quantisation, even in the Sommerfeld formulation. This quantisation leads to discrete levels, the Landau levels. These also result from an exact quantum mechanical calculation.

Problem: Solve the time-independent Schrödinger equation of a particle with charge $(-e)$ which is moving in the x-y plane perpendicular to a constant magnetic field $\boldsymbol{B}$. Do not take the electron spin into account.

Hint: Use the vector potential $\boldsymbol{A}$ in the form $\boldsymbol{A} = (0, B_x, 0)$ and the trial solution

$$\psi(x,y,z) = \mathrm{e}^{iky}\phi(x) .$$

In addition, make use of the fact that $\phi(x)$ satisfies the Schrödinger equation for a displaced harmonic oscillator.

14.2 Show that for the spin operators $\hat{s}_x$ and $\hat{s}_y$ (14.33 a, b), the following relations hold:

$$\hat{s}_x\hat{s}_y + \hat{s}_y\hat{s}_x = 0 , \quad \left(\frac{2}{\hbar}\hat{s}_x\right)^2 = 1 , \quad \left(\frac{2}{\hbar}\hat{s}_y\right)^2 = 1 .$$

Hint: Use the explicit matrix form.

14.3 Demonstrate that the relativistic expression for the energy, (14.132), may be written in the form

$$E \approx m_0c^2 + \frac{1}{2}\frac{p^2}{m_0}$$

provided that

$$\frac{p^2}{2m_0} \ll \frac{1}{2}m_0c^2 .$$

Hint: Expand the square root in a series.

14.4 Show that the (charge) conservation law in the form

$$\frac{d\varrho}{dt} + \operatorname{div} \boldsymbol{j} = 0$$

may be derived from the Klein-Gordon equation.

Hint: Multiply the Klein-Gordon equation (written with $\Box^2$) by Ψ^* and subtract from the result its complex conjugate. Use:

$$\varrho = \frac{\mathrm{i}\hbar}{2m_0c^2}\left(\Psi^*\frac{\partial\Psi}{\partial t} - \Psi\frac{\partial\Psi^*}{\partial t}\right),$$

$$\boldsymbol{j} = \frac{\hbar}{2\mathrm{i}m_0}(\Psi^* \operatorname{grad}\Psi - \Psi \operatorname{grad}\Psi^*)\,.$$

14.5 Show that each component of (14.153) satisfies the Klein-Gordon equation.

Hint: Write the Dirac equation (14.151) in the form

$$\mathrm{i}\hbar\frac{\partial\Psi}{\partial t} = \mathscr{H}\Psi \tag{1}$$

then take $\mathrm{i}\hbar(\partial/\partial t)$ on both sides; use (1) again and rearrange $\mathscr{H}^2$ using the Dirac matrices.

14.6 Solve the Dirac equation for a force-free particle moving in the z direction.

Hint: Substitute the trial solution

$$\Psi(\boldsymbol{r},t) = \begin{pmatrix}\Psi_1\\ \Psi_2\\ \Psi_3\\ \Psi_4\end{pmatrix}\exp(\mathrm{i}kz - \mathrm{i}\omega t)$$

into (14.151) and solve the resulting algebraic equation.

Which energies correspond to the various solutions?

15. Atoms in an Electric Field

15.1 Observations of the Stark Effect

In 1913, *Stark* observed a splitting of the lines of the Balmer series of hydrogen (8.2) in an electric field. He was studying the light emission of H atoms in the field of a condenser (Figs. 15.1, 2). Since then, frequency shifts in optical spectra in the presence of eletric fields have been generally called the Stark effect.

The effect is experimentally more difficult to observe than the Zeeman effect, because it is necessary to generate strong electric fields without sparking over. It has been far less important to experimental atomic physics than the Zeeman effect.

One observes:

- With hydrogen and similar atoms, a splitting of the terms with $l \neq 0$ and the spectral lines associated with them. The splitting is proportional to the field strength F. This so-called *linear* Stark effect is present when the l degeneracy – the degeneracy of states with the same principal quantum number n and different orbital angular momentum quantum numbers l – is lifted by the external electric field, when it is not already lifted by internal atomic fields;
- Displacement and splitting of terms in all other atoms proportional to F^2. This is the *quadratic* Stark effect.

The quadratic Stark effect can be understood qualitatively in an intuitive model. The applied electric field induces an electric dipole moment $p = \alpha F$ in the atom, where

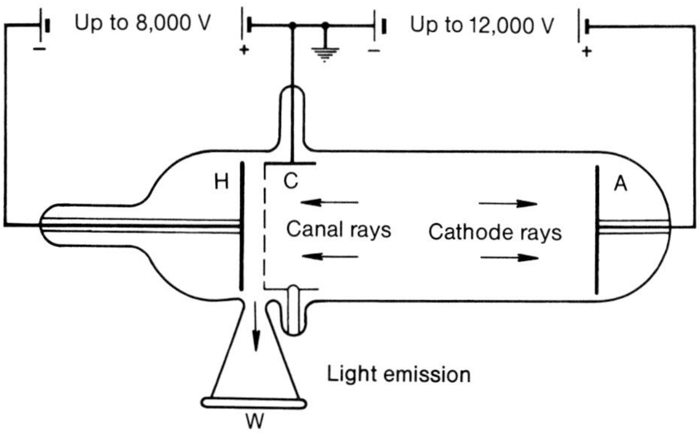

Fig. 15.1. Canal ray tube for investigation of the emission of atoms in an electric field: the Stark effect. The potential between the cathode C and the electrode H can be as high as 8000 V. The resulting splitting of spectral lines is observed through the window W

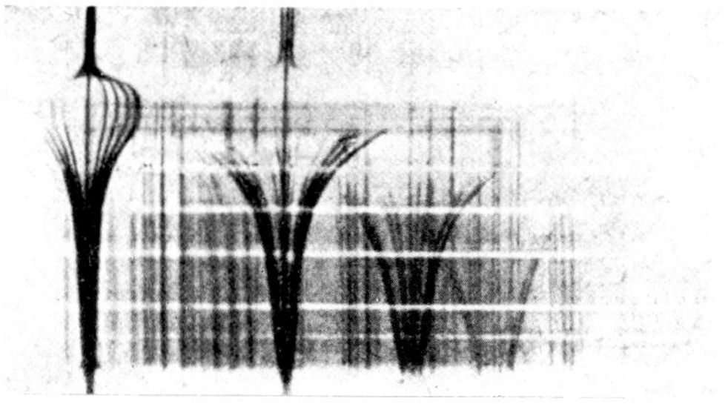

Fig. 15.2. Splitting of the hydrogen atom lines in an electric field. The strength of the field varies along the light source, the image of which is shown after passage through a spectrograph slit. The field is 10^5 V/cm in the region of smaller splitting near the bottom of the figure and rises to a value of $1.14 \cdot 10^6$ V/cm in the region of the greatest splitting. From K. H. Hellwege, *Einführung in die Physik der Atome*, Heidelberger Taschenbücher, Vol. 2, 4th ed. (Springer, Berlin, Heidelberg, New York 1974) Fig. 45

α is the atomic polarisability. The latter is naturally a function of the quantum numbers of the atomic state and is different for each electron configuration.

The electric field acts on this induced dipole moment. The interaction energy is given by

$$V_{el} = \tfrac{1}{2}\boldsymbol{p} \cdot \boldsymbol{F} = \tfrac{1}{2}\alpha F^2 . \tag{15.1}$$

We have thus explained qualitatively the proportionality between the term shifts and the square of the electric field strength.

The linear Stark effect, which is observed in the hydrogen atom, cannot be so easily understood on an intuitive basis. It will be treated in greater detail in Sect. 15.2.

The fundamental difference between the Stark effect and the splitting of spectral lines in a magnetic field is the fact that in an electric field, states with the same absolute value of the magnetic quantum number m_j, i.e. m_j and $-m_j$, behave in an identical manner. This can be easily understood: the effect of an electric field on a "clockwise" and on a "counterclockwise" rotating electron, when the spatial distribution of the electrons is otherwise the same, is, averaged over time, the same. The number of split components is therefore smaller in the Stark effect than in the Zeeman effect: the number of different terms is not $2j+1$, but rather $j+1$ for integral j and $j+1/2$ for half-integral j.

An example is the Stark effect of the Na D lines, shown in Fig. 15.3. The magnitude of the Stark shift is about 0.05 Å for the Na D lines in fields of about 10^7 V/m (10^5 V/cm). It increases with the principal quantum number n, since orbits with a larger principal quantum number also have a larger polarisability. Therefore the Stark effect is extremely important for the investigation of Rydberg atoms (cf. Sect. 8.12).

The Stark effect which is caused by the strong electric fields resulting from the chemical bonding between atoms, is extremely important to the understanding of molecular spectra. It is also important for the clarification of the influence of the crystal electric field in solids on the atomic term diagrams of component atoms, as well as in gases at high densities. In the latter, the Stark effect is the most important source of spectral line broadening.

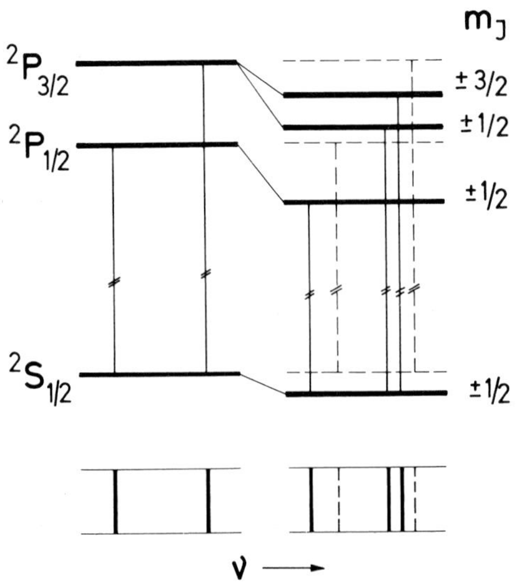

Fig. 15.3. Energy level diagram showing the Stark effect on the sodium doublet $3^2P_{3/2,1/2} - 3^2S_{1/2}$ and the splitting pattern of the D lines

15.2 Quantum Theory of the Linear and Quadratic Stark Effects

15.2.1 The Hamiltonian

We shall treat the quantum theoretical explanation of the Stark effect in some detail here, since we shall, in the process, be able to introduce the general and important methods of perturbation theory.

We wish to investigate how the wavefunctions and the energy levels of an electron are changed, when, in addition to the attractive nuclear potential $V(r)$, a constant electric field is acting on it. We write the Hamiltonian of the complete problem in the form

$$\mathscr{H} = \mathscr{H}_0 + \mathscr{H}^{\mathrm{P}} , \quad (15.2)$$

where

$$\mathscr{H}_0 = -\frac{\hbar^2}{2m_0}\nabla^2 + V(r) \quad (15.3)$$

is the original Hamiltonian without the applied field. In (15.2), as in the following, the upper index "P" indicates "perturbation".

If the electric field has the field strength $\boldsymbol{F}$, the electron is acted on by the force

$$-e\boldsymbol{F} . \quad (15.4)$$

(In order to avoid confusion between the energy E and the electric field strength, we denote the latter here by $\boldsymbol{F}$.) The corresponding potential energy, which results from "force times distance", is then

$$V^{\mathrm{P}} = e\boldsymbol{F} \cdot \boldsymbol{r} , \quad (15.5)$$

provided $\boldsymbol{F}$ is homogeneous.

Since the formalism which we are about to develop may be applied to perturbations which are more general than (15.5), we have written $\mathscr{H}^{\mathrm{P}}$ instead of V^{P} in (15.2). It is found in many cases that the applied electric field produces only a small change in the electron wavefunctions and energies, i.e. it acts as a so-called small perturbation. In order to express the smallness of this perturbation explicitly, we write $\mathscr{H}^{\mathrm{P}}$ in the form

$$\mathscr{H}^{\mathrm{P}} = \lambda \mathscr{H}^1 , \quad (15.6)$$

where λ is a small parameter. In the following, we also assume that the time-independent Schrödinger equation without the external perturbation potential has already been solved:

$$\mathscr{H}_0 \phi_\nu = E_\nu^0 \phi_\nu . \quad (15.7)$$

The indices 0 on $\mathscr{H}_0$ and E_ν^0 indicate that these quantities refer to the unperturbed problem. We shall at first assume that the energies E_ν^0 are all different from one another.

15.2.2 The Quadratic Stark Effect. Perturbation Theory Without Degeneracy*

To be able to solve the Schrödinger equation which also contains the perturbation potential, namely

$$\mathscr{H}\psi = E\psi\,, \tag{15.8}$$

we represent the solution for which we are searching, ψ, as the superposition of the unperturbed solutions ϕ_ν. We expect, indeed, that the electric field will shift and perhaps also change the shapes of the wavefunctions. These modified wavefunctions may be constructed from the unperturbed ones by adding them to wavefunctions belonging to other energy levels (Fig. 15.4). On the basis of such considerations, which also may be justified mathematically in a strict sense, we arrive at the following trial solution for the wavefunctions we are seeking:

$$\psi(\boldsymbol{r}) = \sum_{\nu=1}^{\infty} c_\nu \phi_\nu(\boldsymbol{r})\,. \tag{15.9}$$

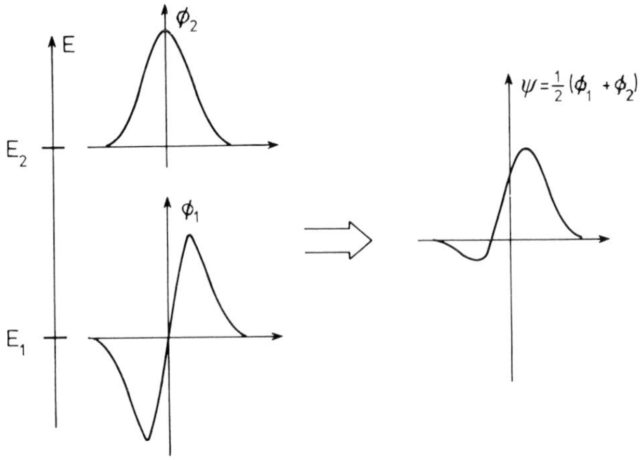

Fig. 15.4. Example of how the superposition of two wavefunctions ϕ_1 and ϕ_2 (*left*) can lead to a new wavefunction with a shifted centre of gravity (centre of charge) (*right*)

Here it is important to note that the wavefunctions ϕ_ν depend upon the position coordinate $\boldsymbol{r}$, but that the coefficients c_ν do *not*. To fix the latter, we insert (15.9) into (15.8) and obtain immediately

$$\mathscr{H}_0 \sum_\nu c_\nu \phi_\nu(\boldsymbol{r}) + \mathscr{H}^{\mathrm{P}} \sum_\nu c_\nu \phi_\nu(\boldsymbol{r}) = E \sum_\nu c_\nu \phi_\nu(\boldsymbol{r})\,. \tag{15.10}$$

In the first term on the left side, we shall use the fact that the wavefunctions ϕ_ν obey (15.7) and we may thus replace $\mathscr{H}_0\phi_\nu$ by $E_\nu^0\phi_\nu$. In order to free ourselves from the $\boldsymbol{r}$ dependence in (15.10), we multiply from the left by ϕ_μ^* and integrate over all space. As we shall show in the appendix, the wavefunctions are orthonormal, i.e. the following relations hold:

$$\int \phi_\mu^* \phi_\nu dV = \delta_{\mu\nu}\,. \tag{15.11}$$

Furthermore, we introduce the abbreviations

$$H_{\mu\nu}^{\mathrm{P}} = \int \phi_\mu^* \mathscr{H}^{\mathrm{P}} \phi_\nu dV . \tag{15.12}$$

Since the parameters $H_{\mu\nu}^{\mathrm{P}}$ carry the two subscripts μ and ν, one often thinks of the $H_{\mu\nu}^{\mathrm{P}}$ arranged in a square array. Such an array is called a "matrix" in mathematics, and the $H_{\mu\nu}^{\mathrm{P}}$ are thus also called "matrix elements", or more exactly, "matrix elements of the perturbation operator $\mathscr{H}^{\mathrm{P}}$". With the help of (15.11) and (15.12), we obtain the following equation from (15.10)

$$(E_\mu^0 - E) c_\mu + \sum_\nu H_{\mu\nu}^{\mathrm{P}} c_\nu = 0 , \tag{15.13}$$

which one must imagine as written out for all indices μ. Thus far, our procedure is completely generally valid and makes no use of the small magnitude of the perturbation. We now assume, however, that the perturbation is small, in that we imagine, according to (15.6), that the parameter λ increases from zero. If the perturbation is exactly equal to zero, the solution sought, (15.9), must naturally be the same as one of the starting solutions ϕ_ν: The coefficients which result for $\lambda = 0$ are indicated by the superscript 0. The initial state is denoted by a subscript κ. We thus obtain the relation

$$c_\nu^0 = \begin{cases} 1 & \text{for} \quad \nu = \kappa \\ 0 & \text{for} \quad \nu \neq \kappa \end{cases} \tag{15.14}$$

or in shorter form,

$$c_\nu^0 = \delta_{\nu\kappa} . \tag{15.15}$$

If we now let λ increase, the coefficients c_ν will also change, of course. We shall expect that as a first approximation, the coefficients c_ν increase proportionally to λ. As the next approximation, we must then take into account changes proportional to λ^2, and so on. The same will naturally also hold for the new energy values E. We thus arrive at

$$c_\nu = \delta_{\nu\kappa} + \lambda c_\nu^{(1)} + \lambda^2 c_\nu^{(2)} + \dots \tag{15.16}$$

and

$$E = E_\kappa^0 + \lambda \varepsilon^{(1)} + \lambda^2 \varepsilon^{(2)} + \dots . \tag{15.17}$$

We subsitute these expressions in (15.13) and thereby obtain

$$(E_\mu^0 - E_\kappa^0 - \lambda \varepsilon^{(1)} - \lambda^2 \varepsilon^{(2)} - \dots)(\delta_{\mu\kappa} + \lambda c_\mu^{(1)} + \dots) + \sum_\nu \lambda H_{\mu\nu}^1 (\delta_{\nu\kappa} + \lambda c_\nu^{(1)} + \dots) = 0 . \tag{15.18}$$

Expressions (15.16) and (15.17) define orders of magnitude, as one can easily convince oneself by setting, e.g. $\lambda = 0.1$. In this case, $\lambda^2 = 0.01$, which is only 10% of λ. Speaking a bit loosely, what we are doing is to solve (15.18) for the different decimal places. In strict mathematical terms, this means that we must multiply out all the terms of (15.18) and arrange them according to powers of λ. We then have to require that the coefficients of the individual powers of λ cancel out independently. For the zeroth power, we obtain

$$(E_\mu^0 - E_\kappa^0)\,\delta_{\mu\kappa} = 0\,, \tag{15.19}$$

which is identically fulfilled. For the 1st power of λ,

$$-\varepsilon^{(1)}\delta_{\mu\kappa} + (E_\mu^0 - E_\kappa^0)\,c_\mu^{(1)} + H_{\mu\kappa}^1 = 0\,. \tag{15.20}$$

For further discussion of this equation, we differentiate between the cases where $\mu = \kappa$ and $\mu \neq \kappa$. For $\mu = \kappa$, (15.20) reduces to

$$\varepsilon^{(1)} = H_{\kappa,\kappa}^1 \equiv \int \phi_\kappa^* \mathscr{H}^1 \phi_\kappa\, dV\,. \tag{15.21}$$

For the perturbed energy, this means according to (15.17) that the 1st order perturbation approximation,

$$E = E_\kappa^0 + H_{\kappa,\kappa}^{\mathrm{P}}\,, \tag{15.22}$$

applies. If we choose $\mu \neq \kappa$, the coefficients of the 1st order perturbation approximation can be calculated from (15.20):

$$c_\mu^{(1)} = \frac{H_{\mu,\kappa}^1}{E_\kappa^0 - E_\mu^0}\,, \qquad \mu \neq \kappa\,. \tag{15.23}$$

So far, the coefficient $c_\kappa^{(1)}$ has not been determined. As can be shown from the normalisation coefficient, it must be set equal to zero:

$$c_\kappa^{(1)} = 0\,. \tag{15.24}$$

If we substitute the coefficients we have calculated into (15.9), the perturbed wavefunction is, in the 1st order perturbation approximation,

$$\psi(\boldsymbol{r}) = \phi_\kappa(\boldsymbol{r}) + \sum_{\mu \neq \kappa} \frac{H_{\mu,\kappa}^{\mathrm{P}}}{E_\kappa^0 - E_\mu^0}\,\phi_\mu(\boldsymbol{r})\,. \tag{15.25}$$

Now we can take into account the terms in second order, i.e. with λ^2. A short calculation yields

$$\varepsilon^{(2)} = \sum_{\nu \neq \kappa} \frac{|H_{\kappa,\nu}^1|^2}{E_\kappa^0 - E_\nu^0}\,. \tag{15.26}$$

With this, the energy in the second order perturbation approximation can be expressed as

$$E = E_\kappa^0 + H_{\kappa,\kappa}^{\mathrm{P}} + \sum_{\nu \neq \kappa} \frac{|H_{\kappa,\nu}^{\mathrm{P}}|^2}{E_\kappa^0 - E_\nu^0}\,. \tag{15.27}$$

We shall now examine the meaning of the formulae (15.25) and (15.27) in the case when an external electric field $\boldsymbol{F}$ is applied. It can be shown (see Sect. 16.1.3 on selection rules), for example that for the hydrogen atom $H^{\mathrm{P}}_{\kappa,\kappa}=0$. The matrix elements which are not equal to zero are, according to (15.5) and (15.12), proportional to the field strength F. Thus the energy E is shifted from the unperturbed energy E^0_κ by an amount, according to (15.27), which is proportional to F^2. One therefore speaks of the *quadratic Stark effect.*

15.2.3 The Linear Stark Effect. Perturbation Theory in the Presence of Degeneracy*

In addition to this quadratic Stark effect, observations have shown a *linear Stark effect*. We approach it as follows. In a purely formal way, we can see from (15.23 – 27) that the method we used above will not work if the denominator, i.e. $E^0_\kappa - E^0_\nu$ cancels and at same time the matrix element in the numerator is not equal to zero. This can actually happen, however, if we are considering degenerate states, such as we have seen in the hydrogen atom. There we have an entire set of different wavefunctions with the same principal quantum number n, but different l and m, which belong to a given energy. To treat this case in the presence of a perturbation, one must fall back on the so-called *perturbation theory in the presence of degeneracy.*

Let us briefly review what was done in the first step of perturbation theory in the absence of degeneracy. We set up the requirement (15.14). In the case of degeneracy, as we know, it is not only the mutually degenerate wavefunctions which can be solutions to the Schrödinger equation for the energy E^0_κ, but also any linear combination of these wavefunctions. Thus if we think of the perturbation being turned off, the perturbed solution can be transformed into a linear combination of unperturbed solutions whose coefficients are not known. The basic idea of perturbation theory in the presence of degeneracy is to find these coefficients, in the zero order approximation, by a systematic procedure. To this end we write

$$\psi(\boldsymbol{r}) = \sum_{\substack{\nu \text{ only over} \\ \text{degenerate} \\ \text{states}}} c_\nu^{(0)} \phi_\nu(\boldsymbol{r}) + \text{corrections}\,, \tag{15.28}$$

where the summation includes *only* the mutually degenerate states. We ignore the correction terms. Equation (15.28) is formally the same as our earlier (15.9), but now we are not summing over all states. However, the coefficients $c_\nu^{(0)}$ can be formally determined if we return to the equation system of the form of (15.13), and allow the set of coefficients $c_\nu^{(0)}$ to replace the coefficients c_ν there. If we have N mutually degenerate states, we now have N equations with N unknown coefficients before us. In order that the homogeneous system of equations be soluble, the determinant of the coefficients must vanish. This provides the condition

$$\begin{vmatrix} (E^0_\kappa - E + H^{\mathrm{P}}_{1,1}) & H^{\mathrm{P}}_{1,2} & \dots & H^{\mathrm{P}}_{1,N} \\ H^{\mathrm{P}}_{2,1} & (E^0_\kappa - E + H^{\mathrm{P}}_{2,2}) & \dots & H^{\mathrm{P}}_{2,N} \\ \vdots & & & \\ H^{\mathrm{P}}_{N,1} & \dots & & (E^0_\kappa - E + H^{\mathrm{P}}_{N,N}) \end{vmatrix} = 0\,. \tag{15.29}$$

The determinant here is also called a secular determinant. If calculated, it yields an Nth degree polynomial in the energy E. When this is set equal to zero, it becomes an algebraic equation for E which has N roots, some of which may be equal to each other.

As a concrete example, let us treat the first excited state of the hydrogen atom with the principal quantum number $n = 2$. The wavefunctions of hydrogen are, as we know, indicated by the quantum numbers n, l and m. To relate these to the present system of indices, we write

$$\underbrace{\phi_{n,l,m}}_{\nu}, \quad n = 2, \tag{15.30}$$

where we use the table

$$\nu = \begin{cases} 1 & \text{for} \quad l = 0, \quad m = 0 \\ 2 & \text{for} \quad l = 1, \quad m = 0 \\ 3 & \text{for} \quad l = 1, \quad m = 1 \\ 4 & \text{for} \quad l = 1, \quad m = -1. \end{cases} \tag{15.31}$$

The trial solution (15.28) thus becomes

$$\psi(\boldsymbol{r}) = c_1^{(0)}\phi_1(\boldsymbol{r}) + c_2^{(0)}\phi_2(\boldsymbol{r}) + c_3^{(0)}\phi_3(\boldsymbol{r}) + c_4^{(0)}\phi_4(\boldsymbol{r}), \tag{15.32}$$

where, to repeat once more, the ϕ's are wavefunctions of the hydrogen atom in the $n = 2$ state, which are all degenerate. The matrix elements (15.12) are then, in concrete terms,

$$H_{\mu\nu}^{\mathrm{P}} = \int \underbrace{\phi_{n,l,m}^{*}}_{\mu}(\boldsymbol{r})\, eFz\, \underbrace{\phi_{n,l',m'}}_{\nu}(\boldsymbol{r})\, dV, \tag{15.33}$$

where it is assumed that the field is applied in the z direction. Using selection rules, it can be shown, as in Sect. 16.1, that all the matrix elements disappear except for

$$H_{1,2}^{\mathrm{P}} = H_{2,1}^{\mathrm{P}}. \tag{15.34}$$

This can be written in the form

$$H_{1,2}^{\mathrm{P}} = H_{2,1}^{\mathrm{P}} = eFd \tag{15.35}$$

because the wavefunctions referred to are real. In the present case, where $N = 4$ and all matrix elements except for (15.35) disappear, (15.13) reduces to

$$(E_2^0 - E)c_1 + eFdc_2 = 0, \tag{15.36}$$

$$eFdc_1 + (E_2^0 - E)c_2 = 0, \tag{15.37}$$

$$(E_2^0 - E)c_3 = 0, \tag{15.38}$$

$$(E_2^0 - E)c_4 = 0. \tag{15.39}$$

It is obvious that this system of equations breaks down into two groups of two: (15.36) and (15.37) in one group, and (15.38) and (15.39) in the other. The determinant for (15.36, 37) is

$$\begin{vmatrix} E_2^0 - E & eFd \\ eFd & E_2^0 - E \end{vmatrix} = 0 . \tag{15.40}$$

This goes to zero when E assumes the values

$$E_\pm = E_2^0 \pm eFd . \tag{15.41}$$

It can be shown that the positive sign is associated with $c_1 = c_2$, and the negative sign with $c_1 = -c_2$. The energy E is increased or decreased with respect to the unperturbed energy, by an amount proportional to the field strength F. Equation (15.38) or (15.39) requires that the perturbed energy be the same as the unperturbed energy. In particular, it turns out that the wavefunctions $\phi_3(r)$ and $\phi_4(r)$ are in each case the "right linear combination". This can also be seen from the fact that for $\phi_3(r)$ and $\phi_4(r)$, the perturbation theory without degeneracy actually does not fail, because for the critical terms (zero energy difference in the denominator), the matrix elements of the numerator also go to zero.

We thus obtain overall the scheme shown in Fig. 15.5 for the energy splitting. The same figure shows the wavefunctions in the presence of the field.

The linear Stark effect discussed here is a special case, in that it is only observed in the hydrogen atom. The reason for this is easy to understand: the matrix element (15.33) differs from zero only when $l \neq l'$. In contrast to hydrogen, the l degeneracy is lifted in other atoms, i.e.

$$E^0_{n,l,m} \neq E^0_{n,l',m'}$$

as was shown in Chap. 11.

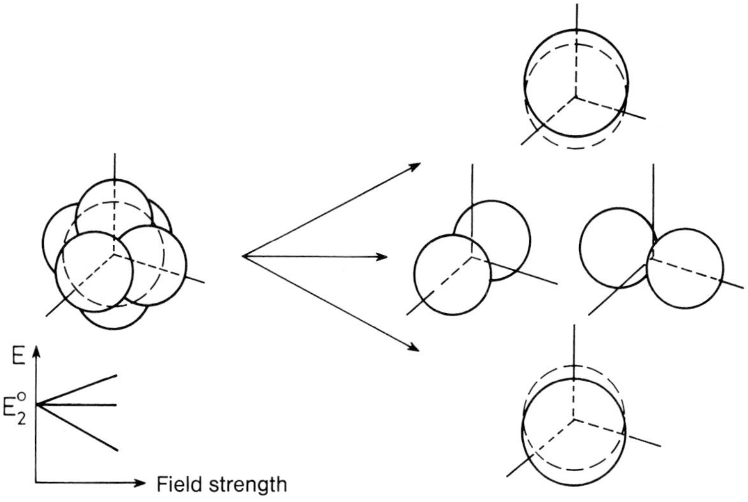

Fig. 15.5. The linear Stark effect. *Lower left:* As the field strength F increases, the energy level E_2^0 is split into three levels. *Upper left:* Representation of four mutually degenerate wavefunctions in a single figure. Dotted line, s functions; solid lines, the p function "dumbbells" in the x, y and z directions. *Upper right:* Superposition of s function and p dumbbell in the z direction causes a shift in the centre of charge of the electron. *Middle right:* The dumbbells in the x and y directions are not affected. *Lower right:* Superposition of s function and p dumbbell (with amplitude in the opposite direction compared to the previous case) leads to a shift of the centre of charge in the negative z direction

15.3 The Interaction of a Two-Level Atom with a Coherent Radiation Field

In the previous chapters on the interaction of a spin with a changing magnetic field, we came upon the interesting phenomenon of spin flipping, which has found numerous applications in physics and chemistry. In this section, we shall show that a two-level atom interacts with a coherent radiation field in a manner which is exactly analogous to spin flipping. Although the spin is a system with exactly two levels, this assumption is only an approximation for an atom. We simply assume that a radiation field induces transitions between two neighbouring levels, and that all the other levels of the atom are energetically so far from the two under consideration that we can neglect the effects of the other states on the two at hand. This is understandable in light of the perturbation theory in the absence of degeneracy which was treated above, because combinations of the wavefunctions associated with distant levels with the wavefunctions belonging to the two close levels will have large energy differences in the denominator, and will thus make only small contributions (15.25).

In the quantum mechanical treatment, we begin with a Schrödinger equation for an electron which is moving in the potential field V of the nucleus and in the additional potential of the radiation field V_a. This equation has the form

$$\left(-\frac{\hbar^2}{2m_0}\nabla^2+V+V_\mathrm{a}\right)\psi(\boldsymbol{r},t)=\mathrm{i}\hbar\frac{d\psi(\boldsymbol{r},t)}{dt}\,. \tag{15.42}$$

To find the explicit form of V_a, let us think of the radiation field in the form of a plane wave:

$$\boldsymbol{F}=\boldsymbol{F}_0\cos(kx-\omega t)\,. \tag{15.43}$$

In order to avoid confusion between the energy and the electric field strength, we again indicate the latter with $\boldsymbol{F}$. We assume, furthermore, that the atom is localised at $\boldsymbol{r}=0$. Since the wavelength of the light $\lambda=2\pi/k$ is in general much larger than the extent of the electron wavefunctions of an atom, we can, to a very good approximation, set $x=0$ in (15.43). We then have a practically homogeneous radiation field over the atom,

$$\boldsymbol{F}=\boldsymbol{F}_0\cos(\omega t)\,. \tag{15.44}$$

We also assume that the radiation field is polarised in the z direction:

$$\boldsymbol{F}_0=(0,0,F_0)\,. \tag{15.45}$$

The force exerted on an electron by an electric field $\boldsymbol{F}$ is given by $-e\cdot\boldsymbol{F}$. The resulting potential energy (negative of the force times the z vector) is

$$V_\mathrm{a}=eF_0 z\cos\omega t\,. \tag{15.46}$$

In the following, we shall use this V_a in (15.42). We assume that we have already solved the Schrödinger equation in the absence of an external field:

$$\left(-\frac{\hbar^2}{2m_0}\nabla^2+V\right)\phi_j=E_j\phi_j\,,\qquad j=1,2\,,\tag{15.47}$$

i.e. we assume that the wavefunctions and energies are known, at least for the indices $j=1,2$.

Since we expect transitions only between the two levels 1 and 2, we cast the wavefunction of (15.42) in the form of a superposition of the unperturbed wavefunctions of (15.47),

$$\psi(\boldsymbol{r},t)=c_1(t)\,\phi_1(\boldsymbol{r})+c_2(t)\,\phi_2(\boldsymbol{r})\,.\tag{15.48}$$

In order to determine the still unknown coefficients c_1 and c_2, we substitute (15.48) into (15.42), multiply from the left by ϕ_1^* or ϕ_2^*, exactly as in Sect. 15.2, and integrate over the total space. By introducing the abbreviation

$$H_{ij}^{\mathrm{P}}=\int\phi_i^*(\boldsymbol{r})\,eF_0z\,\phi_j(\boldsymbol{r})\,dV\cos\omega t\,,\tag{15.49}$$

we obtain the equations

$$\dot{c}_1=\frac{1}{\mathrm{i}\hbar}\left[(E_1+H_{11}^{\mathrm{P}})\,c_1+H_{12}^{\mathrm{P}}c_2\right]\tag{15.50}$$

and

$$\dot{c}_2=\frac{1}{\mathrm{i}\hbar}\left[H_{21}^{\mathrm{P}}c_1+(E_2+H_{22}^{\mathrm{P}})\,c_2\right]\,,\tag{15.51}$$

in analogy with Sect. 15.2.

In many cases, we can assume that H_{11}^{P} and H_{22}^{P} vanish (compare Chap. 16 on symmetries and selection rules). For the solution of (15.50, 51), we attempt the trial function

$$c_j=d_j(t)\,\mathrm{e}^{(-\mathrm{i}/\hbar)E_jt}\,.\tag{15.52}$$

With this equation, (15.50) and (15.51) reduce to

$$\dot{d}_1=\frac{1}{\mathrm{i}\hbar}\left(H_{12}^{\mathrm{P}}d_2\,\mathrm{e}^{\mathrm{i}(E_1-E_2)t/\hbar}\right)\tag{15.53}$$

and

$$\dot{d}_2=\frac{1}{\mathrm{i}\hbar}\left(H_{21}^{\mathrm{P}}d_1\,\mathrm{e}^{-\mathrm{i}(E_1-E_2)t/\hbar}\right)\,.\tag{15.54}$$

Meanwhile we have assumed, according to (15.43), that the radiation field is monochromatic. We now introduce the further assumption that the field is in resonance with the electronic transition. This means that the following relation is valid:

$$E_2-E_1=\hbar\omega\,.\tag{15.55}$$

If we now extract the factor

$$\cos\omega t=\tfrac{1}{2}(e^{i\omega t}+e^{-i\omega t}) \tag{15.56}$$

from (15.49), and multiply by the exponential function in (15.53), we obtain the overall factor

$$\tfrac{1}{2}(1+e^{-2i\omega t})\,. \tag{15.57}$$

As we shall see in the following, d_1 and d_2 vary quite slowly in time compared to the frequency ω, so long as the field strength is not too great. This makes it possible for us to average (15.53) and (15.54) over a time which is long compared to $1/\omega$, but still short compared to the time constant which determines the change in d_j, see (15.61).

The result of this averaging is that the rapidly changing term $\exp(-2i\omega t)$ makes a contribution which is much less than 1 and can therefore be neglected compared to 1. This is called the "rotating wave approximation" in the literature. The expression comes from spin resonance. (In Sect. 14.4 the term $\exp(-i\omega t)$ did not appear, because we had applied a *rotating* magnetic field in the first place.) In (15.54) there is a term with $\exp(+2i\omega t)$ which corresponds to (15.57), and in this case too, it is negligibly small. If we abbreviate the integral in (15.49) by substituting the dipole moment matrix element $(\theta_z)_{ij}\equiv\int\phi_i^*(\boldsymbol{r})ez\phi_j(\boldsymbol{r})dV$, (15.53) and (15.54) reduce to

$$\dot{d}_1=\frac{1}{i\hbar}\tfrac{1}{2}F_0(\theta_z)_{12}d_2 \tag{15.58}$$

and

$$\dot{d}_2=\frac{1}{i\hbar}\tfrac{1}{2}F_0(\theta_z)_{21}d_1\,. \tag{15.59}$$

These equations are strikingly similar to the spin equations (14.94) and (14.95) which we met in Sect. 14.4. It can be shown that $(\theta_z)_{12}=(\theta_z)_{21}^*$ can be chosen to be real. By introducing another abbreviation,

$$\Omega=\frac{1}{2\hbar}F_0(\theta_z)_{12}\,, \tag{15.60}$$

where Ω is to be understood as a frequency, we obtain as a solution to (15.58, 59)

$$d_1=\cos\Omega t\,, \tag{15.61}$$

$$d_2=-i\sin\Omega t\,. \tag{15.62}$$

We have based this on the assumption that at time $t=0$, the electron is known for certain to be in the lower level. Thus the Schrödinger equation (15.42) for a two-level system interacting with an external monochromatic radiation field is solved. The coefficients c_1 and c_2 in (15.48) now obviously have the form

$$c_1=e^{-(i/\hbar)E_1t}\cos\Omega t\,, \tag{15.63}$$

$$c_2=-ie^{-(i/\hbar)E_2t}\sin\Omega t\,. \tag{15.64}$$

As we know, the square of the absolute value of c_j gives the probability of finding the system in state j. $|c_j|^2$ can thus be unterstood as the *occupation number* N_j of the state j. As indicated by the corresponding formulae

$$N_1 \equiv |c_1|^2 = \cos^2 \Omega t\,, \tag{15.65}$$

$$N_2 \equiv |c_2|^2 = \sin^2 \Omega t\,, \tag{15.66}$$

the electron oscillates with the frequency Ω between states 1 and 2. It is instructive to calculate the dipole matrix element according to

$$\theta_z = \int \psi^* ez\, \psi\, dV = (\theta_z)_{12} c_1^* c_2 + (\theta_z)_{21} c_2^* c_1\,. \tag{15.67}$$

The final result of the whole process is

$$\theta_z = -(\theta_z)_{12} \sin(2\,\Omega t) \sin \omega t\,. \tag{15.68}$$

This says that the dipole moment swings back and forth with the rapidly oscillating component $\sin \omega t$, and that its magnitude is also modulated by $\sin 2\Omega t$. The dipole moment is thus largest when the electron has exactly the occupation number $N_1 = N_2 = \frac{1}{2}$, that is, its probability of occupying either level is the same. The result (15.68) and the formulae (15.65) and (15.66) are very closely analogous to the results obtained for spin resonance in Sect. 14.4. We shall examine this analogy more closely in the next chapter. It makes it possible to extend a series of spin experiments to optical transitions between electronic states in atoms. Such experiments require coherent light with a high field strength. The latter, F_0, is necessary so that the transitions can occur in a time $t_0 \sim 1/\Omega \sim 1/F_0$ – [compare (15.60)!] – which is so short that the electron motion is not appreciably perturbed by other effects, e.g. collisions between atoms in gases or the spontaneous emission of light from excited states. Typical values for t_0 lie between 10^{-9} and 10^{-11} s.

15.4 Spin- and Photon Echoes

In this section we shall treat two especially interesting phenomena, spin echo and photon echo. If we compare the results of Sect. 14.4 with those of the preceding section, we see a very close analogy between the behaviour of a spin subjected to both a constant magnetic field and a transverse oscillating magnetic field, and an electron in a two-level atom which is subjected to an oscillating electric field. In both cases we assume that the frequency of the applied field is in resonance with the transition frequency of the spin or of the electron transition from the lower to the upper state. With reference to the analogy which we shall wish to use, we shall first discuss the behaviour of a spin.

As we saw in Sect. 14.4, the application of a coherent, resonant field causes the spin eventually to flip. How far it flips depends on the length of time the external field is applied. If one lets the field work just long enough to flip the spin by $\pi/2$, one speaks of a $\pi/2$ or a 90° pulse. If the field is left on twice as long, the spin will flip completely over. In this case one speaks of a π or 180° pulse. The so-called spin echo is an important application of these ideas. Here one first applies a $\pi/2$ pulse (Fig. 15.6). In a

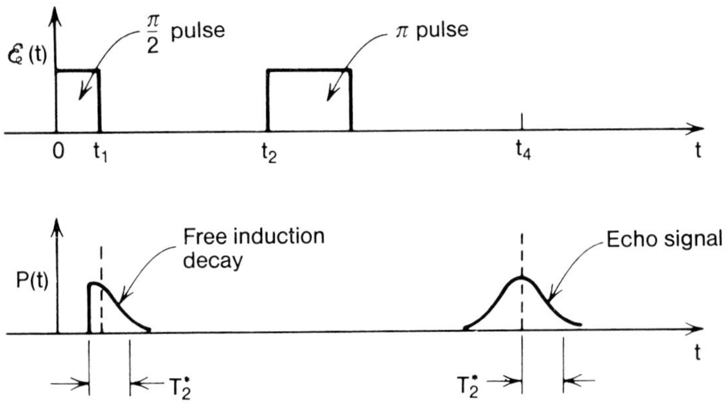

Fig. 15.6. The spin echo (photon echo) experiment. *Above:* the applied pulses of electric field strength (photon echo) or magnetic field (spin echo) as a function of time. *Below:* magnitude of the dipole moment (spin) of the atom as a function of time (schematically)

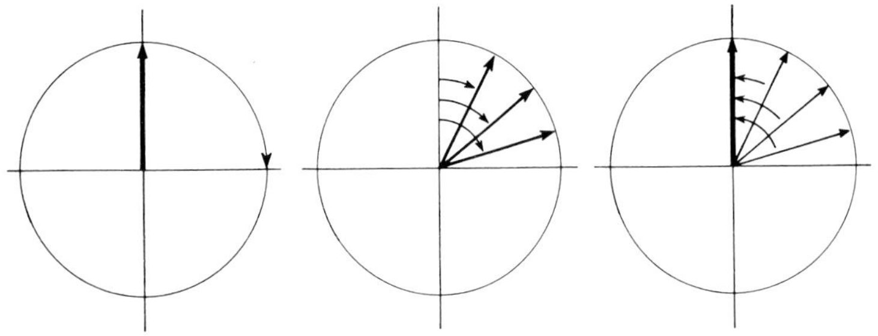

Fig. 15.7. Spreading out and moving together of spins. *Left:* Starting condition, all spins parallel. *Middle:* Spreading out of spins. *Right:* Moving back together after 180° pulse (schematically)

number of practical cases, the spins do not precess with the same velocity, due to spatially varying static magnetic fields. They therefore spread out in the course of time (see Fig. 15.7). Denoting the frequency width of the precession by $\Delta\omega^*$, we may define a mean time, T_2^*, within which the spins spread, by $\Delta\omega^* = 2\pi/T_2^*$. $\Delta\omega^*$ is called "inhomogeneous width". A single precessing spin can emit electromagnetic radiation. Because the spins get out of phase with each other, so do their electromagnetic emissions, which leads to a reduction in the total intensity. If one now applies another pulse, this time a 180° pulse (Fig. 15.6), the spins are flipped. What happens here can best be compared with runners on a track. At the beginning of the race, all the runners are at the same place, the starting line. After the starting gun (the 90° pulse), they have, however, moved different distances away from the starting line because of their different velocities. The effect of the 180° pulse is the same as that of a second gun, which signals the runners to turn around and return to the start at the same speed as before. Obviously, they all reach the start at the same time. For the spins, this means that at a certain time after the 180° pulse, they will again all be in phase, and thus their radiation will be in phase. As a result, the original radiation intensity is reached again. This picture must be somewhat modified, because there are also irreversible phase changes of the spins, which are characterised by the so-called homogeneous linewidth, which we shall discuss in Sect. 16.2. Due to this homogeneous linewidth, the original starting intensity can no longer quite be attained (see Fig. 15.8). If one repeats the 180° pulse, the result is that shown in Fig. 15.9. The decay time T_2, which we met earlier in Sect. 14.5 in the Bloch equations, can be calculated from the decay of the peaks. The homogeneous linewidth can then be determined from T_2: $\Delta\omega = 2\pi/T_2$. The analogy between the spin and the two-level atom has now made it possible to apply the entire

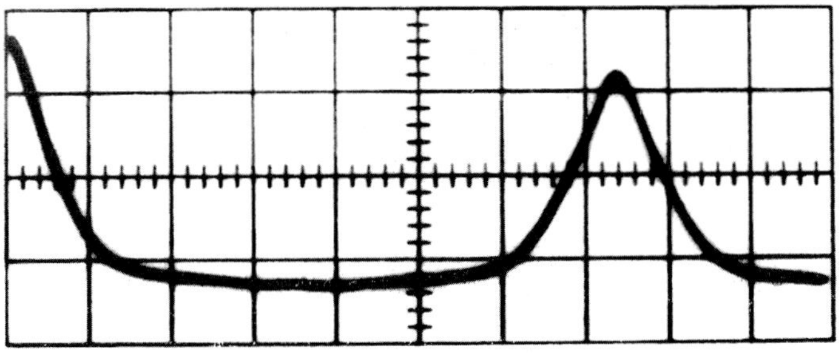

Fig. 15.8. Decay of the spin emission and echo emission of protons in water. From A. Abragam: *The Principles of Nuclear Magnetism* (Oxford 1962)

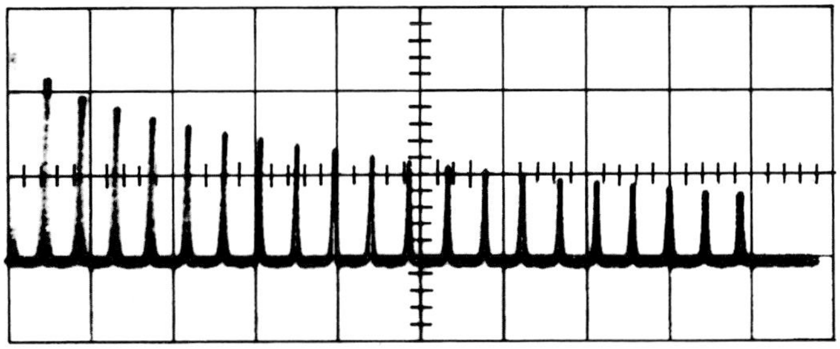

Fig. 15.9. A series of spin echos obtained from protons in ordinary water. In this method, a 90° pulse is applied at time $t=0$, and is followed at times τ, 3τ, $5\tau, \dots$, $(2n-1)\tau$ by a 180° pulse. The echoes are observed at the times 2τ, $4\tau, \dots$, $2n\tau$ (n an integer). It can be shown that the height of the peaks decays according to the formula $f(n)=\exp(-2n\tau/T_2)$. From A. Abragam: *The Principles of Nuclear Magnetism* (Oxford, 1962)

process of spin echo to the radiation field in the case of the photon echo. We leave it to the reader to develop this analogy on the basis of the above.

To this end, let us consider a system of two-level atoms. An example is the ruby, in which chromium ions are included as impurities in a basic lattice of aluminium oxide. To a certain approximation, these ions can be treated as two-level atoms. As we saw in Sect. 15.4, the occupation number of the upper state increases due to the applied resonant oscillating electric field (15.66). The electron of the impurity atom is thus to be found more and more frequently in the upper state, while the occupation of the lower state decreases correspondingly. Eventually a state is reached in which the occupation numbers are equal. Because this is an exact analogy to the state in which the spin has flipped through 90°, or in other words, the occupation numbers of the "spin up" and "spin down" states are equal, one speaks in this case, too, of a $\pi/2$ or 90° pulse. If one applies the external electric field to the electron for twice as long, it goes completely into the upper state, again in analogy to the spin, which is at this point rotated through 180°. Therefore one speaks again, in the two-level atom case, of a 180° or π pulse. As (15.66) and (15.60) show, the "flipping" occurs more rapidly when the applied field strength F_0 is larger.

We now imagine that the electron has been excited by a $\pi/2$ pulse. It can be shown that the dipole moment of the electron then oscillates freely with the frequency of the optical transition, ω. This can be shown by a calculation of the expectation value of the dipole which is analogous to (15.67) and (15.68), although there is no external applied field. According to Maxwell's Theory, however, an oscillating dipole can emit electromagnetic waves, in this case, light waves. This means that the ensemble of impurity atoms emits light after the $\pi/2$ pulse. However, since the atoms are subject to different fields within the crystal, their transition frequencies are not all the same, and some of the oscillating dipoles emit light which is out of phase. The emitted intensity is therefore reduced.

Now, just as in the spin case, we can apply a π pulse which brings the diverging phases of the oscillating dipoles back together. This prediction is justified because of the complete mathematical correspondence between the behaviour of spin and a two-level atom. As the oscillating dipole moments come back into phase, they emit a light pulse which can be seen as the "echo" of the previously applied π pulse. In the optical range, the phases of the dipoles diverge very rapidly, so it is necessary in these experiments to use short pulses of about 10^{-10} s and less.

15.5 A Glance at Quantum Electrodynamics*

15.5.1 Field Quantization

In this section we want to sketch the nonrelativistic theory of the Lamb shift. To this end we first show by means of an example how the light field can be quantized. We start with Maxwell's equations in vacuum which reads:

$$\operatorname{curl} \boldsymbol{E} = -\frac{\partial \boldsymbol{B}}{\partial t}, \tag{15.69}$$

$$\operatorname{curl} \boldsymbol{B} = \varepsilon_0 \mu_0 \frac{\partial \boldsymbol{E}}{\partial t}, \tag{15.70}$$

$$\operatorname{div} \boldsymbol{E} = 0, \tag{15.71}$$

$$\operatorname{div} \boldsymbol{B} = 0 \tag{15.72}$$

where

$$\mu_0 \varepsilon_0 = \frac{1}{c^2} \tag{15.73}$$

and where c is the light velocity in vacuum. Let us consider a standing electric wave with wave vector k and with its electric vector in the z-direction

$$\boldsymbol{E} = (0, 0 E_z), \qquad E_z = p(t) N \sin(kx) \tag{15.74}$$

and where $p(t)$ is a still unknown function of time. To derive the corresponding magnetic induction we insert (15.74) into (15.69). One can convince oneself readily that only the y-component of this equation is non-vanishing.

$$-\frac{\partial E_z}{\partial x} = -\frac{\partial B_y}{\partial t}. \tag{15.75}$$

Since the left-hand side of this equation is proportional to $\cos(kx)$ it suggests that we put B_y proportional to $\cos(kx)$. This leads us to the ansatz

$$B_y = q(t)\frac{N}{c}\cos(kx) \tag{15.76}$$

where we have included the factor $1/c$ for later convenience. This factor gives p and q the same physical dimension. Inserting (15.76) into (15.75) yields

$$\frac{dq}{dt} = \omega p \tag{15.77}$$

where we have used the abbreviation

$$\omega = ck\,. \tag{15.78}$$

Since k is a wave number and c a velocity, ω in (15.78) is a circular frequency. Inserting $\boldsymbol{E}$ (15.74) and $\boldsymbol{B}$ (15.76) into (15.70) yields

$$\frac{dp}{dt} = -\omega q\,. \tag{15.79}$$

When we differentiate (15.77) with respect to time and eliminate p from it by means of (15.79), we obtain

$$\frac{d^2q}{dt^2} + \omega^2 q = 0\,. \tag{15.80}$$

This equation is the well-known equation of a harmonic oscillator with circular frequency ω. Equations (15.77) and (15.79) can be written in a very elegant form by introducing the Hamiltonian

$$\mathscr{H} = \tfrac{1}{2}\omega(p^2+q^2)\,. \tag{15.81}$$

With its aid we can write (15.77) and (15.79) in the form

$$\frac{dq}{dt} = \frac{\partial\mathscr{H}}{\partial p} \tag{15.82}$$

$$\frac{dp}{dt} = -\frac{\partial\mathscr{H}}{\partial q}\,. \tag{15.83}$$

Quite evidently we are dealing here with the Hamiltonian equations of a harmonic oscillator. This then allows us to identify p with the momentum and q with the coordinate of an harmonic oscillator. With this identification we have the key in our hands to quantize the electromagnetic field. This is done by a purely formal analogy. In Sect. 9.4 we saw how to quantize the motion of the harmonic oscillator. Here we want to do exactly the same. To put this analogy between the harmonic oscillator and the electromagnetic field on firm ground we show that $\mathscr{H}$ (15.81) is identical with the energy of the electromagnetic field mode. According to electrodynamics, the field energy in the volume $\mathscr{V} \equiv L^3$ is given by

$$\bar{U} = \int \frac{1}{2}\left(\varepsilon_0 \boldsymbol{E}^2 + \frac{1}{\mu_0}\boldsymbol{B}^2\right) d^3\boldsymbol{x} . \tag{15.84}$$

By inserting (15.74) and (15.76) into the energy expression (15.84) we obtain

$$\bar{U} = \frac{1}{2} N^2 \varepsilon_0 \left\{ L^2 \int_0^L [p^2 \sin^2(kx) + q^2 \cos^2(kx)]\, dx \right\} . \tag{15.85}$$

The integration over x can easily be performed so that we are left with

$$\bar{U} = \tfrac{1}{4} \mathscr{V} N^2 \varepsilon_0 (p^2 + q^2) . \tag{15.86}$$

We find exactly the same function of p and q as occurring in (15.81). However, this identification now allows us to determine the still unknown normalization factor N. Comparing (15.86) with (15.81) we obtain

$$N = \sqrt{\frac{\omega}{\varepsilon_0}} \sqrt{\frac{2}{\mathscr{V}}} . \tag{15.87}$$

Now let us return to the quantization problem. We wish to utilize the analogy between the Hamiltonian (15.81) and that of the harmonic oscillator. It is convenient to use its Hamiltonian in the form

$$\tfrac{1}{2} \hbar \omega (\Pi^2 + \xi^2) . \tag{15.88}$$

The equivalence of (15.81) with (15.88) is achieved by putting

$$p = \sqrt{\hbar}\, \Pi , \qquad q = \sqrt{\hbar}\, \xi \tag{15.89}$$

so that the Hamiltonian (15.81) acquires exactly the same form (15.88). Here, however, we known what the quantum version looks like. We have to replace Π by the operator $\partial / \mathrm{i} \partial \xi$ exactly in analogy to Sect. 9.4. By exploiting that analogy further we introduce creation and annihilation operators by

$$\frac{1}{\sqrt{2}} \left(-\frac{\partial}{\partial \xi} + \xi \right) = b^+ , \tag{15.90}$$

$$\frac{1}{\sqrt{2}} \left(\frac{\partial}{\partial \xi} + \xi \right) = b \tag{15.91}$$

or, solving for p and q

$$p = \mathrm{i} \sqrt{\frac{\hbar}{2}} (b^+ - b) , \tag{15.92}$$

$$q = \sqrt{\frac{\hbar}{2}} (b^+ + b) . \tag{15.93}$$

The creation and annihilation operators b^+ and b obey the commutation relation

$$bb^+ - b^+ b = 1 . \tag{15.94}$$

By using (15.92) and (15.93), we can express the free fields $\boldsymbol{E}$ and $\boldsymbol{B}$ by means of these operators in the form

$$E_z = \mathrm{i}(b^+ - b)\sqrt{\frac{\hbar}{2}}N\sin(kx) \tag{15.95}$$

$$B_y = (b^+ + b)\sqrt{\frac{\hbar}{2}}\,\frac{N}{c}\cos(kx) . \tag{15.96}$$

The normalization factor is given by

$$N = \sqrt{\frac{\omega}{\varepsilon_0}}\sqrt{\frac{2}{\mathscr{V}}}, \qquad \varepsilon_0\mu_0 = \frac{1}{c^2} . \tag{15.97}$$

With the transformations (15.90) and (15.91), the Hamiltonian (15.88) can be expressed by the creation and annihilation operators exactly as in Sect. 9.4.

$$\mathscr{H} = \hbar\omega(b^+ b + \tfrac{1}{2}) . \tag{15.98}$$

We leave it as an exercise to the reader to convince himself that this Hamiltonian could be also derived by inserting (15.95) and (15.96) into (15.84). For a number of problems dealing with the interaction between electrons and the electromagnetic field we need the vector potential $\boldsymbol{A}$. $\boldsymbol{A}$ is connected with the magnetic induction by

$$\boldsymbol{B} = \mathrm{curl}\boldsymbol{A} . \tag{15.99}$$

In our book we choose the "Coulomb gauge"

$$\mathrm{div}\boldsymbol{A} = 0 . \tag{15.100}$$

For $\boldsymbol{B}$ in the form (15.96) the relations (15.99) and (15.100) are fulfilled by

$$A_x = 0 , \quad A_y = 0 , \quad A_z = -(b^+ + b)\sqrt{\frac{\hbar\omega\mu_0}{2}}\sqrt{\frac{2}{\mathscr{V}}}\,\frac{1}{k}\sin(kx) . \tag{15.101}$$

Let us summarize the above results. When we quantize the electromagnetic field, the electric field strength, the magnetic induction, and the vector potential become operators that can be expressed by the familiar creation and annihilation operators b^+, b of a harmonic oscillator. The total energy of the field also becomes a Hamiltonian operator of the form (15.98).

Since the normalization of waves in infinite space provides some formal difficulties (which one may overcome, however), we shall use a well-known trick. We subject the wavefunctions

$$\exp(\mathrm{i}\boldsymbol{k}_\lambda \cdot \boldsymbol{r}) \tag{15.102}$$

to periodic boundary conditions. In order to apply the above formalism in particular to the Lamb shift, two changes must be made:

1) Instead of using a single (standing) wave, the electromagnetic field must be written as a superposition of all possible waves.
2) Instead of standing waves, the use of running waves has some advantages.

Since the derivation of the corresponding relations does not give us any physical insight beyond what we gained above, we immediately write down these relations.

$$\boldsymbol{E}(\boldsymbol{r}) = \sum_\lambda \boldsymbol{e}_\lambda \sqrt{\frac{\hbar\omega_\lambda}{2\varepsilon_0 \mathscr{V}}}\,[\mathrm{i}b_\lambda \exp(\mathrm{i}\boldsymbol{k}_\lambda \cdot \boldsymbol{r}) - \mathrm{i}b_\lambda^+ \exp(-\mathrm{i}\boldsymbol{k}_\lambda \cdot \boldsymbol{r})] \tag{15.103}$$

$$\boldsymbol{B}(\boldsymbol{r}) = \sum_\lambda \hat{\boldsymbol{k}} \times \boldsymbol{e}_\lambda \sqrt{\frac{\hbar\omega_\lambda \mu_0}{2 \mathscr{V}}}\,[\mathrm{i}b_\lambda \exp(\mathrm{i}\boldsymbol{k}_\lambda \cdot \boldsymbol{r}) - \mathrm{i}b_\lambda^+ \exp(-\mathrm{i}\boldsymbol{k}_\lambda \cdot \boldsymbol{r})]\,. \tag{15.104}$$

The individual expressions have the following meaning:
λ index labeling the individual waves
$\boldsymbol{e}_\lambda$ vector of polarization of wave
ω_λ circular frequency
$\boldsymbol{k}_\lambda$ wave vector of wave

$$\hat{\boldsymbol{k}} = \boldsymbol{k}/|\boldsymbol{k}|\,. \tag{15.105}$$

The vector potential $\boldsymbol{A}$ reads

$$\boldsymbol{A} = \sum_\lambda \boldsymbol{e}_\lambda \sqrt{\frac{\hbar}{2\omega_\lambda \varepsilon_0 \mathscr{V}}}\,[b_\lambda \exp(\mathrm{i}\boldsymbol{k}_\lambda \cdot \boldsymbol{r}) - b_\lambda^+ \exp(-\mathrm{i}\boldsymbol{k}_\lambda \cdot \boldsymbol{r})]\,. \tag{15.106}$$

The operators b_λ^+, b_λ again obey the commutation relations

$$b_\lambda b_{\lambda'}^+ - b_{\lambda'}^+ b_\lambda = \delta_{\lambda\lambda'} \tag{15.107}$$

$$b_\lambda b_{\lambda'} - b_{\lambda'} b_\lambda = 0 \tag{15.108}$$

$$b_\lambda^+ b_{\lambda'}^+ - b_{\lambda'}^+ b_\lambda^+ = 0\,. \tag{15.109}$$

The Hamiltonian operator reads

$$\mathscr{H}_{\text{field}} = \sum_\lambda \hbar\omega_\lambda (b_\lambda^+ b_\lambda + \tfrac{1}{2})\,. \tag{15.110}$$

As usual, the Schrödinger equation is obtained by applying the Hamiltonian operator to a wavefunction, which we call Φ:

$$\mathscr{H}_{\text{field}}\,\Phi = E\Phi\,. \tag{15.111}$$

We shall determine the wavefunction and energies in the exercises. For what follows, we need only the ground state defined by $b_\lambda \Phi_0 = 0$ for all λ, and the states which are occupied by a single light quantum (photon).

$$\Phi_\lambda = b_\lambda^+ \Phi_0 \tag{15.112}$$

with quantum energy

$$E_\lambda = \hbar\omega_\lambda (+E_0)\,. \tag{15.113}$$

It has the zero point energy

$$E_0 = \sum_\lambda \tfrac{1}{2}\hbar\omega_\lambda\,. \tag{15.114}$$

This expression is infinite, but it is also unobservable and therefore dropped.

15.5.2 Mass Renormalization and Lamb Shift

We shall now treat the interaction of a hydrogen atom with the quantized light-field. The interaction is brought about by the vector potential $\boldsymbol{A}$ occurring in (14.12). Since $\boldsymbol{A}$ is small in the present case, we shall retain only terms linear in $\boldsymbol{A}$, and assume $\operatorname{div}\boldsymbol{A} = 0$. But in contrast to that former case, the fields are not externally given, but become now by themselves variables of the system. Therefore we must not only write $\boldsymbol{A}$ in the form (15.106), but we must also add to the Hamiltonian $\mathscr{H}_{\text{el}} + \mathscr{H}_{\text{int}}$ that of the quantized field, i.e. $\mathscr{H}_{\text{field}}$. Thus the Schrödinger equation to be solved reads:

$$(\mathscr{H}_{\text{el}} + \mathscr{H}_{\text{field}} + \mathscr{H}_{\text{int}})\,\Psi = E\,\Psi \tag{15.115}$$

where

$$\mathscr{H}_{\text{el}} = -\frac{\hbar^2}{2m_0}\Delta + V(\boldsymbol{r}) \tag{15.116}$$

$$\mathscr{H}_{\text{field}} = \sum_\lambda \hbar\omega_\lambda b_\lambda^+ b_\lambda \tag{15.117}$$

$$\mathscr{H}_{\text{int}} = \frac{e}{m_0}\boldsymbol{A}\cdot\hat{\boldsymbol{p}} \equiv \frac{e}{m_0}\sum_\lambda \boldsymbol{e}_\lambda \sqrt{\frac{\hbar}{2\omega_\lambda\varepsilon_0\mathscr{V}}}\,[b_\lambda \exp(\mathrm{i}\boldsymbol{k}_\lambda\cdot\boldsymbol{r}) - b_\lambda^+ \exp(\mathrm{i}\boldsymbol{k}_\lambda\cdot\boldsymbol{r})]\,\hat{\boldsymbol{p}}$$

where

$$\hat{\boldsymbol{p}} \equiv \frac{\hbar}{\mathrm{i}}\,\text{grad}\,. \tag{15.118}$$

In order to solve the Schrödinger equation (15.115), we shall apply perturbation theory, where

$$\mathscr{H}_0 = \mathscr{H}_{\text{el}} + \mathscr{H}_{\text{field}} \tag{15.119}$$

serves as unperturbed Hamiltonian and $\mathscr{H}_{\text{int}}$ as perturbation. The eigenfunctions Φ_ν of $\mathscr{H}_0$ are products of an eigenfunction of $\mathscr{H}_{\text{el}}$, i.e. ψ and of an eigenfunction of $\mathscr{H}_{\text{field}}$, i.e. Φ. We shall abbreviate the set of quantum numbers n, l, m of ψ by n. In the following we shall be concerned with the vacuum state Φ_0 and one-photon states $\Phi_\lambda \equiv b_\lambda^+ \Phi_0$. Identifying the index ν of ϕ_ν with $(n,0)$, or (n,λ) we may write the unperturbed wavefunction as

$$\phi_\nu = \psi_n(\boldsymbol{r})\,\Phi_\kappa \quad \text{where} \tag{15.120}$$

$$\kappa = 0\ , \quad \text{or} \quad \lambda\ . \tag{15.121}$$

The corresponding energy levels are

$$E^0_{\nu,\text{tot}} = E^0_n \quad (\kappa = 0) \qquad \text{and} \qquad E^0_{\nu,\text{tot}} = E^0_n + \hbar\omega_\lambda \quad (\kappa = \lambda)\ . \tag{15.122}$$

For the perturbation theory we need the matrix elements of $\mathscr{H}_{\text{int}}$ which read in bra and ket notation:

$$H^1_{\mu\nu} = \left\langle \psi_n \Phi_\kappa \left| \frac{e}{m_0} \sum_\lambda \boldsymbol{e}_\lambda \sqrt{\frac{\hbar}{2\,\omega_\lambda\,\varepsilon_0\,\mathscr{V}}}\,[b_\lambda \exp(\mathrm{i}\boldsymbol{k}_\lambda\cdot\boldsymbol{r}) - b_\lambda^+ \exp(-\mathrm{i}\boldsymbol{k}_\lambda\cdot\boldsymbol{r})]\hat{\boldsymbol{p}} \right| \psi_{n'}\,\Phi_{\kappa'} \right\rangle \tag{15.123}$$

which may be rearranged as

$$H^1_{\mu\nu} = \frac{e}{m_0} \sum_\lambda \sqrt{\frac{\hbar}{2\,\omega_\lambda\,\varepsilon_0\,\mathscr{V}}}\,[\langle n|\exp(\mathrm{i}\boldsymbol{k}_\lambda\cdot\boldsymbol{r})\boldsymbol{e}_\lambda\cdot\hat{\boldsymbol{p}}|n'\rangle\langle\Phi_\kappa|b_\lambda|\Phi_{\kappa'}\rangle$$
$$- \langle n|\exp(-\mathrm{i}\boldsymbol{k}_\lambda\cdot\boldsymbol{r})\boldsymbol{e}_\lambda\cdot\hat{\boldsymbol{p}}|n'\rangle\langle\Phi_\kappa|\,b_\lambda^+\,|\Phi_{\kappa'}\rangle]\ . \tag{15.124}$$

As we know from the quantized harmonic oscillator,

$$\langle\Phi_\kappa|b_\lambda^+|\Phi_\kappa\rangle = \langle\Phi_\kappa|b_\lambda|\Phi_\kappa\rangle = 0\ . \tag{15.125}$$

Therefore the perturbation energy in first order vanishes. Starting from the vacuum as unperturbed state, i.e. $\kappa' = 0$, we further have

$$\langle\Phi_\kappa|b_\lambda|\Phi_0\rangle = 0\ , \qquad \langle\Phi_\kappa|b_\lambda^+|\Phi_0\rangle = \delta_{\lambda\kappa} \tag{15.126}$$

In other words, a single photon λ may be generated. We assume that the atom sits at the origin and that the extension of the electronic wave function is small compared to the wavelength of the light waves $\exp[\mathrm{i}\boldsymbol{k}_\lambda\cdot\boldsymbol{r}]$. This allows us to ignore this factor in the matrix elements. In this way $H^1_{\mu\nu}$ is reduced to (with $\kappa = \lambda$!)

$$H^1_{\mu\nu} \equiv H^1_{n,\lambda;n',0} = -\frac{e}{m_0}\sqrt{\frac{\hbar}{2\,\omega_\lambda\,\varepsilon_0\,\mathscr{V}}}\,\langle n|\boldsymbol{e}_\lambda\cdot\hat{\boldsymbol{p}}|n'\rangle\ . \tag{15.127}$$

The expression for the perturbed energy in second order therefore reads

$$\varepsilon^{(2)} = \sum_{n',\lambda} \frac{|H^1_{n,\lambda;n',0}|^2}{E^0_n - E^0_{n'} - \hbar\omega_\lambda}\ . \tag{15.128}$$

In quantum electrodynamics, the processes which lead to $\varepsilon^{(2)}$ are usually visualized as follows:

An electron is in its initial state n, and there is no photon. Then a photon of kind λ is emitted, [creation operation b_λ^+ in (15.124) and (15.128)!] and the electron goes into the state n'. Finally the photon is reabsorbed [annihilation operator b_λ in (15.124) and (15.128)!] and the electron returns to its state n. This process is described by the following "Feynman diagram" (Fig. 15.10).

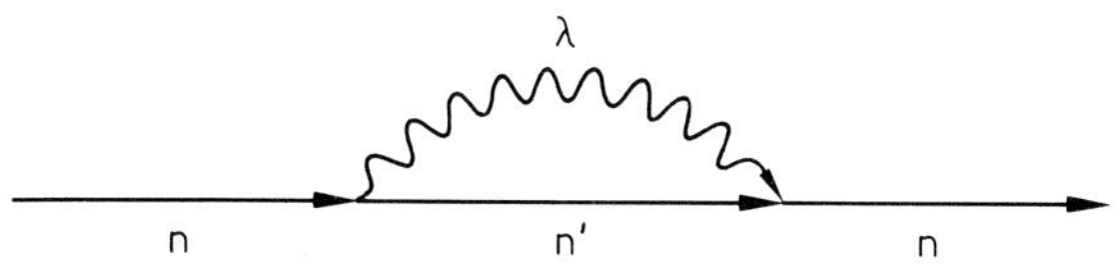

Fig. 15.10. Example of a Feynman diagram: virtual emission and reabsorption of a photon

In the following we have to distinguish the different field modes more carefully. To this end we replace the general mode index λ by the wave vector $\boldsymbol{k}$ and an index j indicating one of the two directions of polarization. Furthermore we use the relation between frequency and wave number

$$\omega_\lambda \equiv \omega_k = ck \ . \tag{15.129}$$

While we initially start with waves normalized in a volume $\mathcal{V}$ we will eventually go over to an integration which is done by the rule

$$\frac{1}{\mathcal{V}} \sum_\lambda \rightarrow \sum_{j=1,2} \int \frac{d^3k}{(2\pi)^3} \ . \tag{15.130}$$

Using (15.127, 129) and (15.130) we may cast (15.128) into the form

$$\varepsilon^{(2)} = \frac{1}{(2\pi)^3} \frac{e^2\hbar}{2m_0^2\varepsilon_0} \int d^3\boldsymbol{k} \frac{1}{\omega_k} \sum_{n',j} \frac{|\langle n|\boldsymbol{e}_j\cdot\hat{\boldsymbol{p}}|n'\rangle|^2}{E_n^0 - E_{n'}^0 - \hbar\omega_k} \ . \tag{15.131}$$

For further evaluation we split the integral over $\boldsymbol{k}$-space into one over the space angle Ω and one over the magnitude of $\boldsymbol{k}$, i.e. k,

$$\int d^3\boldsymbol{k} = \int k^2 dk \int d\Omega \ . \tag{15.132}$$

We then first perform the integration over the space angle and sum up over the two directions of polarizations. Since the evaluation is purely formal we immediately write down the result

$$\int d\Omega \sum_j |\langle n|\boldsymbol{e}_j\cdot\hat{\boldsymbol{p}}|n'\rangle|^2 = 4\pi \tfrac{2}{3} |\langle n|\hat{\boldsymbol{p}}|n'\rangle|^2 \ . \tag{15.133}$$

This leads us to the following result for the self-energy

$$\varepsilon^{(2)} = \frac{1}{(2\pi)^2} \frac{2e^2\hbar}{3m_0^2\varepsilon_0 c^3} \int_0^\infty \omega\, d\omega \sum_{n'} \frac{|\langle n|\hat{\boldsymbol{p}}|n'\rangle|^2}{E_n^0 - E_{n'}^0 - \hbar\omega} \ . \tag{15.134}$$

A detailed discussion of the sum over n' reveals that this sum certainly does not vanish more strongly than $1/\omega$. We thus immediately recognize that the integral over ω in (15.134) diverges which means that the energy shift is infinitely great. This seemingly absurd result presented a great difficulty to theoretical physics. It was overcome by ideas of Bethe, Schwinger and Weisskopf which we will now explain.

When we do similar calculations for free electrons we again find an infinite result, which can be seen as follows. We repeat the whole calculation above but instead of eigenfunctions ψ_n of the hydrogen atom we use the wave functions of free electrons

$$\psi_n(\boldsymbol{r}) \rightarrow \psi_p = N \exp(\mathrm{i}\boldsymbol{p} \cdot \boldsymbol{r}/\hbar) \tag{15.135}$$

Note that in this formula $\boldsymbol{p}$ is a usual vector whereas $\hat{\boldsymbol{p}}$ occurring for instance in (15.127) and (15.131) is the momentum operator $\hbar/\mathrm{i}$ grad. Instead of matrix elements, which were between the eigenstates of the hydrogen atom, we now have to evaluate matrix elements between plane waves. We immediately obtain

$$\langle \boldsymbol{p}' | \hat{\boldsymbol{p}} | \boldsymbol{p} \rangle = N^2 \int \exp(-\mathrm{i}\boldsymbol{p} \cdot \boldsymbol{r}/\hbar) \hat{\boldsymbol{p}} \exp(\mathrm{i}\boldsymbol{p} \cdot \boldsymbol{r}/\hbar) d^3 r \tag{15.136}$$

and

$$\langle \boldsymbol{p}' | \hat{\boldsymbol{p}} | \boldsymbol{p} \rangle = (\hbar/\mathrm{i}) \boldsymbol{p} \, \delta_{p,p'} \ . \tag{15.137}$$

Furthermore, we have to make the substitution

$$E_n^0 - E_{n'}^0 \rightarrow E_p - E_{p'} \tag{15.138}$$

but we immediately find

$$E_p - E_{p'} = 0 \tag{15.139}$$

on account of (15.137). By putting all the results together we obtain the self-energy of a free electron in the form

$$\varepsilon_p^{(2)} = -\frac{1}{(2\pi)^2} \frac{2e^2}{3m_0^2 \varepsilon_0 c^3} p^2 \sum_0^\infty d\omega \ . \tag{15.140}$$

We notice that the self-energy of a free electron of momentum $\boldsymbol{p}$ is proportional to $\boldsymbol{p}^2$. Equation (15.140) can be interpreted as giving rise to a shift in the mass of the electron, which can be seen as follows: The energy of the free electron without interaction with the electromagnetic field ("bare electron") reads

$$E_p = \boldsymbol{p}^2/(2\bar{m}_0) \ . \tag{15.141}$$

In it, $\bar{m}_0$ is the "bare" mass. The energy shift just calculated is

$$\Delta E_p = -\frac{1}{(2\pi)^2} \frac{2e^2}{3m_0^2 \varepsilon_0 c^3} p^2 \int_0^\infty d\omega \ . \tag{15.142}$$

Thus the total energy reads

$$E_p + \Delta E_p = \frac{p^2}{2\,m_0} \tag{15.143}$$

While the mass of the "bare" electron neglecting electromagnetic interaction is $\bar{m}_0$, taking this interaction into account the electron mass is m_0. Note that in this type of consideration one uses m_0 and not $\bar{m}_0$ in (15.142). This follows from the "renormalization" procedure we will describe now.

Since we always make observations on free electrons with the electromagnetic interaction present, (15.143) must be just the expression we normally write down for the energy of a free electron where m_0 is the observed mass. Thus we can make the identification

$$\frac{1}{m_0} = \frac{1}{\bar{m}_0} - \frac{1}{(2\pi)^2}\,\frac{4\,e^2}{3\,m_0^2\,\varepsilon_0\,c^3}\int_0^\infty d\omega \equiv \frac{1}{\bar{m}_0} - 2\tilde{a} \tag{15.144}$$

where $2\tilde{a}$ is merely an abbreviation of the last term of the middle part of (15.144). The electromagnetic self-energy can be interpreted as a shift of the mass of an electron from its "bare" value to its observed value m_0. This shift is called renormalization of the mass.

The argument used in renormalization theory is now as follows. The reason that the result (15.134) is infinite lies in the fact that it included an infinite energy change that is already counted when we use the observed mass in the Hamiltonian rather than the bare mass. In other words, we should in fact start with the Hamiltonian for the hydrogen atom in the presence of the radiation field given by

$$\mathscr{H} = \frac{\hat{p}^2}{2\,\bar{m}_0} - \frac{e^2}{4\,\pi\,\varepsilon_0\,r} + \mathscr{H}_{\text{int}} \;. \tag{15.145}$$

Then using (15.144) we can rewrite $\mathscr{H}$ as

$$\mathscr{H} = \frac{\hat{p}^2}{2\,m_0} - \frac{e^2}{4\,\pi\,\varepsilon_0\,r} + \{\mathscr{H}_{\text{int}} + \tilde{a}\hat{p}^2\} \;. \tag{15.146}$$

Thus if we use the observed free particle mass in the expression for the kinetic energy (which we always do) we should not count that part of $\mathscr{H}_{\text{int}}$ that produces the mass shift, i.e. we should regard

$$\mathscr{H}_{\text{int}} + \tilde{a}\hat{p}^2 \tag{15.147}$$

as the effective interaction of an electron of a renormalized mass m_0 with the radiation field.

Returning then to the calculation of the Lamb shift, we see that to first order in $e^2/\hbar c$ we must add the expectation value of the second term in (15.147) to (15.134) in order to avoid counting the electromagnetic interaction twice, once in m_0 and once in $\mathscr{H}_{\text{int}}$. Thus more correctly the shift of the level n is given by

$$\varepsilon^{(2)} = \frac{1}{(2\pi)^2} \frac{2e^2\hbar}{3m_0^2\varepsilon_0 c^3} \int_0^\infty \omega\, d\omega \left[\sum_{n'} \frac{|\langle n'|\hat{p}|n\rangle|^2}{E_n^0 - E_{n'}^0 - \hbar\omega} + \frac{\langle n|\hat{p}^2|n\rangle}{\hbar\omega} \right] . \tag{15.148}$$

The second term under the integral in (15.148) can be brought into a form similar to the first term under the integral by means of the relation

$$\langle n|\hat{p}^2|n\rangle = \sum_{n'} \langle n|\hat{p}|n'\rangle\langle n'|\hat{p}|n\rangle \tag{15.149}$$

In order not to interrupt the main discussion we will postpone the proof of this relation to the exercises. Using (15.149) and (15.148) we find after a slight rearrangement of terms

$$\varepsilon^{(2)} = \frac{1}{(2\pi)^2} \frac{2e^2}{3m_0^2\varepsilon_0 c^3} \sum_{n'} |\langle n'|\hat{p}|n\rangle|^2 \int_0^\infty d\omega \frac{E_n^0 - E_{n'}^0}{E_n^0 - E_{n'}^0 - \hbar\omega} . \tag{15.150}$$

We note that the integral over ω is still divergent, however, only logarithmically. This divergence is not present in a more sophisticated relativistic calculation. Such a calculation yields a result quite similar to (15.150), but with an integrand falling off more rapidly at high frequencies $\hbar\omega \geq m_0c^2$. We can mimic the result of such a calculation by cutting off the integral at $\omega = m_0c^2/\hbar$. The integral can be immediately performed and yields

$$\varepsilon^{(2)} = \frac{1}{(2\pi)^2} \frac{2e^2}{3m_0\varepsilon_0\hbar c^3} \sum_{n'} |\langle n'|\hat{p}|n\rangle|^2 (E_{n'}^0 - E_n^0) \ln\left|\frac{m_0c^2}{E_{n'}^0 - E_n^0}\right| \tag{15.151}$$

where we have neglected $|E_n^0 - E_{n'}^0|$ compared to m_0c^2. The further evaluation of (15.151) requires some formal tricks which are purely mathematical. Namely,

$$\ln\left|\frac{m_0c^2}{(E_{n'}^0 - E_n^0)}\right|$$

is replaced by an average

$$\ln\frac{|m_0c^2|}{\langle |E_{n'}^0 - E_n^0|\rangle} .$$

Clearly, we may now rewrite (15.151) in the form

$$\varepsilon^{(2)} = \frac{1}{(2\pi)^2} \frac{2e^2}{3m_0^2\varepsilon_0\hbar c^3} \ln\frac{|m_0c^2|}{\langle |E_{n'}^0 - E_n^0|\rangle} \sum_{n'} |\langle n'|\hat{p}|n\rangle|^2 (E_{n'}^0 - E_n^0) . \tag{15.152}$$

To simplify (15.152) further we use the relation

$$\sum_{n'} |\langle n'|\hat{p}|n\rangle|^2 (E_{n'}^0 - E_n^0) = -\tfrac{1}{2}\langle n|[\hat{p}, [\hat{p}, \mathscr{H}_{\mathrm{el}}]]|n\rangle \tag{15.153}$$

which will be proved in the exercises. The double commutator on the right-hand side can be easily evaluated. We assume $\mathscr{H}_{\text{el}}$ in the form

$$\mathscr{H}_{\text{el}} = -\frac{\hbar^2}{2m_0}\Delta + V(r) \ . \tag{15.154}$$

We readily obtain (cf. Sect. 9.3)

$$[\hat{p}_x, \mathscr{H}_{\text{el}}] = \frac{\hbar}{\mathrm{i}} \frac{\partial V(r)}{\partial x} \tag{15.155}$$

and in a similar fashion

$$[\hat{\boldsymbol{p}}, [\hat{\boldsymbol{p}}, \mathscr{H}_{\text{el}}]] = -\hbar^2 \Delta V(r) \ . \tag{15.156}$$

Using for V the Coulomb potential of the electron in the hydrogen atom, $V = -e^2/(4\pi\varepsilon_0|\boldsymbol{r}|)$, we can readily evaluate the right-hand side of (15.156). Using a formula well known from electrostatics (potential of a point charge!) we find

$$\Delta \frac{1}{|\boldsymbol{r}|} = -4\pi\delta(\boldsymbol{r}) \tag{15.157}$$

where $\delta(\boldsymbol{r})$ is Dirac's delta function in three dimensions. Using this result and the definition of bra and kets (see Problem 9.19), we readily obtain

$$(15.153) = \frac{\hbar^2}{2}\left\langle n \left| \Delta \frac{-e^2}{4\pi\varepsilon_0|\boldsymbol{r}|} \right| n \right\rangle = \frac{e^2\hbar^2}{2\varepsilon_0}\int |\psi_n(\boldsymbol{r})|^2 \delta(\boldsymbol{r})\, d^3r \tag{15.158}$$

and making use of the properties of the δ-function

$$(15.158) = \frac{e^2\hbar^2}{2\varepsilon_0}|\psi_n(0)|^2 \ . \tag{15.159}$$

We are now in a position to write down the final formula for the renormalized self-energy shift by inserting the result (15.156) with (15.153) and (15.159) into (15.152). We then obtain

$$\varepsilon^{(2)} = \frac{1}{(2\pi)^2} \frac{2e^2}{3m_0^2\varepsilon_0\hbar c^3} \ln \frac{|m_0c^2|}{\langle |E^0_{n'} - E^0_n| \rangle} \frac{e\hbar^2}{2\varepsilon_0} |\psi_n(0)|^2 \ . \tag{15.160}$$

To obtain final numerical results we have to calculate numerically the average as well as $|\psi_n(0)|^2$. For the hydrogen atom $|\psi_n(0)|^2$ is well known and is nonzero only for s-states. The average value was calculated by Bethe for the $2S$ level. Inserting all the numerical values we eventually find $\varepsilon^{(2)}/\hbar = 1040$ megacycles. According to these considerations a shift between an S and a P level must be expected. Such a shift was first discovered between the $2S_{1/2}$ and $2P_{1/2}$ level of hydrogen by Lamb and Retherford.

At first sight, it may seem strange that it is possible to obtain reasonable results by a subtraction procedure in which two infinitely large quantities are involved. However, it has turned out that such a subtraction procedure can be formulated in the frame of a beautiful theory, called renormalization, and such procedures are now a legitimate part of theoretical physics giving excellent agreement between theory and experiment. Unfortunately it is beyond the scope of our introductory book to cover the details of these renormalization techniques.

Problems

15.1 A harmonic oscillator with mass m, charge e and eigenfrequency ω is subjected to a constant electric field. Calculate the wavefunctions to the first and second approximations by perturbation theory, and compare the result with the exact solution and energy values (see Problem 9.13).
As an example, one might choose the perturbation of the $n=1$ and $n=2$ levels.

Hint: Use the b^+, b formalism in perturbation theory and the results of the corresponding problems from Chap. 9.

15.2 The rotational motion of a two-atom molecule or of a rotating atomic nucleus can be approximately described by the Schrödinger equation

$$-\frac{\hbar^2}{Mr_0^2}\frac{d^2}{d\theta^2}\phi(\theta)=E\phi(\theta)\,,$$

where M is a mass and r_0 is an "effective" distance. Let the wavefunction $\phi(\theta)$ be periodic in the angular coordinate θ: $\phi(\theta+2\pi)=\phi(\theta)$. What are the wavefunctions and energy values of this system? Then assume the system is perturbed by the additional potential $a\cos(2\theta)$. Calculate E and ϕ in this case using perturbation theory with degeneracy.

15.3 Prove that (15.112) fulfills (15.111) and verify (15.113).

Hint: Insert (15.112) into (15.111), use the commutation relations (15.107), (15.109) and $b_\lambda\Phi_0=0$ for all λ's.

15.4 Prove that the general wavefunction

$$\Phi=\frac{1}{\sqrt{n_1!\,n_2!\ldots n_N!}}(b_1^+)^{n_1}(b_2^+)^{n_2}\ldots(b_N^+)^{n_N}\Phi_0 \qquad (*)$$

fulfills (15.111) and determine E.

Hint: Insert (*) into (15.111), use the relations

$$b_\lambda(b_\kappa^+)^n-(b_\kappa^+)b_\lambda=n\,b_\kappa^{n-1}\delta_{\lambda,\kappa}\,,$$

(15.109) and $b_\lambda\Phi_0=0$ for all λ's.

15.5 Prove $\langle n|\hat{p}^2|n\rangle = \sum_{n'} \langle n|\hat{p}|n'\rangle\langle n'|\hat{p}|n\rangle$.

Hint: Write the bra-ket as integrals $\int \psi_n^* \hat{p}\, \psi_{n'}\, dx\, dy\, dz$ and use the completeness relation

$$\sum_{n'} \psi_{n'}(\boldsymbol{r})\, \psi_{n'}^*(\boldsymbol{r}') = \delta(\boldsymbol{r}-\boldsymbol{r}') \qquad (**)$$

where δ is Dirac's function.

15.6 Prove $\sum_{n'} |\langle n'|\hat{p}|n\rangle|^2 (E_{n'}^0 - E_n^0) = -\frac{1}{2}\langle n|[\hat{p}, [\hat{p}, \mathscr{H}_{\mathrm{el}}]]|n\rangle$.

Hint: Start from the expression

$$\sum_{n'} \langle n|\hat{p}|n'\rangle\, \langle n'|\hat{p}|n\rangle E_{n'}^0$$

and use the fact that the ψ_n's are eigenfunctions of $\mathscr{H}_{\mathrm{el}}$. Use the completeness relation (**).

15.7 Show that the Lamb shift can be understood qualitatively by the assumption that the zero-point fluctuations of the field ($\equiv$ set of harmonic oscillators!) cause a shift of the potential energy of the electron.

Hint: Write the electron coordinate as

$$\boldsymbol{r}' = \boldsymbol{r} + \boldsymbol{s}$$

where $\boldsymbol{r}$ corresponds to the unperturbed motion, and $\boldsymbol{s}$ to the field fluctuations.
Expand

$$\langle \psi_n(\boldsymbol{r})\, V(\boldsymbol{r}')\, \psi_n(\boldsymbol{r})\rangle$$

in the first two powers of $\boldsymbol{s}$, where $\langle \boldsymbol{s}\rangle = 0$ and convince yourself, that an expression analogous to (15.158) results.

16. General Laws of Optical Transitions

16.1 Symmetries and Selection Rules

16.1.1 Optical Matrix Elements

Selection rules and symmetry considerations play a basic role in modern physics. This holds for the atomic shells, the nucleus, elementary particles and for many other areas. We shall present only a small but typical sample here, taking up the perturbation theory discussed in Sect. 15.2. We began with the unperturbed wavefunctions ϕ_n. The perturbation Hamiltonian in one example contained the dipole moment ez. In the following, we shall first choose the x coordinate instead of the z coordinate. We became acquainted with matrix elements of the form

$$H_{mn}^{\mathrm{P}} = \int \phi_m^*(r)\, ex\, \phi_n(r)\, dV \tag{16.1}$$

as important parameters. As we observed at that time, there are a number of realistic cases in which matrix elements vanish when the indices m and n are equal, on the basis of symmetry alone. We shall now examine these cases.

16.1.2 Examples of the Symmetry Behaviour of Wavefunctions

As an unperturbed wavefunction, let us consider that of a particle in a box (Sect. 9.1) or of a harmonic oscillator. We recognise (see Figs. 16.1, 2) that these wavefunctions are either "symmetric" or "antisymmetric".

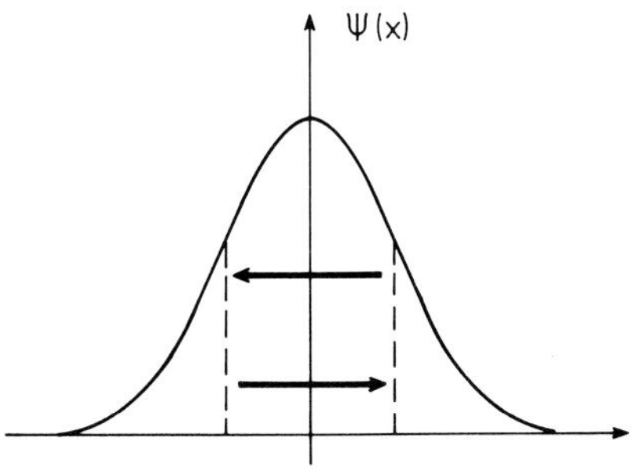

Fig. 16.1. Example of a *symmetric* wavefunction. When reflected through the ψ axis, ψ is converted into itself

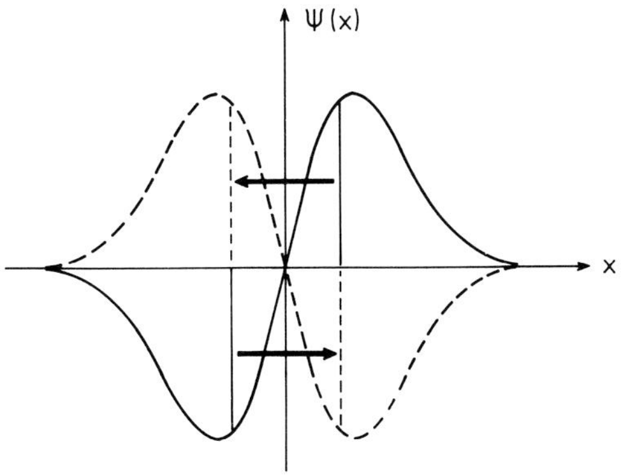

Fig. 16.2. Example of an *antisymmetric* wavefunction. When the wavefunction is reflected through the ψ axis it is transformed into a function with the same magnitude but the opposite sign at each point

A symmetric wavefunction is converted into itself when every x is converted to $-x$. In an antisymmetric wavefunction, this process reverses the sign of the function. We shall now show two things:

1) How this symmetry property of the wavefunction can be derived directly from the Schrödinger equation, without solving it explicitly, and

2) how these symmetry properties can be used to prove that

$$H_{mm}^{\mathrm{P}}=0\,. \tag{16.2}$$

We first consider the symmetry properties of the Hamiltonian and select the harmonic oscillator as an example. If we replace x by $-x$, we obviously obtain

$$x^2\rightarrow(-x)^2=x^2\,. \tag{16.3}$$

x^2 is thus unchanged when x is replaced by $-x$, or in other words, x^2 is "invariant" with respect to the transformation

$$x\rightarrow -x\,. \tag{16.4}$$

Since the potential energy of the harmonic oscillator is proportional to x^2, this invariance property naturally applies to the oscillator potential itself:

$$V(-x)=V(x)\,. \tag{16.5}$$

In analogous fashion, one can show that the second derivative with respect to x is also invariant with respect to the transformation (16.4):

$$\frac{d^2}{dx^2}\rightarrow\frac{d^2}{d(-x)^2}=\frac{d^2}{dx^2}\,. \tag{16.6}$$

We can now assume in general that for a certain one-dimensional problem, the Hamiltonian is invariant with respect to the transformation (16.4). If we take the appropriate Schrödinger equation

$$\mathscr{H}(x)\,\psi(x)=E\,\psi(x) \tag{16.7}$$

and replace each x by $-x$, we naturally obtain

$$\mathscr{H}(-x)\,\psi(-x)=E\,\psi(-x)\,. \tag{16.8}$$

However, since the Hamiltonian is supposed to be invariant with respect to the transformation (16.4), we can replace $\mathscr{H}(-x)$ by $\mathscr{H}(x)$ in (16.8),

$$\mathscr{H}(x)\,\psi(-x)=E\,\psi(-x)\,. \tag{16.9}$$

This tells us that if $\psi(x)$ is an eigenfunction of (16.7), $\psi(-x)$ is also an eigenfunction of (16.7). We now make the simplifying assumption that there is only a single eigenfunction for the energy E. The word "single" should be taken with a grain of salt, in that eigenfunctions can differ from one another by a constant numerical factor.

As we see from (16.7) and (16.8), the eigenfunctions $\psi(x)$ and $\psi(-x)$ belong to the energy E. They can differ at most by a constant factor, which we shall call α, due to our assumption above. We thus have the relation

$$\psi(-x) = \alpha\psi(x) . \quad (16.10)$$

If we replace x by $-x$ on both sides of (16.10), the relation becomes

$$\psi(x) = \alpha\psi(-x) . \quad (16.11)$$

We now replace $\psi(x)$ on the right side of (16.10) by $\psi(-x)$ according to (16.11), obtaining as the overall result

$$\psi(-x) = \alpha\psi(x) = \alpha^2\psi(-x) . \quad (16.12)$$

Since we know that ψ does not vanish identically, we can divide both sides by $\psi(-x)$, yielding

$$\alpha^2 = 1 , \quad (16.13)$$

or, taking the square root,

$$\alpha = \pm 1 . \quad (16.14)$$

Inserting this result now into (16.10), we obtain the relation

$$\psi(-x) = \pm\psi(x) . \quad (16.15)$$

This, however, is just the relation we were searching for. As may be seen in Figs. 16.1 and 16.2, the plus sign means that the wavefunction is symmetric, and the minus sign means that it is antisymmetric. We thus see that from the symmetry of the Hamiltonian, it follows automatically that the wavefunctions will have a particular symmetry behaviour.

The argumentation which we have just carried out may be immediately generalised to three dimensions, by utilising consistently the replacement

$$x \rightarrow \boldsymbol{r} \quad (16.16)$$

and by replacing the transformation (16.14) by the transformation

$$\boldsymbol{r} \rightarrow -\boldsymbol{r} . \quad (16.17)$$

If the Hamiltonian is invariant with respect to (16.17), it follows, in complete analogy to the derivation of (16.15), that

$$\psi(-\boldsymbol{r}) = \pm\psi(\boldsymbol{r}) . \quad (16.18)$$

The transformation behaviour described by (16.18) is denoted as *parity*. If the positive sign is valid, one speaks of positive (or even) parity; if the negative sign holds, the parity is negative (or odd). Correspondingly, the oscillator wavefunctions of Figs. 9.9a, b have positive parity for even $n = 0, 2, 4, \ldots$ and negative parity for odd $n = 1, 3, 5, \ldots$.

Up to now we have assumed that only a single wavefunction (up to a constant factor) belongs to E, or, in other words, that the wavefunctions are not degenerate with one another. It can be shown by the detailed theory that even in the case of degeneracy, the degenerate wavefunctions may be so defined that here, too, the relation (16.18) is fulfilled, as long as the Hamiltonian is invariant with respect to the transformation (16.17).

Before we show by means of symmetry arguments that (16.2) is valid, let us consider another example of a symmetry. Keeping in mind the hydrogen atom problem, which is three dimensional, we investigate symmetry with respect to rotations in three-dimensional space; as a concrete example, we consider a rotation by the angle ϕ around the z axis. We assume, as is for example fulfilled in the hydrogen atom, that the Hamiltonian remains unchanged when we rotate the coordinate axes through the angle $\phi = \phi_1$ about the z axis (Fig. 16.3):

$$\mathscr{H}(r, \phi + \phi_1) = \mathscr{H}(r, \phi) , \quad (16.19)$$

where we have not shown the constant angle θ as an argument.

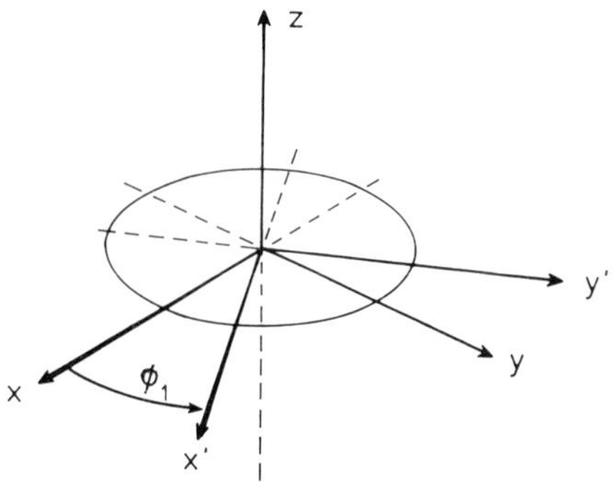

Fig. 16.3. Rotation of the coordinate system around the z axis by an angle ϕ_1

If, in the corresponding Schrödinger equation, we replace ϕ everywhere by $\phi + \phi_1$, we obtain

$$\mathscr{H}(r, \phi + \phi_1)\, \psi(r, \phi + \phi_1) = E\, \psi(r, \phi + \phi_1) , \quad (16.20)$$

or, using rotational invariance,

$$\mathscr{H}(r, \phi)\, \psi(r, \phi + \phi_1) = E\, \psi(r, \phi + \phi_1) . \quad (16.21)$$

Since the wavefunction which occurs in (16.21) must be an eigenfunction of the original Schrödinger equation, we must have (without degeneracy) in analogy to the example treated before,

$$\psi(r, \phi + \phi_1) = \alpha_{\phi_1}\, \psi(r, \phi) . \quad (16.22)$$

Here the constant α can, as we have explicitly indicated, be a function of the angle of rotation ϕ_1. We should like to determine this functional dependence exactly. For this purpose, we write (16.22) again for a second rotation angle ϕ_2:

$$\psi(r, \phi+\phi_2) = \alpha_{\phi_2}\psi(r,\phi) \,. \tag{16.23}$$

We imagine now that ϕ in (16.22) has been replaced by $\phi+\phi_2$. This gives us

$$\psi(r, \phi+\phi_1+\phi_2) = \alpha_{\phi_1}\psi(r,\phi+\phi_2) \,, \tag{16.24}$$

or, using (16.23) on the right side,

$$\psi(r, \phi+\phi_1+\phi_2) = \alpha_{\phi_1}\alpha_{\phi_2}\psi(r,\phi) \,. \tag{16.25}$$

On the other hand, we could have replaced ϕ_1 in (16.22) on both sides of the equation by $\phi_1+\phi_2$, which would have led immediately to

$$\psi(r, \phi+\phi_1+\phi_2) = \alpha_{\phi_1+\phi_2}\psi(r,\phi) \tag{16.26}$$

(Fig. 16.4).

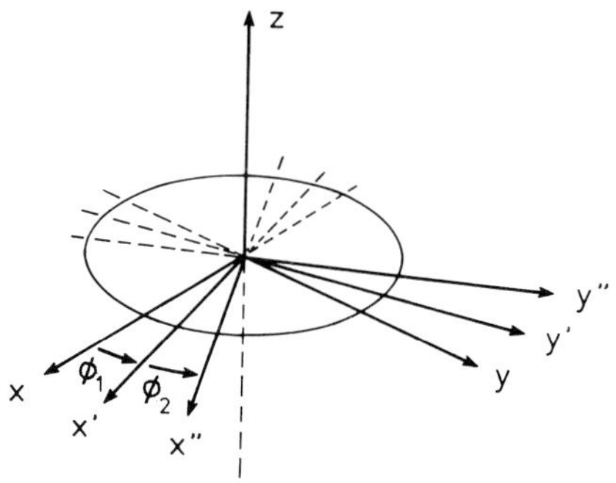

Fig. 16.4. Addition of rotations: two rotations which are carried out one after the other may be replaced by a single rotation

We now compare (16.26) with (16.25). We see immediately that the α's are connected to one another by the relation

$$\alpha_{\phi_1+\phi_2} = \alpha_{\phi_1}\alpha_{\phi_2} \,. \tag{16.27}$$

It may be shown mathematically in an exact way that (16.27) can only be fulfilled by

$$\alpha_\phi = \mathrm{e}^{im\phi} \,, \tag{16.28}$$

where m is still an unknown parameter. (Addition of arguments leads to multiplication of the functions → exponential function.) Now, however, we recognise that every wavefunction must go into itself when we rotate the coordinate system completely around in a circle, that is, when we rotate by an angle of 2π. From this fact, we obtain directly the requirement

$$\mathrm{e}^{im2\pi} = 1 \,. \tag{16.29}$$

This may be fulfilled by requiring m to be a positive or a negative integer.

To utilise this result, we employ (16.22) again, setting $\phi = 0$ in it:

$$\psi(r, 0+\phi_j) = \alpha_{\phi_j}\psi(r,0) \,. \tag{16.30}$$

If we leave off the index j and use (16.28), we finally obtain the relation

$$\psi(r,\phi) = \mathrm{e}^{\mathrm{i}m\phi}\psi(r,0) \,, \qquad m \text{ integral} \,. \tag{16.31}$$

Thus we have found that the wavefunction ψ is dependent on the angle ϕ, and in a way which is in perfect agreement with (10.82), the result which we obtained for the hydrogen atom.

As this example indicates, extremely general conclusions about the structure and the transformation behaviour of wavefunctions may be drawn from symmetry considerations. Naturally, one can also consider rotations around other axes, and rotations around various axes may also be added to give new rotations. The corresponding transformation behaviour of the wavefunctions is treated by what is called the theory of representations of the rotation group, which, however, is beyond the subject matter of this book. We hope, however, that the reader has gained a feeling for the way in which the symmetries of the original problem (i.e. of the Hamiltonian) can lead to a certain transformation behaviour of the wavefunctions.

16.1.3 Selection Rules

We have already met selection rules several times in this book. As we have seen, the coupling of an atom to external fields, especially to the radiation field (light), produces transitions between the electronic states in the atom. However, these transitions only occur (at least in first order) when the corresponding matrix element of the external field, H^{P}_{mn}, is nonzero. If it is zero, the transition is forbidden; we thus obtain selection rules for the occurrence of transitions. We shall now show, using several simple examples, how it follows from the symmetry properties of the wavefunctions that certain matrix elements are identically equal to zero. This is valid independently of the form which the wavefunctions may happen to have.

As a first example we consider an integral which results from leaving out the factor e from the matrix element (16.1). If we replace the functions ϕ_n and ϕ_m by the wavefunction ψ which we are now considering, we have

$$I = \int_{-\infty}^{+\infty} \psi^*(x)\, x\, \psi(x)\, dx \,, \tag{16.32}$$

where we have chosen a one-dimensional example (for an intuitive picture of this integral, see Fig. 16.5). We rename the variable of integration x, in that we replace x by $-x$:

$$x \rightarrow -x \,. \tag{16.33}$$

The value of the integral naturally remains the same. However, the following individual changes are made in the integral (16.32):

$$I = \int_{+\infty}^{-\infty} \psi^*(-x)(-x)\psi(-x)\, d(-x) \,. \tag{16.34}$$

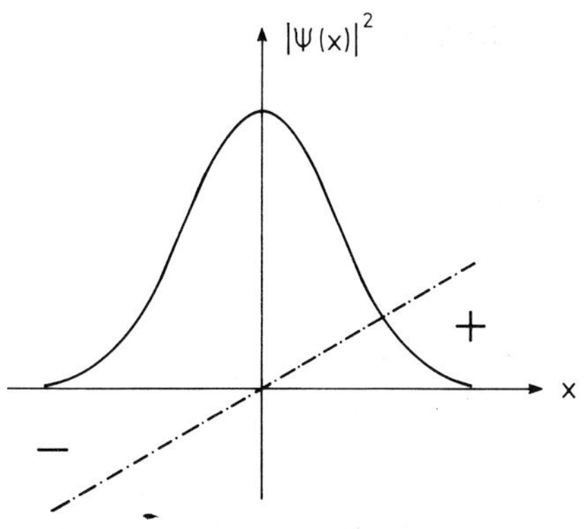

Fig. 16.5. Explanation of the integral (16.32). The integral is obtained by multiplying the value of the function x (− · −) by the value of the function $|\psi|^2$ (——)at each point along the x axis and then integrating the products over all values of x. As can be seen from the diagram, the product for each value of x has the same magnitude as the product for $-x$, but the opposite sign. When added, each pair has a sum of zero, and thus the integral vanishes

By exchanging the limits and accordingly changing the sign of the differential, we obtain

$$\int_{-\infty}^{+\infty} \psi^*(-x)(-x)\,\psi(-x)\,dx\,. \tag{16.35}$$

Now we make use of a symmetry property of the wavefunction, namely the transformation relation (16.15). The transformation (16.4) leaves $\psi^*\psi$ invariant, so that the integral (16.35) differs from the original integral (16.32) only with respect to its sign. We thus obtain the relation

$$I = -I\,, \tag{16.36}$$

which, of course, can only be satisfied by

$$I = 0\,. \tag{16.37}$$

This result is enormously important. It shows us that we have found the value of the integral to be 0 without having done any integrating. We have used only symmetry relationships.

What would have happened if we had calculated a matrix element with different indices m and n? If the perturbation operator were again x, we would then have obtained the following: if ϕ_n and ϕ_m have the same parity, the result is still zero. The integral can only have a non-zero value if the parity of ϕ_n and ϕ_m are different. Here we have the simplest example of a *selection rule*.

As a second example, which is particularly important for atomic shells, we will again consider a dipole matrix element, but now between eigenfunctions of the hydrogen atom. We shall examine a matrix element of the form

$$I_z = \int \psi^*_{n,l,m}(\boldsymbol{r})\,z\,\psi_{n',l',m'}(\boldsymbol{r})\,dV\,. \tag{16.38}$$

In the following, it is useful to employ polar coordinates, in which (16.38) has the form

$$I_z = \int dV\,\psi^*_{n,l,m}(r,\theta,\phi)\,r\cos\theta\,\psi_{n',l',m'}(r,\theta,\phi)\,. \tag{16.39}$$

In order to determine when the matrix element I_z is identical to zero, we again make use of symmetry relationships. This time, however, we consider rotations about the z axis. We allow a rotation through the angle ϕ_0. This converts (16.39) into

$$I_z = e^{-i(m-m')\phi_0} I_z , \tag{16.40}$$

as one can easily see from the transformation property expressed in (16.31). The left side must naturally remain I_z. Equation (16.40) can be satisfied in two ways. Either $I_z = 0$, or $I_z \neq 0$, but in the latter case m must be equal to m'. This is another example of a selection rule. If the perturbation operator contains the dipole moment in the z direction, a matrix element can only be different from 0 if $m = m'$.

We shall now derive a selection rule for the x and y components of the dipole moment, by considering the integrals

$$I_x = \int dV\, \psi^*_{n,l,m}(r,\theta,\phi)\, x\, \psi_{n',l',m'}(r,\theta,\phi) \tag{16.41}$$

and

$$I_y = \int dV\, \psi^*_{n,l,m}(r,\theta,\phi)\, y\, \psi_{n',l',m'}(r,\theta,\phi) . \tag{16.42}$$

We shall multiply the second line by i and add it to the first, and also express x and y in spherical polar coordinates. We then obtain

$$I_x + i I_y = \int dV\, \psi^*_{n,l,m} r \sin\theta e^{i\phi} \psi_{n',l',m'} . \tag{16.43}$$

If we again carry out a rotation through the angle ϕ_0 about the z axis, we obtain, in complete analogy to the preceding case,

$$I_x + i I_y = \exp[-i(m-1-m')\phi_0](I_x + i I_y) . \tag{16.44}$$

From this it immediately follows that

$$I_x + i I_y = 0 \quad \text{for} \quad m \neq m' + 1 \tag{16.45}$$

and similarly by subtraction of (16.41) and (16.42)

$$I_x - i I_y = 0 \quad \text{for} \quad m \neq m' - 1 . \tag{16.46}$$

In summary, if $m \neq m' + 1$ and $m \neq m' - 1$, then

$$I_x = 0 \quad \text{and} \quad I_y = 0 . \tag{16.47}$$

Here we have discovered another selection rule: I_x or I_y can have non-zero values only when $m = m' + 1$ or $m = m' - 1$.

It can be derived from the rules (16.40, 45 and 46) which transitions in an atom can be induced by a radiation field. If the radiation field is polarised in the z direction, only transitions with $m = m'$ are possible. However, if it is polarised in the x or the y direction, the transitions with

$$m = m' \pm 1 \tag{16.48}$$

are possible. These are the π or σ transitions which were introduced in Sect. 13.3.2.

The selection rule of the angular momentum of the electron,

$$l = l' \pm 1 , \tag{16.49}$$

can be derived from the matrix element (16.38) on the basis of considerations which are similar in principle to the above, but require somewhat more extensive mathematics. The basic matrix element occurs just when light, or more exactly, light quanta, are absorbed or emitted (see also Sect. 16.1.4). Since the total angular momentum of electron and light quantum is conserved, it can be deduced from (16.49) that a quantum of light has the angular momentum (spin) $\hbar$.

We can summarise what has been learned in this chapter as follows: The invariance properties of the Hamiltonian operator lead to certain transformation properties of the wavefunctions. From the transformation properties of the wavefunctions and of the perturbation operator (in the present case this is the dipole operator) it can be rigorously determined which matrix elements vanish identically and which can in principle differ from zero. However, these considerations give no information about the magnitude of the matrix elements, and it can happen that still other matrix elements vanish for other reasons. As it turns out, symmetry considerations allow quite exact predictions, and it is therefore not surprising that such considerations, appropriately generalised, have a basic role in the physics of elementary particles.

16.1.4 Selection Rules and Multipole Radiation*

In this chapter we have been primarily concerned with the matrix element

$$\int \phi_n^*(\boldsymbol{r})\, ex\, \phi_m(\boldsymbol{r})\, dV . \tag{16.50}$$

We met this earlier, in Sect. 15.3, where we examined the effect of a light wave on an atom. Conversely, the matrix element (16.50) is also found to apply to the production of light by atomic transitions. Within the limits of this introduction we cannot go into details, but we will describe the basic ideas. As we know from classical electrodynamics, an oscillating dipole generates electromagnetic waves. The dipole is mathematically described by the dipole moment

$$\boldsymbol{P} = -e\boldsymbol{r}(t) \tag{16.51}$$

where $\boldsymbol{r}$ is the vector from the positive charge to the oscillating negative charge. As usual, we assume that the oscillation is purely harmonic, $\boldsymbol{r}(t) = \boldsymbol{r}_0 \sin\omega t$. The dipole moment (16.51) then appears in the Maxwell equations for electromagnetism as a "source term". (For the more interested reader, we can formulate this somewhat more exactly: In the Maxwell equation $\operatorname{curl}\boldsymbol{H} = \boldsymbol{j} + d\boldsymbol{D}/dt$, the polarisation $\boldsymbol{P}$ appears in $\boldsymbol{D}$ if matter, in the present case atoms, is present. In the classical case, this can be expressed by (16.51) or, if several atoms are present, by a summation of several expressions of the form of (16.51).) The question now arises whether there is a quantum theoretical analogue to this source term of the form of (16.51). Let us recall the translation table in Sect. 9.3.4, and accordingly, assign to the classical observable "dipole moment $-e\boldsymbol{r}(t)$" an operator $-e\boldsymbol{r}$ and, finally, the expectation value

$$\int \psi^*(\boldsymbol{r},t)(-e\boldsymbol{r})\,\psi(\boldsymbol{r},t)\, dV . \tag{16.52}$$

What does the wavefunction $\psi(\boldsymbol{r},t)$ mean here? To obtain an insight, let us imagine an atom with two energy levels E_1 and E_2 and the related wave functions $\phi_1(\boldsymbol{r})$ and $\phi_2(\boldsymbol{r})$. To generate an oscillation, we form a superposition in the form of a "wave packet",

$$\psi(\boldsymbol{r},t) = \frac{1}{\sqrt{2}} e^{-iE_1t/\hbar}\phi_1(\boldsymbol{r}) + \frac{1}{\sqrt{2}} e^{-iE_2t/\hbar}\phi_2(\boldsymbol{r}) , \qquad (16.53)$$

where the factors $1/\sqrt{2}$ serve to normalise $\psi(\boldsymbol{r})$. Let us substitute (16.53) in (16.52) and multiply out the individual terms. The result is a sum over expressions of the form of (16.50), in which m and n can take on the values 1 and 2. Let us assume, as above, that

$$\int \phi_1^* e\boldsymbol{r}\,\phi_1 dV \quad \text{and} \quad \int \phi_2^* e\boldsymbol{r}\,\phi_2 dV$$

vanish. Then (16.52) takes the form

$$\tfrac{1}{2}\int \phi_1^*(\boldsymbol{r})(-e\boldsymbol{r})\,\phi_2(\boldsymbol{r})\,dV\cdot e^{-i\omega t} + \text{c.c.}^1 , \qquad (16.54)$$

where we have also introduced the abbreviation $\omega = (E_2 - E_1)/\hbar$. The expectation value of the dipole moments thus actually oscillates as a classical dipole and generates the corresponding classical electromagnetic field. In this way the relationship between a dipole matrix element and dipole radiation becomes clear.

Now as we know from classical physics, radiation fields are generated not only by oscillating dipoles, but also by other oscillating charge or even current distributions. A loop conducting an electric current acts like a magnetic dipole. If we vary the current in the loop, for example in a sine wave, the magnetic dipole moment oscillates and generates "magnetic dipole radiation". Can such a radiation field also be generated by an *atomic* transition? This is in fact possible, but the matrix element (16.52) or (16.54) is no longer sufficient to describe it mathematically. Instead we must base it on a more exact theory of the interaction between light and electrons.

With the exception of constants, the corresponding part of the Schrödinger equation is

$$e\boldsymbol{p}\cdot\boldsymbol{A} , \qquad (16.55)$$

where $\boldsymbol{p} = -i\hbar\nabla$ is the momentum operator and $\boldsymbol{A}$ is the vector potential of the light field. We have already encountered the interaction expression (16.55) in Sect. 14.1, e.g. in (14.9). There the vector potential referred to a constant magnetic field; here it describes the electromagnetic field. If we express $\boldsymbol{A}$ in terms of plane waves (in complex notation), we have instead of (16.55) expressions of the form

$$e\boldsymbol{p}\cdot\boldsymbol{e}\,e^{i\boldsymbol{k}\cdot\boldsymbol{r}} \qquad (16.56)$$

in which $\boldsymbol{e}$ is the polarisation vector of the light wave with the wavenumber vector $\boldsymbol{k}$. The matrix element which now appears in perturbation theory,

$$e\int \phi_n^*(\boldsymbol{r})\boldsymbol{p}\cdot\boldsymbol{e}\,e^{i\boldsymbol{k}\cdot\boldsymbol{r}}\phi_m(\boldsymbol{r})\,dV , \qquad (16.57)$$

---

[1] Here and in the following, c.c. indicates the complex conjugate of the preceding expression

takes the place of (16.50). Since the wavelength λ is generally large compared to the extent a of the wave function ϕ, we can assume $\boldsymbol{k}\cdot\boldsymbol{r}=2\pi r/\lambda \ll 1$ and expand the exponential function

$$e^{i\boldsymbol{k}\cdot\boldsymbol{r}} \approx 1+i\boldsymbol{k}\cdot\boldsymbol{r}+\dots \; . \tag{16.58}$$

If we substitute this in (16.57), the first term is

$$\int \phi_n^*(\boldsymbol{r})\, e\boldsymbol{p}\cdot\boldsymbol{e}\, \phi_m(\boldsymbol{r})\, dV \, . \tag{16.59}$$

In the special case that $\boldsymbol{e}$ is parallel to the x axis and $n=1$, $m=2$, we have

$$\int \phi_1^*(\boldsymbol{r})\, e p_x \phi_2(\boldsymbol{r})\, dV \, . \tag{16.60}$$

As is demonstrated exactly in quantum mechanics, (16.60) is identical to

$$-\mathrm{i}m\omega \int \phi_1^*(\boldsymbol{r})\, e x\, \phi_2(\boldsymbol{r})\, dV \quad \text{where} \quad \omega=(E_2-E_1)/\hbar \, , \quad m\text{: particle mass,} \tag{16.61}$$

or, except for a numerical factor, the familiar dipole matrix element. There are cases of atomic transitions, however, where (16.61) vanishes. (These are "forbidden electric dipole transitions".) Then the matrix element derived from the second term in (16.58),

$$\mathrm{i}e \int \phi_1^*(\boldsymbol{r})(\boldsymbol{p}\cdot\boldsymbol{e})(\boldsymbol{k}\cdot\boldsymbol{r})\, \phi_2(\boldsymbol{r})\, dV \, , \tag{16.62}$$

becomes important. This can also be rearranged (which will not be demonstrated here) and becomes (except for the factor $\omega m \boldsymbol{e}$)

$$\int \phi_1^*(\boldsymbol{r})[-e\boldsymbol{r}(\boldsymbol{k}\cdot\boldsymbol{r})]\, \phi_2(\boldsymbol{r})\, dV \, . \tag{16.63}$$

To establish the connection with classical physics, let us think of the integral over $\boldsymbol{r}$ being replaced by a discrete sum over points $\boldsymbol{r}_j$, each with the charge $(-e_j)$. Then

$$(16.63) \to \sum_j (-e_j)\boldsymbol{r}_j(\boldsymbol{k}\cdot\boldsymbol{r}_j) \, . \tag{16.64}$$

However, just this sum appears as the "source term" in the classical theory of electromagnetic fields (it appears as the "Hertz vector"). It is known from this theory that the sum represents a superposition of the electric quadrupole moment and the magnetic dipole moment, the oscillations of which produce the corresponding radiation fields. (In the case of the magnetic dipole, use is made of the fact that $\dot{\boldsymbol{r}}=\mathrm{i}\omega\boldsymbol{r}$.) Thus we recognise a one-to-one correspondence between quantum theory and classical physics with regard to the radiation properties of atoms and classical sources, just as the Bohr correspondence principle requires.

The radiation fields of electric and magnetic dipoles are shown in Figs. 16.6 and 16.7. The theory sketched above, in which the radiation field is treated classically, according to the Maxwell equations, but the "source term" is introduced as a quantum mechanical expectation value, is often referred to in the literature as the "semiclassical" theory.

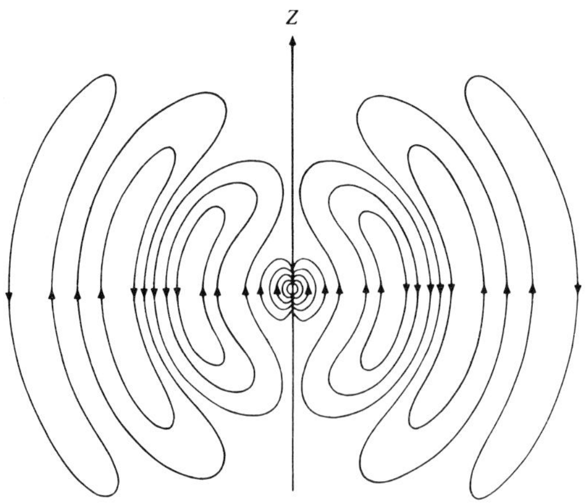

Fig. 16.6. Radiation field for electric dipole radiation. The dipole oscillates in the z direction. The lines connect elements of the electric field $\boldsymbol{E}$ which have the same phase. The radiation field of magnetic dipole radiation is formally the same as that for electric dipole radiation if we replace the electric dipole moment $\boldsymbol{p}$ by the magnetic dipole moment μ and simultaneously make the substitution $\boldsymbol{E} \rightarrow \boldsymbol{B}$ and $\boldsymbol{B} \rightarrow \boldsymbol{E}$

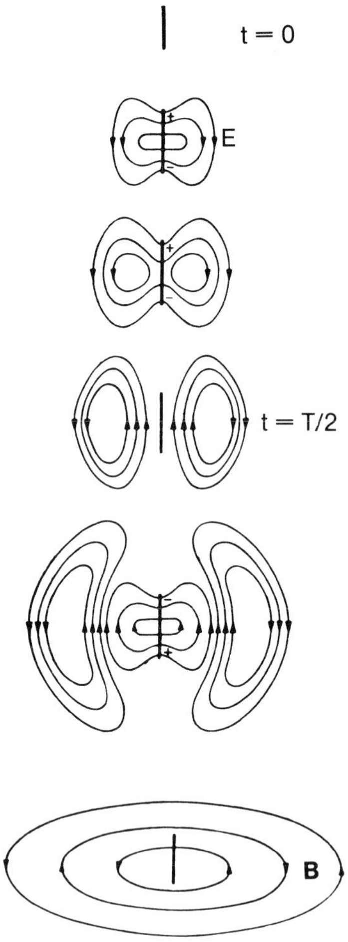

Fig. 16.7. Visualisation of the time dependence of the radiation of a Hertzian dipole. T is the period of oscillation

16.2 Linewidths and Lineshapes

As we showed in the Schrödinger theory, electrons can assume certain energy states in the atom. If an electron is in an excited state, it can go from there to an energetically lower state by emitting a quantum of light. The result of this is that the lifespan of the excited state is no longer infinite. In classical electrodynamics, it is shown that the energy of a Hertzian oscillator (= oscillating dipole) decays exponentially with time. Measurements on excited atoms show (Fig. 16.8) that their radiation intensity also decays exponentially. Such behaviour is to be expected, according to the correspondence principle (Sect. 8.11). In fact, the quantum mechanical treatment of radiation, which cannot be given here in detail, shows that the number N of excited atoms decreases according to

$$N = N_0 \, e^{-2\gamma t} . \tag{16.65}$$

$1/(2\gamma)$ is the time t_0 in which the number N decays to N/e. t_0 is called the lifetime of the state. In the sense of the statistical interpretation of quantum mechanics, (16.65) refers

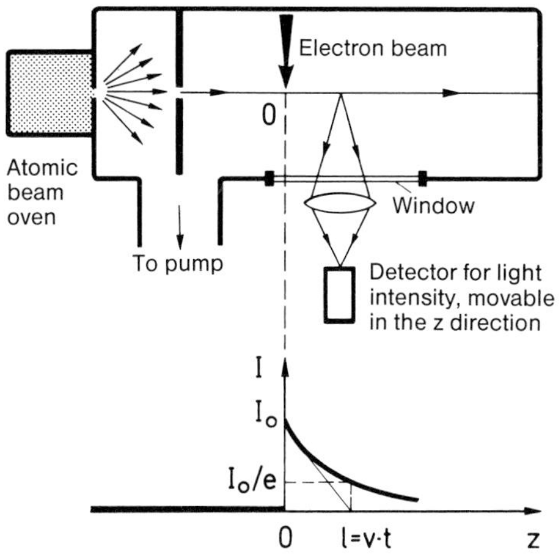

Fig. 16.8. The decay times of excited atomic states can be measured, for example with the arrangement shown here. Atoms in a neutral beam are lifted into defined excited states by irradiation with electrons of the appropriate kinetic energy or with laser light. A movable device for measuring radiation intensity along the line of flight is used. The velocity of the atoms is known, so that the radiative decay of the excited state during the flight through the chamber can be used to calculate the decay time, or lifetime, of the excited state

to the quantum mechanical average which describes the behaviour of many atoms, i.e. of the ensemble. For a single atom, the emission occurs at a completely random time. In analogy to (16.65), the amplitude of the emitted light decreases exponentially. The radiation field amplitude F therefore has the form (Fig. 16.9) in complex notation

$$F(t) = F_0(e^{-\gamma t + i\omega_0 t} + \text{c.c.}) , \quad t > 0 . \tag{16.66}$$

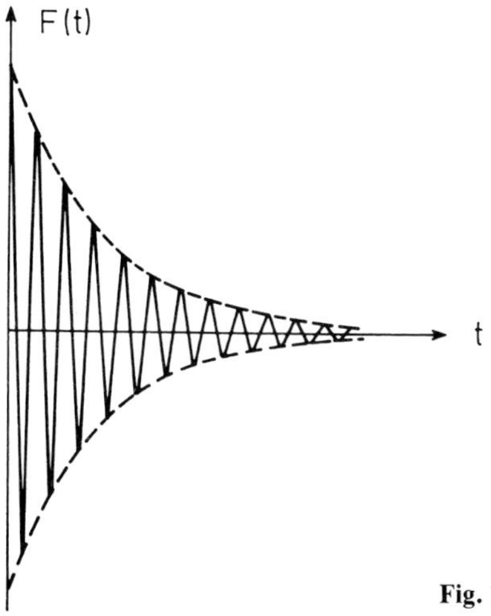

Fig. 16.9. Exponential decay of the radiation field amplitude F (16.66)

Here ω_0 is given by $(E_o - E_f)/\hbar$, where E_o and E_f are the energies of the initial and final atomic states, respectively. F_0 is a real amplitude.

We can imagine that the excitation occurs at time $t = 0$, so that for $t < 0$ there is no light wave. If we use a spectrometer to examine the emitted light, we find that this light is composed of monochromatic components, i.e. waves of the form

$$c(\omega)\, e^{i\omega t} \quad \text{(in complex notation)} ,$$

where $\omega = 2\pi c/\lambda$ (λ: wavelength, c: velocity of light). The radiation field amplitude (16.66) can be represented as a superposition of such waves:

$$F(t) = \frac{1}{2\pi} \int_{-\infty}^{\infty} c(\omega)\, e^{i\omega t} d\omega \,. \qquad (16.67)$$

This decomposition is known in mathematics as a Fourier transformation. It is shown in Fourier theory that the Fourier coefficients $c(\omega)$ are given by

$$c(\omega) = \int_{-\infty}^{+\infty} F(t)\, e^{-i\omega t} dt \,. \qquad (16.68)$$

The intensity of monochromatic light with frequency ω is then given by

$$|c(\omega)|^2 \,. \qquad (16.69)$$

For an exponentially decaying light field (16.66), the spectral distribution is thus

$$c(\omega) = -F_0 \left(\frac{1}{i(\omega_0 - \omega) - \gamma} + \frac{1}{i(-\omega_0 - \omega) - \gamma} \right) . \qquad (16.70)$$

Since $(\omega_0 - \omega) \ll (\omega_0 + \omega)$ and $\gamma \ll (\omega_0 + \omega)$, $(\omega > 0)$, the second term in (16.70) is much smaller than the first and can be neglected. The intensity distribution is then

$$|c(\omega)|^2 = F_0^2 \frac{1}{(\omega - \omega_0)^2 + \gamma^2} \,. \qquad (16.71)$$

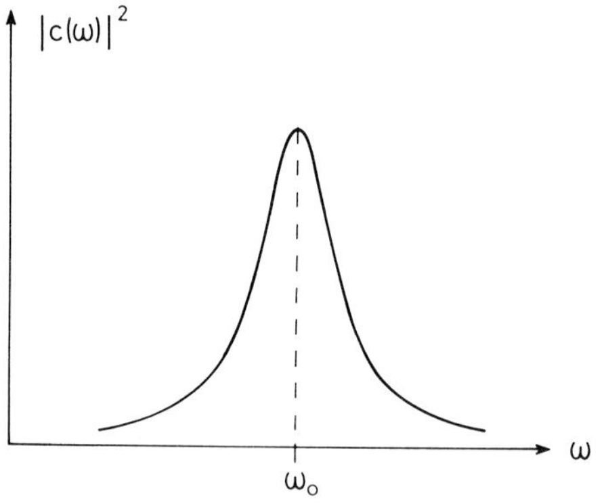

Fig. 16.10. Lorentzian line

A spectral line with the intensity distribution given in (16.71) and shown in Fig. 16.10 is called a Lorentzian line. Its width at half height is given by the atomic decay time t_0 according to $\gamma = 1/(2t_0)$. The linewidth introduced here is also called the *natural* linewidth. Typical numerical values are $t_0 \approx 10^{-8} - 10^{-9}$ s (10^{-8} s $\triangleq 5 \cdot 10^{-4}$ cm^{-1} $\triangleq$ 15 MHz), and thus $\gamma = 10^8 - 10^9$ s^{-1}.

It is intuitively reasonable that the electron orbitals of gas atoms can also be disturbed by collisions between the atoms. This constantly causes changes in the resulting light emissions, resulting in a line broadening which is called collision broadening. If atoms are incorporated into solids, they interact constantly with the lattice oscillations, which again disturb the electron orbitals and lead to line broadening. In these cases, the atoms are all completely identical, and the resulting line broadening is "*homogeneous*". "*Inhomogeneous line broadening*" occurs when individual atoms, which were originally identical, become distinguishable through additional physical conditions. For example, atoms in solids can occupy different types of positions within the lattice, so that the energies of individual electrons are differently displaced. These displacements often lie along a continuum and their intensities assume a Gaussian distribution.

Another example of inhomogeneous line broadening is *Doppler broadening* in gases. According to the Doppler principle, the frequency of light emitted by an atom moving toward the observer with velocity v differs from that emitted by an atom at rest by

$$\omega = \omega_0(1 - v/c)\,, \tag{16.72}$$

where c is the velocity of light. The frequency is increased when the atoms are moving toward the observer, and decreased when they are moving away. Let us now imagine a gas at thermal equilibrium. According to the Boltzmann distribution (2.8) the number of gas atoms whose velocity components v *in the direction of the observed light wave* lie in the interval $(v, v+dv)$ is given by

$$n(v)\,dv = N\left(\frac{2kT}{\pi m_0}\right)^{1/2} \exp\left(-\frac{m_0 v^2}{2kT}\right) dv\,. \tag{16.73}$$

N is the total number of atoms, k is the Boltzmann constant, T the absolute temperature and m the atomic mass. Since, according to (16.72), a frequency displacement is associated with the velocity v, we obtain an intensity distribution given by

$$I(\omega) = \text{const} \cdot \exp\left(-\frac{m_0 c^2(\omega_0 - \omega)^2}{2kT\omega_0^2}\right), \tag{16.74}$$

as is shown in Fig. 16.11. The total linewidth at half height is given by

$$\varDelta\omega_{\mathrm{D}} = \frac{2\omega_0}{c}\left(2\ln 2 \cdot \frac{kT}{m_0}\right)^{1/2}\,. \tag{16.75}$$

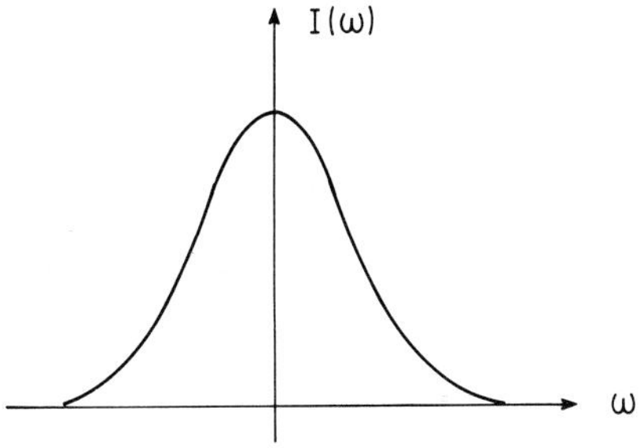

Fig. 16.11. Gaussian line shape

According to (16.75), the Doppler width of a spectral line, $\Delta\omega_D = 2\pi\Delta\nu_D$ is thus proportional to both the frequency and the square root of the temperature. For the yellow D line of sodium, at $T = 500$ K, the Doppler width is $\Delta\nu_D = 1700$ MHz, or $(1/\lambda)_D = 0.056\ \mathrm{cm}^{-1}$. For optical spectral lines, this width is generally significantly greater than the natural linewidth γ [see (16.71)].

In contrast, the Doppler broadening of microwave or radiofrequency transitions between excited atomic states (which can be studied by double resonance methods, see Sect. 13.3.7) is generally smaller than the natural linewidth. At frequencies $\nu < 10^{10}$ Hz, the Doppler width at the same temperature is, according to (16.75), $< 10^4$ Hz. The spectral resolution of such double resonance methods is thus no longer limited by Doppler broadening.

Doppler broadening also occurs in atoms which are incorporated into solids and are vibrating there at high temperatures.

As we saw in Chaps. 8 and 12, there are a number of interesting line shifts and splittings. In order to measure these exactly, the linewidth must be small compared to these shifts or splittings. Therefore, in Chap. 22 we shall discuss methods by which line broadening, and in particular Doppler broadening, can be avoided, thus permitting the experimenter to carry out Doppler-free spectroscopy. A simple method of reducing the Doppler broadening consists of cooling the source of the atomic beam which is being investigated. In order to reduce collisional broadening, one must ensure that the mean time between two collisions of one atom with other atoms is longer than the mean lifetime of the excited state. This is done by reducing the pressure in the experimental apparatus.

17. Many-Electron Atoms

17.1 The Spectrum of the Helium Atom

The simplest many-electron atom is the helium atom. In the ground state, its two s electrons exactly fill the innermost shell with the principal quantum number $n=1$. There is no room for more electrons in this shell, as we shall see below.

In the excited state, one electron remains behind in the half-filled first shell, while the other is excited into a higher shell. Therefore we have

electron 1	in state $n=1$	$l=0$,	and
electron 2	in state $n>1$,	$l=0\ldots n-1$.	

In the previous chapters, we treated the spectra of atoms in which the quantum numbers of only *one* electron were sufficient to characterise the terms. Other electrons were either not present – as in the H atom – or they were all in so-called closed shells or closed subshells. That means, as we shall also see presently, that they make no contribution to the total angular momentum or magnetic moment of the atom.

The experimentally derived term scheme for helium (Fig. 17.1) is similar in some respects to those of the alkali atoms. It differs from them, however, in that there are two term systems which do not combine with each other, as if there were two kinds of helium atoms: a singlet and a triplet system. These names come from the fact that in the singlet system, all the terms are single, while in the triplet system, they are generally split into triplets.

What is important for practical applications, such as gas discharges and lasers, is that both the lowest state of the triplet system (2^3S in Fig. 17.1) and the second-lowest state of the singlet system (2^1S in Fig. 17.1) are metastable in the helium atom. "Metastable" means that the lifetime of the system is long compared to 10^{-8} s, which is the usual lifetime of a state which can be emptied by an allowed optical transition. An atom which is excited to one of these states can thus radiate its energy of about 20 eV only in a time that is long compared to 10^{-8} s.

Helium in the singlet state is also called parahelium. Unlike the alkali atoms, it has no fine structure. All its lines are single. The lowest term is given the symbol $1\,^1S$. Here the first 1 stands for the principal quantum number, the superscript 1 for the multiplicity (here singlet), and the letter S for the total orbital angular momentum, which in this case is zero. The higher terms are $2\,^1S$, $2\,^1P$, $3\,^1S$, $3\,^1P$, $3\,^1D$, etc. From the lack of fine structure, one can conclude that the spins of the two electrons are antiparallel, and add vectorially to a total spin $S=0$. The same holds for the magnetic moment, $\mu_S=0$. Here the upper case letters indicate quantum states which result from the coupling of many (here 2) electrons.

Triplet helium, which in contrast to the singlet system has fine structure, is called orthohelium. Its lowest level is $2\,^3S$. Here the 2 stands for the excited electron with

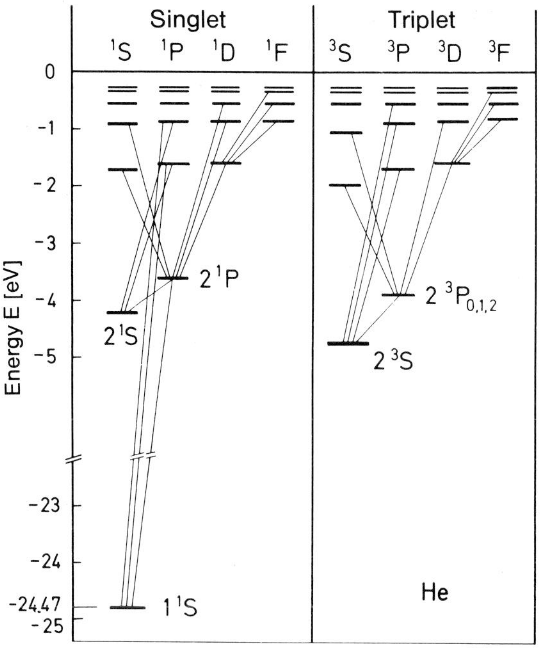

Fig. 17.1. Term scheme of the He atom. Some of the allowed transitions are indicated. There are two term systems, between which radiative transitions are forbidden. These are the singlet and triplet systems. The transitions in the singlet system span an energy range of 25 eV, while those in the triplet system span only 5 eV

$n = 2$, the superscript 3 for the multiplicity (triplet), and the letter S for $L = 0$. Unfortunately, both the total spin quantum number and a term with the total orbital angular momentum $L = 0$ are represented by the same letter, S. The reader must be aware of this difference.

In this system, the spins of the two electrons are parallel to each other. The quantum number of the total spin is $s_1 + s_2 = S = 1$. The magnetic moment $\boldsymbol{\mu}_S = \boldsymbol{\mu}_{s_1} + \boldsymbol{\mu}_{s_2}$ is different from zero. The resulting total spin has three possible orientations with respect to an internal magnetic field $\boldsymbol{B}_l$ which is coupled to the orbital angular momentum of the electrons. The spin-orbit coupling resulting from this leads to a triple fine structure splitting of the terms which have non-vanishing total orbital angular momentum.

The spectrum of parahelium lies mostly in the ultraviolet, while that of orthohelium is in the infrared and visible. Combinations between the systems are not observed, i.e. there are no optical transitions between the singlet and triplet systems. If one compares corresponding configurations in the two non-combining systems, one discovers considerable differences in energy, especially for the low quantum numbers. The $2\,^1S_0$ state lies about 0.8 eV above the $2\,^3S_1$, and the $2\,^1P_1$ state about 0.25 eV higher than the $2\,^3P_2$ (the notation in this sentence is explained in Sect. 17.3.2). This energy difference is a result of the differences in the electrostatic interactions of the two electrons in parallel and antiparallel spin orientations. It is also called the symmetry energy, because it arises from the difference in the average distance between two electrons with symmetrical and antisymmetrical wavefunctions (see Sect. 17.2). For P states, this can also be seen in Fig. 17.9.

17.2 Electron Repulsion and the Pauli Principle

Compared to the one-electron atom, the new factor in the helium atom is the repulsive interaction of the two electrons. For the total binding energy of the He atom, one must therefore write

$$E = -\underbrace{\frac{Ze^2}{4\pi\varepsilon_0 r_1}}_{\substack{\text{Nucleus – electron 1}\\ \text{attraction at}\\ \text{distance } r_1}} - \underbrace{\frac{Ze^2}{4\pi\varepsilon_0 r_2}}_{\substack{\text{Nucleus – electron 2}\\ \text{attraction at}\\ \text{distance } r_2}} + \underbrace{\frac{e^2}{4\pi\varepsilon_0 r_{12}}}_{\substack{\text{Electron 1 – electron 2}\\ \text{repulsion at mutual}\\ \text{distance } r_{12}}} . \tag{17.1}$$

The repulsion energy of the two electrons naturally depends on the n, l states they occupy, because the spatial distribution of the electrons depends on the quantum numbers. This repulsion energy thus lifts the l degeneracy to a considerable degree.

The Schrödinger equation of this relatively simple two-electron problem cannot be solved exactly. As a first approximation, in the independent-particle model, one neglects the third term and sets the total energy equal to the sum of two H-atom terms. The binding energy is then

$$E = -\left(\frac{RhcZ^2}{n^2}\right)_1 - \left(\frac{RhcZ^2}{n^2}\right)_2 \tag{17.2}$$

where the indices 1 and 2 refer to the two electrons. One would then expect the energy of the ground state to be

$$E_{\mathrm{He}} = 2\cdot(-54.4)\,\mathrm{eV} = -108.8\,\mathrm{eV}\,.$$

The experimental value is distinctly different from this. The total work of removing both electrons is 79 eV, 24.6 eV for removing the first electron (ionisation of the He to the singly charged positive ion He^+) and 54.4 eV for the removal of the second electron (ionisation of the singly charged He^+ to the doubly charged positive ion He^{2+}). The second value is the same as one would expect from a comparison with the hydrogen atom. There the ionisation energy is 13.6 eV. For helium, one would expect the energy to be four times as great, because the nucleus is doubly charged. The work of removing the first electron, however, is much smaller. The model for the binding energy must therefore be refined by taking into account the energy of the interaction of the two electrons. We shall present an approximation process in Sect. 19.4.

The observation that the helium atom has a $1\,^1S$ state but not a $1\,^3S$ state was the starting point for the Pauli principle (*Pauli* 1925). In its simplest form, it says:

The electronic states of an atom can only be occupied in such a way that no two electrons have exactly the same set of quantum numbers.

The electrons must therefore differ in at least one quantum number. In addition to the orbital quantum numbers n, l and m_l, the spin quantum number s or m_s is taken into account here. In the 1 ^3S configuration, both electrons would have exactly the same quantum numbers, as we shall show in detail below. For more details see Sect. 19.4.

This principle is the generalisation of the previously stated empirical rule for all atoms with more than one electron: There is always a unique ground state which has the lowest principal quantum number. It has the lowest multiplicity which is compatible with the principal quantum number.

17.3 Angular Momentum Coupling

17.3.1 Coupling Mechanism

We have already learned that in the one-electron system, the individual angular momenta $\boldsymbol{l}$ and $\boldsymbol{s}$ combine to give a resultant angular momentum $\boldsymbol{j}$. There is a similar coupling between the angular momenta of different electrons in the same atom. We have already inferred, as an important result of empirical spectral analysis, that the total angular momentum of filled shells is equal to zero. This follows directly from the observation that the ground state of all noble gas atoms is the $^1S_0$ state. In calculating the total angular momentum of an atom, it is therefore necessary to consider only the angular momenta of the valence electrons, i.e. the electrons in non-filled shells. These angular momenta are coupled by means of magnetic and electric interactions between the electrons in the atom. They combine according to specific quantum mechanical rules to produce the total angular momentum $\boldsymbol{J}$ of the atom. These quantum rules are those which have already been discussed. The vector model provides insight into the composition of the angular momentum. There are two limiting cases in angular momentum coupling: the LS or Russell-Saunders coupling, and jj coupling.

17.3.2 *LS* Coupling (Russell-Saunders Coupling)

If the interactions $(\boldsymbol{s}_i \cdot \boldsymbol{l}_i)$ between the spin and orbital angular momenta of the individual electrons i are smaller than the mutual interactions of the orbital or spin angular momenta of different electrons [coupling $(\boldsymbol{l}_i \cdot \boldsymbol{l}_j)$ or $(\boldsymbol{s}_i \cdot \boldsymbol{s}_j)$], the orbital angular momenta $\boldsymbol{l}_i$ combine vectorially to a total orbital angular momentum $\boldsymbol{L}$, and the spins combine to a total spin $\boldsymbol{S}$. $\boldsymbol{L}$ couples with $\boldsymbol{S}$ to form the total angular momentum $\boldsymbol{J}$.

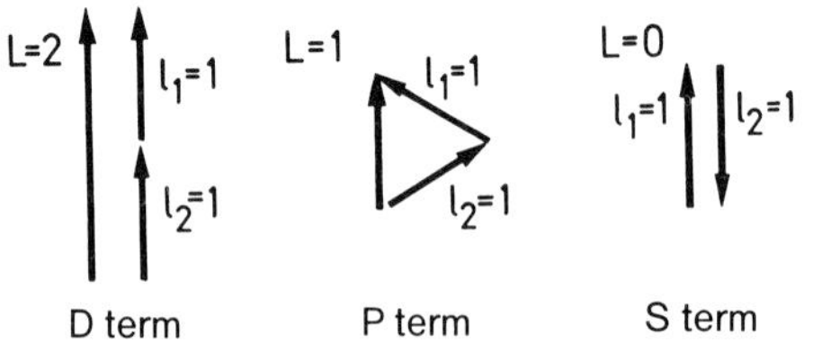

Fig. 17.2. Coupling of the orbital angular momenta of two electrons l_1 and l_2 to the total angular momentum L. Coupling to D, P and S, corresponding to $L = 2, 1$ and 0

For a two-electron system like the He atom, the resulting behaviour is shown in Fig. 17.2. The orbital angular momentum $\boldsymbol{L}$ of the atom is the sum of the two electron orbital angular momenta

$$\boldsymbol{L} = \boldsymbol{l}_1 + \boldsymbol{l}_2 . \tag{17.3}$$

For the absolute value of L it again holds that $|\boldsymbol{L}| = \sqrt{L(L+1)}\,\hbar$ with the quantum number L, for which the following values are possible:

$$L = l_1 + l_2, l_1 + l_2 - 1 \dots l_1 - l_2, \quad \text{where} \quad l_1 \geqq l_2 .$$

The quantum number L determines the term characteristics:

$L = 0, 1, 2 \dots$ indicates $S, P, D \dots$ terms .

It should be noted here that a term with $L = 1$ is called a P term but this does not necessarily mean that in this configuration one of the electrons is individually in a p state.

For optical transitions the following selection rules hold:

$\Delta l = \pm 1$ for the single electron ,

$\Delta L = 0, \pm 1$ for the total system .

$\Delta L = 0$ means here that the quantum states of two electrons change simultaneously, and in opposite directions. This is only possible when the coupling is strong, which is the case in heavy atoms.

Furthermore, for the total spin,

$$\boldsymbol{S} = \boldsymbol{s}_1 + \boldsymbol{s}_2 \quad \text{with} \quad |\boldsymbol{S}| = \sqrt{S(S+1)}\,\hbar . \tag{17.4}$$

The total spin quantum numbers S can take one of two values here, either

$$S = 1/2 + 1/2 \quad \text{or} \quad S = 1/2 - 1/2 , \quad \text{i.e.} \quad S = 0 \quad \text{or} \quad S = 1 .$$

The selection rule for optical dipole radiation is $\Delta S = 0$. This means that combinations between states with different total spins are not allowed, or in other words, that spin flipping is not associated with optical dipole radiation.

Finally, the interaction between $\boldsymbol{S}$ (or its associated magnetic moment μ_S), and the magnetic field $\boldsymbol{B}_L$, which arises from the total orbital angular momentum $\boldsymbol{L}$, results in a coupling of the two angular momenta $\boldsymbol{L}$ and $\boldsymbol{S}$ to the total angular momentum $\boldsymbol{J}$:

$$\boldsymbol{J} = \boldsymbol{L} + \boldsymbol{S} , \quad |\boldsymbol{J}| = \sqrt{J(J+1)}\,\hbar . \tag{17.5}$$

The quantum number J can have the following values:

for $S = 0$, $J = L$,

for $S = 1$, $J = L + 1, L, L - 1$,

in which case all terms are triplets.

The individual angular momenta are combined according to exactly the same quantum rules with which we became acquainted in Chap. 13.

In the general case of a several-electron system, there are $2S + 1$ possible orientations of S with respect to L, i.e. the multiplicity of the terms is $2S + 1$ (if $S < L$). As an example, Fig. 17.3 shows the possible couplings in the case of $S = 1$, $L = 2$.

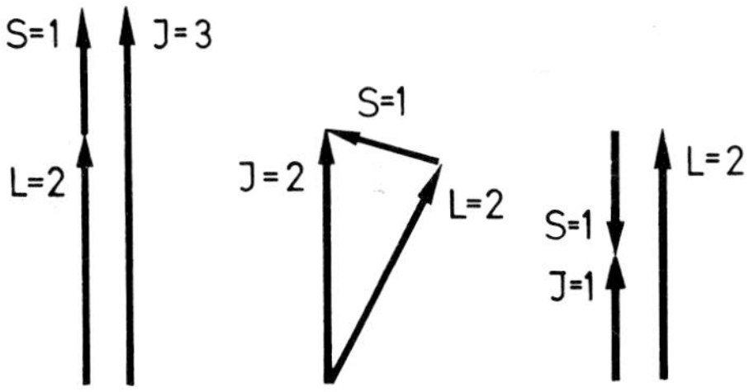

Fig. 17.3. Combination of spin (S) and orbital (L) angular momenta to form the total (J) angular momentum. If $L=2$ and $S=1$, J can assume the values 3, 2 or 1

Using the helium atom as an example, we shall show once more which atomic terms can be derived from the given electron configurations. The lowest states of the helium atom have the following term symbols and quantum numbers:

– If both electrons are in the lowest shell, the electron configuration is $1s\ 1s$, or in the common notation $1s^2$. They then have the following quantum numbers:

$$n_1 = n_2 = 1\,, \quad l_1 = l_2 = 0\,, \quad s_1 = 1/2\,, \quad s_2 = 1/2\,.$$

The resulting quantum numbers for the atom are then either

$$L = 0\,, \quad S = 0 \quad \text{for} \quad m_{s_1} = -m_{s_2}\,, \quad J = 0\,,$$

the singlet ground state $^1S_0$; or

$$L = 0\,, \quad S = 1 \quad \text{for} \quad m_{s_1} = m_{s_2}\,, \quad J = 1\,,$$

the triplet ground state $^3S_1$. Only the singlet ground state is actually observed. The triplet ground state is forbidden by the Pauli Principle, because in it the two electrons would not only have the same quantum numbers n and l, but they would also have the same spin orientation m_s.

– However, when one electron remains in the shell with $n = 1$, while the other is raised into the state with $n = 2$, that is in the electron configuration $1s\,2s$, we have the following quantum numbers:

$$n_1 = 1\,, \quad n_2 = 2\,, \quad l_1 = l_2 = 0\,, \quad s_1 = 1/2\,, \quad s_2 = 1/2\,.$$

This yields either $L = 0$, $S = 0$, $J = 0$, the singlet state $^1S_0$; or $L = 0$, $S = 1$, $J = 1$, the triplet state $^3S_1$. Both states are allowed and are observed.

– In the same way, the states and term symbols can be derived for all electron configurations. More on this subject is to be found in Sect. 19.1.

We finally arrive at the term scheme with the allowed optical transitions by taking into account the selection rules. Intercombination lines with $\Delta S = 1$ are forbidden, because there is no spin flipping with optical dipole radiation. This is the reason for the existence of non-combining term systems like that in Fig. 17.1.

The complete nomenclature for terms or energy states of atoms, which we have already used, is then

$$n^{2S+1}L_J\,.$$

One first writes the principal quantum number n of the most highly excited electron, which is called the *valence* electron. The superscript is the multiplicity $2S+1$. It is followed by the alphabetical symbol S, P, D... for the total orbital angular momentum L; the subscript to this symbol is the quantum number J for the total angular momentum of the atom.

For many-electron systems, this must be expanded over the possible multiplicities as shown below:

For two electrons	$S=0$ Singlet	$S=1$ Triplet		
For three electrons	$S=1/2$ Doublet	$S=3/2$ Quartet		
For four electrons	$S=0$ Singlet	$S=1$ Triplet	$S=2$ Quintet	
For five electrons	$S=1/2$ Doublet	$S=3/2$ Quartet	$S=5/2$ Sextet	etc.

According to the rules for coupling of angular momenta, which have been treated in this section, the possible atomic states can now be easily derived from known electron configurations. We shall only explain a simple example here. In the configuration

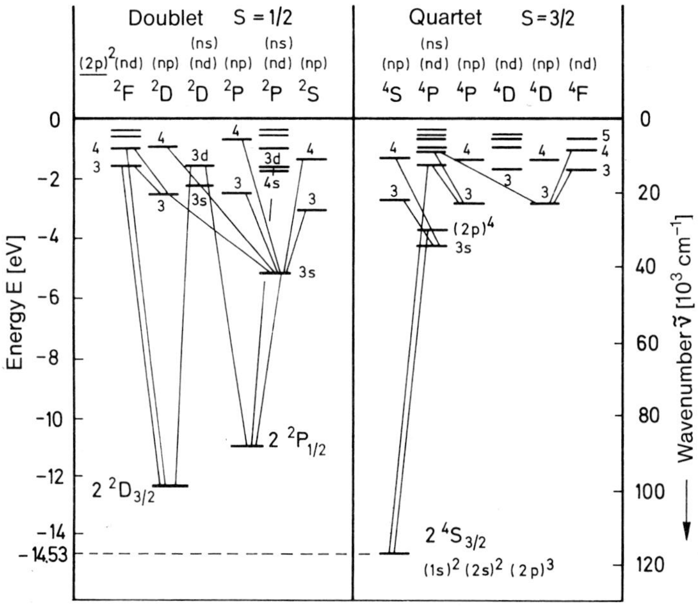

Fig. 17.4. Term diagram for the nitrogen atom (only simple terms, no J splitting). Nitrogen as a doublet and a quartet system. The electronic configuration of the valence electrons is given at the top

Fig. 17.5. Term diagram for the carbon atom (only simple terms, no J splitting). Carbon has a singlet and a triplet system. The configuration of the valence electrons is given at the top ▶

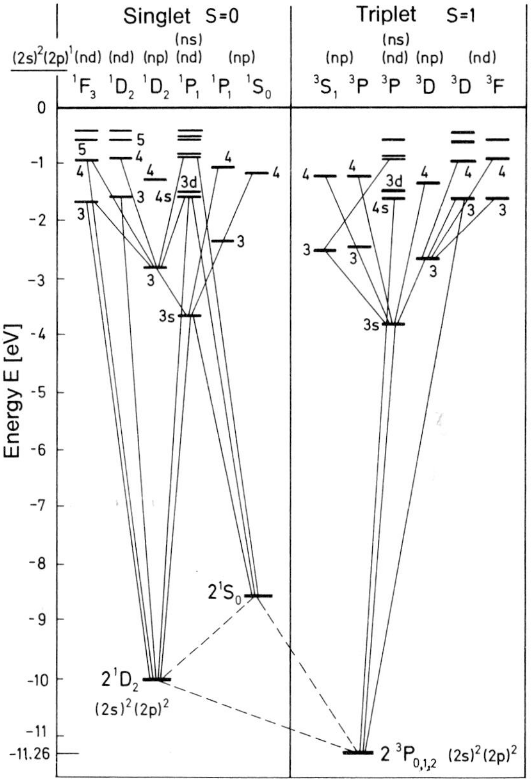

$ns\,n's$, i.e. with two non-equivalent s electrons, the total orbital momentum is necessarily $L=0$, because $l_1=l_2=0$. The spins of the two electrons can be either parallel or antiparallel. This means that the total spin quantum number $S=1$ or $S=0$. In either case, the total angular momentum $J=S$. The possible terms of the configuration $ns\,n's$ are therefore the triplet term $^3S_1$ and the singlet term $^1S_0$.

We shall discuss other examples in Chap. 19, when we know somewhat more about the energetic order of these possible terms.

To make the above somewhat more clear, Figs. 17.4 and 17.5 show the Grotrian diagrams for the nitrogen and carbon atoms. Here, only those terms are taken into account which result from the excitation of a single electron, the so-called valence electron. Of the remaining electrons, only those need be considered which are in non-closed subshells. The remaining electrons form the spherically symmetric core of the electronic shell. The terms and term symbols resulting from LS coupling and from the quantum states of the individual electrons can now be immediately understood, and can be derived from the electron configuration, which is also given. We shall return to these many-electron atoms in Chap. 19.

17.3.3 jj Coupling

The second limiting case for coupling of electron spin and orbital angular momenta is the so-called jj coupling, which occurs only in heavy atoms, because the spin-orbit coupling for each individual electron increases rapidly with the nuclear charge Z.

In jj coupling, the spin-orbit interaction $(\boldsymbol{l}_i \cdot \boldsymbol{s}_i)$ for a single electron is large compared to the interactions $(\boldsymbol{l}_i \cdot \boldsymbol{l}_j)$ and $(\boldsymbol{s}_i \cdot \boldsymbol{s}_j)$ between different electrons. This type of coupling is shown schematically in Fig. 17.6b. For comparison, LS coupling is shown in Fig. 17.6a.

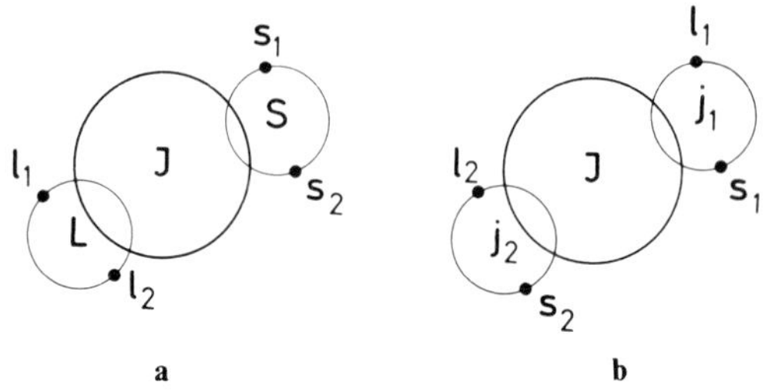

Fig. 17.6. a) Schematic representation of LS coupling between two electrons. **b)** jj coupling between two electrons. In jj coupling, the angular momenta L and S are not even defined

In jj coupling, the angular momenta of individual electrons couple according to the pattern

$$\boldsymbol{l}_1+\boldsymbol{s}_1 \rightarrow \boldsymbol{j}_1\,, \quad \boldsymbol{l}_2+\boldsymbol{s}_2 \rightarrow \boldsymbol{j}_2\,,$$

and so on to give individual total angular momenta $\boldsymbol{j}$. These then combine vectorially to give the total angular momentum $\boldsymbol{J}$ of the atom. Here $\boldsymbol{J}=\Sigma \boldsymbol{j}_i$ and

$$|\boldsymbol{J}|=\sqrt{J(J+1)}\,\hbar\,.$$

In this type of coupling, the quantum number J arises from a generalised quantum mechanical vector model. A resultant orbital angular momentum $\boldsymbol{L}$ is not defined. There are therefore no term symbols S, P, D, etc. One has to use the term notation (j_1, j_2) etc., where j are the angular momentum quantum numbers of the individual electrons. It can easily be shown that the number of possible states and the J values are the same as in LS coupling.

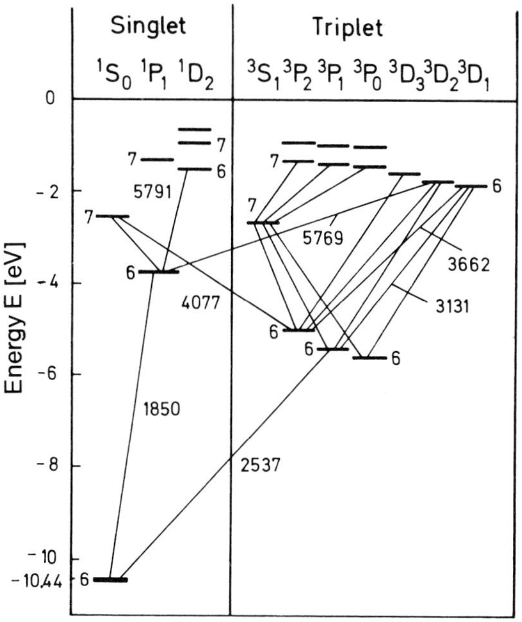

◀ **Fig. 17.7.** Simplified energy diagram of the mercury atom as an example of a heavy atom. The wavelengths [Å] of a few of the more important lines are given. The strongest line in the spectrum of a mercury lamp is the line at 2537 Å which results from the intercombination of the $6\,^1S_0$ and the $6\,^3P_1$ states. Intercombination between terms of different multiplicity is strictly forbidden in light atoms. In heavy atoms it is possible

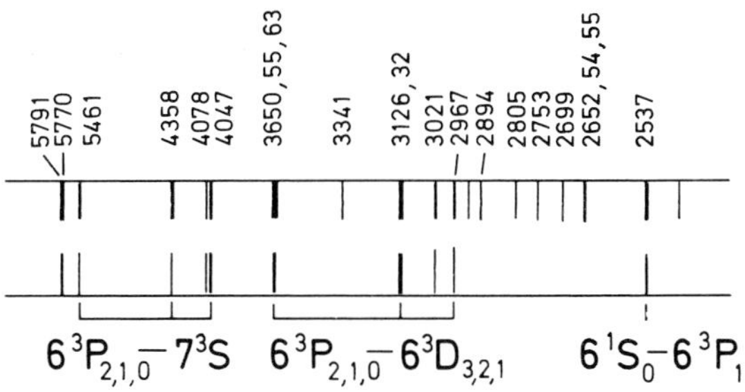

Fig. 17.8. Photographically recorded spectrum of a low-pressure mercury lamp: segment between 2500 and 5800 Å. Due to the superposition of the different series in the spectrum, which are shown in the energy diagram Fig. 17.7, a series structure cannot be *immediately* recognised in the spectrum of a heavy, several-electron atom, as might be inferred from the energy diagram

Purely jj coupling is only found in *very* heavy atoms. In most cases there are intermediate forms of coupling, in such a way, for example, that the intercombination between terms of different multiplicity is not so strictly forbidden. This is called intermediary coupling. The most prominent example of this is the strongest line in the spectrum of mercury high-pressure lamps, $\lambda = 253.7$ nm (compare the energy diagram in Fig. 17.7 and the photograph of the spectrum of a low-pressure mercury lamp in Fig. 17.8). In the frequently used high-pressure mercury lamps, the intensity distribution of the spectrum is different from that of a low-pressure lamp. The 253.7 nm line is relatively most strongly emitted, in addition broadened, and reabsorbed. This is an intercombination line between the singlet and triplet systems. The selection rule for optical transitions is $\Delta J = 0, \pm 1$, and a transition from $J = 0$ to $J = 0$ is forbidden. An example for the transition from LS to jj coupling is shown in Fig. 17.9.

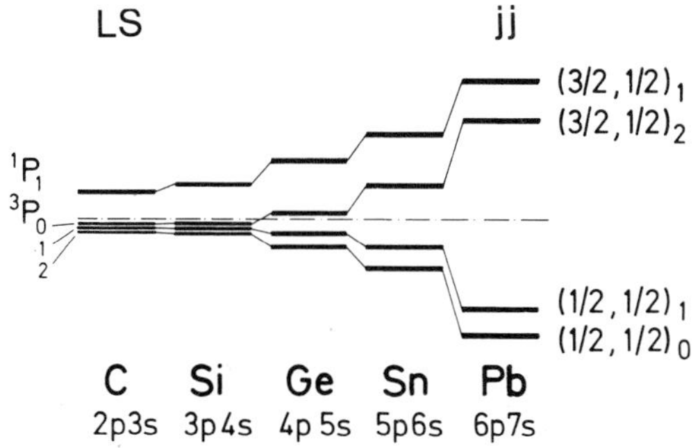

Fig. 17.9. Transition from LS coupling in light atoms to jj coupling in heavy atoms in the series C – Si – Ge – Sn – Pb. In carbon the $^3P_0 - {}^3P_1$ and $^3P_1 - {}^3P_2$ distance are 20 and 40 cm^{-1}, while the $^1P_1 - {}^3P_2$ distance is 1589 cm^{-1}. The term nomenclature is that of LS coupling for carbon, and of jj coupling for lead. The quantum symbols of the two outermost electrons are given. The symbol $(3/2, 1/2)_1$ means $j_1 = 3/2, j_2 = 1/2, J = 1$

17.4 Magnetic Moments of Many-Electron Atoms

Having calculated the total angular momentum of many-electron atoms, we shall now also calculate the total magnetic moment. The treatment exactly follows that of the one-electron system in Chap. 13. In the case of LS coupling, the magnetic moment is composed of

$$\boldsymbol{\mu}_L + \boldsymbol{\mu}_S = \boldsymbol{\mu}_J .$$

Here $\boldsymbol{\mu}_L$ is antiparallel to $\boldsymbol{L}$ and $\boldsymbol{\mu}_S$ is antiparallel to $\boldsymbol{S}$, but because of the different g factors of orbital and spin magnetism, $\boldsymbol{\mu}_J$ and $\boldsymbol{J}$ are not antiparallel to each other. Their directions do not coincide; instead the total moment $\boldsymbol{\mu}_J$ precesses around the direction of $\boldsymbol{J}$. As was mentioned in Chap. 13 and illustrated in Fig. 13.9, the observable magnetic moment is only that component of $\boldsymbol{\mu}_J$ which is parallel to $\boldsymbol{J}$. For this component $(\boldsymbol{\mu}_J)_J$, as was shown in Sect. 13.5,

$$|(\boldsymbol{\mu}_J)_J| = \frac{3J(J+1)+S(S+1)-L(L+1)}{2\sqrt{J(J+1)}} \cdot \mu_\mathrm{B} = g_J\sqrt{J(J+1)}\,\mu_\mathrm{B} \tag{17.6}$$

with the Landé factor

$$g_J = 1 + \frac{J(J+1)+S(S+1)-L(L+1)}{2J(J+1)} . \tag{17.7}$$

In one chosen direction z, the only possible orientations are quantised and they are described by whole or half-integral values of the quantum number m_J, depending on the magnitude of J.

$$(\mu_J)_{J,z} = -m_J g_J \mu_\mathrm{B} \tag{17.8}$$

with

$$m_J = J, J-1, \ldots -J .$$

The contents of this chapter are the quintessence of years of spectroscopic work: measurement of spectra, setting up of term schemes, determination of quantum numbers, and so on. Measurements in magnetic fields have also been an essential tool. If one applies the insights discussed in Chap. 13 to many-electron atoms, one can determine the magnetic quantum numbers of the atomic states from measurements of the splitting of spectral lines. The same considerations which were discussed in Chap. 13 on the behaviour of atoms in magnetic fields apply to many-electron atoms. Here too, the normal and anomalous Zeeman effects and the Paschen-Back effect are important limiting cases. The LS coupling can be broken in sufficiently strong magnetic fields, and in very strong fields, even the jj coupling breaks down.

17.5 Multiple Excitations

Let it be only briefly mentioned here that observed spectra can be made much more complicated by multiple excitation processes, such as those in which several electrons are involved and change their states in the atom. This is especially likely in systems where there is strong mutual interaction between the electrons. In ionisation processes, for example, it is possible that a second electron is simultaneously excited. Excitation energies higher than the ionisation limit can be obtained if a light quantum simultaneously removes one electron (ionisation) and raises a second one to a discrete excitation level. This makes the analysis of heavy-atom spectra much more difficult.

Problems

17.1 The energy levels of helium-like atoms with one electron in the ground state ($n = 1$) and the other in an excited state ($n > 1$) can be expressed as

$$E = -RhcZ^2 - \frac{Rhc(Z-1)^2}{n^2} .$$

This expression is based on the assumption that the ground state electron completely shields one unit of nuclear charge. Discuss the plausibility of the expression. Calculate the energy levels for helium with $n = 2$, 3 and 4 and compare them with the experimental results. Why does the accuracy of the above expression for E increase as n increases?

17.2 Show that the sum $\sum (2J+1)$ over all possible values of J for a given pair of quantum numbers L and S is equal to the product $(2L+1)(2S+1)$. What is the physical meaning of this product?

17.3 Discuss a two-electron system with a $2p$ and a $3d$ electron for the case of jj coupling and show that the number of possible states and their total angular momentum J are the same as in LS coupling.

17.4 a) Ignoring spin-orbit coupling, determine the number of possible terms of an excited carbon atom with the electronic configuration $1s^2 2s^2 2p\, 3d$.

b) Calculate the effective magnetic moment of an atom in the ground state with the configuration $1s^2 2s^2 2p^6 3s^2 4s^2 3d^3$, assuming that L has the largest possible value consistent with Hund's rule (Sect. 19.2) and the Pauli principle.

c) Calculate the ground state of the atoms with the electronic configurations $4d\, 5s^2$ (Y) or $4d^2 5s^2$ (Zr). [The closed shells are not given. L is determined as under b).]

d) The manganese atom ($Z = 25$) has in its ground state a subshell which is exactly half-filled with 5 electrons. Give the electronic configuration and the ground state of the atom.

17.5 a) Calculate the maximum components of the magnetic moments in the direction of the magnetic field for vanadium (^4F), manganese (^6S) and iron (^5D), if beams of these atoms are split into 4, 6 or 9 parts in a Stern-Gerlach experiment.

b) What is the term symbol of the singlet state with a total splitting of $\bar{\nu} = 1.4\ \mathrm{cm}^{-1}$ in a magnetic field $B_0 = 0.5$ tesla?

18. X-Ray Spectra, Internal Shells

18.1 Introductory Remarks

Up to this point, the discussion has been concerned with the energy states and spectra of the most weakly bound electrons. In the lighter atoms, these are usually the outermost or valence electrons. If we now turn to x-ray spectra, we shall be concerned with the energy states of electrons in inner shells. We shall see, however, that x-ray spectra can for the most part be treated as one-electron spectra, although they belong to many-electron atoms.

Historically, it was the x-ray spectra which led to the theory of the shell structure of the atom (*Kossel* 1914). At present, however, we shall assume that the shell structure is familiar as we turn to the x-ray spectra.

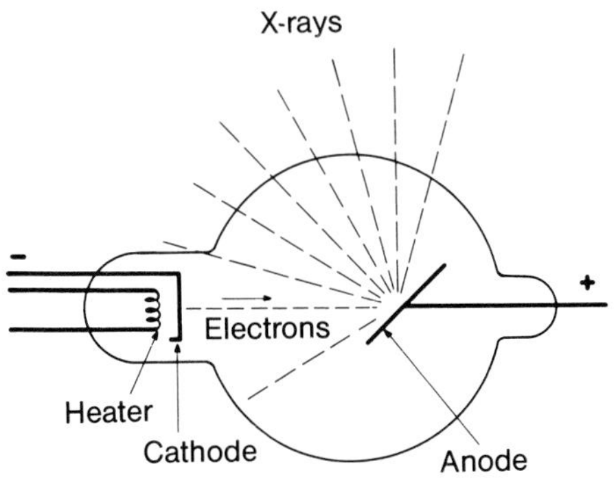

Fig. 18.1. An x-ray tube, schematically

X-rays are usually generated by irradiating an anode, which is often called the anticathode, with fast electrons (Fig. 18.1). The x-rays are detected by photographic plates, film, counting tubes, or more recently, by semiconductor detectors. The latter are made as silicon or germanium diodes. Absorption of x-rays in the "space-charge" zone releases charge carriers. These can be measured as in an ionisation chamber. Pulse-height analysis permits the apparatus to be used both to measure the energy of the x-ray quanta and as a simple spectrometer. For higher resolution spectroscopy and wavelength measurements, one still uses the crystal spectrometer described in Sect. 2.4.5. One can also obtain a rough estimate of the wavelength of x-rays by measuring their "hardness" – their ability to penetrate solids.

18.2 X-Radiation from Outer Shells

By "x-rays", we usually mean electromagnetic radiation (light) which has a wavelength shorter than that of ultraviolet light – though there is no sharp boundary. The range is usually considered to be 0.1 to 10 Å, which corresponds to quantum energies of 1 – 100 keV. The x-ray region is attained according to the series formula (Sect. 8.2)

$$\bar{\nu} = RZ^2(1/n^2 - 1/n'^2)$$

for hydrogen-like atoms, i.e. atoms with only one electron, if the nuclear charge is large enough. For $Z = 20$, the quantum energies are already 400 times as large as the energies of corresponding transitions in the hydrogen atom with $Z = 1$. It is generally not possible to generate such "Balmer series" for highly ionised atoms in the laboratory, but these spectra can be observed in stellar atmospheres.

18.3 X-Ray Bremsstrahlung Spectra

If an anticathode is bombarded with electrons which have passed through an accelerating voltage V_0, x-rays are generated. Spectral analysis of these reveals that

- there is always a continuum, the *x-ray bremsstrahlung* (Fig. 18.2),
- and under certain conditions, there is in addition a line spectrum, the *characteristic spectrum* (Fig. 18.3).

If the intensity is plotted against the frequency rather than the wavelength (as in Fig. 18.2), the bremsstrahlung spectrum for an accelerating voltage V_0 is described, to a good approximation (outside the low-energy range) by

$$I(\nu) = \text{const} \cdot Z(\nu_{max} - \nu) , \qquad (18.1)$$

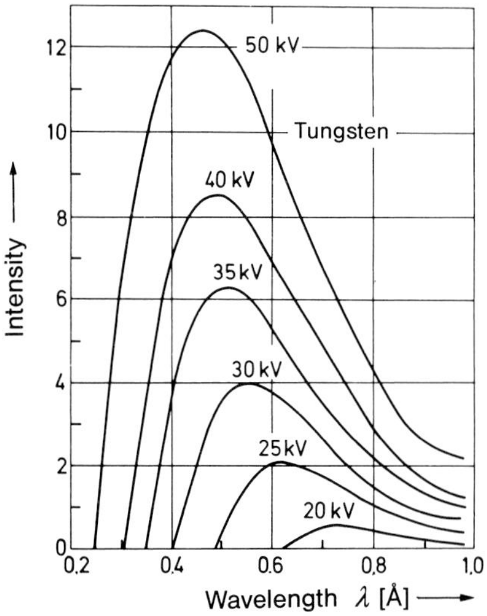

Fig. 18.2. X-ray bremsstrahlung. Spectral energy distribution of the x-rays emitted from a tungsten anticathode at various accelerating voltages for the bombarding electrons. The intensity is given in arbitrary units

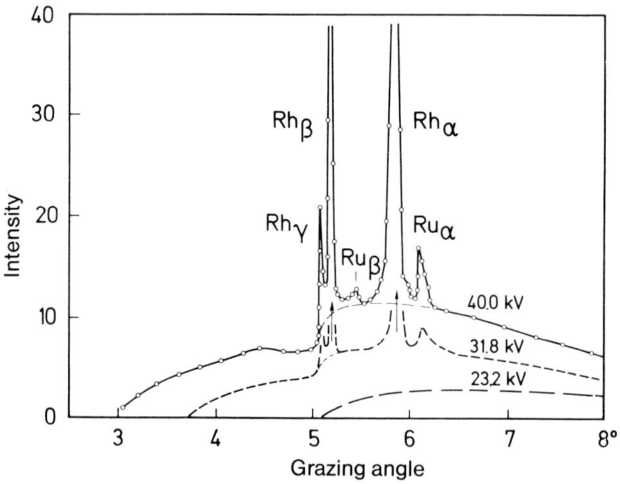

Fig. 18.3. Line spectrum of a Rh anticathode doped with Ru impurity. The lines are superimposed on the bremsstrahlung spectrum. The intensity is plotted against the grazing angle of the crystal spectrometer instead of the wavelength

where I is the intensity of the radiation (energy per time and frequency interval and solid angle) and Z is the atomic number of the anticathode material. The limiting frequency ν_{max} is given by

$$h\nu_{max} = e \cdot V_0 . \tag{18.2}$$

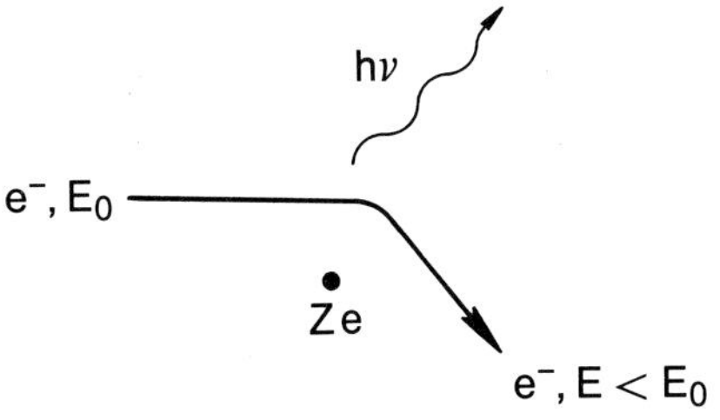

Fig. 18.4. Origin of x-ray bremsstrahlung. An electron with energy E_0 is deflected in the field of a nucleus and slowed down. In the process a quantum of light is emitted

This means that the high-energy or short-wavelength limit of the x-ray spectrum ν_{max} is given by the energy equivalent eV_0. The bremsstrahlung spectrum is a result of the fact that when electrons pass close to the atomic nuclei, they are deflected and slowed down (Fig. 18.4). A positive or negative accelerated charge will, according to classical electrodynamics, emit electromagnetic radiation. This is "white" or continuous x-ray bremsstrahlung. In terms of quantum theory, this can be understood as follows: for each braking incident, a quantum of light $h\nu = E_0 - E$ is emitted. However, since the beginning and end states are not quantised – the electrons are free, not bound – a "white" spectrum arises when there are many individual events.

The reaction equation is

$$\text{Atom} + e^-(\text{fast}) \rightarrow \text{Atom} + e^-(\text{slow}) + h\nu .$$

In the limiting case, the entire energy of the electron is emitted in a single quantum in the course of a single braking event. This x-ray quantum then has the energy $h\nu_{max} = eV_0$. The measurement of this short-wave limit is one method of determining Planck's constant h with great precision. However, one must be careful about such precision measurements, because the work of escaping the solid, and the band structure of the solid, lead to uncertainties or corrections of a few electron volts in the energies at the short-wavelength limit of the bremsstrahlung spectrum.

The spatial distribution of the radiation can also be explained in terms of the classical view of bremsstrahlung. With a thin anticathode, in which multiple events are less probable, and with energies eV_0 which are not too high, the distribution is the same as with the classical Hertzian dipole. The maximum is perpendicular to the direction

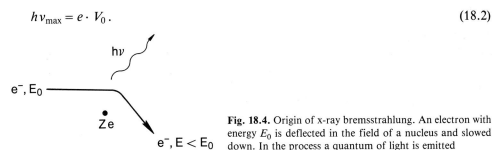

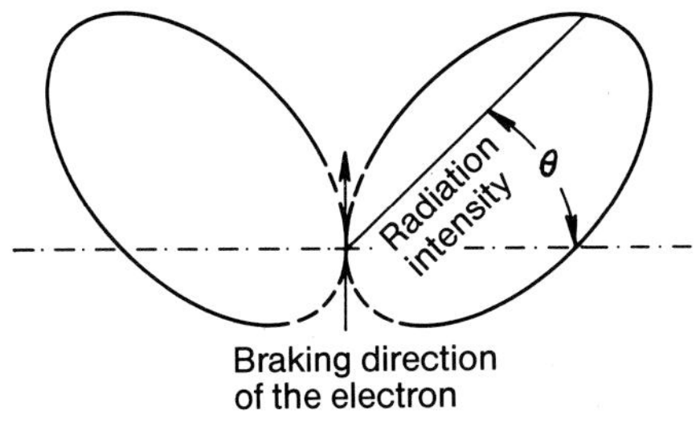

Fig. 18.5. Spatial distribution of the bremsstrahlung

from which the electrons are coming, and thus to the direction in which braking is occurring. The minimum of radiation intensity is in the direction of electron travel (Fig. 18.5). At higher accelerating voltages V_0, the electron's momentum plays a stronger rôle and the radiation pattern folds "ahead", i.e. in the beam direction.

18.4 Emission Line Spectra: Characteristic Radiation

Characteristic radiation consists of a relatively small number of lines. Figure 18.3 gives an example of this. The lines are again grouped into series, which converge to a short-wavelength limit, which is called an "edge". With a rhodium anticathode, for example, one can observe the following lines and series by increasing the accelerating voltage on the electrons in steps:

For accelerating voltages $V_0 > 0.5$ kV, the lines of the *M* series,
For accelerating voltages $V_0 > 3.0$ kV, the *L* series also,
For accelerating voltages $V_0 > 23$ kV, the *K* series as well.

The lines of the *K* series are doublets.

In general it holds for characteristic spectra that while optical spectra contain a large number of lines which depend on the nuclear charge *Z* in a rather complicated way, and which are strongly influenced by chemical bonding, x-ray spectra include a limited number of lines which can be grouped into a few series. There is also a clear relationship to the nuclear charge (Fig. 18.6). Corresponding lines and edges are found at increasing quantum energies as the nuclear charge increases. The series are designated by the letters *K, L, M, N*, ... and the lines within the series by Greek lower case letters beginning with α. The fine structure splitting of the lines is indicated by numbers written as subscripts.

To a good approximation, the first line of the *K* series, the line K_α, can be described for atoms with different nuclear charge numbers *Z* by the expression

$$\bar{\nu}_{K_\alpha} = \tfrac{3}{4}R(Z-1)^2 \equiv R(Z-1)^2(1/1^2 - 1/2^2)\,. \tag{18.3}$$

The first lines of the *L* series (L_α) are described by

$$\bar{\nu}_{L_\alpha} = \tfrac{5}{36}R(Z-7.4)^2 \equiv R(Z-7.4)^2(1/2^2 - 1/3^2)\,. \tag{18.4}$$

A linear relationship between $\sqrt{\nu}$ and the nuclear charge number *Z* for analogous x-ray lines or edges in the spectra of different elements was discovered in 1913 by *Moseley* (Fig. 18.7). Comparison with the Balmer formula for hydrogen suggests that for the *K* line the nuclear charge is screened by one unit of charge, while for the *L* line, it is screened by almost eight units.

Chemical bonding of an atom has only a slight influence on its x-ray spectrum. However, exact measurement of this effect does provide important information about the behaviour of electrons in chemical bonds.

The emission of x-rays can be elicited not only by bombarding an anticathode with electrons, but also by irradiation of atoms, molecules or solids with x-rays. This is called x-ray fluorescence.

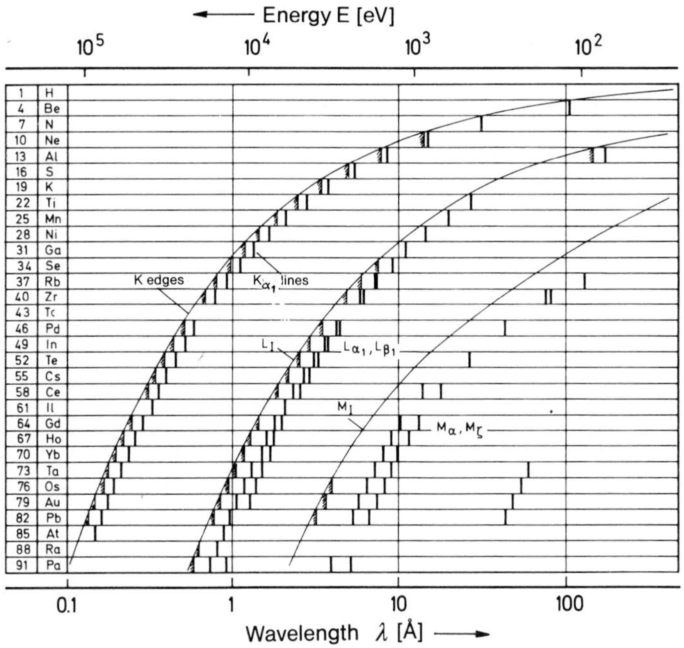

◀ **Fig. 18.6.** Spectral positions of the characteristic x-ray emission lines and the absorption edges of the elements. The quantum energies increase with increasing nuclear charge number

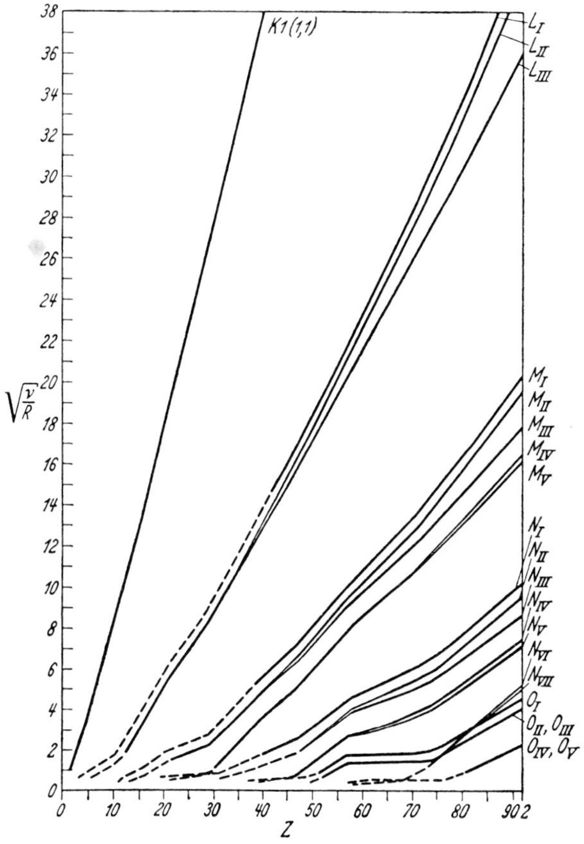

Fig. 18.7. Moseley diagram of the absorption edges. For the edge frequencies ν, $\sqrt{\nu/R} = (Z-s)/n$, where s is the screening number. n and s are different for different shells, and this is why the observed Moseley lines are not parallel. The lines for the spin doublets, e.g. $L_{\text{II},\text{III}}$ diverge at the top because spin-orbit coupling increases with Z. [From K. H. Hellwege: *Einführung in die Physik der Atome*, Heidelberger Taschenbücher, Vol. 2, 4th ed. (Springer, Berlin, Heidelberg, New York 1974) Fig. 72]

The wavelength of the x-radiation is greater than, or at least equal to, that of the exciting light, but other than that, it is independent of the wavelength of the exciting radiation within certain limits. The lines of a series appear in a fluorescence spectrum, and then all of them at once, only when the quantum energy of the exciting radiation is at least as great as the quantum energy of the highest-energy, or shortest-wavelength line in the characteristic spectrum. It is the same with excitation of x-radiation by electron bombardment: the kinetic energy of the elctrons eV_0 must be at least as great as the quantum energy of the shortest-wavelength line of the series before this series appears in the emission spectrum. Thus emission of the K_α line cannot be excited by the quantum energy of K_α; instead it is necessary to supply the energy of the K edge. This is the energy to which the lines of the K series converge, the series limit. From this and other observations, it was concluded that x-ray lines correspond to states of "inner" electrons which are bound in filled shells, in contrast to the more loosely bound outer electrons, which give rise to the optical spectra.

In 1916, *Kossel* interpreted the generation of the x-ray line spectra as follows: first the exciting electron must remove an atomic electron from an inner shell. The resulting hole is filled by outer electrons, and their binding energy is released in the form of characteristic light quanta. All transitions which end on the same inner shell occur together, and form a series (Fig. 18.8).

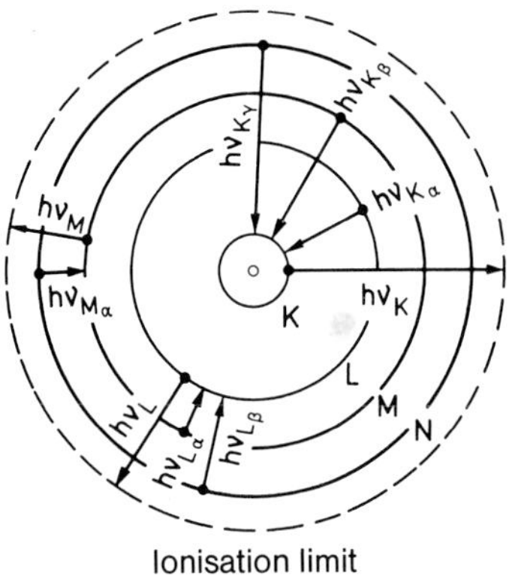

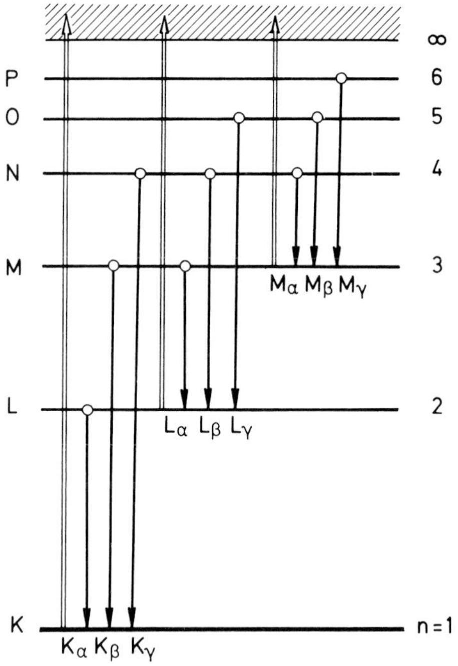

Fig. 18.8. Schematic explanation of the K, L and M series in x-ray spectra. *Left:* An electron hole is formed by ionisation of an inner shell (outward-pointing arrow). This is filled by an electron from a shell wich is farther out. The binding energy is emitted as an x-ray quantum (inward-pointing arrow). *Right:* The same in the form of a term scheme. The ionisation limit is shaded in at the top. Unfortunately, in practice and by historical reasons, the use of the greek letters in the L, M, N... series is less systematic than indicated here, cf. Fig. 18.9

The transitions involving inner shells are much more energetic than those in the outermost shell, because the nuclear charge is shielded only by those electrons in still lower shells. This results in screening to a charge $(Z-1)$ for the K_α lines, and to $(Z-7.4)$ for the L_α lines. The field strength in the interior of a sphere with a uniformly charged surface is zero, so the external electrons make no contribution to the field experienced by the inner ones.

18.5 Fine Structure of the X-Ray Spectra

The x-ray transitions indicated by Greek letters, K_α, K_β, L_α, L_β, etc. thus start from terms with different principal quantum numbers n. To understand the "fine structure" of x-ray spectra, that is the occurrence of several components in a given transition, one must also take into account the orbital angular momentum and spin of the electrons.

For electrons in inner shells, orbital degeneracy (l degeneracy) is naturally lifted. The reason for this, the different degrees of screening for electrons with different orbital angular momenta and the associated differences in the Coulomb potential, has already been discussed in the case of the spectra of alkali atoms (Sect. 11.2). Furthermore, we must also take into account the actual fine structure due to spin-orbit coupling. The energy of this coupling increases rapidly with nuclear charge, as Z^4 (Sect. 12.8). In heavy atoms such as uranium the spin-orbit splitting amounts to as much as 2 keV! One can understand the structure of x-ray spectra if one realises that a missing electron, or a hole, in an otherwise full shell is equivalent to a single electron in an otherwise empty shell. Naturally this equivalence goes only as far as the sign: to remove an electron from the atom we must apply energy. If we consider the binding energy of the electron to be negative, then we must consider the energy required to generate a hole to be positive.

X-ray spectra can thus be understood, similarly to the spectra of alkali atoms, as one-electron (or one-hole) spectra. The terms may be characterised, as in the alkalis, by the quantum numbers of *one* electron; we thus arrive at a term diagram of the type shown in Fig. 18.9.

In the K shell, $n = 1$, l can have only the value 0, j is equal to 1/2, and the state is denoted by the symbol $^2S_{1/2}$.

For optical transitions, the selection rules are $\Delta l = \pm 1$ and $\Delta j = 0, \pm 1$. The longest-wavelength lines of the K series, $K_{\alpha 1}$ and $K_{\alpha 2}$, are thus produced in a manner analogous to the two sodium D lines. They connect the state $n = 1$, $^2S_{1/2}$ with the states $n = 2$, $^2P_{1/2}$ and $^2P_{3/2}$, which are split by the spin-orbit interaction.

Correspondingly, we can understand all of the fine structure of x-ray spectra. The shells which are characterised by the quantum numbers n are also split up into subshells. The latter are numbered using Roman numerals (e.g. L_I, L_II, L_III in Fig. 18.9). A subshell is characterised by a triplet of quantum numbers n, l, and j. The energy splitting between the edges L_I, L_II, and L_III has, as indicated in Fig. 18.9, a variety of origins. The spacing between L_II and L_III, i.e. between $^2P_{1/2}$ and $^2P_{3/2}$, is the well-known doublet splitting, which increases with increasing Z. The spacing between

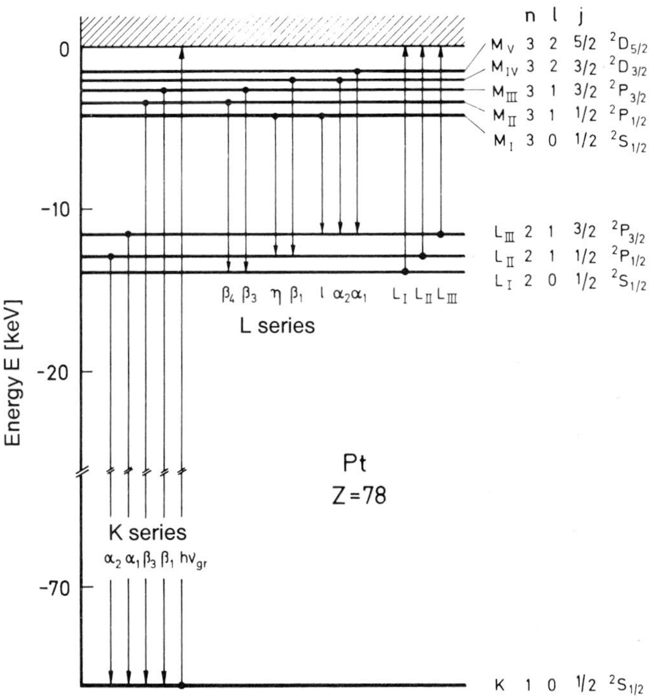

Fig. 18.9. Fine structure diagram for the x-ray spectra of a platinum anode with $Z = 78$. The notations for the series, the lines, and the quantum numbers are shown. Arrows pointing upwards mean absorption, those pointing downwards mean emission. The use of Greek letters to denote the individual lines is not systematic and is not uniform in the literature. – The spacings between the L subshells L_I, L_II, L_III and the M subshells $M_\mathrm{I} - M_\mathrm{V}$ are not shown to scale. For a given l-value, they result from the normal doublet splitting; otherwise, they are produced by differing screening of the nuclear charge and are therefore not all equal

L_I and L_{II}, on the other hand, results from variations in screening. The latter is produced only by the inner electrons and is less dependent on Z. This is clarified in the Mosely diagrams (Figs. 18.6 and 7).

18.6 Absorption Spectra

X-rays, like any other electromagnetic radiation, are absorbed and scattered on passing through matter. The primary experimentally determined quantity is the extinction coefficient μ, defined by the equation $I = I_0 \exp(-\mu x)$, where x is the thickness of the material irradiated, I_0 is the incident intensity, and I is the transmitted intensity. The result of a measurement is often given as the half-absorption thickness, $d = \mu^{-1} \ln 2$. The half-absorption thickness depends, in general, on the material irradiated and on the energy of the x-ray quanta. Table 18.1 gives some numerical data.

Table 18.1. Half-absorption thickness [cm] for x-rays in aluminium and lead

V_0 [kV]	Al	Pb
10	$1.1 \cdot 10^{-2}$	$7.5 \cdot 10^{-4}$
100	1.6	$1.1 \cdot 10^{-2}$

The extinction is the sum of scattering – which does not interest us here – and absorption. The dependence of the absorption coefficient on quantum energy, i.e. the spectral distribution in absorption spectra, is represented schematically in Fig. 18.10, where the emission spectra at various excitation energies are also shown for comparison.

X-ray absorption spectra typically display a large decrease in the absorption coefficient with increasing quantum energy, and absorption "edges", which are quantum energies at which the absorption coefficient jumps to a higher value. These "edges" correspond to the series limits for the $K, L, M, \ldots$ series, and they are correspondingly labelled. The subshells also appear as edges, for example L_I, L_{II}, and L_{III} in Fig. 18.10.

The position of the K edge for lead at 88 keV (Fig. 18.11) means that the work of removing an electron from the K shell, where it experiences the field of the nearly unshielded nuclear charge of the lead nucleus, is 88 keV. The screening for lead ($Z = 82$) can be calculated. For the innermost electron, the work of separation is $Z_{\text{eff}}^2 \cdot 13.6$, where $Z_{\text{eff}} \equiv Z - s$ is the effective nuclear charge and 13.6 eV is the ionisation energy for the hydrogen atom (Sect. 8.4). From $(82 - s)^2 \cdot 13.6$ eV $= 88$ keV, $s = 1.61$.

In order for an atom to absorb x-radiation, an electron must be excited from an inner shell into a less strongly bound state. Since the neighbouring shells are already occupied, discrete absorption lines due to transitions from one shell into another are scarcely observable. There is, however, a continuum of free states on the other side of the series limit into which the absorbing electron can be lifted. Therefore, absorption is usually associated with ionisation, and the absorption spectra are the superimposed series-limit continua of the various shells and subshells. This is shown in Fig. 18.10.

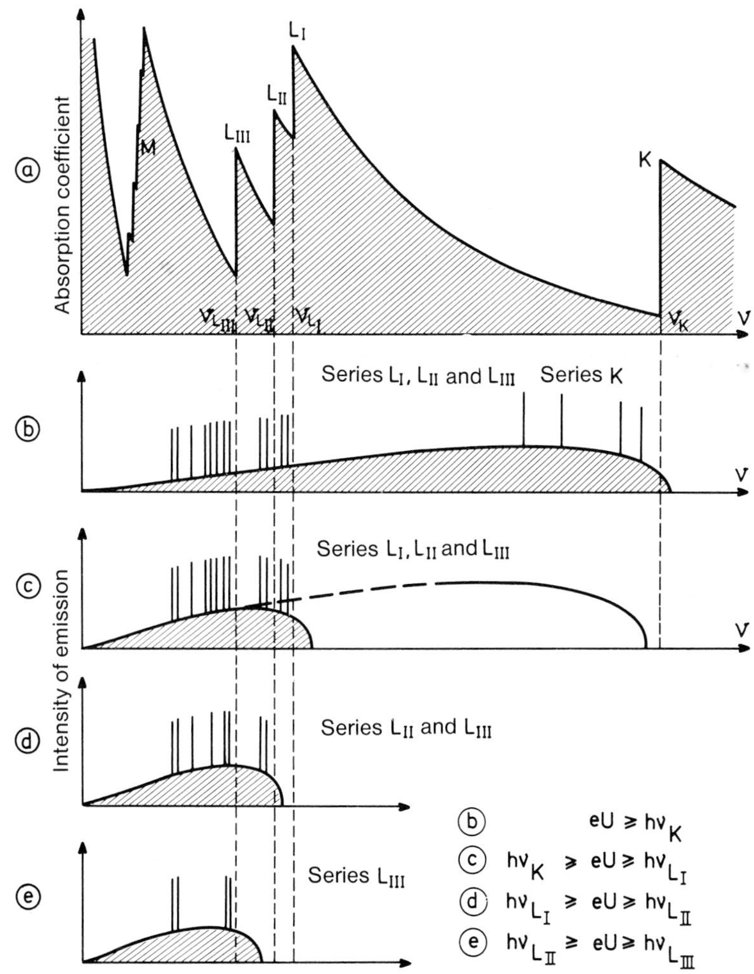

Fig. 18.10a–e. Comparison of x-ray absorption and emission spectra. **a)** Absorption coefficient, e.g. of platinum for x-rays, as a function of frequency, schematically. The spectrum consists primarily of superimposed absorption edges. **b–e)** Emission spectrum of platinum at various excitation energies. All the line series are excited in **b**, in **c** the K series is lacking, in **d** the K and L_I series are absent, and in e the K, L_I and L_{II} series are missing

Going from lower to higher frequencies, absorption edges, or jumps in the absorption coefficient, are located at those points where the energy of the x-ray quantum is just sufficient to allow an absorptive transition from a new (lower) shell into the limiting continuum.

At lower frequencies, the quantum energy $h\nu$ is only sufficient to release electrons from outer shells. As $h\nu$ increases, an energy is reached which is sufficient to release even K electrons, and at this point the absorption coefficient increases abruptly. The fine structure of the absorption edges is further evidence for the existence of shells and subshells: there is one K edge, but $3\,L$ edges, $5\,M$ edges, and so on.

In addition, if the spectral resolution is good enough, it is possible to detect effects of chemical bonding on the energies and fine structures of the absorption edges.

Aside from the edges, the frequency dependence of the absorption coefficient is essentially expressed by

$$\mu_{\mathrm{abs}} \cong Z^x/(h\nu)^3 \quad \text{with} \quad 3 \lesssim x \lesssim 4 \tag{18.5}$$

or $\mu_{\mathrm{abs}} \cong \lambda^3 Z^x$.

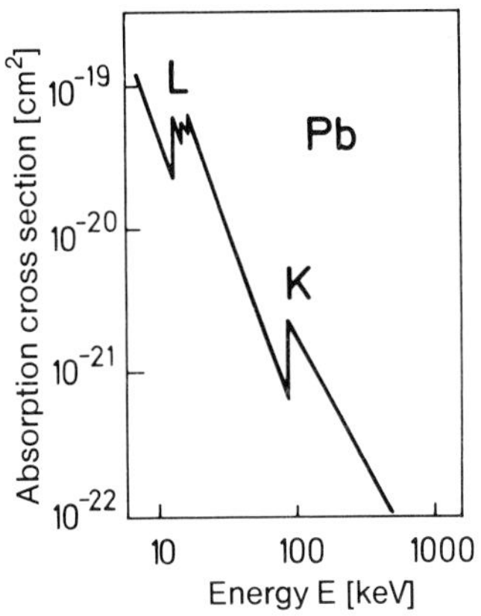

Fig. 18.11. Absorption cross section of lead for x-rays in the region of the L and K edges. The absorption coefficient is expressed in terms of the absorption cross section of an atom

The hardness or penetrating ability of the x-rays thus increases as the wavelength decreases, or as the accelerating voltage increases. Figure 18.11 shows the frequency dependence of the atomic absorption cross section (see Sect. 2.4.2) of lead in the region of the K and L edges.

18.7 The Auger Effect (Inner Photoeffect)

Not all atoms from which electrons have been removed from inner shells by electron bombardment or other forms of energy transfer return to the ground state by emitting x-rays. Instead, the observed quantum yield for x-ray emission is frequently less than 1. It is defined here as

$$\eta = \frac{\text{Number of x-ray emitting atoms}}{\text{Number of } K, L \ldots \text{ ionised atoms}} \, .$$

Thus it must be possible for the atoms to return to the ground state without emitting radiation. The probability of such non-radiative processes, which compete with x-ray emission, has been found to decrease with increasing nuclear charge. In light atoms, the non-radiative processes far outweigh emittive processes (Fig. 18.12).

The non-radiative return to ground state is accomplished by the *Auger effect.* After an electron has been removed from an inner shell, the excess energy can be released either in the form of an x-ray quantum or through the emission of an electron from a shell farther out. The Auger effect is thus similar to an "inner photoeffect". If one electron falls into a lower shell and another is simultaneously emitted, the Coulomb interaction between the two must be responsible for the process.

The Auger effect is represented schematically in Fig. 18.13. First the K shell is ionised. An L electron falls from the L to the K shell, and fills the hole there. The energy released by this is used to remove a second L electron from the L shell, and this one escapes from the atom. The result is that the L shell loses two electrons. These are then replaced by electrons from shells farther out, namely the M shell and so on.

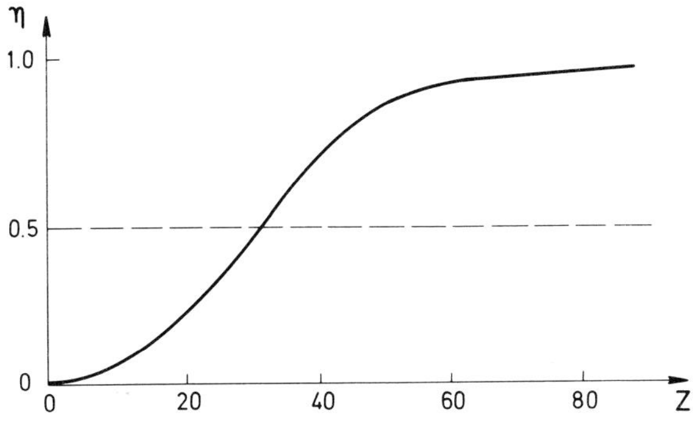

Fig. 18.12. Auger coefficient as a function of the Z number (schematically)

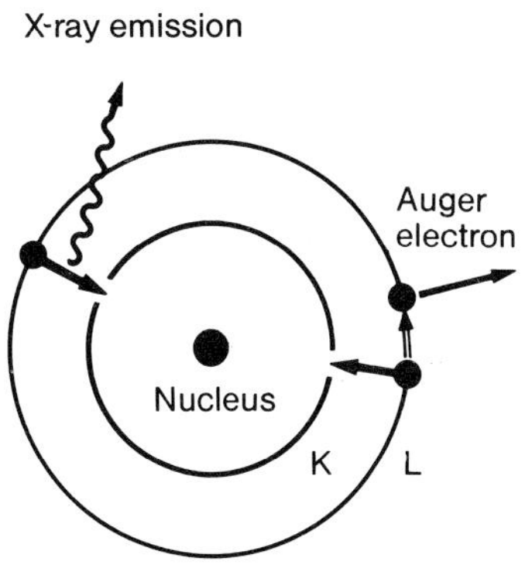

Fig. 18.13. Auger electron emission (*right*) competes with x-ray emission (*left*)

The kinetic energy of the Auger electron is given by

$$E_{\text{kin}} = h\nu_{K_\alpha} - E_L = E_K - E_L - E_L = E_K - 2E_L\,, \tag{18.6}$$

where E_L and E_K are the binding energy of the electrons in the L and K shells, respectively.

Auger electrons can also be observed directly in a cloud chamber. Their energy can be determined from the length of their tracks in the cloud chamber, or by means of a braking field method, and thus can be used to verify the explanation of the Auger effect.

Finally, a numerical example: Silver is bombarded with K_α radiation from a tungsten anticathode (energy = 59.1 keV). Electrons with the following energies are observed:

1) 55.8 keV — Interpretation: Photoelectrons from the Ag L shell
Because: the ionisation energy of the Ag L shell is $E_{\text{ion}L} = 3.34$ keV
Therefore: $59.1 - 3.34 = 55.76$ keV.

2) 33.8 keV — Interpretation: Photoelectrons from the Ag K shell
Because: $E_{\text{ion}K} = 25.4$ keV
Therefore: $59.1 - 25.4 = 33.7$ keV.

3) 21.3 keV — Interpretation: Auger electrons
Because: $E_{K_\beta}(\text{Ag}) - E_{\text{ion}L} = 24.9 - 3.34 = 21.56$ keV.

4) 18.6 keV — Interpretation: Auger electrons
Because: $E_{K_\alpha}(\text{Ag}) - E_{\text{ion}L} = 22.1 - 3.34 = 18.76$ keV.

18.8 Photoelectron Spectroscopy (XPS), ESCA

A relatively new method for investigating the energy states of the inner electrons of an atom is *photoelectron spectroscopy*. This technique is a modern application of the photoelectric effect (Sect. 5.3).

Electrons are ejected from the shells of an atom by exciting them with light of a known quantum energy. The kinetic energies of the photoelectrons correspond to the difference between the quantum energy of the photons and the binding energies of the electrons in the atom, following the energy-balance equation of the photoelectric effect, $E_{kin} = h\nu - E_{bind}$. The principle is illustrated in Fig. 18.14. Figure 18.15 gives an example of a measurement. With this method, the binding energies of inner-shell electrons can be determined directly, in contrast to x-ray absorption spectroscopy, which gives only the energy of the absorption edge.

The measurement of the kinetic energy of the photoelectrons is performed with high-resolution analysers, which allow a precise determination of the velocities of the electrons using the principle of the determination of e/m (deflection in electric and magnetic fields) described in Sect. 6.4. (Simultaneous measurement over a wide range of electron kinetic energies is also possible using time-of-flight analysers, with somewhat poorer energy resolution.) The best energy resolutions currently available allow determination of the electron energies to about 20 meV.

The light source provides either UV light of short wavelengths, e.g. the resonance lines from the Ne and the He spectra in the region between 15 and 50 eV, or else, for the investigation of states with higher binding energies, the characteristic x-ray lines, e.g. the K_α lines from Cu (8048 eV), from Al (1487 eV), or from Mg (1254 eV). Synchrotron radiation, which has a continuously variable photon energy in the whole UV and x-ray region (cf. Sect. 5.1), is a particularly suitable light source.

The binding energies of the electrons are characteristic of the particular atoms being investigated, so that the method can also be used for *chemical analysis* of a sample.

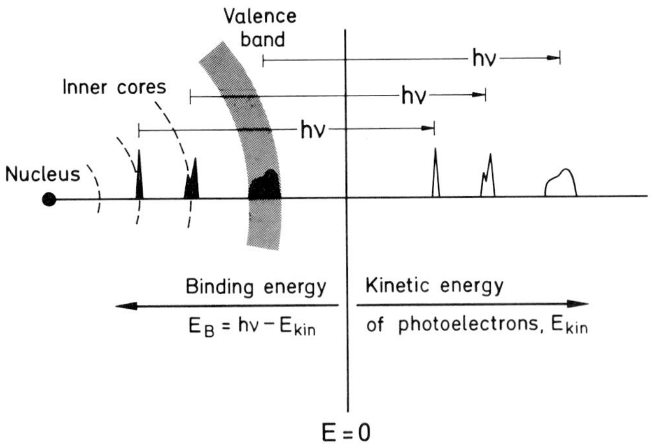

Fig. 18.14. Illustration of photoelectron spectroscopy. The kinetic energy of the photoelectrons is the difference between the quantum energy $h\nu$ of the excitation photons and the binding energy of the electrons in an atom or a solid. The *dashed lines* represent the orbitals of the electrons. One should note, however, that here the binding energies, and not the distance from the nucleus, are indicated

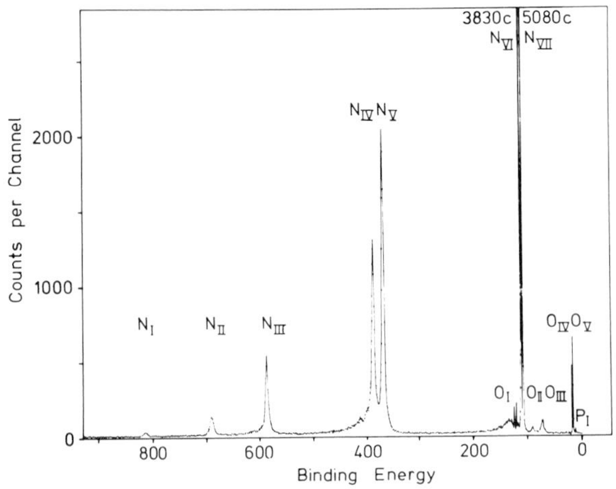

Fig. 18.15. The photoelectron spectrum of Hg vapour, after K. Siegbahn (overview of the spectrum). All the levels which can be excited with the K_α radiation from Al are to be seen. These are the N and O shells, as well as P_I

Furthermore, the chemical bonding between atoms in molecules or in solids leads to a redistribution of the valence electrons. This, in turn, changes the entire bonding potential in the atoms, whereby even the inner electrons are affected. The resulting small shifts in the inner-shell binding energies due to changes in the state of the outer (valence) electrons are termed *chemical shifts* and can be measured, e.g. between atoms in different ionisation states. The corresponding analytical method, called ESCA (*E*lectron *S*pectroscopy for *C*hemical *A*nalysis), was developed in particular by *K. Siegbahn* and coworkers. It has become an important experimental technique in chemistry and in molecular and solid state physics.

Problems

18.1 What is the shortest possible wavelength for bremsstrahlung observed when an electron which has been accelerated through a potential difference of 40 kV is stopped by the anticathode of an x-ray tube? In what region of the electromagnetic spectrum does this wavelength lie?

18.2 The K_α line of cobalt is at 1.785 Å. What is the energy difference between the $1s$ and $2p$ orbitals in cobalt? Compare this result with the energy difference between the $1s$ and $2p$ orbitals in hydrogen (that is, the first Lyman line). Why is the difference much greater for cobalt than for hydrogen?

18.3 The most intense line in the x-ray spectrum arises from the transition in which an electron goes from the shell with $n = 2$ to the shell with $n = 1$. The transition is described by Moseleys rule (cf. Sect. 18.4). What is the wavelength of this line for copper?

18.4 The maximum energy of the characteristic x-rays emitted by a sample of unknown composition corresponds to the wavelength 2.16 Å. Of what element does the sample consist?

18.5 X-rays are allowed to pass through aluminium foils, each 4×10^{-3} m thick. A Geiger counter registers 8×10^3, 4.7×10^3, 2.8×10^3, 1.65×10^3 and 9.7×10^2 events/min when the rays have passed through 0, 1, 2, 3 and 4 foils, respectively. Calculate the linear absorption coefficient of aluminium.

18.6 Gamma rays with energies of 0.05, 0.3 and 1 MeV, but the same intensities, fall onto a lead absorber. The linear absorption coefficients for these energies are 8×10^3, 5×10^2 and 78 m^{-1}.

a) Calculate the thickness of lead required to reduce the intensity of each beam of gamma rays to one tenth of its original value.
b) What is the relation of the total intensity (at each photon energy) at a depth of 5 mm to the total incident intensity?

18.7 How many times the half-absorption thickness of a material is required to reduce the intensity of an x-ray beam to (a) 1/16, (b) 1/20, or (c) 1/200 of the incident intensity?

18.8 a) The K absorption edge of tungsten is at 0.178 Å, and the wavelengths of the lines of the K series are (ignoring the fine structure) $K_\alpha = 0.210$ Å, $K_\beta = 0.184$ Å, and $K_\gamma = 0.179$ Å. Sketch the energy levels of W and give the energies of the K, L, M and N shells.

b) What is the minimum energy required to excite the L series in tungsten? What is the wavelength of the L line?

18.9 The L_{I} absorption edge in tungsten is at 1.02 Å. Assume that a K_α photon is "absorbed" by one of the $2s$ electrons by an Auger process. Determine the velocity of the photoelectron released.

19. Structure of the Periodic System. Ground States of the Elements

19.1 Periodic System and Shell Structure

It is one of the goals of atomic physics to understand the ordering and the properties of the chemical elements in the periodic system. The empirically determined physical and chemical properties of the atoms and their dependence on the atomic number Z ought to be explainable starting from the electronic structures of the atoms. After having, in the previous chapters, discussed one-electron spectra, many-electron spectra, and the spectra of the inner shells in detail, we are now in a position to understand, at least in principle, the spectra of any atom in any state of excitation. We will outline and extend this understanding in the present chapter.

In particular, in the discussions of alkali atoms and x-ray spectra, we have met with some important experimental facts which led to the concept of the shell structure of

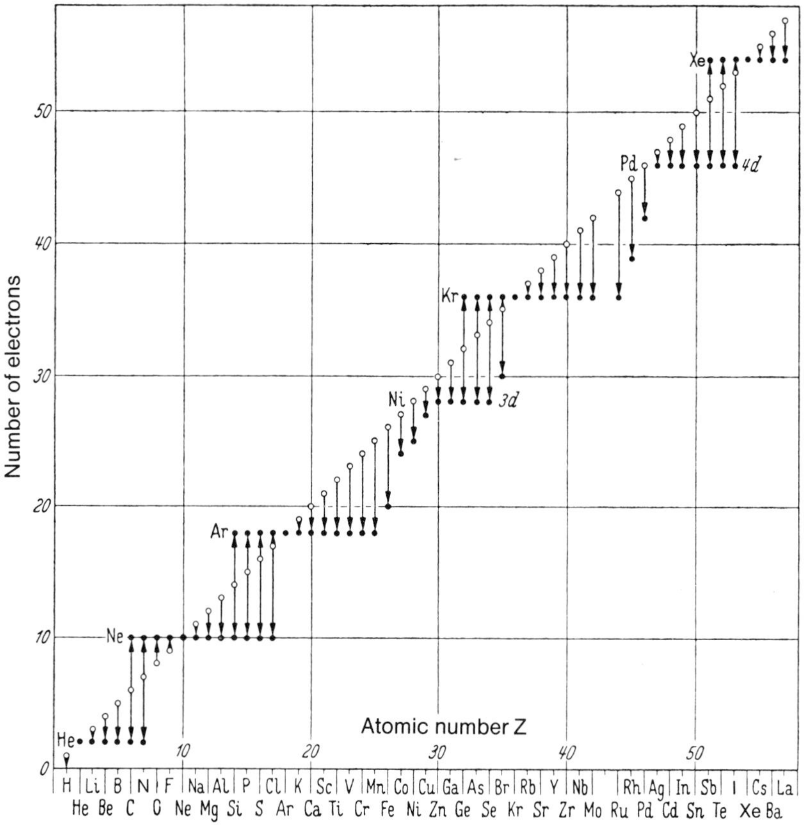

Fig. 19.1. Electronic shells of the ions of atoms, following *Kossel*, 1916. The number of electrons is plotted against the nuclear charge number for atoms and for the ions which are preferentially formed from those atoms in chemical bonding. The electronic shells of the atoms are depleted or filled up to the electron numbers indicated by arrows in the diagram when ions are formed. With the exceptions of nickel and palladium, the preferred electron numbers are those of the noble gases. [From K. H. Hellwege: *Einführung in die Physik der Atome*, Heidelberger Taschenbücher, Vol. 2, 4th ed. (Springer, Berlin, Heidelberg, New York 1974) Fig. 61]

atoms. It is known that the noble gases are chemically particularly stable. Furthermore, if the electrochemical valency of an ion is compared to the distance of the corresponding neutral element from the nearest noble gas in the periodic table, it is found that the atoms have been ionised to such an extent that the ions have assumed the electron number of a neutral noble gas. The observed ionisation levels of the atoms are shown in a way that makes this clear in Fig. 19.1. Thus when an ion is formed, those electrons which are in excess of the electron number of a noble gas are most easily lost, or the electrons which are lacking from the electron number of the next heavier noble gas are most easily acquired. Consider, for example, the elements around the noble gas neon ($Z = 10$). The next element in the periodic table is sodium ($Z = 11$), which preferentially forms singly charged positive ions. The next element, magnesium ($Z = 12$), forms doubly charged positive ions. The next lighter element than neon, fluorine ($Z = 9$), forms singly charged negative ions (by acquisition of an electron) and so on.

It can be seen from Fig. 19.1 that nickel and palladium also have preferred electron numbers, because the atoms of neighbouring elements tend to assume these numbers when they become ionised. The stable electron configurations are thus not limited to the noble gases. This is made clear in the discussion of Table 19.3.

We know from the spectroscopic studies of atoms discussed in earlier chapters that the unusually stable electron configurations are characterised by complete mutual compensation of all angular momenta and magnetic moments. Furthermore, the shell and subshell structure of the electron distribution of atoms appears especially clearly in x-ray spectra as a system of absorption edges.

We now turn to the question of which electronic configurations are possible in atoms, which are particularly stable, and how the electrons of an atom are distributed among the possible quantum states.

To understand this, one first needs the quantum numbers with which each atomic electron can be characterised:

the principal quantum number n,
the orbital angular momentum quantum number $l = 0, 1 \ldots n-1$,
the magnetic quantum number $m_l = 0, \pm 1 \ldots \pm l$,
the magnetic spin quantum number $m_s = \pm 1/2$.

One also needs the Pauli principle. In non-mathematical terms, this says that only those atomic states can exist in nature in which any two electrons differ, in at least one of their quantum numbers. This results in limitations in the possible combinations of quantum numbers in an atomic state.

Strictly speaking, in order to define the quantum numbers of an electron in the atom, one would have to solve the Schrödinger equation for a many-particle problem, namely for all the electrons in the atom. The problem is solved approximately using the Hartree-Fock technique, which is based on the model of independent particles (Sect. 19.4). The basic idea is that instead of trying to calculate the interactions of $N-1$ electrons with the Nth electron, one replaces the Coulomb attraction of the nucleus for the Nth electron by an effective potential. One then calculates the eigenstates and eigenvalues of the Nth electron in this potential field.

Figure 19.2 shows, as the result of such calculations, the shell structure of the hydrogen atom and the positive ions of lithium, sodium and potassium. As the figure shows, the shells overlap, and thus have no unique geometric significance. As the nuclear charge increases, they are drawn closer to the nucleus. Another example of a calculated electron density distribution was discussed in Fig. 11.8.

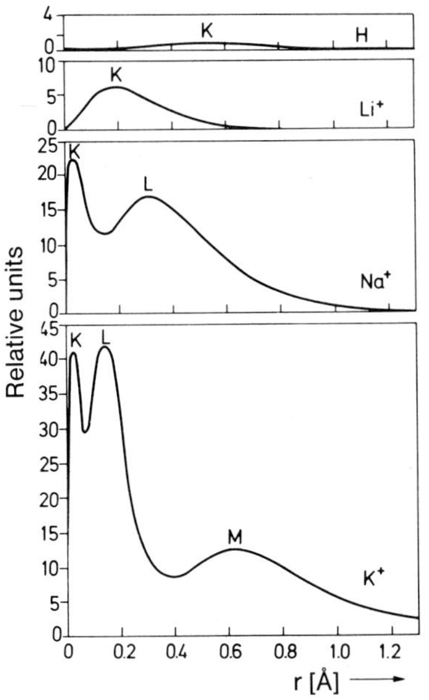

Fig. 19.2. Radial density distribution of electrons in the hydrogen atom, and the singly charged positive ions of lithium, sodium and potassium. It is especially noticeable how the *K* shell electrons move closer to the nucleus as the nuclear charge increases

In the absence of an external magnetic field, the magnetic quantum numbers would seem to lose their function, because of degeneracy. In order to apply the Pauli principle to atomic terms in such cases, we make use of of *Ehrenfest*'s adiabatic invariance principle, which states that if a parameter changes continuously, the states of an atom also change in a continuous and uniquely determined way. In the present case, this means that the states of the atom in the absence of a magnetic field must derive from those in the presence of a strong field – where the m degeneracy is lifted – in a continuous and uniquely determined way as the field is slowly turned off.

With the Pauli principle it is easy to count the maximum number of electrons with a given principal quantum number n which can be bound to an atom:

- For a given value of the principal quantum number n there are n different values for the orbital angular momentum quantum number l.
- For every value of l there are $2l+1$ different values of the magnetic quantum number m_l.
- For each pair of quantum numbers l and m_l there are two different values of the spin quantum number m_s.
- Thus for each pair of numbers n, l there are at most $2(2l+1)$ electrons.

The maximum number of electrons in a shell with a given value of n is then

$$\sum_{l=0}^{n-1} 2(2l+1) = 2n^2 . \qquad (19.1)$$

Table 19.1 gives the possible combinations of quantum numbers for $n = 1, 2$ and 3.

If a shell is defined as the group of all electrons with the same principal quantum number n, then the atoms with closed shells are helium ($n = 1$), neon ($n = 2$), nickel ($n = 3, Z = 28$) and neodymium ($n = 4, Z = 60$). Nickel and neodymium are neither

Table 19.1. Possible quantum numbers and numbers of electrons in the shells with $n = 1, 2, 3$

n	l	m_l	m_s	Number of electrons		Configuration	Shell
1	0	0	$\pm 1/2$	2		$1s^2$	*K* shell
2	0	0	$\pm 1/2$	$1 \cdot 2$	= 8	$2s^2p^6$	*L* shell
	1	1	$\pm 1/2$				
		0	$\pm 1/2$				
		-1	$\pm 1/2$	$3 \cdot 2$			
3	0	0	$\pm 1/2$	$1 \cdot 2$	= 18	$3s^2p^6d^{10}$	*M* shell
	1	1	$\pm 1/2$				
		0	$\pm 1/2$				
		-1	$\pm 1/2$	$3 \cdot 2$			
	2	2	$\pm 1/2$				
		1	$\pm 1/2$				
		0	$\pm 1/2$				
		-1	$\pm 1/2$				
		-2	$\pm 1/2$	$5 \cdot 2$			

noble gases nor chemically very inactive. The simple equation of a closed shell with a noble gas configuration breaks down for higher electron numbers.

Instead, it is observed that the closure of a partial shell, that is occupation of all states with the same value of l for a given value of n, leads to especially stable electron configurations. This is understandable in light of the fact that even for fully occupied partial shells, the angular momentum and magnetic moments add up to zero, so that the atom is outwardly spherically symmetric. In fact, this is the case with the third noble gas, argon, which has the electronic configuration $1s^2\,2s^2\,2p^6\,3s^2\,3p^6$. In other words, all of the s and p electrons are present in the third shell, but there are no d electrons. The cases of the noble gases krypton and xenon are similar, as is shown in Table 19.2.

Table 19.2. Electronic configuration for the highest occupied shells or partial shells of the noble gases. Only helium and neon have completely filled highest shells. The particular stability of the other noble gases is due to completion of partial shells

Highest occupied state	Z	Element	1st ionisation potential [eV]
$(1s)^2$	2	He	24.58
$(2s)^2(2p)^6$	10	Ne	21.56
$(3s)^2(3p)^6$	18	Ar	15.76
$(4s)^2(3d)^{10}(4p)^6$	36	Kr	14.00
$(5s)^2(4d)^{10}(5p)^6$	54	Xe	12.13
$(6s)^2(4f)^{14}(5d)^{10}(6p)^6$	86	Rn	10.75

Having seen that each partial or complete shell can only accommodate a certain maximum number of electrons, we can turn to the configurations of all the atoms of the periodic system. Table 19.3 a and b shows all the elements with their electronic configurations. Closed complete and partial shells are shaded in. The table also shows the first ionisation energy for each element.

The transition elements scandium to nickel owe their unusual properties, such as colour and paramagnetism, to their partially filled inner $3d$ shells. Their chemical reactivities are determined by the outer valence electrons. Because the valence electrons are energetically very close to the inner $3d$ electrons, the $3d$ electrons of the transition elements tend to shift from one shell to the other, and these elements have several valencies.

The $4d$ transition elements $_{39}$Y to $_{46}$Pd have similar behaviour, as do the $4f$ transition elements $_{57}$La to $_{70}$Yb (the "rare earths", in which the $4f$ shell is filled up after the $6s$ subshell is occupied), the $5d$ elements $_{71}$Lu to $_{78}$Pt and the $5f$ elements $_{91}$Pa to $_{103}$Lw.

The rare earths are the most impressive example of the unusual properties of elements in which the energetically higher, but more inwardly located shells are being filled up. They are chemically very similar, because their outermost configurations in the $6s$ subshell are identical or similar. Their colour and paramagnetism are due to the inner $4f$ electrons. The screening of the latter from the outside by the $6s$ electrons is responsible for the fact that the optical spectra of rare earths have very sharp lines, even in solids. Rare earth atoms or their ions in solids are therefore very suitable materials for lasers (Chap. 21). Neodymium has been most commonly used for this purpose.

Palladium, with $Z = 46$ is also especially interesting. It has a completely closed subshell configuration. The fact that it is nevertheless not a noble gas can be understood by comparing it with the neighbouring element rhodium. The $5s$ electron present in rhodium is shifted to the $4d$ shell in palladium. However, it requires very little energy to raise it back into the $5s$ shell, and for this reason, palladium is neither chemically inactive nor a noble gas.

Figure 19.3 summarises the energetic sequence of the subshells in the successive construction of the atoms and their shell structure. At several points the distance to the next level is especially large. These points are the particularly stable configurations of the noble gases.

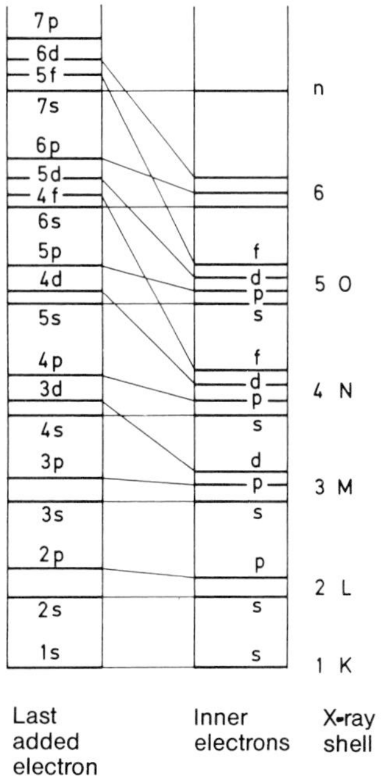

Fig. 19.3. Shell structure of the atomic energy levels and order of the levels for the last electron added and for inner electrons. The particularly stable configurations are those which are followed by an especially large distance to the next level

Table 19.3a. Periodic Table with electron configurations, ground state terms, and ionisation energies. The filled shells and subshells are shaded

Atomic number Z	Element		Shells K $n=1$ s	L $n=2$ s	p	M $n=3$ s	p	d	N $n=4$ s	p	d	O $n=5$ s	p	LS configuration of the ground state	First ionisation potential [eV]	
1	Hydrogen	H	1											$^2S_{1/2}$	13.60	
2	Helium	He	2											$^1S_0$	24.58	
3	Lithium	Li	2	1										$^2S_{1/2}$	5.39	
4	Beryllium	Be	2	2										$^1S_0$	9.32	
5	Boron	B	2	2	1									$^2P_{1/2}$	8.30	
6	Carbon	C	2	2	2									$^3P_0$	11.26	
7	Nitrogen	N	2	2	3									$^4S_{3/2}$	14.54	
8	Oxygen	O	2	2	4									$^3P_2$	13.61	
9	Fluorine	F	2	2	5									$^2P_{3/2}$	17.42	
10	Neon	Ne	2	2	6									$^1S_0$	21.56	
11	Sodium	Na	2	2	6	1								$^2S_{1/2}$	5.14	
12	Magnesium	Mg	2	2	6	2								$^1S_0$	7.64	
13	Aluminium	Al	2	2	6	2	1							$^2P_{1/2}$	5.98	
14	Silicon	Si	2	2	6	2	2							$^3P_0$	8.15	
15	Phosphorous	P	2	2	6	2	3							$^4S_{3/2}$	10.55	
16	Sulphur	S	2	2	6	2	4							$^3P_2$	10.36	
17	Chlorine	Cl	2	2	6	2	5							$^2P_{3/2}$	13.01	
18	Argon	Ar	2	2	6	2	6							$^1S_0$	15.76	
19	Potassium	K	2	2	6	2	6		1					$^2S_{1/2}$	4.34	
20	Calcium	Ca	2	2	6	2	6		2					$^1S_0$	6.11	
21	Scandium	Sc	2	2	6	2	6	1	2					$^2D_{3/2}$	6.56	Transition elements
22	Titanium	Ti	2	2	6	2	6	2	2					$^3F_2$	6.83	
23	Vanadium	V	2	2	6	2	6	3	2					$^4F_{3/2}$	6.74	
24	Chromium	Cr	2	2	6	2	6	5	1					$^7S_3$	6.76	
25	Manganese	Mn	2	2	6	2	6	5	2					$^6S_{5/2}$	7.43	
26	Iron	Fe	2	2	6	2	6	6	2					$^5D_4$	7.90	
27	Cobalt	Co	2	2	6	2	6	7	2					$^4F_{9/2}$	7.86	
28	Nickel	Ni	2	2	6	2	6	8	2					$^3F_4$	7.63	
29	Copper	Cu	2	2	6	2	6	10	1					$^2S_{1/2}$	7.72	
30	Zinc	Zn	2	2	6	2	6	10	2					$^1S_0$	9.39	
31	Gallium	Ga	2	2	6	2	6	10	2	1				$^2P_{1/2}$	6.00	
32	Germanium	Ge	2	2	6	2	6	10	2	2				$^3P_0$	7.88	
33	Arsenic	As	2	2	6	2	6	10	2	3				$^4S_{3/2}$	9.81	
34	Selenium	Se	2	2	6	2	6	10	2	4				$^3P_2$	9.75	
35	Bromine	Br	2	2	6	2	6	10	2	5				$^2P_{3/2}$	11.84	
36	Krypton	Kr	2	2	6	2	6	10	2	6				$^1S_0$	14.00	
37	Rubidium	Rb	2	2	6	2	6	10	2	6		1		$^2S_{1/2}$	4.18	
38	Strontium	Sr	2	2	6	2	6	10	2	6		2		$^1S_0$	5.69	
39	Yttrium	Y	2	2	6	2	6	10	2	6	1	2		$^2D_{3/2}$	6.38	Transition elements
40	Zirconium	Zr	2	2	6	2	6	10	2	6	2	2		$^3F_2$	6.84	
41	Niobium	Nb	2	2	6	2	6	10	2	6	4	1		$^6D_{1/2}$	6.88	
42	Molybdėnum	Mo	2	2	6	2	6	10	2	6	5	1		$^7S_3$	7.13	
43	Technetium	Tc	2	2	6	2	6	10	2	6	6	1		$^6D_{9/2}$	7.23	
44	Ruthenium	Ru	2	2	6	2	6	10	2	6	7	1		$^5F_5$	7.37	
45	Rhodium	Rh	2	2	6	2	6	10	2	6	8	1		$^4F_{9/2}$	7.46	
46	Palladium	Pd	2	2	6	2	6	10	2	6	10			$^1S_0$	8.33	
47	Silver	Ag	2	2	6	2	6	10	2	6	10	1		$^2S_{1/2}$	7.57	
48	Cadmium	Cd	2	2	6	2	6	10	2	6	10	2		$^1S_0$	8.99	
49	Indium	In	2	2	6	2	6	10	2	6	10	2	1	$^2P_{1/2}$	5.79	
50	Tin	Sn	2	2	6	2	6	10	2	6	10	2	2	$^3P_0$	7.33	
51	Antimony	Sb	2	2	6	2	6	10	2	6	10	2	3	$^4S_{3/2}$	8.64	
52	Tellurium	Te	2	2	6	2	6	10	2	6	10	2	4	$^3P_2$	9.01	
53	Iodine	J	2	2	6	2	6	10	2	6	10	2	5	$^2P_{3/2}$	10.44	
54	Xenon	Xe	2	2	6	2	6	10	2	6	10	2	6	$^1S_0$	12.13	

Table 19.3b. Periodic Table with electron configurations, ground state terms, and ionisation energies. The filled shells and subshells are shaded. (The subshells $5g$ and $6f$, $6g$, $6h$ are not shown, since there are no atoms which have electrons in these shells in their ground states)

Atomic number Z	Element		Shells N $n=4$ s	p	d	f	O $n=5$ s	p	d	f	P $n=6$ s	p	d	Q $n=7$ s	LS configuration of the ground state	First ionisation potential [eV]	
55	Cesium	Cs	2	6	10		2	6			1				$^2S_{1/2}$	3.89	
56	Barium	Ba	2	6	10		2	6			2				$^1S_0$	5.21	
57	Lanthanum	La	2	6	10		2	6	1		2				$^2D_{3/2}$	5.61	
58	Cerium	Ce	2	6	10	2	2	6			2				$^3H_4$	5.6	
59	Praseodymium	Pr	2	6	10	3	2	6			2				$^4I_{9/2}$	5.46	
60	Neodymium	Nd	2	6	10	4	2	6			2				$^5I_4$	5.51	
61	Promethium	Pm	2	6	10	5	2	6			2				$^6H_{5/2}$		
62	Samarium	Sm	2	6	10	6	2	6			2				$^7F_0$	5.6	
63	Europium	Eu	2	6	10	7	2	6			2				$^8S_{7/2}$	5.67	Rare earths
64	Gadolinium	Gd	2	6	10	7	2	6	1		2				$^9D_2$	6.16	
65	Terbium	Tb	2	6	10	9	2	6			2				—	5.98	
66	Dysprosium	Dy	2	6	10	10	2	6			2				$^5I_8$	6.8	
67	Holmium	Ho	2	6	10	11	2	6			2				$^4I_{15/2}$		
68	Erbium	Er	2	6	10	12	2	6			2				$^3H_6$	6.08	
69	Thulium	Tm	2	6	10	13	2	6			2				$^2F_{7/2}$	5.81	
70	Ytterbium	Yb	2	6	10	14	2	6			2				$^1S_0$	6.22	
71	Lutetium	Lu	2	6	10	14	2	6	1		2				$^2D_{3/2}$	6.15	
72	Hafnium	Hf	2	6	10	14	2	6	2		2				$^3F_2$	5.5	
73	Tantalum	Ta	2	6	10	14	2	6	3		2				$^4F_{3/2}$	7.7	
74	Tungsten	W	2	6	10	14	2	6	4		2				$^5D_0$	7.98	
75	Rhenium	Re	2	6	10	14	2	6	5		2				$^6S_{5/2}$	7.87	Transition elements
76	Osmium	Os	2	6	10	14	2	6	6		2				$^5D_4$	8.7	
77	Iridium	Ir	2	6	10	14	2	6	9						$^2D_{5/2}$	9.2	
78	Platinum	Pt	2	6	10	14	2	6	9		1				$^3D_3$	9.0	
79	Gold	Au	2	6	10	14	2	6	10		1				$^2S_{1/2}$	9.22	
80	Mercury	Hg	2	6	10	14	2	6	10		2				$^1S_0$	10.43	
81	Thallium	Tl	2	6	10	14	2	6	10		2	1			$^2P_{1/2}$	6.11	
82	Lead	Pb	2	6	10	14	2	6	10		2	2			$^3P_0$	7.42	
83	Bismuth	Bi	2	6	10	14	2	6	10		2	3			$^4S_{3/2}$	7.29	
84	Polonium	Po	2	6	10	14	2	6	10		2	4			$^3P_2$	8.43	
85	Astatine	At	2	6	10	14	2	6	10		2	5				9.5	
86	Radon	Rn	2	6	10	14	2	6	10		2	6			$^1S_0$	10.75	
87	Francium	Fr	2	6	10	14	2	6	10		2	6		1		4	
88	Radium	Ra	2	6	10	14	2	6	10		2	6		2		5.28	
89	Actinium	Ac	2	6	10	14	2	6	10		2	6	1	2			
90	Thorium	Th	2	6	10	14	2	6	10		2	6	2	2			
91	Protactinium	Pa	2	6	10	14	2	6	10	2	2	6	1	2			
92	Uranium	U	2	6	10	14	2	6	10	3	2	6	1	2			
93	Neptunium	Np	2	6	10	14	2	6	10	4	2	6	1	2			
94	Plutonium	Pu	2	6	10	14	2	6	10	6	2	6		2			
95	Americium	Am	2	6	10	14	2	6	10	7	2	6		2			
96	Curium	Cm	2	6	10	14	2	6	10	7	2	6	1	2			Actinides
97	Berkelium	Bk	2	6	10	14	2	6	10	8	2	6	1	2			
98	Californium	Cf	2	6	10	14	2	6	10	10	2	6		2			
99	Einsteinium	Es	2	6	10	14	2	6	10	11	2	6		2			
100	Fermium	Fm	2	6	10	14	2	6	10	12	2	6		2			
101	Mendelevium	Md	2	6	10	14	2	6	10	13	2	6		2			
102	Nobelium	No	2	6	10	14	2	6	10	14	2	6		2			
103	Lawrencium	Lw	2	6	10	14	2	6	10	14	2	6	1	2			
104	Kurchatovium		2	6	10	14	2	6	10	14	2	6	2	2			
105	Hahnium		2	6	10	14	2	6	10	14	2	6	3	2			

Let it be re-emphasised here that the shell structure which we have been discussing, and the one which is important for the properties of the periodic system, is that of the last added electron. As more and more electrons are added, in atoms of higher nuclear charge, the relative energies of the shells and subshells change because the electrons in them experience the potential of other electrons. This is also shown in Fig. 19.3. The position of the subshells on the energy scale is then determined solely by the principal quantum number n. This is the shell structure with which we have been acquainted from x-ray spectroscopy, and it is also being investigated by the newer methods of photoelectron spectroscopy. As discussed in Chap. 18, this is a technique for determining the binding energies of the inner electrons of an atom.

19.2 Ground States of Atoms

We are now familiar with the electronic configurations of the atoms, as far as the quantum numbers n and l are concerned. It remains to discuss the energetic order of the states with different values of m_l and m_s and the combination of the angular momenta of individual electrons to form the total angular momentum of the atom.

Figure 19.4 shows the ground states and the electronic configurations of the first 11 elements. For the beryllium and carbon atoms the lowest excited state is also given. Hydrogen and helium, with the ground states $^2S_{1/2}$ and $^1S_0$ have already been discussed at length. In lithium, the second shell is started, with a $2s$ electron. Beryllium, with a closed $2s$ subshell, has a $^1S_0$ ground state. It can easily be excited to the $2p$ configuration, which is energetically very close to the ground state. The occupation of the $2p$ subshell begins with boron; from its spectrum it is clear that this element has a $^2P_{1/2}$

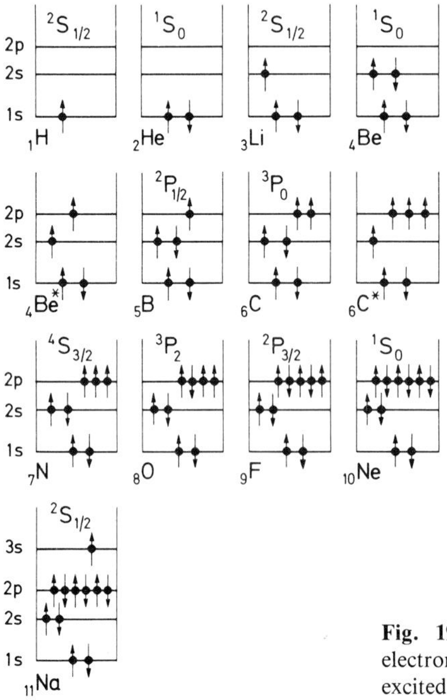

Fig. 19.4. Electronic configurations and arrangement of the electron spins in the ground states of the atoms H to Na. The lowest excited state is also given for beryllium and carbon

ground state, i.e. that its orbital and spin angular momenta point in opposite directions. The spectrum of carbon indicates that the spins of the two $2p$ electrons are parallel, so the ground state is $^3P_0$. The excited state C^* shown in Fig. 19.4, in which there are one $2s$ and three $2p$ electrons, is responsible for the valence of 4 which carbon displays in organic chemistry. The four electrons in the second shell are coupled in this state so that they are energetically equivalent. This leads to what is called sp^3 hybridisation (cf. Sect. 23.7) which effectively determines the character of the chemical bonding.

It turns out that in nitrogen, the three p electrons have parallel spins, so that they form a $^4S_{3/2}$ ground state. Spin saturation of the p electrons then begins with oxygen. Fluorine lacks only one electron in the p subshell. With neon, the noble gas ground state $^1S_0$ is reached again. The occupation of the M shell with $n = 3$ begins with the alkali atom sodium.

There are several rules for the energetic ordering of the electrons within the subshells, which hold in addition to the Pauli principle. In LS coupling, i.e. in all light atoms, the angular momenta in the ground state are governed by *Hund's Rules.* They are:

1) Full shells and subshells contribute nothing to the total angular momenta $\boldsymbol{L}$ and $\boldsymbol{S}$. We have already shown this earlier.

2) The electrons having the same value of l which are divided among the corresponding m_l subshells – called equivalent electrons – are placed into the ground state in such a way that the resulting total spin $\boldsymbol{S}$ is maximised. States with the highest multiplicities thus lie energetically lowest, e.g. triplet states are lower than singlet states. This is a consequence of the Pauli principle, which requires that the total wavefunction be antisymmetric (see Sects. 19.4, where this "antisymmetry" will be discussed in more detail). The higher the multiplicity, the more parallel spins there are, which are thus completely symmetrical in their symmetry properties. Therefore their spatial functions must be antisymmetric. As a result, their binding energy is maximised, because the mutual Coulomb repulsion of the electrons is lowest for antisymmetric spatial functions.

Let us consider the nitrogen ground state as an example. Nitrogen has three electrons in the outermost subshell. It therefore has a doublet and a quartet system, i.e. $S = 1/2$ and $S = 3/2$ (Fig. 17.4). Given the possible combinations of the quantum numbers m_l for the three p electrons of the configuration $1s^2 2s^2 2p^3$, the overall state can be either ^2P, ^2D or ^4S. Of these, the one with the lowest energy is ^4S, which is the state with the highest multiplicity. The other two states actually occur at somewhat higher energy in the doublet part of the nitrogen term scheme (Fig. 17.4).

3) When the highest value of the quantum number S is reached, the Pauli principle requires that the electrons be distributed among the substates m_l in such a way that $L_z = \sum m_l \hbar = m_L \hbar$ is maximised. The resulting angular momentum quantum number L is then equal to $|m_L|$. For a given multiplicity S, the higher values of L give states of lower energy.

4) Finally, when the spin-orbit coupling is also taken into account, the terms with the smallest quantum numbers J have the lowest energy in "normal" multiplets, but otherwise the converse holds. "Normal" means here that the subshells are less than half full.

This rule is a consequence of the fine structure calculation which was carried out earlier for the one-electron atom (Sect. 12.8). As the negative charge circulates around the positive nucleus, the orbital magnetic field at the position of the electron is directed in such a way that an antiparallel orientation of $\boldsymbol{L}$ and $\boldsymbol{S}$ corresponds to a minimum energy. If the shell is more than half full, however, each electron which would be needed to fill the shell is equivalent to a positively charged "hole". The sign of the magnetic field $\boldsymbol{B}_L$ then changes, and the state with the highest J has the lowest energy. In Fig. 19.4 this is shown by comparison of the one-electron atom lithium (ground state $^2S_{1/2}$) and the "one-hole" atom fluorine (ground state $^2P_{3/2}$).

The carbon atom can serve as an example here. Its term scheme is shown in Fig. 17.5. In the ground state, the carbon atom has two electrons in the outermost subshell. It therefore has a singlet and a triplet system with $S=0$ and $S=1$. The possible ground states for the electronic configuration $1s^2 2s^2 2p^2$ are ^1S, ^1D and ^3P. Because it has the largest multiplicity, ^3P has the lowest energy, as can also be seen in Fig. 17.5. If the angular momentum quantum number J is included, the possibilities are $^3P_0$, $^3P_1$ and $^3P_2$, i.e. $J=0, 1$ and 2. In a "normal" order of terms, the state $^3P_0$ has the lowest energy, by rule 4. The singlet terms are also observed, at somewhat higher energy, as the term scheme in Fig. 17.5 and Fig. 19.5 shows.

In the oxygen atom, however, the possible ground states for the $1s^2 2s^2 2p^4$ configuration are ^1S, ^1D and ^3P. According to rule 2, ^3P has the lowest energy. Now, however, with an inverted term order, the state with the largest value of J (rule 4) has the very lowest energy. The ground state is therefore $^3P_2$.

19.3 Excited States and Complete Term Scheme

Several complete term schemes for atoms have already been shown in the figures. Each of the energy terms indicated there corresponds to a particular electron configuration and to a certain type of coupling of the electrons in non-filled shells. The energetic positions of these terms are uniquely determined by the energies of interaction between the nucleus and electrons and between the electrons, themselves. Quantitative calculations are extremely difficult, because atoms with more than one electron are complicated "many-particle" systems.

However, using a few examples, we shall consider how many different terms are possible for a given electronic configuration and how these are arranged energetically.

Our first example is an atom with two p electrons in unfilled shells, in the configuration $(np)^1(n'p)^1$. If $n=n'$, the two electrons are equivalent and the configuration is np^2. Carbon is a concrete example of this case.

Figure 19.5 shows the terms discussed already in Sect. 19.2 which are possible if the two p electrons are coupled. First of all, the two spins can be either parallel or antiparallel. The position functions associated with these alignments differ with respect to Coulomb repulsion, so that according to Hund's rule, the state with the multiplicity $S=1$ has a lower energy. Thus one obtains an $S=1$ (triplet) and an $S=0$ (singlet) term scheme, with the triplet scheme having lower energies. In addition, the orbital angular momenta $l_1=1$ and $l_2=1$ can couple to give $L=2, 1$ and 0. This produces a D, a P and an S state. The state with the highest value of L has the lowest energy. This, like the other Hund rules, is not adequately understandable by intuition alone.

For the triplet terms ^3P and ^3D, the spin-orbit interaction leads to a further splitting into three states each. The singlet terms and the ^3S term are not subject to fine structure splitting, as we have seen earlier. Thus we obtain, in all, 10 spectral terms. In an external magnetic field, all those terms which have a total angular momentum $\boldsymbol{J}$ not equal to zero are further split into the m_J states. The terms indicated by dashed lines in Fig. 19.5 are not possible for equivalent electrons (i.e. in the np^2 configuration) due to

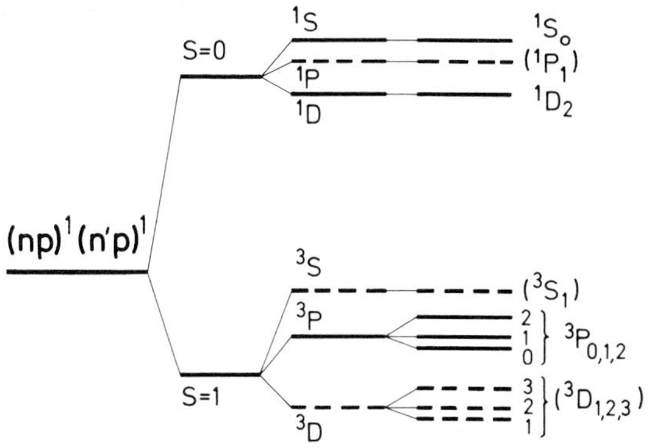

Fig. 19.5. Coupling of two p electrons. If the two electrons are equivalent (np^2 configuration), the terms indicated by dashed lines are forbidden by the Pauli principle. The presence of an external magnetic field lifts the J degeneracy and leads to further splitting of the terms, which is not shown here

the Pauli principle. Since the quantum numbers n and l are the same for the two electrons in these configurations, they must differ in their m_l or m_s quantum numbers. In this case, the terms ^1P, ^3S and ^3D would disobey the Pauli principle, and they are thus forbidden for equivalent electrons. For these terms, it is impossible to construct an antisymmetrical wavefunction for two p electrons with the same principal quantum number n.

For two non-equivalent s electrons, i.e. for the configuration $(ns)^1(n's)^1$, the only possible spectral terms are $^3S_1$ and $^1S_0$. For the configuration $(nd)^1(n'p)^1$, $^1P_1$, $^1D_2$, $^1F_3$, $^3P_{0,1,2}$, $^3D_{1,2,3}$ and $^3F_{2,3,4}$ are possible. It is left to the reader to confirm this. For heavy atoms, LS coupling is replaced by jj coupling.

All those terms which arise from allowed electronic configurations of the atom are given in the term schemes, for example those in Fig. 17.4, 5 Together with the selection rules for optical transitions

$$\begin{aligned}
&\Delta J = 0, \pm 1 \quad [\text{except for } (J=0) \rightarrow (J=0)] \\
&\Delta m_J = 0, \pm 1 \quad [\text{except for } (m_J=0) \rightarrow (m_J=0) \text{ where } \Delta J = 0] \\
&\left.\begin{aligned}&\left.\begin{aligned}&\Delta S = 0\\ &\Delta L = 0, \pm 1\end{aligned}\right\} \text{for the atom}\\ &\Delta l = \pm 1 \text{ for the electron changing its configuration}\end{aligned}\right\} \text{in } LS \text{ coupling} \\
&\Delta j = 0, \pm 1 \text{ for one of the electrons in } jj \text{ coupling ,}
\end{aligned}$$

the term schemes and the spectra of all atoms are explained.

19.4 The Many-Electron Problem. Hartree-Fock Method*

19.4.1 The Two-Electron Problem

In Chap. 10, we were able to solve exactly the hydrogen problem, in which only a single electron orbits the nucleus. Unfortunately, there is no exact solution for any other atom with more than one electron. In spite of this, it is possible, usually to a very good approximation, to calculate wavefunctions and energies. In order to demonstrate the problems which arise here, we shall consider an atom with two electrons, for example the helium atom or a multiply ionised atom in which only two electrons remain

(Fig. 19.6). We distinguish the coordinates of the two electrons by using the subscripts $j=1$ or $j=2$. If only one electron were present, the Hamiltonian would be

$$\mathscr{H}_j = -\frac{\hbar^2}{2m_0}\nabla_j^2 - \frac{Ze^2}{4\pi\varepsilon_0 r_j}\,, \tag{19.2}$$

where the Laplace operator is defined by

$$\nabla_j^2 = \frac{\partial^2}{\partial x_j^2} + \frac{\partial^2}{\partial y_j^2} + \frac{\partial^2}{\partial z_j^2}\,. \tag{19.3}$$

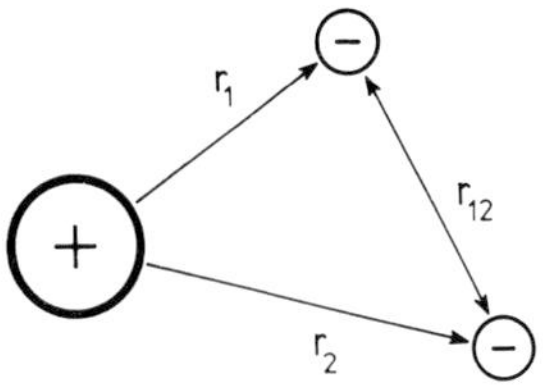

Fig. 19.6. Atom with two electrons

If we ignore the interactions between the electrons, in a *classical* treatment the energy of the whole system of two electrons is simply equal to the energy of the individual electrons. The Hamiltonian of the whole system is thus equal to the sum of the Hamiltonians of the two electrons. If we make use of the translation rule of quantum theory, which says that the kinetic energy is always to be replaced by the Laplacian (19.3), multiplied by $-\hbar^2/2m_0$, we obtain for the two-electron system the Hamiltonian

$$\mathscr{H}^0 = \mathscr{H}_1 + \mathscr{H}_2\,, \tag{19.4}$$

where the Hamiltonian (19.2) is used. Actually, however, there is a direct interaction between the electrons. The most important part of this is the Coulomb interaction energy, so that we must replace (19.4) by

$$\mathscr{H} = \mathscr{H}_1 + \mathscr{H}_2 + \frac{e^2}{4\pi\varepsilon_0 r_{12}}\,, \tag{19.5}$$

where r_{12} is the distance between the two electrons. Our problem is to find an exact solution to the Schrödinger equation belonging to (19.5). As we remarked above, this is not possible in closed form. Therefore, we first consider the simpler problem in which the Coulomb interaction between the two electrons is left out and the Hamiltonian (19.4) is used in the Schrödinger equation. We make use of the fact that the single-particle Schrödinger equation with the Hamiltonian (19.2) has already been solved. In order to stay closer to reality, we include the spin of the electron and introduce the spin variable along with the position vector $\boldsymbol{r}_k$. As an abbreviation, we use the variable $\boldsymbol{R}_k$, which is defined as

$$\boldsymbol{R}_k = (\boldsymbol{r}_k, \text{spin variable}) \equiv (\boldsymbol{r}_k, k)\,. \tag{19.6}$$

As we saw in Chap. 10, the wavefunction of the one-electron problem is characterised by the quantum numbers n, l, m_l, and m_s: Ψ_{n,l,m_l,m_s}. The energy of this function is E_{n,l,m_l,m_s}. We now consider the Schrödinger equation for the total Hamiltonian (19.4):

$$\mathscr{H}^0 \Psi = E_t \Psi \,, \tag{19.7}$$

where E_t is an abbreviation for E_{tot}. Since the Hamiltonian $\mathscr{H}^0$ (19.4) includes the two variables $\boldsymbol{R}_1$ and $\boldsymbol{R}_2$, the wavefunction must naturally depend on these variables: $\Psi(\boldsymbol{R}_1, \boldsymbol{R}_2)$. In order to avoid writing out the quantum numbers explicitly in each of the following expressions, we introduce the abbreviation Q for the total set of quantum numbers,

$$Q = (n, l, m_l, m_s) \equiv (q, m_s) \,. \tag{19.8}$$

The Schrödinger equation (19.7) is solved by the wavefunction

$$\Psi(\boldsymbol{R}_1, \boldsymbol{R}_2) = \Psi_{Q_1}(\boldsymbol{R}_1)\, \Psi_{Q_2}(\boldsymbol{R}_2) \,, \tag{19.9}$$

as can be immediately demonstrated by substitution. The Ψ's on the right side of the equation are solutions for the one-particle Schrödinger equations. The total energy is, as can also be easily demonstrated,

$$E_t = E_{Q_1} + E_{Q_2} \,, \tag{19.10}$$

where the energies E on the right side are the one-particle energies associated with the quantum numbers Q_1 and Q_2. It would thus appear that we have reproduced, with no limitations, the above-mentioned classical result, that the one-particle energies are simply additive, if we ignore the electron interactions.

The solution (19.9) also appears to allow the case of $Q_1 = Q_2$, which would mean that the two electrons would have exactly the same four quantum numbers. According to the Pauli principle, however, this case is excluded. Theoretical physicists have therefore considered whether a wavefunction can be found which automatically excludes this case. To find it, we make use of the fact that not only (19.9), but also the wavefunction

$$\Psi_{Q_2}(\boldsymbol{R}_1)\, \Psi_{Q_1}(\boldsymbol{R}_2) \tag{19.11}$$

satisfies the Schrödinger equation (19.7), and yields exactly the same energy (19.10) as (19.9). As we know, any linear combination of wavefunctions which have the same energy is also a solution of the Schrödinger equation and has the same energy. The linear combination which automatically fulfils the Pauli principle is a difference between (19.9) and (19.11), namely

$$\Psi(\boldsymbol{R}_1, \boldsymbol{R}_2) = \frac{1}{\sqrt{2}} [\Psi_{Q_1}(\boldsymbol{R}_1)\, \Psi_{Q_2}(\boldsymbol{R}_2) - \Psi_{Q_2}(\boldsymbol{R}_1)\, \Psi_{Q_1}(\boldsymbol{R}_2)] \,. \tag{19.12}$$

This wavefunction vanishes identically if $Q_1 = Q_2$. The factor $1/\sqrt{2}$ serves to normalise the whole wavefunction. If we exchange the coordinates $\boldsymbol{R}_1$ and $\boldsymbol{R}_2$ in

(19.12), the wavefunction is obviously converted to its additive inverse, or in other words, the wavefunction is *antisymmetric*. This is now a formulation which makes possible a statement of the Pauli principle even when there is interaction between the electrons: *The wavefunction must be antisymmetric with respect to the coordinates* $\boldsymbol{R}_1$ *and* $\boldsymbol{R}_2$. It should be remembered that the variables $\boldsymbol{R}_j$ include the spin variables.

Let us now investigate the rôle of the spin variables and of the corresponding spin wavefunctions.

To this end, we separate the total set of quantum numbers Q_k defined in (19.8) into

$$Q_k = (q_k, m_{s,k}) \ , \tag{19.13}$$

as well as separating the variables $\boldsymbol{R}_k$ as in (19.6).

We now write the wavefunction $\Psi_{Q_k}(\boldsymbol{R}_k)$ as the product of a wavefunction $\psi_{q_k}(\boldsymbol{r}_k)$ for the orbital motion (space function) and a spin wavefunction $\phi_{m_{s,k}}(k)$ (spin function). We then have instead of (19.12)

$$\begin{aligned}\Psi(\boldsymbol{R}_1, \boldsymbol{R}_2) = \frac{1}{\sqrt{2}} & (\psi_{q_1}(\boldsymbol{r}_1)\, \psi_{q_2}(\boldsymbol{r}_2)\, \phi_{m_{s,1}}(1)\, \phi_{m_{s,2}}(2) \\ & - \psi_{q_1}(\boldsymbol{r}_2)\, \psi_{q_2}(\boldsymbol{r}_1)\, \phi_{m_{s,1}}(2)\, \phi_{m_{s,2}}(1)).\end{aligned} \tag{19.14}$$

For simplicity, we characterise the spin quantum numbers $m_s = \pm\frac{1}{2}$ by $\uparrow$ (spin up) or $\downarrow$ (spin down). We then find the following possibilities:

$$1)\quad m_{s,1} = \uparrow \ , \qquad m_{s,2} = \uparrow \tag{19.15}$$

$$2)\quad m_{s,1} = \downarrow \ , \qquad m_{s,2} = \downarrow \tag{19.16}$$

$$3)\quad m_{s,1} = \uparrow \ , \qquad m_{s,2} = \downarrow \tag{19.17}$$

$$4)\quad m_{s,1} = \downarrow \ , \qquad m_{s,2} = \uparrow \ . \tag{19.18}$$

In the wavefunctions belonging to 1) and 2), we can factor out the spin functions and obtain

$$1)\ \Psi_{\uparrow\uparrow}(\boldsymbol{R}_1, \boldsymbol{R}_2) = \frac{1}{\sqrt{2}} (\psi_{q_1}(\boldsymbol{r}_1)\, \psi_{q_2}(\boldsymbol{r}_2) - \psi_{q_1}(\boldsymbol{r}_2)\, \psi_{q_2}(\boldsymbol{r}_1)) \times \phi_{\uparrow}(1)\, \phi_{\uparrow}(2) \tag{19.19}$$

and

$$2)\ \Psi_{\downarrow\downarrow}(\boldsymbol{R}_1, \boldsymbol{R}_2) = \frac{1}{\sqrt{2}} (\psi_{q_1}(\boldsymbol{r}_1)\, \psi_{q_2}(\boldsymbol{r}_2) - \psi_{q_1}(\boldsymbol{r}_2)\, \psi_{q_2}(\boldsymbol{r}_1)) \times \phi_{\downarrow}(1)\, \phi_{\downarrow}(2) \ . \tag{19.20}$$

The wavefunction for two electrons may be written as a product of the space function and the spin function:

$$\Psi(\boldsymbol{R}_1, \boldsymbol{R}_2) = \psi(\boldsymbol{r}_1, \boldsymbol{r}_2)\, \Phi(1,2) \ , \qquad \text{where} \tag{19.21}$$

$$\Phi_{\uparrow\uparrow}(1,2) = \phi_{\uparrow}(1)\, \phi_{\uparrow}(2) \qquad \text{and} \tag{19.22}$$

$$\Phi_{\downarrow\downarrow}(1,2) = \phi_{\downarrow}(1)\, \phi_{\downarrow}(2) \ . \tag{19.23}$$

Let us consider the precise meaning of the spin wavefunctions Φ. Since they refer to two electrons, it seems reasonable of inquire into the total spin of the pair of electrons. The z component of the latter corresponds to the operator

$$\Sigma_z \equiv \sigma_{z,1} + \sigma_{z,2} \tag{19.24}$$

where $\sigma_{z,k}$ is the operator for the z component of the spin of electron k. As one can quickly verify, the following relations hold:

$$\Sigma_z \phi_\uparrow(1)\,\phi_\uparrow(2) = \hbar\,\phi_\uparrow(1)\,\phi_\uparrow(2) \quad \text{and} \tag{19.25}$$

$$\Sigma_z \phi_\downarrow(1)\,\phi_\downarrow(2) = -\hbar\,\phi_\downarrow(1)\,\phi_\downarrow(2)\ , \tag{19.26}$$

i.e. the spin functions (19.22, 23) are eigenfunctions of (19.24). Likewise, one can verify (cf. Problem 19.8) that they are simultaneously eigenfunctions of

$$\Sigma^2 \equiv (\boldsymbol{\sigma}_1 + \boldsymbol{\sigma}_2)^2 \tag{19.27}$$

with the eigenvalue $\frac{3}{4}\hbar^2$.

Let us now return to the possible spin arrangements (19.15 – 18) and write down the remaining wavefunctions which correspond to (19.17, 18):

$$3)\ \Psi(\boldsymbol{R}_1, \boldsymbol{R}_2) = \frac{1}{\sqrt{2}}(\psi_{q_1}(\boldsymbol{r}_1)\,\psi_{q_2}(\boldsymbol{r}_2)\,\phi_\uparrow(1)\,\phi_\downarrow(2) - \psi_{q_1}(\boldsymbol{r}_2)\,\psi_{q_2}(\boldsymbol{r}_1)\,\phi_\uparrow(2)\,\phi_\downarrow(1)) \tag{19.28}$$

$$4)\ \Psi(\boldsymbol{R}_1, \boldsymbol{R}_2) = \frac{1}{\sqrt{2}}(\psi_{q_1}(\boldsymbol{r}_1)\,\psi_{q_2}(\boldsymbol{r}_2)\,\phi_\downarrow(1)\,\phi_\uparrow(2) - \psi_{q_1}(\boldsymbol{r}_2)\,\psi_{q_2}(\boldsymbol{r}_1)\,\phi_\downarrow(2)\,\phi_\uparrow(1))\ . \tag{19.29}$$

These may evidently *not* be written in the form of products (19.21). However, as we know, we may take linear combinations of wavefunctions which belong to the same energy; in fact we must do so when applying perturbation theory to degenerate states. We thus take the sum of 3) and 4) and obtain after rearranging

$$\begin{aligned} 5) = 3) + 4) &= \frac{1}{\sqrt{2}}(\psi_{q_1}(\boldsymbol{r}_1)\,\psi_{q_2}(\boldsymbol{r}_2) - \psi_{q_1}(\boldsymbol{r}_2)\,\psi_{q_2}(\boldsymbol{r}_1)) \times (\phi_\uparrow(1)\,\phi_\downarrow(2) + \phi_\uparrow(2)\,\phi_\downarrow(1)) \\ &= \Psi_{\uparrow\downarrow+\downarrow\uparrow}(\boldsymbol{R}_1, \boldsymbol{R}_2) \end{aligned} \tag{19.30}$$

and

$$\begin{aligned} 6) = 3) - 4) &= \frac{1}{\sqrt{2}}(\psi_{q_1}(\boldsymbol{r}_1)\,\psi_{q_2}(\boldsymbol{r}_2) + \psi_{q_1}(\boldsymbol{r}_2)\,\psi_{q_2}(\boldsymbol{r})) \times (\phi_\uparrow(1)\,\phi_\downarrow(2) - \phi_\uparrow(2)\,\phi_\downarrow(1)) \\ &= \Psi_{\uparrow\downarrow-\downarrow\uparrow}(\boldsymbol{R}_1, \boldsymbol{R}_2)\ . \end{aligned} \tag{19.31}$$

These new wavefunctions thus indeed have the desired form (19.21) and are furthermore eigenfunctions of Σ_z and Σ^2:

$$\Sigma_z(\phi_\uparrow(1)\,\phi_\downarrow(2) \pm \phi_\uparrow(2)\,\phi_\downarrow(1)) = 0 \tag{19.32}$$

$$\Sigma^2(\phi_\uparrow(1)\,\phi_\downarrow(2)+\phi_\uparrow(2)\,\phi_\downarrow(1)) = \tfrac{1}{4}\hbar^2(\phi_\uparrow(1)\,\phi_\downarrow(2)+\phi_\uparrow(2)\,\phi_\downarrow(1)) \tag{19.33}$$

$$\Sigma^2(\phi_\uparrow(1)\,\phi_\downarrow(2)-\phi_\uparrow(2)\,\phi_\downarrow(1)) = 0 \ . \tag{19.34}$$

Clearly, the wavefunction

$$\Phi = \phi_\uparrow(1)\,\phi_\downarrow(2)-\phi_\uparrow(2)\,\phi_\downarrow(1) \tag{19.35}$$

corresponds to a singlet state ($S=0$), while the spin wave-functions

$$\Phi = \begin{cases} \phi_\uparrow(1)\,\phi_\uparrow(2) & (19.36)\\ \phi_\uparrow(1)\,\phi_\downarrow(2)+\phi_\uparrow(2)\,\phi_\downarrow(1) & (19.37)\\ \phi_\downarrow(1)\,\phi_\downarrow(2) & (19.38)\end{cases}$$

correspond to a triplet state ($S=1$) with $M_{s,z}=1,0,-1$.

The corresponding space functions, which are the wavefunctions for the orbital motion, have differing symmetries with respect to the space coordinates in the singlet and the triplet states.

The triplet state has antisymmetric space functions; thus, the probability density for both electrons being in the same place vanishes, and it becomes small when the electrons approach each other. The positive Coulomb repulsion energy between the electrons is thus lower for the triplet state than for the singlet state, which has a space function symmetric in the space coordinates. This forms the theoretical basis for Hund's rules (see Sect. 19.2).

19.4.2 Many Electrons Without Mutual Interactions

The considerations which led us to the wave function (19.12) can be generalised. We shall give the results here without proof. In the general case, we are concerned with the variables $\boldsymbol{R}_1, \dots \boldsymbol{R}_N$ of N electrons. In the absence of interactions among the electrons, the Hamiltonian is a sum of one-electron Hamiltonians:

$$\mathcal{H}^0 = \sum_{j=1}^{N} \mathcal{H}_j \,. \tag{19.39}$$

The solution of the Schrödinger equation

$$\mathcal{H}^0 \Psi = E_t \Psi \tag{19.40}$$

associated with (19.39) is naturally a function of the coordinates $\boldsymbol{R}_1, \dots \boldsymbol{R}_N$:

$$\Psi(\boldsymbol{R}_1, \boldsymbol{R}_2, \dots \boldsymbol{R}_N) \,. \tag{19.41}$$

Retaining the notation introduced above, we can immediately show that the Schrödinger equation (19.40) is solved by the product

$$\Psi(\boldsymbol{R}_1, \boldsymbol{R}_2, \dots \boldsymbol{R}_N) = \Psi_{Q_1}(\boldsymbol{R}_1)\,\Psi_{Q_2}(\boldsymbol{R}_2)\dots\Psi_{Q_N}(\boldsymbol{R}_N) \,. \tag{19.42}$$

This approach is often called the Hartree method. The energy of the solution is given by

$$E_t = E_{Q_1} + E_{Q_2} + \ldots + E_{Q_N}, \tag{19.43}$$

where the energies on the right side are again the energies of the individual electrons.

The solution (19.42) is not yet compatible with the Pauli principle, because it allows solutions $Q_i = Q_k$ for the pair i, k. The solution which is compatible with the Pauli principle is given by a determinant of the form

$$\Psi(\boldsymbol{R}_1, \boldsymbol{R}_2, \ldots \boldsymbol{R}_N) = \frac{1}{\sqrt{N!}} \begin{vmatrix} \Psi_{Q_1}(\boldsymbol{R}_1)\, \Psi_{Q_1}(\boldsymbol{R}_2) & \ldots \Psi_{Q_1}(\boldsymbol{R}_N) \\ \Psi_{Q_2}(\boldsymbol{R}_1)\, \Psi_{Q_2}(\boldsymbol{R}_2) & \ldots \\ \vdots & \\ \Psi_{Q_N}(\boldsymbol{R}_1)\, \Psi_{Q_N}(\boldsymbol{R}_2) \ldots & \Psi_{Q_N}(\boldsymbol{R}_N) \end{vmatrix} . \tag{19.44}$$

A determinant changes its sign if two rows or two columns are exchanged. If we exchange two variables $\boldsymbol{R}_i$ and $\boldsymbol{R}_k$, this is equivalent to exchanging two columns in (19.44). Thus (19.44) guarantees the antisymmetry of the wavefunction. Since a determinant is zero if any two rows or columns are identical, we see that (19.44) vanishes if $Q_i = Q_k$ for any pair i, k.

We shall now discuss a special case of (19.44), namely the case that all the electrons have parallel spins. Then the spin quantum number m_s is e.g. 1/2 for all j. If we also ignore spin-orbit coupling, then the wavefunction of an individual electron can be written as a product (where we write just m for the quantum number m_l):

$$\Psi_{Q_k}(\boldsymbol{R}_k) = \psi_{n_k, l_k, m_k}(\boldsymbol{r}_j) \cdot \phi_\uparrow(k) \tag{19.45}$$

where ψ refers to the orbital motion and ϕ to the spin. As can be shown using elementary rules for determinants, substitution of (19.44) in (19.45) yields

$$\Psi(\boldsymbol{R}_1, \ldots \boldsymbol{R}_N) = \frac{1}{\sqrt{N!}} \begin{vmatrix} \psi_{n_1, l_1, m_1}(\boldsymbol{r}_1) \ldots \psi_{n_1, l_1, m_1}(\boldsymbol{r}_N) \\ \vdots \\ \psi_{n_N, l_N, m_N}(\boldsymbol{r}_1) \end{vmatrix} \cdot \phi_\uparrow(1) \ldots \phi_\uparrow(N), \tag{19.46}$$

i.e. the total wavefunction is now a product of a spin function which is symmetric (because the spin quantum numbers of the electrons are identical) and an antisymmetric wavefunction, which is represented by a determinant as in (19.46) and depends only on the position variable $\boldsymbol{r}_j$.

If we set the coordinates for a pair i, k equal, $\boldsymbol{r}_i = \boldsymbol{r}_k$, in the determinant, it will of course vanish. This means that two electrons with parallel spins cannot occupy the same position. Since the wavefunction ψ is continuous, so is Ψ. This means that the probability of finding two electrons at the same position goes to zero continuously if the coordinates of the two become equal. The Pauli principle thus provides automatically for a certain distance between electrons with parallel spins.

19.4.3 Coulomb Interaction of Electrons. Hartree and Hartree-Fock Methods

We now turn to the actual problem, in which the Coulomb interaction among the electrons is taken into account. The energy of the Coulomb interaction between the pair of electrons j, k is given by $e^2/(4\pi\varepsilon_0 r_{jk})$ (r_{jk} is the distance between the two electrons), so the Hamiltonian is

$$\mathcal{H} = \sum_{j=1}^{N} \mathcal{H}_j + \sum_{j<k} \frac{e^2}{4\pi\varepsilon_0 r_{jk}}. \tag{19.47}$$

The summation rule $j<k$ prevents the interaction energy between pairs of electrons being counted twice. If instead of using this rule, we sum over all indices j and k, with the limitation that $j \neq k$, we must set a factor 1/2 in front of the interaction sum:

$$\mathcal{H} = \sum_{j=1}^{N} \mathcal{H}_j + \frac{1}{2} \sum_{j \neq k} \frac{e^2}{4\pi\varepsilon_0 r_{jk}}. \tag{19.48}$$

Our next task is to solve the Schrödinger equation

$$\mathcal{H}\Psi = E_t \Psi \tag{19.49}$$

for the Hamiltonian (19.48). Since there is no exact solution we follow this train of thought: First, since the one-electron problem has already been solved, the wavefunctions of the individual electrons, in the form

$$\Psi_Q(\boldsymbol{R}_j) \tag{19.50}$$

are already known. There is a charge density distribution associated with the wavefunction (19.50),

$$\varrho(\boldsymbol{r}_j) = e\,|\Psi_Q(\boldsymbol{R}_j)|^2. \tag{19.51}$$

As we know from electrostatics, there is an interaction energy between a charge at position $\boldsymbol{r}$ and the charge distribution given by (19.51). This energy is given by the product of the charge and the electrostatic potential. The latter can be calculated from the charge distribution ϱ. In all, the Coulomb interaction energy is expressed by

$$V(\boldsymbol{r}) = \frac{1}{4\pi\varepsilon_0} \int \frac{e\varrho(\boldsymbol{r}_j)}{|\boldsymbol{r}-\boldsymbol{r}_j|} d\tau_j, \tag{19.52}$$

where the integral covers the total volume (the volume element is indicated here by $d\tau$, in order to distinguish it from the interaction energy V). If we substitute (19.51) in (19.52), we obtain

$$V(\boldsymbol{r}) = \frac{1}{4\pi\varepsilon_0} \int \frac{e^2 |\Psi_Q(\boldsymbol{R}_j)|^2}{|\boldsymbol{r}-\boldsymbol{r}_j|} d\tau_j. \tag{19.53}$$

The main idea of the Hartree and Hartree-Fock methods, which are discussed below, is to reduce the many-electron problem to a one-electron problem. Let us consider a single electron. It is moving not only in the field of the atomic nucleus, but also in the field of all the other electrons. The simplifying assumption is then made that the electron density distribution of all the other electrons can be given, as a first approximation, by one-electron wavefunctions (19.50). In order to calculate the wavefunction of the chosen electron, we must solve a Schrödinger equation in which both the Coulomb potential of the nucleus and the interaction energy with all the other electrons appear. If the chosen electron has the subscript k, and thus the coordinate $\boldsymbol{R}_k$, the Schrödinger equation is

$$\left[-\frac{\hbar^2}{2m_0}\nabla_k^2-\frac{Ze^2}{4\pi\varepsilon_0 r_k}+V_k^{(0)}(\boldsymbol{r}_k)\right]\psi_k^{(1)}(\boldsymbol{R}_k)=E\psi_k^{(1)}(\boldsymbol{R}_k)\,. \tag{19.54}$$

$V_k^{(0)}$ is the Coulomb interaction energy with all the other electrons, and is obtained from

$$V_k(\boldsymbol{r})=\sum_{j=1}^{N}{}'\int\frac{e^2|\psi_{Q_j}(\boldsymbol{R}_j)|^2}{4\pi\varepsilon_0|\boldsymbol{r}-\boldsymbol{r}_j|}d\tau_j \tag{19.55}$$

by substituting, as a first approximation, the wavefunction $\psi_{Q_j}^{(0)}$ for ψ_{Q_j} in (19.55) [1]. The superscript (0) means that we use a given (or guessed) wavefunction to start the whole procedure. In particular, $\psi_{Q_j}^{(0)}$ may be a wavefunction belonging to a given potential. Similarly, the superscript (1) indicates that this wavefunction for the electron k was obtained from (19.54) in the first step of an iterative process. In the second step, we now use the wavefunction $\psi^{(1)}$ for ψ in (19.55). This yields the new wavefunction $\psi^{(2)}$ in a Schrödinger equation analogous to (19.54). The process is repeated until there are no more significant changes in the wavefunctions $\psi^{(j)}$, or in other words, until the method converges. Seen schematically, we have

$$\psi^{(0)}\to V^{(0)}\to\psi^{(1)}\to V^{(1)}\to\psi^{(2)}\to V^{(2)}\ldots\psi^{(j)}\to\psi\,. \tag{19.56}$$

The method described above is somewhat heuristic, of course. It is desirable to set it on a firm mathematical basis, which is possible, but we cannot go into the details here because of space limitations. It can be shown that the Schrödinger equation (19.49) may be solved by using a variational principle. According to this principle, the expression

$$\int\ldots\int\psi^*\mathscr{H}\psi\,d\tau_1\ldots d\tau_N \tag{19.57}$$

must be equal to an extremum (maximum or minimum) with the secondary condition that the wavefunctions are normalised:

$$\int\Psi^*\Psi\,d\tau_1\ldots d\tau_N=1\,. \tag{19.58}$$

If we use a wavefunction Ψ of the form (19.42), i.e. a product wavefunction in such a variational procedure, we find a set of Schrödinger equations for the individual wavefunctions:

$$\left[-\left(\frac{\hbar^2}{2m_0}\right)\nabla_k^2-\frac{Ze^2}{4\pi\varepsilon_0 r_k}+V_k(\boldsymbol{r}_k)\right]\psi_{Q_k}(\boldsymbol{R}_k)=E\psi_{Q_k}(\boldsymbol{R}_k)\,, \tag{19.59}$$

where $V_k(\boldsymbol{r}_k)$ is defined by (19.55). The Hartree method just consists of solving these Schrödinger equations (19.59) iteratively in the way indicated above.

[1] The prime sign ($'$) on the sum in (19.55) indicates that $j\neq k$

The disadvantage of the Hartree method is obvious, in that it uses product trial functions for Ψ, which, as we know, violate the Pauli principle. The key to expanding the Hartree method to include the Pauli principle lies in utilising the determinant trial functions (19.44) for Ψ in the variation equations (19.57) and (19.58). The corresponding calculations are rather long; we shall therefore simply give the result here. For the individual wavefunctions, we find a set of Schrödinger equations of the form

$$\left[-\left(\frac{\hbar^2}{2m_0}\right)\nabla_k^2-\frac{Ze^2}{4\pi\varepsilon_0 r_k}+V_k(\boldsymbol{r}_k)\right]\psi_{Q_k}(\boldsymbol{R}_k)$$
$$-\sum_j{}'\frac{1}{4\pi\varepsilon_0}\int\psi^*_{Q_j}(\boldsymbol{R}_j)\frac{e^2}{|\boldsymbol{r}_k-\boldsymbol{r}_j|}\psi_{Q_k}(\boldsymbol{R}_j)\,d\tau_j\cdot\psi_{Q_j}(\boldsymbol{R}_k)=E\psi_{Q_k}(\boldsymbol{R}_k)\,. \qquad (19.60)$$

The term in brackets on the left-hand side, and the right-hand side of (19.60) agree with the Hartree equation (19.59). The additional term containing the sum over j is new; it is referred to as an exchange term. The reason for this terminology is the following: if one compares $V_k\cdot\psi_{Q_k}$ from (19.60) with this exchange term, one recognises that the wavefunction ψ_{Q_k} has exchanged rôles with the wavefunction ψ_{Q_j}, since the electron coordinates $\boldsymbol{R}_k$ and $\boldsymbol{R}_j$ have been exchanged. This exchange term is a direct result of the determinant approach, i.e. of the antisymmetry of the wavefunctions. Intuitively explained, it means that the Coulomb interaction energy between electrons with parallel spins is reduced relative to that between electrons with antiparallel spins. Since the interaction potentials V themselves depend on the wavefunctions ψ, a solution to (19.60) is difficult to find. For a solution, one therefore resorts again to the Hartree-Fock method, according to which (19.60) is solved iteratively following the scheme (19.56).

Problems

19.1 How many electrons do those atoms have in which the following shells are filled in the ground state:

a) the K and the L shells, the $3s$ subshell and half the $3p$ subshell?
b) the K, L and M shells and the $4s$, $4p$ and $4d$ subshells?

What are the two elements in (a) and (b)?

19.2 Show that for a closed nl shell, $L=S=0$.

19.3 a) Two equivalent p electrons have strong spin-orbit coupling. Calculate the possible values of the total angular momentum quantum number, if the coupling is purely jj. Remember to take the Pauli principle into account.

b) Consider the same problem in the case of weak spin-orbit coupling, so that LS coupling of the two p electrons can be assumed. Do the same values of J occur with equal frequency in the two cases?

Hint: Since the particles cannot be distinguished, configurations which are the same except for the exchange of electron indices may only be counted once. In case (a), the

Pauli principle can be taken into account in this way, because the two electrons cannot have identical sets of quantum numbers. In case (b), the Pauli principle is taken into acount by the requirement that the wavefunction of the overall state must change its sign when the particle indices are exchanged (antisymmetry under particle exchange).

19.4 Give the terms for the following configurations and indicate in each case which term has the lowest energy:
(a) ns, (b) np^3, (c) $(np)^2(n's)$, (d) np^5, (e) $(nd^2)(n'p)$, (f) $(nd)(n'd)$.

19.5 In a diagram, show the occupied electronic states of the Si, Cl and As atoms when the atoms are in the ground state configurations.

In each case, write the electronic configuration and the terms of the ground state.

19.6 Give the ground state configuration and the number of unpaired electrons in the following atoms:

a) S
b) Ca
c) Fe
d) Br .

19.7 Calculate the terms for the np^3 configuration. Give the values of S, L and J for the terms

$${}^1S_0,\ {}^2S_{1/2},\ {}^1P_1,\ {}^3P_2,\ {}^3F_4,\ {}^5D_1,\ {}^1D_2 \text{ and } {}^6F_{9/2}\,.$$

Decide which of the terms correspond to the nd^2 configuration. Use your result to determine the ground and first excited states of titanium.

19.8 Demonstrate that the relations (19.25, 26), and (19.32) are correct. Show also that (19.35 – 38) are eigenfunctions of $\boldsymbol{\Sigma}^2$ and confirm the eigenvalues given in the chapter.

Hint: Write $\boldsymbol{\Sigma}^2$ as $(\sigma_1+\sigma_2)^2=\sigma_1^2+\sigma_2^2+2\sigma_1\sigma_2$.

19.9 Is there a triplet state when the quantum numbers of the two electrons are identical? Justify your answer.

Hint: Consider the symmetry of the spatial wavefunction.

20. Nuclear Spin, Hyperfine Structure

20.1 Influence of the Atomic Nucleus on Atomic Spectra

The atomic nucleus generates the central field in which the electrons of the atom move. The nuclear charge number (atomic number) therefore appears in all the equations for spectral energy terms of atoms. In addition, there are other influences of the nucleus on the electronic spectra, the measurement of which generally requires an especially high spectral resolution. Therefore, one refers to a *hyperfine structure* of the spectral lines. By contrast, the effects of the interactions of the electrons among themselves as well as those of electronic spins with the orbital moments, which we have treated in the preceding chapters, are referred to as fine structure.

To measure the hyperfine structure, one needs spectral apparatus of the highest resolving power. Until a few years ago, this meant interference spectrometers with resolving powers of the order of 10^6. In recent years, laser-spectroscopic methods have been developed, which allow still higher resolving powers to be attained. More details will be given in Chap. 22.

The three most important influences of the nucleus on the electronic spectra, aside from the effect of the Coulomb field of the nuclear charge $+Ze$, are the following:

- The *motion* of the nucleus around the common centre of gravity of the nucleus and the electrons leads (in the model for *Bohr* and *Sommerfeld*) to a dependence of the Rydberg constant on the nuclear mass. As a result, isotopes, that is, atoms with the same nuclear charge Z but with differing masses, have energy terms which differ somewhat. Their spectral lines thus are also different. This isotope effect, which we have already treated (Sect. 8.5), can, in the case of heavy elements, only be measured with a high spectral resolution, since the relative mass differences and thus the differences in the energy terms of different isotopes are then minimal.
- The differing *volumes* of isotopic nuclei, having the same charge Z but different neutron numbers and therefore different masses, also lead to an isotope effect in the spectra, the so-called volume effect. Since the density of nuclear matter is constant, and therefore the nuclear volume is simply proportional to the number of nucleons in the nucleus, different isotopes of a particular chemical element have differing nuclear sizes. From this follows a somewhat differing interaction between

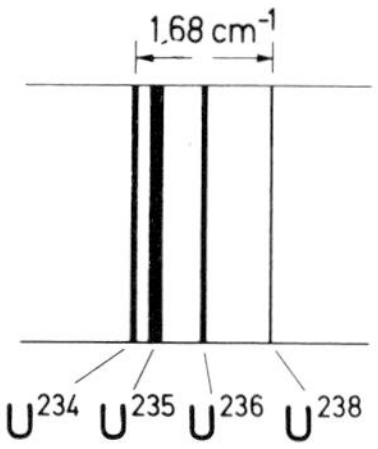

Fig. 20.1. Experimental example of the isotope effect: the isotopic splitting of the 4244.4 Å line in the uranium spectrum, observed in the 6th order in a 9 m grating spectrograph. [After H. G. Kuhn: *Atomic Spectra* (Longmans, London 1962, Plate 18)]

the nucleus and the electrons. These two isotope effects permit a determination of the number, masses, and abundances of the isotopes of a given element. Figure 20.1 shows an experimental example.

- The term *hyperfine structure* in the precise sense refers, finally, to another kind of structure in atomic spactra, which results from the existence of spins and magnetic moments of the atomic nuclei. The presence of these properties in nuclei was first postulated by *Pauli* in 1924 as a means of explaining spectroscopic observations. In the year 1934, *Schüler* further postulated the existence of electric quadrupole moments in nuclei. The interactions of these nuclear moments with the electrons lead to an additional splitting of the spectral lines, namely the above-mentioned hyperfine structure. This will be the topic of the present chapter.

20.2 Spins and Magnetic Moments of Atomic Nuclei

Atomic nuclei possess a mechanical angular momentum

$$|\boldsymbol{I}| = \sqrt{I(I+1)}\,\hbar\,. \tag{20.1}$$

Here, the quantum number I may be integral or half-integral. Stable atomic nuclei are known with I-values between 0 and 15/2. One says, briefly, "The nucleus has a spin I". This means that the largest measurable component of the nuclear angular momentum $\boldsymbol{I}$ has the value $I\hbar$. As we have already seen in the case of electronic angular momenta, only the component in a special direction z, for example in the direction of an applied magnetic field $\boldsymbol{B}$, is observable. The x and y components have time averages equal to zero. For the z component, we have

$$(\boldsymbol{I})_z = m_I\hbar \quad \text{with} \quad m_I = I, I-1, \ldots -I\,. \tag{20.2}$$

There are thus $2I+1$ possible orientations of the nuclear angular momentum relative to the special direction, corresponding to the possible values of the nuclear magnetic quantum number m_I; for example, for the hydrogen nucleus – the proton – we find $I=\frac{1}{2}$, $|\boldsymbol{I}| = (\sqrt{3}/2)\,\hbar$, $I_z = \pm\frac{1}{2}\hbar$ (see also Fig. 20.2).

A magnetic moment $\boldsymbol{\mu}_I$ is connected with the nuclear angular momentum; the two are proportional:

$$\boldsymbol{\mu}_I = \gamma \boldsymbol{I}\,. \tag{20.3a}$$

The constant of proportionality γ is known as the gyromagnetic ratio. The unit of the nuclear magnetic moment is the *nuclear magneton*, $\mu_N = e\hbar/2m_P$, where m_P is the proton mass. The nuclear magneton is thus smaller than the Bohr magneton by the mass ratio of the electron and the proton:

$$\mu_N = \mu_B/1836 = 0.5050824 \cdot 10^{-27}\,\text{Am}^2 \text{ or JT}^{-1}\,.$$

With this definition, we may write the magnetic moment as

$$\boldsymbol{\mu}_I = \frac{g_I \mu_N}{\hbar}\boldsymbol{I}\,. \tag{20.3b}$$

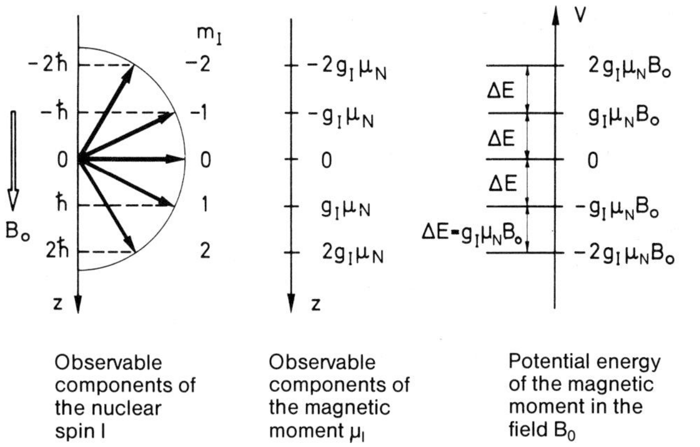

Fig. 20.2. The nuclear spin $\boldsymbol{I}$ (here for the case $I = 2$) has discrete orientations relative to a special direction, here an externally applied magnetic field $\boldsymbol{B}_0$ (*left*). Correspondingly, the measurable component of the magnetic moment (*middle*) and the potential energy (*right*) have discrete values

The nuclear g factor defined in (20.3 b) is a dimensionless number $g_I = \gamma\hbar/\mu_N$ (cf. Sect. 12.2). In contrast to the g_J factor for the electrons, the nuclear g factor cannot, as yet, be calculated from other quantum numbers. Otherwise, the relations between spins and magnetic moments in nuclei and in the electronic shells are completely analogous.

Following the quantisation rules for the angular momentum, one can also only observe the component of the nuclear magnetic moment along a special direction z. From (20.2), this component is

$$(\boldsymbol{\mu}_I)_z = \gamma(\boldsymbol{I})_z = \gamma\hbar m_I = g_I\mu_N m_I . \tag{20.4 a}$$

The maximum possible value for m_I is I, Fig. 20.2. According to (20.2), the maximum observable value of $\boldsymbol{\mu}_I$ is

$$|\boldsymbol{\mu}_I|_{\max} = g_I\mu_N I . \tag{20.4 b}$$

One thus says, simplifying, that the nucleus has the magnetic moment $\mu_I = g_I\mu_N I$. This is the experimental quantity which appears in tabulations.

To clarify these concepts, a few further examples: For the hydrogen nucleus, the proton, we have

$$\mu_I(^1\mathrm{H}) = +2.79\,\mu_N ; \quad I = \tfrac{1}{2} ; \quad g_I = 5.58 ;$$

for the potassium nucleus with mass number 40,

$$\mu_I(^{40}\mathrm{K}) = -1.29\,\mu_N ; \quad I = 4 ; \quad g_I = -0.32 .$$

The sign of μ_I is also contained in the g_I factor: $g_I > 0$ means that $\boldsymbol{\mu}_I > 0$ also, i.e. the directions of the spin $\boldsymbol{I}$ and of the moment $\boldsymbol{\mu}_I$ are parallel.

An example for a positive sign of μ_I is the proton. It has a positive charge; therefore, it is intuitively understandable that the factor connecting the angular momentum with the magnetic moment of the proton has a sign opposite to that in the case of the electron (Sect. 12.4).

There are also numerous nuclei with vanishing spins, $I = 0$. These nuclei do not contribute to the hyperfine structure. Examples of this type of nucleus are

$$^{4}_{2}\mathrm{He},\ ^{12}_{6}\mathrm{C},\ ^{16}_{8}\mathrm{O},\ ^{40}_{20}\mathrm{Ca},\ ^{56}_{26}\mathrm{Fe},\ ^{88}_{38}\mathrm{Sr},\ ^{114}_{48}\mathrm{Cd},\ ^{180}_{72}\mathrm{Hf},\ ^{208}_{82}\mathrm{Pb},\ ^{238}_{92}\mathrm{U}\,.$$

A detailed theoretical description of moments, spins and g-factors of the nucleus is in the province of nuclear physics and will not be gone into here.

The quantities $\boldsymbol{I}$ or I can be calculated from experimental data, such as the number of the hyperfine components of spectral lines. The gyromagnetic ratio γ, or the g_I factor, is measured experimentally by nuclear magnetic resonance, which will be the subject of the next sections.

Since the magnetic moments of nuclei are several orders of magnitude smaller than those of electrons, the interactions to be expected with external or internal fields are likewise three orders of magnitude smaller; therefore, the name hyperfine structure is fitting.

20.3 The Hyperfine Interaction

We now wish to calculate the interaction energy between a nuclear magnetic moment and the magnetic field which the electrons produce at the site of the nucleus.

In solving the analogous problem of the fine structure interaction between the spin of the orbital moments of electrons, we proceeded roughly as follows.

The magnetic field $\boldsymbol{B}_L$ which is produced by the orbital motion of the electron interacts with the electron's magnetic moment and orients its spin $\boldsymbol{s}$. The spin and the orbital angular momentum combine according to the rules in Sect. 12.7, 8 to yield a total angular momentum $\boldsymbol{j}$. In an atom with several electrons and in LS coupling, the corresponding vectors $\boldsymbol{L}$, $\boldsymbol{S}$, and $\boldsymbol{J}$ are to be used. In the vector model, the angular momenta $\boldsymbol{L}$ and $\boldsymbol{S}$ precess about the space-fixed vector $\boldsymbol{J}$ (Fig. 20.3 a). An additional magnetic energy results from the interaction between the magnetic moment $\boldsymbol{\mu}_S$ and the electronic field $\boldsymbol{B}_L$ (Fig. 20.4 a):

$$V_{\mathrm{FS}} = -\boldsymbol{\mu}_S \cdot \boldsymbol{B}_L\,, \tag{20.5}$$

where the indices FS stand for fine structure.

The hyperfine interaction energy is calculated in an exactly analogous way. It is three orders of magnitude smaller than the fine structure energy because, this is the ratio of the magnetic moments of nuclei and electrons, respectively. Therefore the magnetic coupling of the atomic electrons among themselves is not affected by the hyperfine interaction. At the position of the nucleus, there is a magnetic field $\boldsymbol{B}_J$ (Fig. 20.4 b). This influences the magnetic moment of the nucleus and orients the nuclear spin. The result of this interaction is a coupling of the angular momenta of the electrons ($\boldsymbol{J}$) and the nucleus ($\boldsymbol{I}$) to a new total angular momentum $\boldsymbol{F}$. In analogy to LS coupling for electrons (Fig. 20.3 b):

$$\boldsymbol{F} = \boldsymbol{J} + \boldsymbol{I} \tag{20.6}$$

where the absolute value of the total angular momentum $|\boldsymbol{F}| = \sqrt{F(F+1)}\,\hbar$. The quantum number F of the total angular momentum $\boldsymbol{F}$ can have the values

$$F = J+I,\ J+I-1 \ldots J-I\,.$$

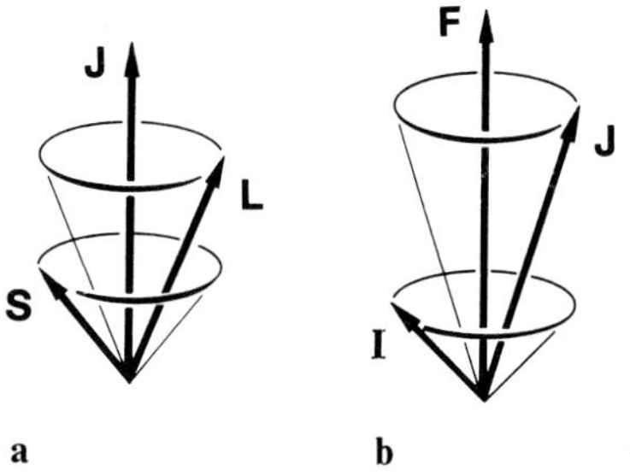

Fig. 20.3. **a)** Coupling of the S (spin) and L (orbital angular momentum) vectors to a total electron angular momentum vector $\boldsymbol{J}$ for the electron shell. **b)** Coupling of the J (electron angular momentum) and $\boldsymbol{I}$ (nuclear angular momentum) vectors to the total angular momentum $\boldsymbol{F}$ of the atom

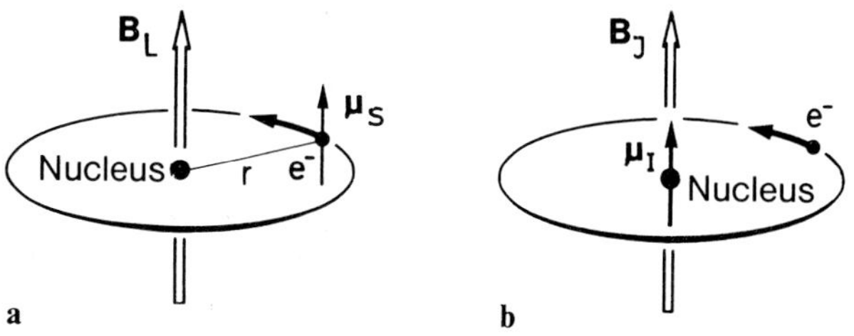

Fig. 20.4. **a)** Calculation of the spin-orbit interaction: the magnetic moment μ_S of the electron interacts with the magnetic field of the orbital motion $\boldsymbol{B}_L$. **b)** Calculation of the hyperfine interaction: the magnetic moment μ_I of the nucleus interacts with the magnetic field of the electron shells $\boldsymbol{B}_J$

This amounts to $(2I+1)$ or $(2J+1)$ possibilities, depending on whether I is smaller or larger than J.

In the vector model, the vectors $\boldsymbol{I}$ and $\boldsymbol{J}$ precess in quantised positions around the spatially fixed vector $\boldsymbol{F}$, Fig. 20.3b. The number of hyperfine levels is thus uniquely determined by the quantum numbers J and I. If J is known, I can thus be measured.

The additional magnetic energy due to the hyperfine interaction is

$$V_{\mathrm{HFS}} = -\boldsymbol{\mu}_I \cdot \boldsymbol{B}_J , \qquad (20.7)$$

where $\boldsymbol{B}_J$ is the magnetic field generated at the position of the nucleus by the electron shell (Fig. 20.4b). (HFS is the abbreviation for "hyperfine structure".)

Further calculation of the hyperfine interaction is quite analogous to calculation of the fine structure interaction. Each of the possible orientations of the nuclear spin in the field $\boldsymbol{B}_J$ corresponds to a definite potential energy. According to (20.7), it is

$$\begin{aligned} V_{\mathrm{HFS}} &= -\mu_I B_J \cos(\boldsymbol{\mu}_I, \boldsymbol{B}_J) \\ &= g_I \mu_{\mathrm{N}} \sqrt{I(I+1)}\, B_J \cos(\boldsymbol{I}, \boldsymbol{J}) , \end{aligned} \qquad (20.8)$$

if we substitute $g_I \mu_{\mathrm{N}} \sqrt{I(I+1)}$ for $|\boldsymbol{\mu}_I|$, according to (20.3b). The reversed sign in the second line of (20.8) is a result of the fact that the vectors $\boldsymbol{B}_J$ and $\boldsymbol{J}$ are antiparallel.

Furthermore, in the quantum mechanical formulation,

$$\cos(\boldsymbol{I}, \boldsymbol{J}) = \frac{F(F+1) - I(I+1) - J(J+1)}{2\sqrt{J(J+1)}\sqrt{I(I+1)}} \qquad (20.9)$$

according to the law of cosines, as shown in Sect. 13.3.5. This can be written as

$$F(F+1) = J(J+1) + I(I+1) + 2\sqrt{J(J+1)}\sqrt{I(I+1)} \cos(\boldsymbol{I}, \boldsymbol{J}) .$$

With (20.9), one finally arrives at the energy of the hyperfine interaction,

$$\Delta E_{\mathrm{HFS}} = \frac{a}{2}[F(F+1) - I(I+1) - J(J+1)] , \qquad (20.10)$$

with

$$a = \frac{g_I \mu_N B_J}{\sqrt{J(J+1)}} .$$

Here, we have again defined a measurable quantity, the hyperfine constant or interval constant a. Figure 20.5 shows an example of (20.10), the hyperfine splitting of a term with $J = 3/2$ and $I = 3/2$.

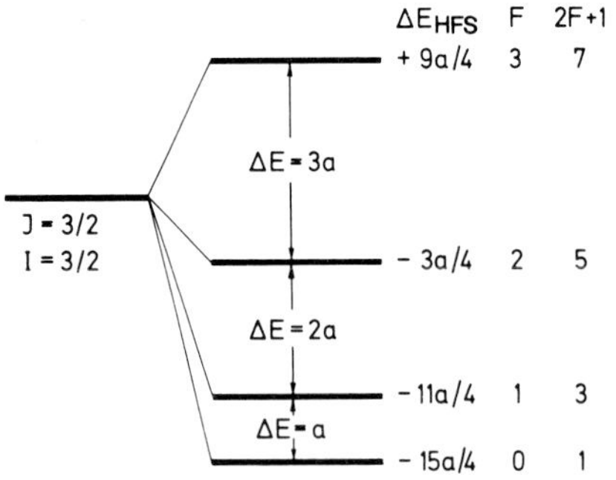

Fig. 20.5. Hyperfine splitting and the interval rule: hyperfine splitting of a state with $J = 3/2$ and $I = 3/2$. The numerical values of the quantum number F and the degree of degeneracy $2F+1$ (with respect to m_F) are given on the right

The field strength $\boldsymbol{B}_J$ produced by the electrons at the nucleus cannot, in general, be measured independantly of the magnetic properties of the nucleus itself. In order to extract $\boldsymbol{B}_J$ from a measurement of the hyperfine splitting of spectral lines, one must know the nuclear moment or g_I, as seen from (20.10). An exact calculation of $\boldsymbol{B}_J$ is possible only in simple particular cases, e.g. for the hydrogen atom, because the wavefunctions are then precisely known (see Chap. 10 and Sect. 12.10).

In the case of s-electrons, the magnetic field $\boldsymbol{B}_J$ is mainly due to the nonvanishing electron density at the nucleus. The magnetic moments of the electrons interact directly with the nuclear moment, which is distributed over the finite nuclear volume. This interaction is known as the (Fermi) contact interaction; it is isotropic. For an s electron ($^2S_{1/2}$, $J = 1/2$), it may be calculated exactly following *Fermi*, thus for example for the ground states of alkali atoms. One obtains the following expression for the hyperfine constant a:

$$a = \tfrac{2}{3}\mu_0 g_e \mu_B g_I \mu_N |\psi(0)|^2 . \tag{20.11}$$

Here, $\psi(0)$ is the wavefunction of the electron at the nucleus. For an s electron in a hydrogen atom (cf. Chap. 10), we have $|\psi(0)|^2 = 1/\pi r_H^3$, with r_H being the radius of the first Bohr orbit. The numerical calculation gives $a = 1420$ MHz or $0.0475\ \mathrm{cm}^{-1}$ or $5.9 \cdot 10^{-6}$ eV (see also Fig. 20.7). In general, for s electrons in hydrogen-like atoms, we have as a good approximation

$$|\psi(0)|^2 = Z^3/(\pi n^3 r_H^3) .$$

For electrons with orbital angular momentum $l > 0$, i.e. for $p, d, f, \ldots$ electrons, the probability density at the nucleus vanishes. Here, the field $\boldsymbol{B}_J$ is calculated from the dipole-dipole interaction between the electronic moment and the nuclear moment. The hyperfine constant a becomes anisotropic.

Even for p electrons, one obtains values of $\boldsymbol{B}_J$ which are an order of magnitude smaller. A term diagram of the hydrogen atom including the hyperfine interaction is shown in Fig. 20.6.

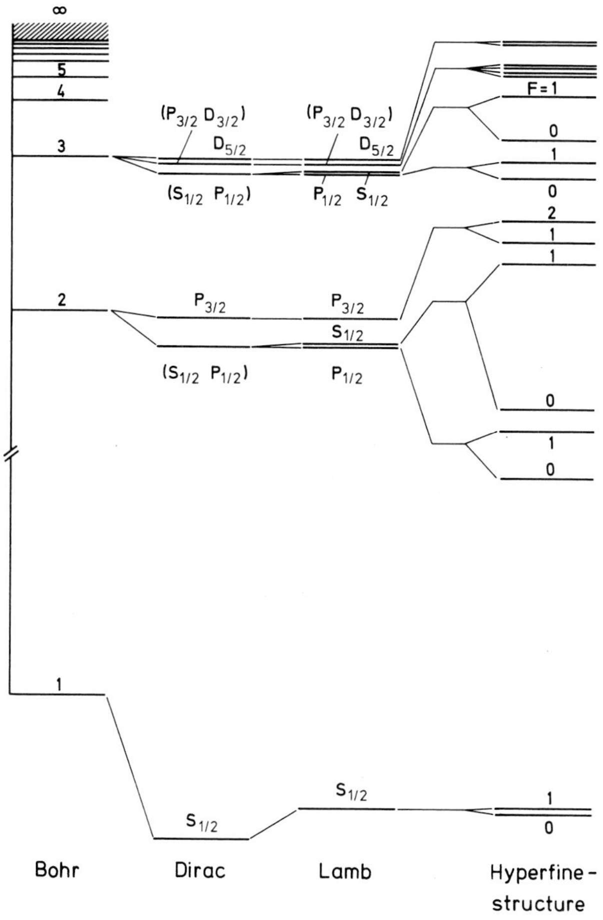

Fig. 20.6. Term diagram of the H atom including the hyperfine splitting. All fine structure terms (see Fig. 12.20) are doubly split by the interaction with the proton. The magnitude of the splitting is not drawn to scale here. In particular, the scale was enlarged on going from the Bohr energy levels to the fine structure and again in going to the hyperfine structure, in order to show the energy differences which become smaller and smaller from left to right. The scale increases from below to above in the case of the hyperfine splitting; the latter is largest in the ground state, $n = 1$

For other atoms, the only possibility is to use approximate methods. In general, however,

- $\boldsymbol{B}_J$ and thus V_{HFS} are large when the electrons are close to the nucleus, i.e. for large Z, small n and small l.
- The number of hyperfine components split apart is the same for a given type of atom (i.e. for a fixed nuclear spin quantum number I) in all its excitation states, so long as $J > I$.
- For a vanishing nuclear spin ($I = 0$) or vanishing electron angular momentum ($J = 0$), $V_{\mathrm{HFS}} = 0$, so there is no hyperfine splitting.

Table 20.1 lists some numerical values for B_J.

Table 20.1. Magnetic field $\boldsymbol{B}_J$ produced at the nucleus by the outermost electron, for various terms (after Segrè, fields in Tesla)

	n	$^2S_{1/2}$	$^2P_{1/2}$	$^2P_{3/2}$
Na	3	45	4.2	2.5
K	4	63	7.9	4.6
Rb	5	130	16	8.6
Cs	6	210	28	13

The magnetic field B_J at the nucleus can be determined from measurements of the hyperfine splitting of spectral lines, if the nuclear moments are known. In many atoms, they can be obtained by measurements in the presence of external fields (see Chap. 13). The magnitude of the hyperfine splitting of spectral lines is, for example 0.027 cm^{-1} for the lowest $^2S_{1/2}$ term of lithium and 0.3 cm^{-1} for the lowest $^2S_{1/2}$ term of cesium.

The measurement of the hyperfine splitting V_{HFS} of spectral terms is thus one method by which it is possible to calculate a product of the nuclear property (g_I) and the electron shell property (B_J). One must still take into account that the observed spectral lines are transitions between two terms, which are in general both subject to hyperfine splitting. The selection rule for optical transitions is $\Delta F = 0, \pm 1$. This will not be derived here.

The evaluation of hyperfine spectra is simplified by the interval rule. It follows from (20.10) that the relative distance between the hyperfine terms for the quantum numbers F and $F+1$ is

$$\Delta E_{F+1} - \Delta E_F = a(F+1) . \tag{20.12}$$

The distance between two terms in a hyperfine structure multiplet is thus proportional to the larger of the two F values, and the spacings within the multiplet are in the ratio $F:(F-1):(F-2)$, etc. This is shown in Fig. 20.5.

Another important matter is the sequence of the levels. If the nuclear moment is positive, so is the hyperfine constant a, so that V_{HFS} increases as F increases. This offers an experimental method of determining the sign of the nuclear moment $\boldsymbol{\mu}_I$.

Finally, the intensities of the spectral lines composing a hyperfine multiplet are also characteristic. In the absence of an external magnetic field, the terms are still $(2F+1)$-fold degenerate with respect to m_F. (The degree of degeneracy is shown in Fig. 20.5.) Terms with different values of F therefore have different statistical weights, and their intensities are proportional to $(2F+1)$. To be sure, it is generally the case that both of the levels between which the transition occurs have hyperfine structure. This complicates the pattern to be expected.

In all, hyperfine spectra are characterised by very typical ratios of intensities and distances between the hyperfine components.

20.4 Hyperfine Structure in the Ground States of the Hydrogen and Sodium Atoms

Here, we shall consider the hyperfine splitting of atomic terms, which was treated in the previous section, in more detail. In the ground state of the hydrogen atom, the proton and the electron spins can only be oriented parallel or antiparallel to one another. This yields the two possible values for F: 1 or 0. The energy difference between these two configurations is (Fig. 20.7)

$$\Delta\bar{\nu} = 0.0475\ \mathrm{cm}^{-1} \quad \text{or}$$

$$\Delta\nu = 1420\ \mathrm{MHz} \quad \text{or}$$

$$\lambda = 21\ \mathrm{cm} .$$

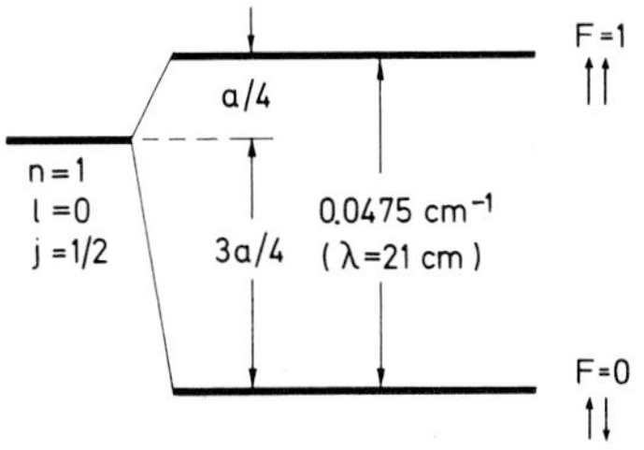

Fig. 20.7. Hyperfine structure of the hydrogen ground state. The spins of electrons and protons can be parallel or antiparallel to each other. The hyperfine splitting is six times smaller than the Lamb shift of the levels (Sect. 12.11)

This energy difference can be determined either by direct absorption of high-frequency radiation of the appropriate frequency, or from the splitting of the spectral lines which end on the $n = 1$, $l = 0$ level, if sufficient resolution is available.

Since the energy difference $\Delta\bar{\nu}$ can be exactly calculated if the magnetic moments of the proton and electron are known (cf. Sect. 20.11), and since the moment of the proton is known to great precision from nuclear resonance measurements (Sect. 20.6), an exact measurement of $\Delta\bar{\nu}$ in hydrogen showed, for the first time, the deviation of the g factor of the electron from the value $g = 2$. The measured value of $\Delta\bar{\nu}$ differed slightly from the value calculated with $g = 2$, and from the measurement, the g factor was calculated to be $g = 2.0023$. This and the measurement of the Lamb shift were instrumental in stimulating the development of quantum electrodynamics.

In interstellar space, hydrogen is present in extremely low concentrations, but due to the immense volume of the universe, the total amount of it is vast. Therefore it is to be expected that radio-frequency radiation at 21 cm wavelength would be emitted and absorbed there in measurable amounts. This radiation was actually discovered in 1951 with large radio telescopes. Since then, the 21 cm line of hydrogen has played an important part in radio-astronomy. For example, there is an especially large amount of this radiation in the region of the Milky Way. The motion of parts of the Milky Way relative to the earth can be deduced from the Doppler shifts of this line, and analysis of this motion has confirmed that the Milky Way is a spiral galaxy. This is probably the most spectacular application of the study of hyperfine structure to astrophysics.

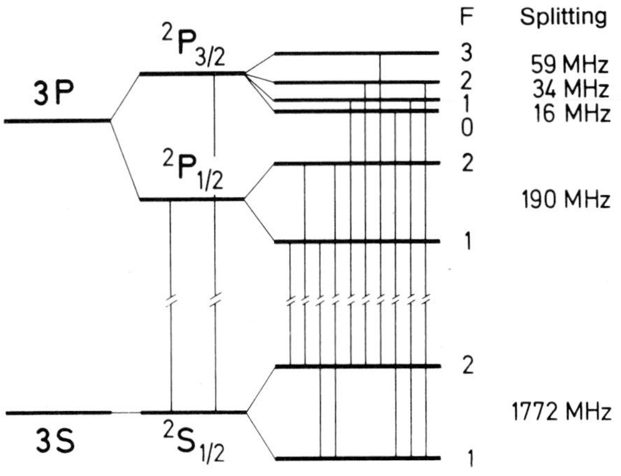

Fig. 20.8. Hyperfine splitting of the lowest terms of the sodium atom with the allowed transitions. The numerical values for the hyperfine splitting are given here in MHz

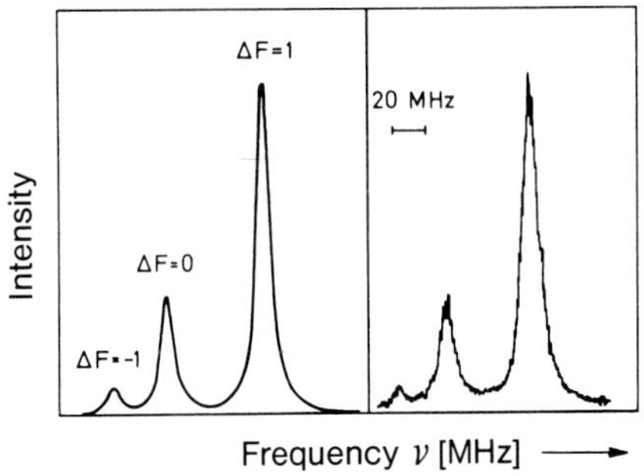

Fig. 20.9. Three components of the hyperfine-split D_2 line of sodium (Fig. 20.7), measured by recording the resonance radiation of sodium atoms in an atomic beam irradiated by light from a narrow-band, variable-frequency dye laser. The ΔF values for the ground and excited state are given. The curve on the left is calculated from the natural linewidth, the curve on the right was observed experimentally. The measured (15 MHz) and calculated natural linewidths are very close [from Lange et al.: Opt. Commun. **8**, 157 (1973)]

The hyperfine splitting of three levels of the sodium atom, $^2S_{1/2}$, $^2P_{1/2}$ and $^2P_{3/2}$ is shown in Fig. 20.8. The nuclear spin of the atom is $I = 3/2$. There are four values of the quantum number F for $P_{3/2}$, $F = 3, 2, 1$ and 0; for $J = 1/2$, $F = 2$ or 1. Together with the selection rule for optical transitions. $\Delta F = 0, \pm 1$, these values yield the spectral lines indicated in Fig. 20.8. Because the linewidth is finite, it is usually possible to see only two components of the line D_1 with $\Delta = 0.023$ Å, and two components of the D_2 line with $\Delta = 0.021$ Å. Figure 20.9 shows a modern spectrum with the highest available resolution, obtained by exciting resonance fluorescence in a beam of sodium atoms with a very narrow-band, variable frequency dye laser. The method is further discussed in Chap. 22.

20.5 Hyperfine Structure in an External Magnetic Field, Electron Spin Resonance

An important tool for the study of the hyperfine interaction is the measurement of the splitting of the lines of the optical spectrum in the presence of an external magnetic field $\boldsymbol{B}_0$. This external field $\boldsymbol{B}_0$ adds to the internal field $\boldsymbol{B}_J$, and the resulting term splitting depends on the relative sizes of the two fields. If the external field is so small that the magnetic potential energy of the atom in it is small compared to the energetic separation of the hyperfine terms, one speaks of the Zeeman effect of the hyperfine structure.

In this case, the coupling of the angular momentum vectors $\boldsymbol{I}$ and $\boldsymbol{J}$ to $\boldsymbol{F}$ remains intact. If the external field $\boldsymbol{B}_0$ is strong enough, this coupling is lifted and one speaks of the Paschen-Back effect of the hyperfine structure. The transition from the Zeeman to the Paschen-Back effect occurs at much smaller fields than in the case of the fine structure, due to the small magnitude of hyperfine splitting. With respect to hyperfine structure, 0.1 T is usually already a "strong" field.

The case of the hyperfine Zeeman effect is represented in Fig. 20.10. The angular momentum vectors $\boldsymbol{I}$ and $\boldsymbol{J}$ remain coupled. The resulting total angular momentum vector $\boldsymbol{F}$ precesses around $\boldsymbol{B}_0$. The precession frequency of $\boldsymbol{I}$ and $\boldsymbol{J}$ about $\boldsymbol{F}$ is much higher than that of $\boldsymbol{F}$ about $\boldsymbol{B}_0$. There are $2F+1$ possibilities for the orientation of the vector $\boldsymbol{F}$ relative to $\boldsymbol{B}_0$, which are given by the quantum numbers m_F. As usual, the relation $|\boldsymbol{F}|_z = m_F\hbar$ with $m_F = F, \; F-1, \ldots -F$ holds. For optical transitions, the selection rules $\Delta F = 0, \pm 1$ and $\Delta m_F = 0, \pm 1$ apply.

The size of the Zeeman splitting in the presence of the hyperfine interaction can be calculated from $V_{\mathrm{HFS}} = -\boldsymbol{\mu}_F \cdot \boldsymbol{B}_0$, quite analogously to the calculation of the Zeeman

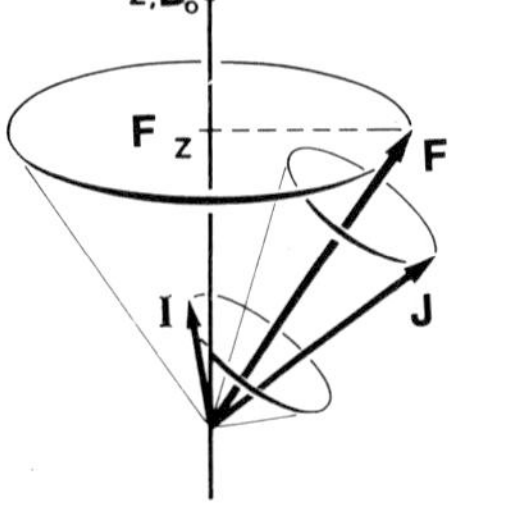

Fig. 20.10. Hyperfine structure in a magnetic field. The vector diagram illustrates the Zeeman effect: the total angular momentum $\boldsymbol{F}$, composed of the angular momentum vectors $\boldsymbol{J}$ and $\boldsymbol{I}$, possesses quantised orientations relative to an applied magnetic field $\boldsymbol{B}_0$. Only the z component of $\boldsymbol{F}$ is observable

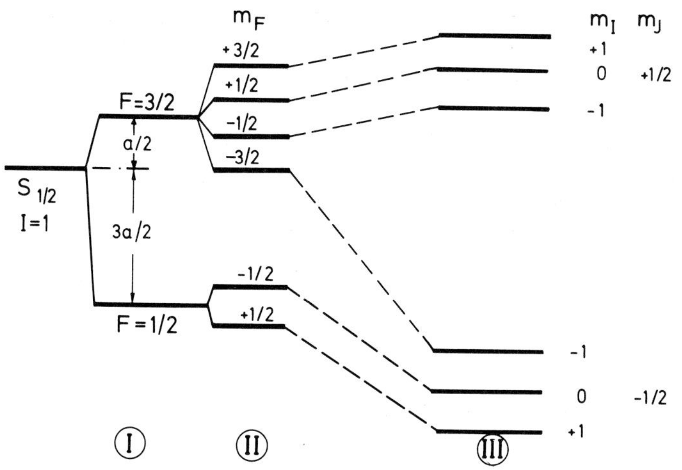

Fig. 20.11. Hyperfine structure of an $S_{1/2}$ state with $I=1$, without applied field (I); in a weak applied field, corresponding to the Zeeman effect with hyperfine structure (II); in a strong applied field, the Paschen-Back effect with hyperfine structure (III)

effect with fine structure in Sect. 13.3. In this case, $\boldsymbol{\mu}_F$ is equal to $\boldsymbol{\mu}_J + \boldsymbol{\mu}_I$. We shall thus restrict ourselves to stating the result: the shift of the atomic terms due to the hyperfine splitting in a magnetic field is given by (weak field case)

$$\Delta E_{\mathrm{HFS}} = g_F \mu_{\mathrm{B}} B_0 m_F \tag{20.13}$$

with

$$g_F = g_J \frac{F(F+1)+J(J+1)-I(I+1)}{2F(F+1)} - g_I \frac{\mu_{\mathrm{N}}}{\mu_{\mathrm{B}}} \frac{F(F+1)+I(I+1)-J(J+1)}{2F(F+1)} .$$

The second term can be neglected relative to the first term because of the factor $\mu_{\mathrm{N}}/\mu_{\mathrm{B}} = 1/1836$. The term splitting in a weak field then yields $2F+1$ equidistant components; see Fig. 20.11.

If the applied field is increased in strength, the first effect is a decoupling of the angular momenta $\boldsymbol{I}$ and $\boldsymbol{J}$. The LS coupling is stronger and remains in effect, since it is determined by the magnetic moments of electrons, while the IJ coupling results from an electronic moment and a much weaker nuclear moment. Therefore, a relatively small applied field suffices to break up the IJ coupling; the hyperfine Zeeman effect then is replaced by the hyperfine Paschen-Back effect. The angular momentum vector $\boldsymbol{J}$ of the electrons precesses about the z direction defined by the field $\boldsymbol{B}_0$. The nuclear moment $\boldsymbol{\mu}_I$ is primarily affected by the electronic field $\boldsymbol{B}_J$, since this is, in general, much larger at the nuclear site than the applied field; the magnitude of B_J is 10 to 100 T. The vector $\boldsymbol{I}$ thus precesses about the direction of $\boldsymbol{J}$. Since, however, the precession of $\boldsymbol{J}$ about $\boldsymbol{B}_0$ is much more rapid, the nuclear moment $\boldsymbol{\mu}_I$ experiences a constant component of $\boldsymbol{B}_J$ in the direction of $\boldsymbol{B}_0$. The rapidly oscillating components in the x and y directions average to zero; the final result is that the nuclear angular momentum vector $\boldsymbol{I}$ is also oriented in the direction of $\boldsymbol{B}_0$. The precession frequencies of $\boldsymbol{I}$ and $\boldsymbol{J}$ about the z direction differ, however. The corresponding vector diagram is shown in Fig. 20.12.

The quantum number F is no longer defined. The orientation quantum numbers m_I and m_J are valid for the vectors $\boldsymbol{I}$ and $\boldsymbol{J}$. The splitting energy of the atomic terms is the sum of three components. One of these is the shell moment, $g_J \mu_{\mathrm{B}} B_0 m_J$, which was treated before under the Zeeman effect without hyperfine structure (Sect. 13.3). Each

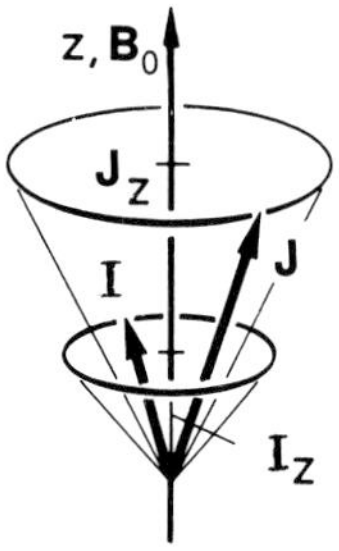

Fig. 20.12. Hyperfine structure in a magnetic field. Vector diagram of the Paschen-Back effect. A total angular momentum F is no longer defined

of these levels is split into $(2I+1)$ hyperfine levels, corresponding to the possible values of m_I: $m = I, I-1, \ldots -I$. The quantum number I can thus be immediately determined by counting, as is made clear in the right side of Fig. 20.11. The magnitude of this second splitting can be easily calculated using the vector model (Fig. 20.12). To a first approximation, it is $a m_I m_J$, where

$$a = \frac{g_I \mu_N B_J}{\sqrt{J(J+1)}} \tag{20.14}$$

is the hyperfine constant defined in (20.10).

If the field is strong enough, the effect of the external field on the nucleus is no longer negligible compared to the field of the shells. Therefore the Zeeman energy of the nucleus, $-g_I \mu_N m_I B_0$, is included in (20.15) as the third term on the right.

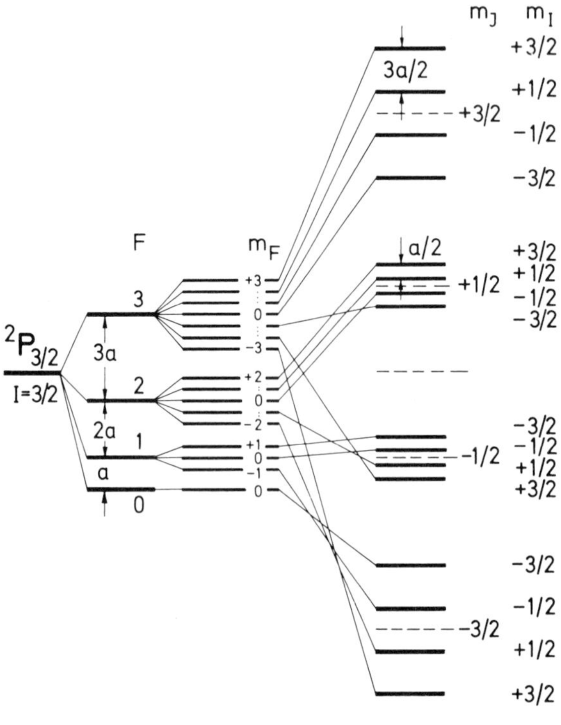

Fig. 20.13. Hyperfine splitting (schematic) for $J = 3/2$, $I = 3/2$ in a weak, medium and strong field

In all, the splitting energy in a strong field is

$$\Delta E_{\mathrm{HFS}} = g_J \mu_{\mathrm{B}} m_J B_0 + a m_I m_J - g_I \mu_{\mathrm{N}} m_I B_0 . \tag{20.15}$$

The region of transition between the limiting cases of strong and weak fields is usually very difficult to calculate, and can only be approximated. Figure 20.13 shows schematically the complicated behaviour of the terms in the intermediate field range for a state with the quantum numbers $J = 3/2$ and $I = 1/2$. A further remark concerning (20.15): it would be consistent to treat the electronic g-factor g_J (and naturally also g_S and g_L, Sect. 12.4, 12.2, and 13.3.5) as negative, since the electron spins are directed opposite to the corresponding magnetic moments. In this case, a minus sign is to be inserted in (20.15) in front of g_J.

In ordering the levels with different possible values of m_I, one must be careful. If the electronic field $\boldsymbol{B}_J$ at the nucleus is larger than the external field $\boldsymbol{B}_0$, the energies of the nuclear states depend on the orientation of the nucleus with respect to the electron. However, the quantum numbers refer to the direction of the external field $\boldsymbol{B}_0$, which is not the same as that of the electronic field for all states. This is illustrated by Fig. 20.14, which shows the splitting of the ground state of the hydrogen atom in a magnetic field (compare Fig. 20.7 and Eq. 20.15). From left to right, the Zeeman energy of the electron in the field $\boldsymbol{B}_0$, the hyperfine interaction energy between the electron and the proton, and the Zeeman energy of the proton are ordered according to decreasing interaction energy. The arrows indicate the spin directions. We note again that for the electron, the spin and the magnetic moment are antiparallel to one another, but are parallel for the proton.

Up to now, we have mainly treated the observation of the hyperfine splitting in the *optical* spectral region; however, in Fig. 20.14 the magnetic dipole transitions which are observable by *electron spin resonance* (ESR) are indicated. A single resonance frequency for the free electron (at the left in Fig. 20.14) becomes two transitions in the hydrogen atom due to the hyperfine and Zeeman interaction with the proton. Their splitting

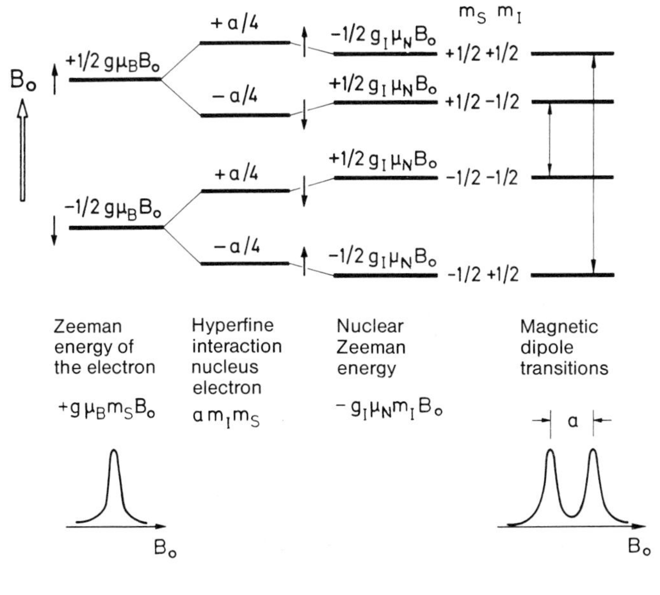

Fig. 20.14. The hyperfine structure of the hydrogen atom in a strong magnetic field $\boldsymbol{B}_0$, and the hyperfine structure observable in electron spin resonance (ESR). From left to right, the three contributions to the splitting according to (20.15) are illustrated, along with the magnetic quantum numbers and the allowed (magnetic) dipole transitions. At the lower left, the ESR spectrum of a free electron is shown; at the lower right, that of an electron bound to a proton (H atom). From the splitting in the ESR spectrum, one obtains the hyperfine interaction constant a (in the limiting case of a strong applied field $\boldsymbol{B}_0$)

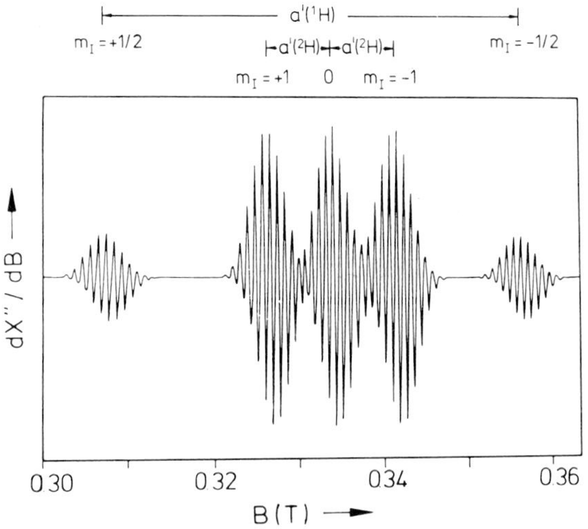

Fig. 20.15. The ESR spectrum of hydrogen atoms which are stored in interstitial sites in a KCl crystal. The two outer groups of lines belong to the transitions with $\Delta m_s = 1$ and $m_I = +1/2$ or $m_I = -1/2$ of the light hydrogen isotope (mass number 1, ${}^1_1$H), having a splitting $a({}^1\mathrm{H})$ of 500 Gauss; cf. Fig. 20.14. Each of these transitions is split into 13 equidistant lines, due to what is termed the superhyperfine structure, i.e. the interaction with the nuclear moments of the four nearest-neighbor Cl^- anions ($I = 3/2$). The three inner groups of lines belong to the transitions with $\Delta m_s = 1$ and $m_I = 1, 0,$ or -1 of heavy hydrogen (${}^2_1$H, deuterium, with $I = 1$ and $g_I = 0.857$). The splitting constant $a({}^2\mathrm{H})$ is smaller by the ratio of the g_I factors than that of ^{1}H (see J. M. Spaeth: Phys. stat. sol. **34**, 71 (1969). In the limiting case of a strong external field $\boldsymbol{B}_0$, the hyperfine coupling constant a' measured by this method is identical to the isotropic hyperfine constant a as in (20.11)

corresponds to about 0.05 T (500 G). The electron thus experiences a field of about 500 G from the proton. By comparison, the nucleus, according to (12.26), is subjected to a field of the order

$$0.05 \cdot \frac{\mu_{\mathrm{electron}}}{\mu_{\mathrm{proton}}} \cong 30\ \mathrm{T} \ .$$

The transition region between a weak and a strong applied field can only be calculated in closed form for the case $F = I \pm 1/2$ by using the Breit-Rabi formula. We shall not treat this region further here.

The electron spin resonance of free hydrogen atoms is difficult to observe in the gas phase due to the thermal motion of the atoms and their low density. However, hydrogen atoms may be stored in alkali-halide crystals in interstitial lattice sites; using such crystals, one does, in fact, obtain the ESR spectrum expected from Fig. 20.14. An example is given in Fig. 20.15. It also shows the ESR spectrum of heavy hydrogen (deuterium), ${}^2_1$H. Due to the nuclear spin $I = 1$, one here observes three ESR transitions; because of the smaller g_I factor ($g_I = 0.857$), a and thus the hyperfine splitting are smaller. Figure 20.15 exhibits a further splitting, the so-called superhyperfine structure. This is due to the interaction of the electronic spin with the nuclei of host-crystal atoms on adjacent lattice sites. Further details are given in the figure caption.

20.6 Direct Measurements of Nuclear Spins and Magnetic Moments, Nuclear Magnetic Resonance

We have seen in the previous sections that the examination of the hyperfine structure of lines of optical spectra makes possible the calculation of nuclear spin. However, the magnetic moment can only be measured in connection with the electronic field at the nucleus by this method. Since the electronic field cannot be measured independently of the nuclear spin, other methods are required to determine the magnetic moment.

The Stern-Gerlach method (see Sect. 12.6) of measuring magnetic moments by deflection in an inhomogeneous magnetic field can be used to measure the nuclear moment if the electron shell of the atom is diamagnetic. Otherwise the deflection due to the nucleus will be completely obscured by the approximately thousand-fold greater effect of the electron-shell moment. In 1933, *Stern*, *Frisch* and *Estermann* used this method to obtain the correct order of magnitude of the proton magnetic moment in hydrogen molecules. They were not able to make a precise measurement, however. In their apparatus, the molecules had a flight path of 150 cm, and a deflection of 10^{-2} mm. The inhomogeneity of the field was 80000 G/cm or 800 T/m.

The atomic or molecular beam resonance method of *Rabi* (1937) is much more precise. Here, as in electron spin resonance (Sect. 13.2), one measures the Larmor frequency of the nuclear spin in an external magnetic field, or in other words, the ratio μ_I/I. For this method, too, the field due to electronic shells must vanish at the nucleus. This condition is fulfilled for some atoms, including Hg, Cu, C and S, and for some molecules, e.g. H_2O, CaO, LiCl, CO_2, H_2 and NH_3, because in these atoms and molecules, the magnetic moments of the electronic shells add up to zero.

For the potential energy of the nuclear moment in the field $\boldsymbol{B}_0$,

$$\begin{aligned} V &= -\boldsymbol{\mu}_I \cdot \boldsymbol{B}_0 \\ &= -g_I\sqrt{I(I+1)}\,\mu_N B_0 \cos(\boldsymbol{I},\boldsymbol{B}_0) \\ &= g_I \mu_N B_0 m_I . \end{aligned} \tag{20.16}$$

The energy difference between two neighbouring orientations in the field $\boldsymbol{B}_0$, i.e. for $\Delta m_I = \pm 1$ is then (Fig. 20.2)

$$\Delta E = g_I \mu_N B_0 . \tag{20.17}$$

If the sample is irradiated with radiation at a frequency corresponding to this energy difference,

$$\nu = \frac{g_I \mu_N}{h} B_0 \quad \text{or}$$

$$\omega_L = \frac{g_I \mu_N}{\hbar} B_0 = \gamma B_0 , \tag{20.18}$$

i.e. the Larmor frequency of the nucleus in the field $\boldsymbol{B}_0$, one induces transitions with $\Delta m_I = \pm 1$ and the nuclear spin changes its orientation. If numerical values are substituted, the resonance condition is

$$\nu\,[\mathrm{Hz}] = 762.3 \frac{\mu_I}{I} B_0[\mathrm{G}] . \tag{20.19}$$

Resonance is detected in the molecular beam experiments by the deflection of the atoms or molecules in inhomogeneous magnetic fields. The apparatus is represented schematically in Fig. 20.16. The apparatus is adjusted so that when the resonance field $\boldsymbol{B}_1$ is not turned on, the particle beam passes through the three magnets A, B and C and is detected at the detector D. The deflection in the inhomogeneous field of magnet A is

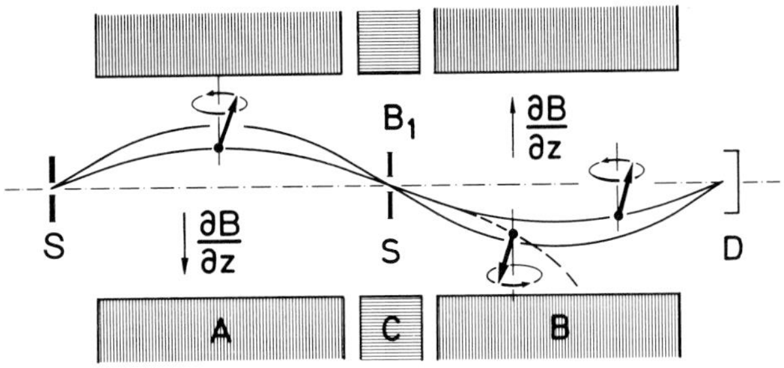

Fig. 20.16. Atomic beam resonance according to *Rabi*. The magnetic fields $\boldsymbol{A}$ and $\boldsymbol{B}$ are inhomogeneous, while magnet C has a homogeneous field $\boldsymbol{B}_0$. A high-frequency field $\boldsymbol{B}_1$ is applied perpendicular to $\boldsymbol{B}_0$ (S, source; D, detector)

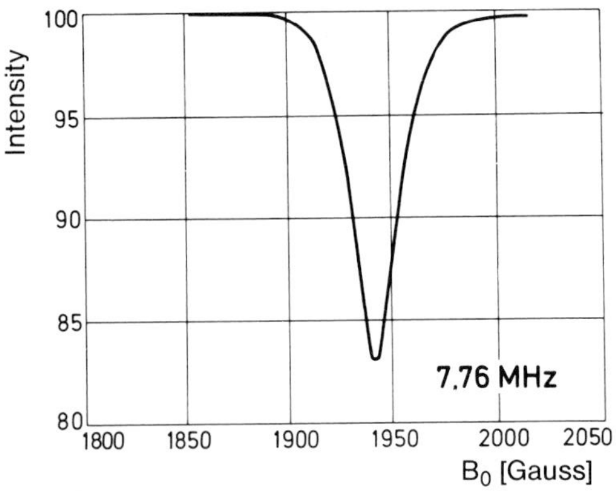

Fig. 20.17. Data curve from the Rabi atomic beam resonance experiment. The intensity at the detector is at a minimum (for a constant frequency of the field $\boldsymbol{B}_1$) when the homogeneous field $\boldsymbol{B}_0$ of magnet C fulfils the resonance condition. The curve shown here is for fluorine nuclei

thus exactly compensated by deflection in magnet B. However, if the resonance field $\boldsymbol{B}_1$ within the homogeneous magnet C ($\boldsymbol{B}_1$ is perpendicular to $\boldsymbol{B}_0$) causes some nuclei to flip from one orientation to another, then the displacement in the field of magnet B is no longer symmetric to that in the field of magnet A. The affected particles thus are no longer picked up by the detector D. The current in the detector D is measured as a function of the frequency of the field $\boldsymbol{B}_1$, or if $\boldsymbol{B}_1$ has a constant frequency, then as a function of the field strength of the magnet C. When the resonance condition is fulfilled, the current in the detector is a minimum, Fig. 20.17. To calculate the nuclear moment μ_I itself, one must have the nuclear spin I from other measurements, most often from the hyperfine lines in the optical spectrum (Sect. 20.3).

The Rabi atomic or molecular beam resonance method permits the measurement of ratios of resonance frequencies to great precision. The exact value of the field $\boldsymbol{B}_0$ does not have to be known. The ratio of the resonance frequencies of the electron and the proton, for example, was measured by letting particles with pure electron spin properties pass through the same apparatus as the molecular beam with pure nuclear spin properties. The ratio of the moments

$$\frac{\mu_{\mathrm{el}}}{\mu_{\mathrm{p}}} = g_{\mathrm{electron}}\mu_{\mathrm{B}}/g_{\mathrm{proton}}\mu_{\mathrm{N}}$$

could then be determined. Because the g factor of the electron is very precisely known from other experiments, it is possible to measure that of the proton, and from it the proton magnetic moment, very precisely. For protons, *Rabi* obtained the moment $\mu_I = (2.875 \pm 0.02)\mu_{\mathrm{N}}$. This value can now be used as a secondary standard for other nuclei or for the exact measurement of magnetic fields $\boldsymbol{B}_0$.

In 1946, *Purcell* and *Bloch* showed both experimentally and theoretically that the precessional motion of the nuclear spin is largely independent of the translational and rotational motion of the nucleus, and that the method of *nuclear spin resonance* can be applied not only to free atoms, but to atomic nuclei in liquids and solids.

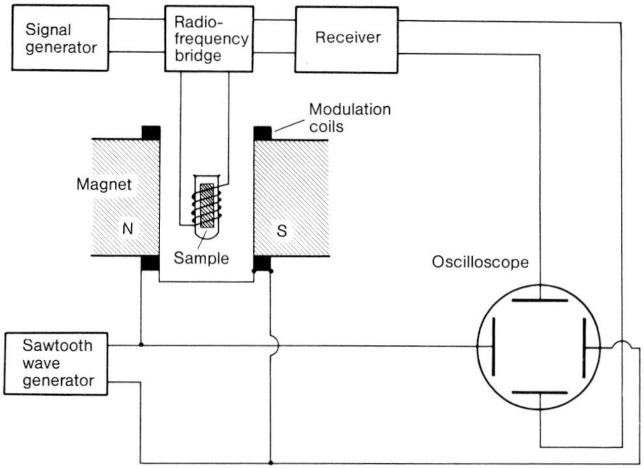

Fig. 20.18. Diagram of a simple nuclear spin resonance apparatus. The sample is placed in a glass tube between the poles of a homogeneous magnet. The high-frequency $\boldsymbol{B}_1$ field is applied across a bridge and an induction coil. For better detection of resonance, the $\boldsymbol{B}_0$ field can also be modulated by an additional coil

The principle of nuclear magnetic resonance in the condensed phase is explained in Fig. 20.18. With the applied magnetic field $\boldsymbol{B}_0$ fixed, one varies the frequency of a signal generator which produces the radio-frequency field $\boldsymbol{B}_1$. When the resonance condition (20.19) is fulfilled, i.e. when the frequency of the generator is identical to that of the nuclear Larmor precession, energy is absorbed from the radio-frequency field by the sample: this energy is taken from the induction coil containing the sample (*Purcell*). A power meter, used as receiver, indicates this energy absorption. The resonance may also be detected by observing a resonance signal induced in a second coil placed perpendicularly to the first (*Bloch*). Often, a fixed frequency is used, and the magnetic field is varied in order to sweep through the resonance.

For protons, with $g_I = 5.58$ (Sect. 20.2), the resonance frequency in a field $\boldsymbol{B}_0 = 1\,\mathrm{T}$ is calculated from (20.18) or (20.19) to be $\nu = 42.576\,\mathrm{MHz}$. This corresponds to a quantum of energy equal to $\Delta E = h\nu = 1.8 \cdot 10^{-7}\,\mathrm{eV}$. Because of this small energy splitting, all of the Zeeman levels of the nuclear spin system are nearly equally populated according to the Boltzmann distribution at room temperature in thermal equilibrium. Thus, in resonance, nearly the same number of absorption and emission transitions are induced, and only a small nett effect remains. At room temperature, the population difference between two nuclear Zeeman levels in a field of 1 T is about 10^{-6} of the populations themselves. This extremely small population difference is responsible for the nuclear resonance signal. In order to obtain high sensitivities, at present the strongest possible magnetic fields are employed, using superconducting coils; e.g. 12 T, giving $\nu = 500\,\mathrm{MHz}$ for a proton spin.

Relaxation processes (compare Sect. 13.3.7) re-establish the population differences; otherwise, the resonance would quickly become saturated, i.e. the levels would all be equally populated and no further signal would be observable.

In modern nuclear magnetic resonance spectrometers, the stationary or continuous wave methods described above are no longer used; instead, pulse methods, as described in Sect. 14.4 and 15.4, are employed. In these methods, one observes the time evolution of the (nuclear) magnetisation $I(t)$ following a short, intense high frequency pulse, and calculates the frequency dependence of the magnetisation, $I(\omega)$, from the observed temporal behavior.

This is illustrated in Fig. 20.19 for a simple case. After a 90° pulse (cf. Sect. 14.4), which turns the magnetisation away from the z direction and into the xy plane, the y

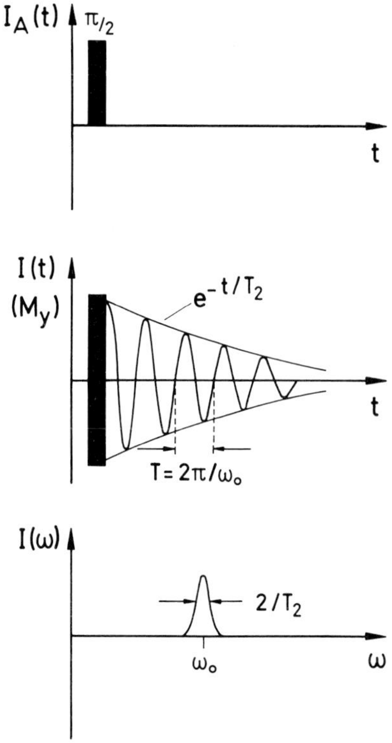

Fig. 20.19. Pulsed nuclear magnetic resonance. (*Upper part*) High-frequency pulses of frequency ω_0 are used to excite the resonance, e.g. 90° pulses. The measurement is repeated many times and summed (signal averaging). (*Middle part*) The time evolution $I(t)$ of the y component of the nuclear magnetisation, M_y, is observed. The envelope decays with the time constant T_2 when, as here assumed, only one species of equivalent nuclear spins (e.g. protons) are present in the sample. This time constant, which we have denoted by T_2 for simplicity, is more precisely called T_2^*, since it may contain T_1 in addition to the transverse relaxation time T_2 defined in Sect. 14.5. (*Lower part*) Through Fourier transformation of the observed time spectrum $I(t)$, one obtains the nuclear magnetic resonance spectrum $I(\omega)$, which is a single line in the case of one equivalent species of nuclear spins

component of the magnetisation exhibits an exponential decay with the time constant T_2. The Fourier transform of this decay is given by

$$I(\omega) = \frac{1}{2\pi} \int_{-\infty}^{+\infty} I(t)\,\mathrm{e}^{-\mathrm{i}\omega t}\,dt$$

and represents the nuclear resonance spectrum $I(\omega)$. If several inequivalent types of nuclei are present, the time evolution $I(t)$ becomes more complex.

20.7 Applications of Nuclear Magnetic Resonance

The precision with which one can determine NMR resonance frequencies is often limited by precise knowledge of the magnetic field $\boldsymbol{B}_0$. The magnetic moment of the proton is now known to very high precision, so that, conversely, NMR can be used for very precise *magnetic field measurements.* Accuracies of better than 10^{-8} are attainable with this method.

There are numerous other applications of NMR. In *nuclear physics*, it is used to determine gyromagnetic ratios and thus, when the nuclear spin I is known, magnetic moments μ_I. The most important applications of NMR are, however, in *chemistry*.

When an atom is chemically bonded, the influence of $\boldsymbol{B}_0$ on the electronic shells of the bonding partners causes a characteristic shift of the NMR frequency of its nucleus.

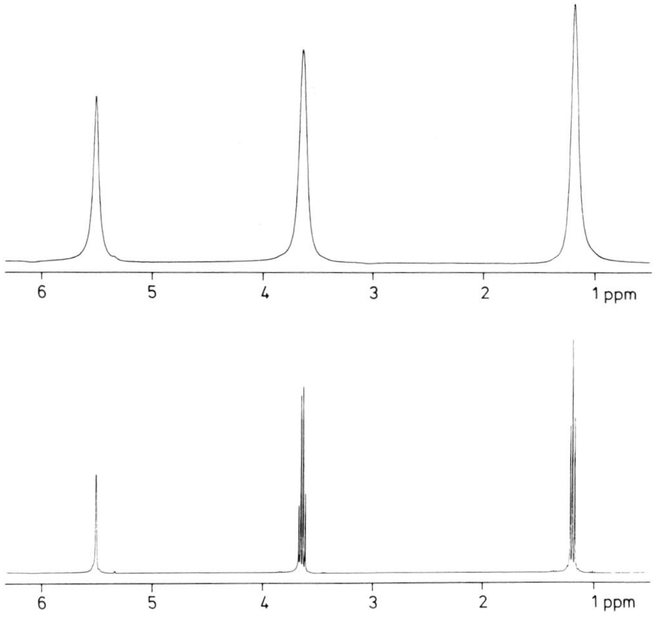

Fig. 20.20. Nuclear magnetic resonance (NMR) of ethyl alcohol, C_2H_5OH. The signal intensity at the receiver is plotted against the transition frequency. The latter is measured in ppm (parts per million) relative to the frequency of a suitable standard. In the upper portion of the figure, one sees three lines having the intensity (area) ratio 1 : 2 : 3. They are due to proton spins in the $-OH$, the $-CH_2$, and the $-CH_3$ groups, respectively, having 1, 2, and 3 protons. The resonance frequencies of these protons differ by a few ppm due to the differing chemical bonding in the three groups. In the lower portion, the same spectrum, but with higher experimental resolution, is shown. The $-CH_2$ signal is now split into a quartet of lines by the interaction with the $-CH_3$ protons, while the signal from the latter is split into a triplet by the indirect nuclear spin interaction with the $-CH_2^-$ protons. The protons of the $-OH$ groups are rapidly exchanged between different molecules, which averages out the indirect spin interactions and leaves a single resonance line

The electrons, depending on the bonding, shield the applied field to varying degrees. The change in the resonance frequency due to this diamagnetic shielding is called the *chemical shift*. It is measured in ppm (parts per million). Thus, two resonance lines are shifted by 1 ppm in a spectrometer with $\boldsymbol{B}_0 = 1\,\mathrm{T}$ and $\nu = 42.5$ MHz if their frequencies differ by 42.5 Hz.

As an example, Fig. 20.20 shows the proton NMR signal from ethanol. Even at low resolution (upper portion), one obtains three resonance maxima, since there are three different chemical shifts corresponding to protons bonded in the $-CH_3$, $-CH_2$, and $-OH$ groups of C_2H_5OH. The areas under the resonance curves are proportional to the number of nuclei at that resonance frequency. With extremely homogeneous magnetic fields and well-defined high-frequency signals, a considerably higher resolution can be attained (lower part of Fig. 20.20). One then observes that the three resonance lines are split into several resolved sublines. This structure is a result of indirect nuclear-nuclear interactions (coupled through the electrons) among all the protons in the molecule (called J coupling). This splitting will not be further treated here; it contains information about the structure and the bonding in the molecule under study.

NMR is thus of great importance for the elucidation of structures in chemistry and in solid state physics. Other applications are found for example in the investigation of motion in liquids and solids, in reaction kinetics, and in analytical chemistry.

In recent years it has become possible to obtain images of the interior of a (nonmetallic) body using NMR. One thus obtains information about the spatial distribution of certain atomic nuclei, e.g. protons. Furthermore, spatial inhomogeneities in the nuclear spin relaxation times which may be present can also be determined. This has opened a promising field in diagnostic medicine and in-vivo NMR of biological systems. Under the name *Magnetic Resonance Tomography,* this method has recently come to be applied in practical medicine, where it complements and enlarges the possibilities of x-ray diagnostic methods.

Spatially resolved NMR is based upon (20.18, 19), according to which the resonance frequency ω of a particular nuclear species is a unique function of the applied magnetic field $\boldsymbol{B}_0$. To obtain an image of spatial structures, one requires NMR signals which contain spatial information in coded form. For this purpose, magnetic field gradients are used; this is illustrated by the following discussion, using proton NMR as an example.

In a homogeneous magnetic field $\boldsymbol{B}_0$, the resonance condition $\omega = \gamma \boldsymbol{B}_0$ is fulfilled for each proton at the same frequency ω, independently of the location of that proton within the sample. The NMR signal therefore contains no spatial information. How-

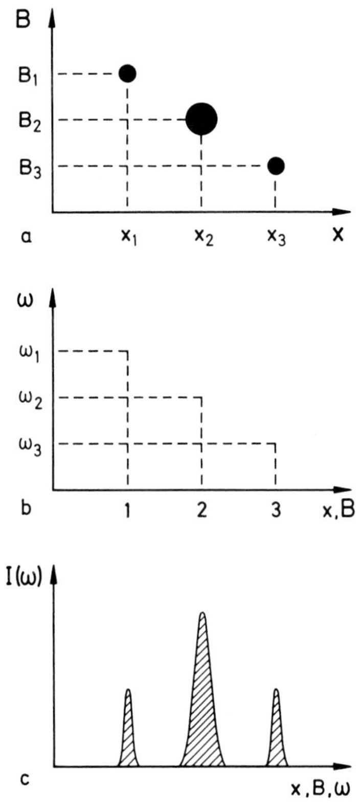

Fig. 20.21a – c

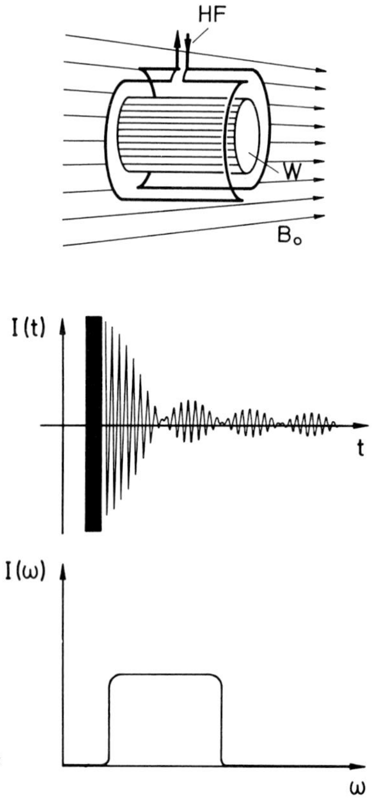

Fig. 20.22

Fig. 20.21a – c. Nuclear resonance of three samples with the same nuclei, e.g. protons, in a field gradient; illustrating position-resolved nuclear magnetic resonance. (**a**) In a field gradient dB/dx, each position x_1, x_2, and x_3 corresponds to a different field strength B_1, B_2, and B_3. Samples *1, 2*, and *3* (with *2* larger than *1* and *3*) are thus in regions of differing field strength. (**b**) Through the resonance condition $\omega = \gamma B_0$, each position x_1, x_2, or x_3 corresponds to a particular resonance frequency ω_1, ω_2, or ω_3. (**c**) The intensity of the NMR signal is a measure of the number of nuclei with the given resonance frequency (ω_1, ω_2, ω_3). A unique correspondence between the measured spectrum and the spatial arrangement of the nuclei exists

Fig. 20.22. The principle of position-resolved NMR. (*Upper part*) A cylinder *W* filled with water, in an inhomogeneous magnetic field $\boldsymbol{B}_0$, and surrounded by input and receiver coils, HF. (*Centre part*) The free induction decay (FID) of the protons in the cylinder *W* following a high-frequency pulse. (*Lower part*) The Fourier transform of the FID signal $I(t)$ yields the spectrum $I(\omega)$. When the proton density in the cylinder is constant, one thus obtains information about the geometry of the sample, here a section along the length of the cylinder

ever, when a field gradient is superposed on the homogeneous field $\boldsymbol{B}_0$, so that the field strength and thus the resonance frequency varies from place to place (Fig. 20.21, upper part), then protons at different locations will have different resonance frequencies (Fig. 20.21, centre). The spatial information is translated into a frequency scale, and the proton NMR signal contains information about the location of the protons (Fig. 20.21, lower part). An image of the spatial structure becomes possible if one knows the relationship between the field strength $\boldsymbol{B}$ and the location within the sample, and measures the signal amplitude as a function of frequency.

If, for example, a cylinder filled with water is placed in a field gradient $d\boldsymbol{B}/dx$ (Fig. 20.22, upper part), then the resonance frequency of the protons in layers perpendicular to $\boldsymbol{x}$ increases with increasing $\boldsymbol{x}$ from left to right. It has proved expedient in practice to measure not the resonance signal as a function of frequency, $f(\omega)$, but rather the time evolution of the signal following a 90° pulse, i.e. the free induction signal (compare Sect. 15.4). The measured function $I(t)$ (Fig. 20.22, centre) can be converted by Fourier transformation to the desired function, $I(\omega)$:

$$I(\omega) = \frac{1}{2\pi} \int_{-\infty}^{+\infty} I(t) \exp(-\mathrm{i}\,\omega t)\, dt \ .$$

The result is indicated in Fig. 20.22 (lower part). The observed free-induction decay following a 90° pulse contains the information about the time behaviour of the proton spins with frequencies in the range $\omega \pm \Delta\omega$ fulfilling the resonance condition (20.18), where $\Delta\omega$ is the frequency spread of the high frequency field $\boldsymbol{B}_1$ of centre frequency ω. The amplitude is proportional to the proton density. If, as in Fig. 20.22, the proton density is constant, the shape of the cylinder, projected onto the plane perpendicular to the $\boldsymbol{x}$ direction, will be imaged.

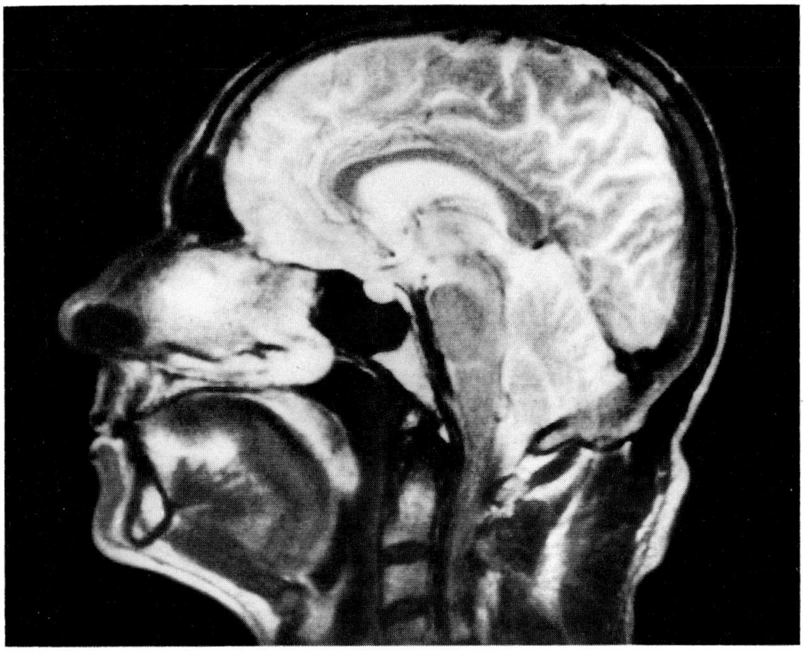

Fig. 20.23. Magnetic resonance tomography of a human head. (Kindly placed at our disposal by Prof. K. H. Hausser, Heidelberg)

In order to investigate the spatial structures of objects with more complex shapes and inhomogeneous proton densities, one requires many projections in different directions. From these, the desired image can be reconstructed with the aid of a computer. The principle of the measurement can thus be simply understood; in practice, somewhat different measurement procedures have been adopted. We shall not treat these here.

In biological samples, in general not only the densities of particular nuclei vary from place to place, but also their relaxation times T_1 and T_2 (see Sect. 14.5). They depend, for example, on the local temperature and the local state of motion of the structures being studied. By introducing the relaxation times as an additional indicator in position-resolved NMR and by using pulse methods for the measurement, it is possible to obtain images with strong contrast even from organs which show little contrast in x-ray images, e.g. in soft tissues.

An example from diagnostic medicine is shown in Fig. 20.23.

A further important modern application of resonance techniques is the *caesium atomic clock*, used as a time and frequency standard. ^{133}Cs is the only stable isotope of caesium. It has a nuclear spin $I = 7/2$, and thus in the atomic ground state with $J = 1/2$ the total angular momenta $F = 4$ and $F = 3$.

With the help of an atomic beam resonance apparatus (see Fig. 20.16), one observes transitions between the hyperfine components $F = 4$, $m_F = 0$ and $F = 3$, $m_F = 0$ in a weak applied field $\boldsymbol{B}_0$, Fig. 20.24. The transition frequency is practically independent of $\boldsymbol{B}_0$.

All the essential components of an atomic clock are similar to those of an atomic beam resonance apparatus following *Rabi* (cf. Sect. 20.6): a furnace which generates a Cs atomic beam; a polarisation magnet which separates the atoms in the $F = 3$ state from those in the $F = 4$ state; a resonator, in which atoms are made to undergo transitions from the $F = 4$, $m_F = 0$ state to the $F = 3$, $m_F = 0$ state by induced emission, caused by irradiation at the resonance frequency of 9.192631770 GHz; a second magnet which serves as analyser and passes only atoms in the $F = 4$, $m_F = 0$ state; and a detector. On resonance, one observes a minimum at the detector.

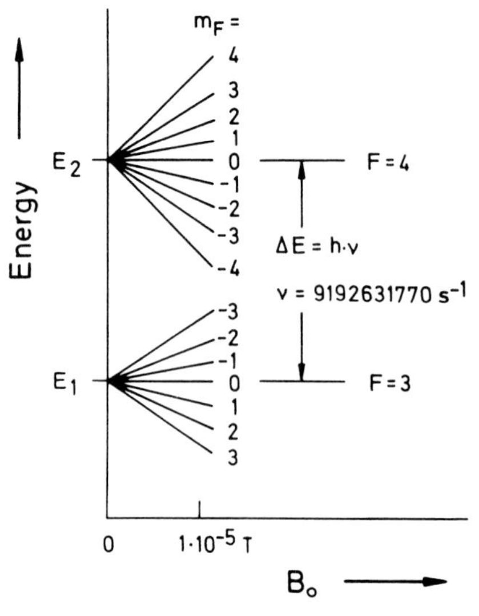

Fig. 20.24. A portion of the term scheme of the Cs atom in the ground state as a function of a weak applied magnetic field $\boldsymbol{B}_0$. The transition frequency used for the Cs atomic clock corresponds to the transition between the states $F = 3$, $m_F = 0$, and $F = 4$, $m_F = 0$

The frequency ν which is fed into the resonator is stabilised to the minimum in the detector. This frequency may be reproduced with an accuracy of ca. 10^{-13}. The most precise atomic clocks, which are used as primary frequency and time standards, are to be found in the laboratories of the Physikalisch-Technische Bundesanstalt in Braunschweig, F. R. Germany; of the National Bureau of Standards in the USA, and of the National Research Council in Canada.

20.8 The Nuclear Electric Quadrupole Moment

In the year 1935 *Schüler* and *Schmidt* found, in studying the hyperfine structure of the optical spectra of the isotopes ^{151}Eu and ^{153}Eu, that this structure cannot be explained through the magnetic interaction between the nuclear dipole moment and the inner and applied magnetic fields alone. They found deviations from the interval rule (20.12) which are caused by an *electrostatic* interaction between the non-spherically symmetric atomic nucleus and inner electric fields, more precisely by the interaction between the nuclear *electric quadrupole moment* and the electric field of the electronic shells.

Up to now, we have regarded the nucleus as spherically symmetric. If that is the case, it is surrounded by a spherically symmetric force field. The next less simple, possible charge distribution is that of an electric dipole; a static electric dipole is, however, not found in nuclei (due to conservation of parity in the nuclear states).

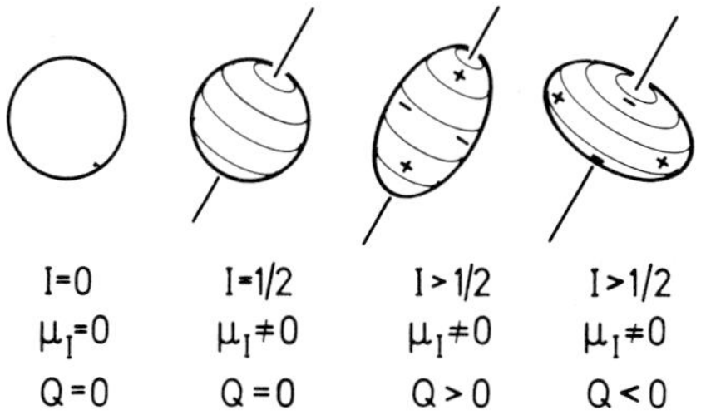

Fig. 20.25. The charge distribution of nuclei without spin, with spin 1/2, and with spin $I > 1/2$. The quadrupole moment is only nonzero for $I > 1/2$. It is positive for cigar-shaped and negative for pincushion-shaped charge distributions

In contrast, there are many nuclei – in fact, all nuclei with spins $I \geqq 1$ – whose form is clearly not spherical, but rather is an ellipsoid of rotation (Fig. 20.25). The electric potential of a homogeneously charged ellipsoid of rotation is likewise no longer spherically symmetric; it is, instead, described by an electric quadrupole moment (in a first approximation).

The electric quadrupole moment of an arbitrary axially symmetric charge distribution $\varrho(r)$ is defined as integral over the volume τ

$$eQ = \int_\tau (3\zeta^2 - r^2)\varrho(r)\,d\tau\,, \tag{20.20}$$

where r is the distance from the centre of the charge distribution and ζ is the coordinate in the direction of the axis of rotational symmetry. If such a charge distribution is brought into an inhomogeneous electric field, described by the potential V ($E = -\operatorname{grad} V$), the energy levels are shifted as a result of the electrostatic interaction by

$$\Delta E_Q = \frac{1}{16\pi\varepsilon_0} eQ \frac{\partial^2 V}{\partial z^2} (3\cos^2\theta - 1/2) , \tag{20.21}$$

where θ is the angle between the z axis of the potential (likewise assumed to be axially symmetric) and the rotational-symmetry axis of the charge distribution.

The quadrupole moment obviously vanishes for a spherically symmetric charge distribution. $Q > 0$ corresponds to a prolate (cigar-shaped) ellipsoid and $Q < 0$ to an oblate (pincushion-shaped) ellipsoid (Fig. 20.25).

In order to use (20.21) in the case of a free atom, we must take account of the fact that in this case, the rotational-symmetry axes of the quadrupole moment and of the potential are collinear with $\boldsymbol{I}$ and with $\boldsymbol{J}$, respectively. Therefore, the angle θ can be expressed in the vector model, again with the help of the law of cosines, analogously to the considerations we have used in the cases of fine structure and hyperfine structure. The result is

$$\Delta E_Q = \frac{eQ}{4\pi\varepsilon_0} \frac{\partial^2 V}{\partial z^2} \frac{(3/4)C(C+1) - I(I+1)J(J+1)}{2I(2I-1)J(2J-1)} \tag{20.22}$$

with the abbreviation $C = F(F+1) - I(I+1) - J(J+1)$. The hyperfine terms are shifted further (Fig. 20.26) and the interval rule is therefore violated.

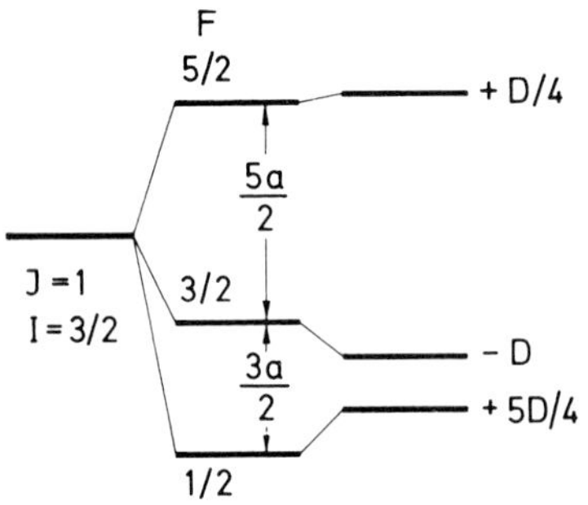

Fig. 20.26. Quadrupole shift of the hyperfine levels for a state with $J = 1$ and $I = 3/2$. The shift indicated corresponds to a positive quadrupole moment. The constant D is equal to the factor $(eQ/4\pi\varepsilon_0)(\partial^2 V/\partial z^2)$ in (20.22). The expected splitting pattern, given by the interval rule, is modified by the quadrupole interaction

A quantitative treatment of quadrupole shifts is difficult, for the most part. To determine the quadrupole moments of nuclei, one needs to know the field gradient $\partial^2 V/\partial z^2$. Only rather large field gradients, as are found in molecules or in solids (but which are larger than can be produced in the laboratory), yield term shifts which are sufficiently large to allow an exact measurement. More precise determinations of nuclear quadrupole moments may be obtained from the direct measurement of nuclear quadrupole resonance in molecules and in solids, by scattering of charged particles from nuclei (Coulomb excitation), by high-frequency resonance in an atomic or molecular beam, and by the study of muonic atoms (Sect. 8.7). The quadrupole shift must be taken into account in the analysis of hyperfine structure in high-resolution atomic spectroscopy in crystals: there, information can be obtained in particular about the internal electric field gradients, and thus about the distribution of the electrons between the atoms or molecules.

Problems

20.1 ^{209}Bi has an excited state with the electronic configuration $^2D_{5/2}$ which is split into 6 hyperfine levels. The distances between each pair of levels are 0.236, 0.312, 0.391, 0.471 and 0.551 cm^{-1}. How large is the nuclear spin quantum number I and what is the hyperfine structure constant a? Sketch the position of the hyperfine levels relative to the unsplit line in units of a.

20.2 Calculate the magnetic field generated at the nucleus by the $1s$, $2s$ or $3s$ electron of the hydrogen atom. How large is the energy difference between parallel and anti-parallel orientations of the proton and electron spins in these states?

20.3 From the data on the hyperfine splitting of the D_2 line in the sodium spectrum, Fig. 20.8, calculate the hyperfine constant a and the field B_J at the nucleus for the $^2P_{3/2}$ state.

20.4 Assume that hydrogen in the ground state is in a magnetic field of 0.3 tesla. Calculate according to Sect. 20.5 and Fig. 20.14 the Zeeman energy of the electron, the hyperfine interaction energy and the Zeeman energy of the nucleus.

20.5 ^{25}Mg atoms in the ground state ($^1S_0$) are examined by the Rabi method. In a field $B_0 = 0.332$ tesla, a resonance frequency $\nu = 3.5$ MHz is measured. ^{25}Mg nuclei have spin $I = 5/2$. Calculate the gyromagnetic ratio γ, the g factor and the largest component of the magnetic moment of the nucleus in the direction of the field (in nuclear magnetons).

20.6 In an atomic beam experiment similar to the Stern-Gerlach experiment, a beam of ^{23}Na ($^2S_{1/2}$) atoms is passed through a strong, inhomogeneous field B_1 ($\mu_B B_1 \gg E_{HFS}$). The beam is split into 8 parts. What is the nuclear spin quantum number of ^{23}Na? Into how many parts would the beam split in a weak inhomogeneous field?

21. The Laser

21.1 Some Basic Concepts for the Laser

The laser has become a light source without which modern spectroscopy could not exist (see also Chap. 22). In addition, the processes which lead to emission of laser light are a beautiful example of the application of basic knowledge about interactions between light and atoms. For both reasons we shall treat the laser in detail here.

The word "laser" is an acronym for "*l*ight *a*mplification by *s*timulated *e*mission of *r*adiation". As we shall see below, however, a laser not only amplifies light, but in most cases acts as a very special kind of light source, which emits light with properties unique to lasers, namely:

1) A particularly high degree of monochromaticity (temporal coherence). Linewidths $\Delta\nu$ of the order of ten hertz can be attained. This means, in the visible region, that the relative linewidth $\Delta\nu/\nu$ is $\lesssim 10^{-15}$. The coherent wave train then has a length l of 300000 km (the coherence length is calculated from $l = c\Delta t = c/\Delta\nu$), in contrast to the light from conventional sources, which consists of a "spaghetti" of light wave trains about 1 m long.

2) A very pronounced bunching of the light, which is practically limited only by diffraction effects at the exit window of the laser (spatial coherence).

3) A high radiation intensity, which can be as high as $10^{12}-10^{13}$ W in pulsed operation. Together with 1) and 2), this results in a very high photon flux density in a very narrow spectral range.

4) The possibility of generating ultra-short light impulses (down to 10^{-14} s) of high intensity.

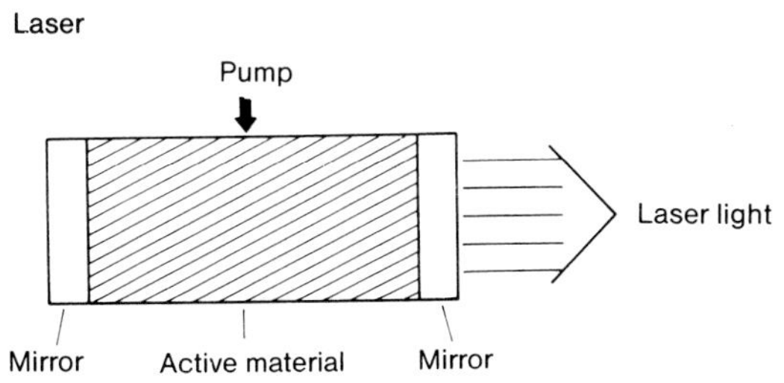

Fig. 21.1. Elements of a laser

The action of a laser can best be understood by considering its elements (Fig. 21.1). It consists, in principle, of a piece of "laser-active" material, which can be a solid, such as a ruby (i.e. a crystal of Al_2O_3 doped with Cr ions). The chromium ions, which are responsible for the red colour, undergo the optical transitions which make the laser process possible. Other examples are glasses or garnets which contain neodymium as

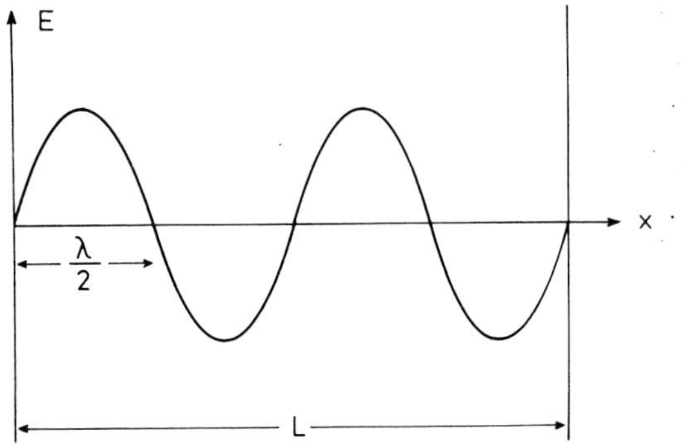

Fig. 21.2. Standing wave between the mirrors of the laser

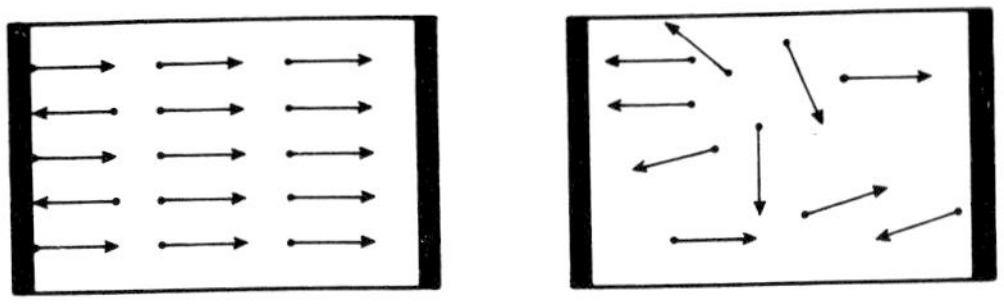

Fig. 21.3. Photons in the laser. See text

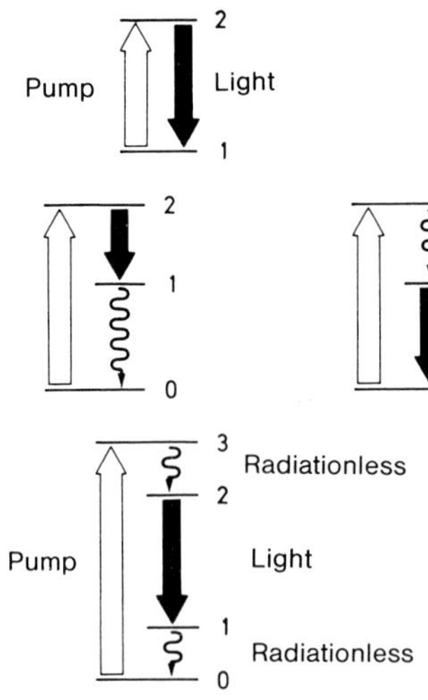

Fig. 21.4. Typical pumping schemes. *Above*, atom with two levels. *Middle*, atom with three levels: *left*, the upper (optical) transition is laser active; *right*, the lower (optical) transition is laser active. *Below*, atom with 4 levels. The middle (optical) transition is laser active

the active lasing atoms. Laser processes can also take place in gases or in certain dye solutions. Dye lasers are especially important for practical applications because they can be continuously tuned. The laser is bounded at two ends by mirrors. If we think of the light as a wave, we recognize that standing waves can only form between two mirrors if a multiple of one half the wavelength exactly fits between them (Fig. 21.2). On the other hand, if we consider the photon nature of light, then the photons which move in the axial direction are reflected back and forth between the mirrors, while those moving in any other direction quickly leave the laser (Fig. 21.3).

Finally, the laser must be "pumped" by addition of energy from outside. Different pumping methods can be used, depending on the atoms. In the 3-level scheme shown in Fig. 21.4, the laser atoms are excited to state 2 by optical pumping, i.e. by irradiation with light. From there, the atoms can fall to state 1 by emission of a light quantum. The same figure shows other simple possibilities for pumping. The helium-neon laser demonstrates a particularly interesting method of bringing about the occupation of excited states. The pumping process in this case is explained in the captions to Figs. 21.5 and 21.6. Figure 21.7 shows an example of a construction plan for a helium-neon laser.

The basic idea of a laser is as follows: We imagine that a considerable number of atoms have been brought into the excited state by optical pumping. Then photons can be spontaneously emitted. If the photon interacts with another excited atom, it can induce this atom to emit another photon. Through repetition of this process, an avalanche of photons builds up. The process cannot go on ad infinitum, since more and more of the atoms go into the ground state. If energy is constantly supplied by pumping, an equilibrium is reached. We must now examine the requirements for excitation conditions and radiation properties which must be met in order for the atom to be laser active. For this activity, the spontaneous emission of light, called fluorescence, must be replaced by induced, collective emission, which results in the typical characteristics of laser light.

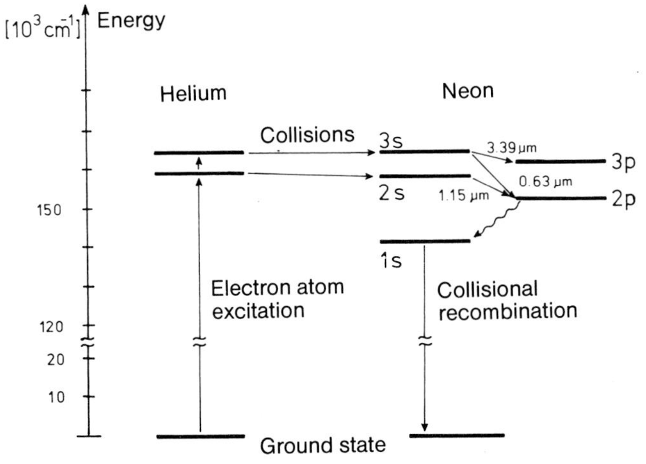

Fig. 21.5. Excitation and recombination processes in the helium-neon laser (schematic). The symbols 1*s*, 2*s*,... used in the neon spectrum on the right are arbitrary technical terms (Paschen symbols) explained in the legend of Fig. 21.6. The helium atoms are excited to the ^1S or ^3S state by collisions with electrons in a gas-discharge tube. Because the excitation energies correspond to the "2*s*" and "3*s*" states of neon, this energy can be transferred from helium to neon by atomic collisions. The electron in the 3*s* or 2*s* state in the neon then goes into the 3*p* or 2*p* state, and laser light emission may be associated with this process (solid arrows). The electron then drops back to the 1*s* state by emission of another light quantum. Finally the neon atom loses the remainder of the energy and drops back to the ground state by collisions with other atoms. In reality, the term scheme of neon is much more complicated than shown here (Fig. 21.6). Note that the energy of the ground state was chosen as the zero point of energy here

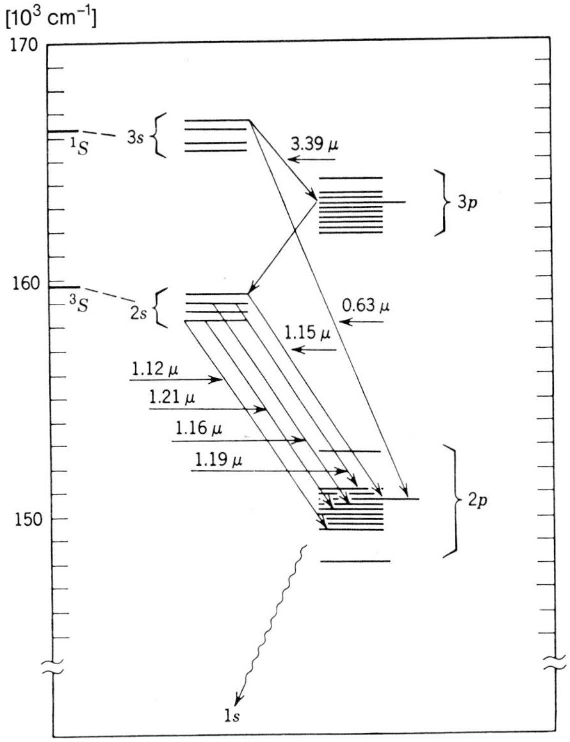

Fig. 21.6. Splitting and transition scheme for the energy levels of helium (*left ordinate*) and neon (*centre*). The ground state of neon, which is not shown here, has the electron configuration $1s^2 2s^2 2p^6$. In the states shown, one of the six 2*p* electrons is excited. If we give only the configuration of the six outermost electrons, then the abbreviations 1*s*, 2*s*, and 3*s* stand for $2p^5 3s$, $2p^5 4s$ and $2p^5 5s$, respectively. Each of these states consists of 4 terms. The abbreviations 2*p* and 3*p* stand for $2p^5 3p$ and $2p^5 4p$, each of which has 10 terms

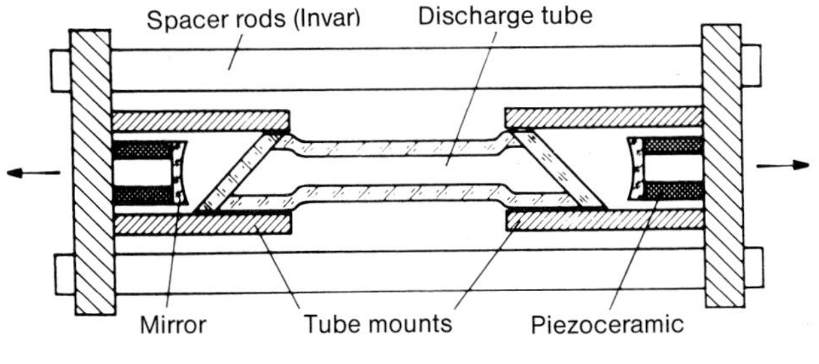

Fig. 21.7. Example of a gas laser. The most important part is the discharge tube which contains the helium-neon mixture. The surfaces at the end are placed to form an angle less than the Brewster angle. The mirrors, which are curved to give the light modes more stability, are located outside the tube. The rest of the parts serve as supports

In the following, we shall discuss the important case in which only a single type of photon is generated by the laser process, namely photons which move in the axial direction and exactly fit between the mirrors as a standing wave. It is also assumed that the basic frequency of the light is in resonance with a transition frequency of the atoms.

A laser which has these properties is called a "single-mode" laser. In the next section we shall develop the theory of this laser in terms of the *photon model*, and then in Sect. 21.3, in terms of the *wave model.*

21.2 Rate Equations and Lasing Conditions

First we consider the rate equations of the laser. It is most useful here to think of the light as photons. As we have already observed, there are different types of photons, e.g. those moving parallel or obliquely to the axis of the laser. We would expect the axially oriented photons to be most likely to contribute to the laser process, since they remain between the mirrors for the longest time and thus have the greatest chance of inducing emission. We let n be the number of photons under consideration and observe how this number changes with time. The number of photons increases as a result of the induced emission. According to *Einstein*, the rate of increase is proportional to the number of atoms in the excited state N_2 and to the number n itself (see Sect. 5.2.3). It is also proportional to a transition probability per second, W. Of course, atoms in the lower state of the optical transition can absorb photons, thus removing them from circulation. The photon number is reduced by WN_1n. The rate of the spontaneous emission is independent of the number of photons present, but it is proportional to the number of excited atoms: WN_2. Finally, the photons can escape through the mirror or be lost by scattering. The rate of loss is proportional to the number n of the photons. If we assume that the lifetime of a photon in the laser is t_0, the loss rate is $-n/t_0$. We thus obtain the scheme:

Process	Rate
Stimulated emission	WN_2n
Absorption	$-WN_1n$
Spontaneous emission	WN_2
Losses	$-n/t_0$

(21.1)

Summarising these expressions, we obtain the rate equation for the number of photons:

$$\frac{dn}{dt} = W(N_2 - N_1)n + WN_2 - \frac{n}{t_0}. \tag{21.2}$$

Using his idea about photons, *Einstein* showed in the derivation of the Planck radiation formula that W is given by

$$W = \frac{1}{VD(\nu)\Delta\nu \cdot \tau}. \tag{21.3}$$

Here V is to be interpreted as the volume of the lasing material, $\Delta\nu$ is the linewidth of the electron transition between levels 2 and 1 in the atom, and τ is the lifetime in level 2.

$D(\nu)\Delta\nu$ is the number of standing waves in a unit volume in the frequency range $\nu \dots \nu+\Delta\nu$. $D(\nu)$ is given explicitly by

$$D(\nu) = 8\pi \frac{\nu^2}{c^3}, \tag{21.4}$$

where ν is the frequency of the atomic transition. With (21.3), it is possible to derive the laser conditions. As we have mentioned, a photon avalanche must be generated. This means that only those terms which are proportional to n in (21.2) are of interest. Furthermore, a more exact mathematical analysis shows that spontaneous emission, represented by the term WN_2, is not correlated with the actual laser light, and only gives rise to noise.

To derive the laser condition, we therefore leave out this term on the right side of (21.2). We now obtain the laser condition by demanding that the rate of photon generation is greater than zero:

$$\frac{dn}{dt} = W(N_2 - N_1)n - \frac{n}{t_0} > 0. \tag{21.5}$$

Since the photon number is assumed not to be equal to zero, (21.5) leads directly to a relationship for the necessary "inversion" per unit volume,

$$\frac{N_2 - N_1}{V} > \frac{8\pi\nu^2 \Delta\nu\tau}{c^3 t_0}. \tag{21.6}$$

Equation (21.6) is the laser condition. In order to obtain laser activity, we must therefore bring enough atoms into the excited state by optical pumping, so that the difference in the occupation densities $(N_2 - N_1)/V$ satisfies the condition (21.6). The smaller the right side of (21.6), the more easily laser activity can be obtained. The right side sets conditions on the atoms to be used. In order to get by with the smallest possible pumping power, the atoms must have as narrow a linewidth as possible. Furthermore, the laser condition becomes more and more difficult to satisfy as the laser frequency ν increases. In addition, the lifetime t_0 of the photons in the laser must be made as large as possible by using the best available mirrors. t_0 can be estimated as follows: $1/t_0$ can be taken as the rate per second at which photons leave the laser. This rate is naturally proportional to the velocity of light. Because the time of flight between the mirrors is proportional to the distance between them, the probability of loss is inversely proportional to the length L. Finally, the probability of exit is inversely related to the reflectivity R of the mirror. This yields the factor $(1-R)$. In this way, we arrive at the formula

$$\frac{1}{t_0} = \frac{c}{L}(1-R) \tag{21.7}$$

for the lifetime of the photons.

The overall dynamics of the laser radiation can only be described when, in addition to (21.2), we have similar equations describing the occupation numbers for the atomic states. These are based on the pumping scheme shown in the middle section of

Fig. 21.4. The rate of change of the occupation number N_2 is given by the following equation, which we shall justify intuitively:

$$\frac{dN_2}{dt} = \underbrace{-WN_2n + WN_1n}_{\text{coherent interaction}} + \underbrace{w_{20}N_0 - w_{12}N_2}_{\text{pumping and recombination}}. \tag{21.8}$$

The first two terms on the right side describe the changes in the occupation number N_2 due to stimulated emission and absorption. The third term on the right side shows how this occupation number is increased by pumping from the ground state (level 0), and the last term represents changes in the occupation number caused by competing processes which do not contribute to the laser process, such as non-radiative recombination from state 2 to state 1. (By the way, the expression "w_{20}" is read "$w_{\text{two zero}}$", not "w_{twenty}".)

One is now easily convinced that the rate equations for the other two levels have the form

$$\frac{dN_1}{dt} = WN_2n - WN_1n + w_{12}N_2 - w_{01}N_1 \tag{21.9}$$

and

$$\frac{dN_0}{dt} = -w_{20}N_0 + w_{01}N_1, \tag{21.10}$$

assuming that we ignore direct recombination from state 2 to state 0. We can obtain an essential clue to the solutions of these equations by considering a special case. We assume that the transition from state 1 to state 0 is extremely rapid, so that level 1 is practically unoccupied, and we can set $N_1 = 0$. Of the equations (21.8 – 10), we are then interested only in (21.8), which we write in the form

$$\frac{dN_2}{dt} = -WN_2n + w_{12}(\bar{N}_2 - N_2) \quad \text{with} \quad \bar{N}_2 = \frac{w_{20}}{w_{12}}N_0. \tag{21.11}$$

Here $\bar{N}_2$ is the value of N_2 which would be obtained through pumping and relaxation processes alone, i.e. if $n \equiv 0$. The expression for the photons also simplifies to

$$\frac{dn}{dt} = WN_2n - \frac{n}{t_0}. \tag{21.12}$$

We now convince ourselves that a laser can attain a steady state, i.e. that it can emit light continuously. This requires that we set

$$\frac{dN_2}{dt} = 0, \quad \frac{dn}{dt} = 0. \tag{21.13}$$

From (21.12), we obtain the relationship

$$N_2 = \frac{1}{Wt_0} \equiv N_{2,\text{th}}. \tag{21.14}$$

This says that a definite occupation number $N_{2,\text{th}}$ (th means threshold) is established during the laser process, and that it remains constant, even if we pump in more energy. The pump energy must therefore be converted primarily into photon energy, as we actually find if we solve (21.11) for n:

$$n = \frac{w_{12}}{W}\left(\frac{\bar{N}_2}{N_{2,\text{th}}} - 1\right). \tag{21.15}$$

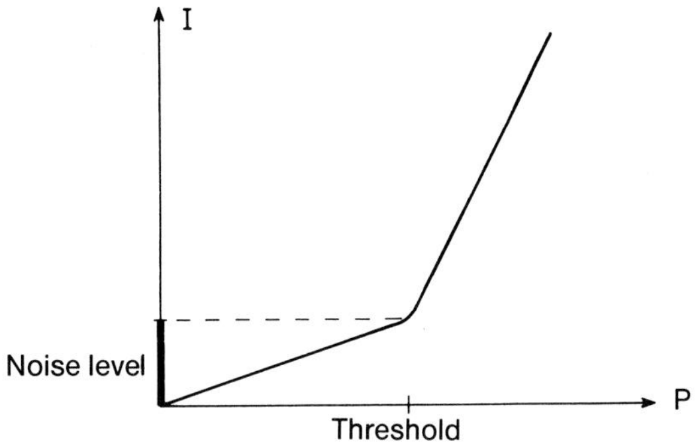

Fig. 21.8. The emission behaviour of a single-mode laser. Abscissa: pumping power input. Ordinate: emission power (in a single mode). Below the threshold, i.e. the critical pumping power, the light consists of spontaneously generated wave trains ("noise"). If this is ignored, the emitted radiation is equal to zero. Laser activity begins at the threshold

If the average number $\bar{N}_2$ is increased by increasing the pump rate w_{20}, then according to (21.15) the number of photons increases. Since the photon number cannot be negative, the laser process does not begin until a critical value of the inversion or occupation number $N_{2,\text{th}}$ is reached. This accounts for the curve given in Fig. 21.8. Below a certain pumping power, or inversion, laser light is not emitted. Above this power, the number of photons emitted increases linearly. In actual data curves, there is emission below the threshold, but this is due to the fact that the atoms can emit light spontaneously, and this light is "noise" as far as the laser is concerned.

21.3 Amplitude and Phase of Laser Light

In the preceding section we treated the single-mode laser in terms of the photon model. This made it possible to derive the lasing conditions and to determine the number of photons present in the laser. We know, however, that light is characterised not only by its intensity, but by its phase relationships, which account for such phenomena as interference. It has been shown experimentally that there are limits to the ability of wave trains to interfere with each other. Let us consider a finite wave train, which we divide into two by means of a half-silvered mirror. We allow the two beams to traverse different paths, and then allow them to interfere. Obviously this is only possible if the difference in the lengths of the paths is less than the length of the wave train. Let us now consider the light from an ordinary lamp, in which each atom emits its wave train independently of the others. Since the emission time (= lifetime of the excited state) is finite, the wave train has only a finite length. The light field generated by the lamp thus consists of individual, uncorrelated wave trains.

The "coherence length" of the light is the average length of a wave train. In the following, we shall examine the question of whether the light from a laser fits this description, only with much longer wave trains, or if it has basically different properties. To answer this question we construct the following laser model, which can be rigorously grounded.

Let us consider a light wave between the two mirrors. It has the form of a standing wave, and the light field strength F can be written

$$F = E_{tot}(t) \sin kx \tag{21.16}$$

where

$$k = \frac{\pi m}{L}. \tag{21.17}$$

Here L is the distance between the mirrors and m is an integer. In a purely harmonic oscillation, E_{tot} would have the form

$$E_{tot}(t) = 2E_0 \cos \omega t, \tag{21.18}$$

where ω is the atomic transition frequency. Actually, however, we must expect that phase shifts arise in the course of time, and that the amplitude E_0 also is not constant. For example, we know that in the case of spontaneous emission from a single atom, E_0 decays exponentially. For this reason, instead of (21.18) we write

$$E_{tot} = \underbrace{E_0 e^{i\phi(t)}}_{E(t)} e^{-i\omega t} + \text{c.c.} \tag{21.19}$$

and thus obtain for (21.16) the general trial solution

$$F = [E(t) e^{-i\omega t} + \text{c.c.}] \sin kx. \tag{21.20}$$

An equation for the complex amplitude $E(t)$ is derived in laser theory. We attempt to arrive at this equation by an intuitive process as follows. The light field amplitude E is increased by the process of induced emission, with the amplification being proportional to the light field amplitude E and to the number of excited atoms N_2. The proportionality factor is called g. The light field amplitude is also subject to a symmetrical decrease, due to absorption, which occurs at the rate $-gN_1E$.

Furthermore, the light can leave the laser through the mirror or be scattered in some way. The result is a loss rate for the light field amplitude expressed by $-\kappa E$. Finally, the light field strength is also subject to constant change due to spontaneous emission events. This is expressed as the so-called "fluctuating forces" $f(t)$, which reflect the statistically fluctuating emission events. We thus obtain the equation

$$\frac{dE}{dt} = g(N_2 - N_1)E - \kappa E + f(t). \tag{21.21}$$

As we saw in the preceding section, the inversion $(N_2 - N_1)$ is changed by the laser process. On the one hand, $(N_2 - N_1)$ is brought to a value of $(N_2 - N_1)_{pumped}$ by the processes of pumping and relaxation – this is called the *unsaturated inversion*. Atoms are constantly being removed from the upper level by the laser process, so that the actual inversion (N_2-N_1) is less than the unsaturated inversion. If the laser intensity is not too high, one can assume that this reduction is proportional to the intensity $|E|^2$ of the light, so that

$$N_2-N_1=(N_2-N_1)_{\text{pumped}}-\text{const}\,|E|^2\,. \tag{21.22}$$

Equation (21.22) is called the *saturated inversion*. If we substitute (21.22) in (21.21), we obtain the fundamental laser equation

$$\frac{dE}{dt}=(G-\kappa)E-C|E|^2E+f(t)\,. \tag{21.23}$$

Here $G=g(N_2-N_1)_{\text{pumped}}$ is a gain factor, and $C=g\cdot\text{const}$.

To arrive at an intuitive understanding of the meaning of (21.23), we can take refuge in mechanics and think of E as a coordinate q of a particle. If we introduce an acceleration term $m\ddot{q}$ on the left side of (21.23), we obtain

$$m\ddot{q}+\dot{q}=K(q)+f\,, \tag{21.24}$$

where we can think of m as being so small that it can be ignored. Equation (21.24), however, is the equation for the damped oscillation of a particle subject to the forces K and f. We can calculate K from a potential V: $K=-dV/dq$. The potential field is shown in Fig. 21.9. For $G<\kappa$, i.e. for low pumping power, the upper curve applies. Our fictitious particle is repeatedly pushed up the potential curve by the fluctuating force f, and then falls back to the equilibrium position $q=0$ under the influence of the force K. If we identify q with E, this means that each push represents the beginning of an emission event in the atom. After the event, the field strength again decays to zero. Since the pushes are directed to the left and right with equal frequency, the average position of the particle is equal to zero, or in other words, the average field strength E is equal to zero.

Now let us consider the case in which $G>\kappa$. In this case the lower curve of Fig. 21.9 applies. It is obvious that the original position $q=0$ has become unstable and is to be replaced by a different q. On the other hand, however, the pushes exerted on our fictitious particle drive it around among these new equilibrium positions, so that q has the form $q=r_0+\varrho(t)$. Now comes an important point. We spoke before of a real particle coordinate. In reality, however, the field amplitude in (21.23) is complex.

This means that in order to complete the picture, we must think of the potential field being rotated about the V axis (Fig. 21.10). The fictitious particle in this picture is rolling in a valley, in which it is pushed either in a radial or a tangential direction by the fluctuating forces. Because there is no restoring force in the tangential direction, the particle carries out a sort of diffusion motion. If we apply this to laser light, we obtain the following picture:

The complex amplitude can be broken down according to the rule

$$E=[r_0+\varrho(t)]\,e^{i\phi(t)}\,, \tag{21.25}$$

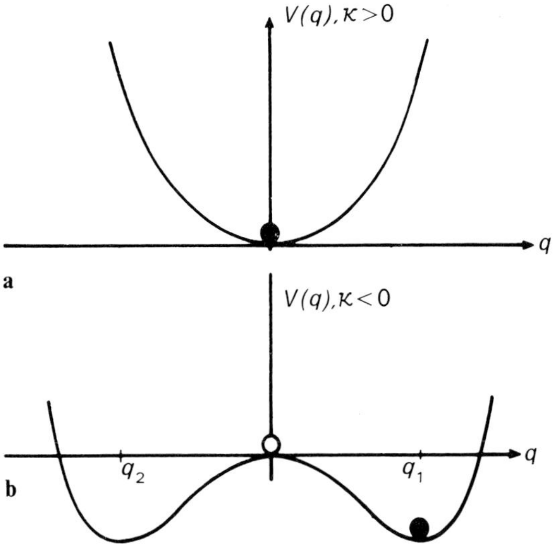

Fig. 21.9a, b. The "effective potential" for the laser amplitude $q \equiv E$. **a)** Potential for pumping power below the threshold. **b)** The same above the threshold

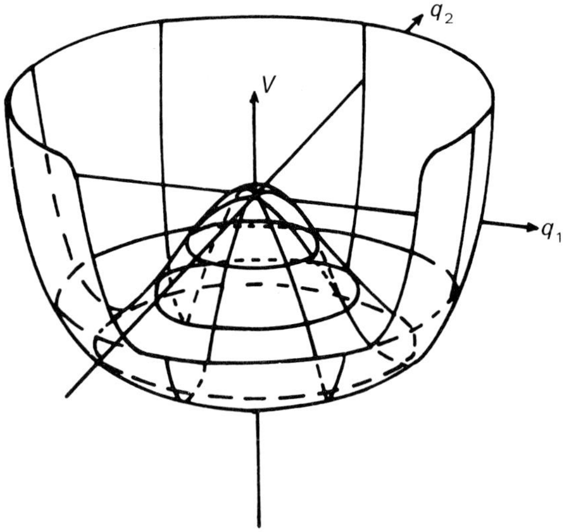

Fig. 21.10. The "effective potential" for the complex laser amplitude $E = q_1 + \mathrm{i}q_2$ above the threshold

where r_0 is the new stable amplitude, ϱ is the amplitude fluctuation, and the factor $\exp[\mathrm{i}\phi(t)]$, which describes the tangential diffusional motion mentioned above, is the phase diffusion. As is shown in laser theory, the average value of ϱ^2 decreases with increasing laser intensity I according to $\langle \varrho^2 \rangle \sim 1/I$. The equation (21.25) represents an infinitely long wave train, which merely undergoes small fluctuations in phase and amplitude. Phase diffusion is responsible for the finite linewidth, which also decreases with $1/I$. We recognise from this that the properties of laser light are fundamentally different from those of light from normal lamps. Laser light has a *stable amplitude* and its *linewidths* are very *sharp*. The theoretical values naturally depend on the laser materials, but for typical materials they are of the order of 1 Hz. The linewidths observed with precision lasers are about 10 Hz.

Problems

21.1 Calculate W for a ruby laser with the following parameters:

$$\begin{aligned} V &= 62.8\ \mathrm{cm}^3 \\ \nu &= 4.32 \cdot 10^{14}\ \mathrm{Hz} \\ \Delta\nu &= 2.49 \cdot 10^{13}\ \mathrm{Hz} \\ c &= 2.9979 \cdot 10^{10}\ \mathrm{cm/s} \\ \tau &= 3.0\ \mathrm{ms} \end{aligned}$$

21.2 Calculate t_0 for the following cases:

Resonator length $L = 1$, 10 or 100 cm;
Reflection coefficient $R = 99$, 90 or 10%.

21.3 Using the rate equation (21.2), calculate the exponential rise time for the parameters calculated in Problems 21.1 and 21.2. The population difference $(N_2 - N_1)$ should be taken to be constant with the value 10^{14}.

21.4 Using the laser conditions, calculate the critical inversion density for ruby. The following data apply:

$V = 62.8\ \mathrm{cm}^3$
$\nu = 4.32 \cdot 10^{14}\ \mathrm{Hz}$
$\Delta\nu = 2.49 \cdot 10^{13}\ \mathrm{Hz}$
$\tau = 3.0\ \mathrm{ms}$
$c' = 1.70 \cdot 10^{10}\ \mathrm{cm\,s^{-1}}$ (velocity of light in the medium)
$R = 99\%$

22. Modern Methods of Optical Spectroscopy

22.1 Classical Methods

We have come to know optical spectroscopy in the preceding chapters as the most important method for the investigation of the electronic shells of atoms and for understanding atomic structure. We have seen that in many cases, extremely small splittings or shifts of the spectral lines must be determined with a high spectral resolution. Progress in our knowledge of atomic structure and of the fundamentals of quantum mechanics has only been possible through a continuous process of improvements in experimental techniques.

In this chapter, we shall first treat the question of how a high spectral resolving power can be attained. If $\Delta\lambda$ is the wavelength difference which two closely spaced spectral lines may have and still be recognised as two lines – i.e. resolved – then the quotient $\lambda/\Delta\lambda$ is called the spectral resolving power of the apparatus. (We note that a large value of this quotient implies a "high" resolving power in the above sense.)

With prism and grating spectrographs, which were developed nearly to perfection in the first quarter of this century, it is difficult to obtain a resolving power significantly above a few $\times 10^5$. Diffraction limits the resolving power of a grating spectrograph to the product Nm, where N is the total number of grating rulings and m is the spectral order. Both numbers cannot be increased without limit while maintaining a measurable light intensity. Nevertheless, an extremely high resolving power may be reached by observing in a high order with a correspondingly shaped ("blazed") grating, to be sure in a limited range of wavelengths. If resolving powers of 10^6 and above are required, an interferometer must be used. The most important type is the parallel-plate interferometer of *Fabry* and *Pérot* (Fig. 22.1).

This interferometer consists of two half-silvered glass plates which are set up parallel to each other at a distance of several cm. Through multiple reflections, inter-

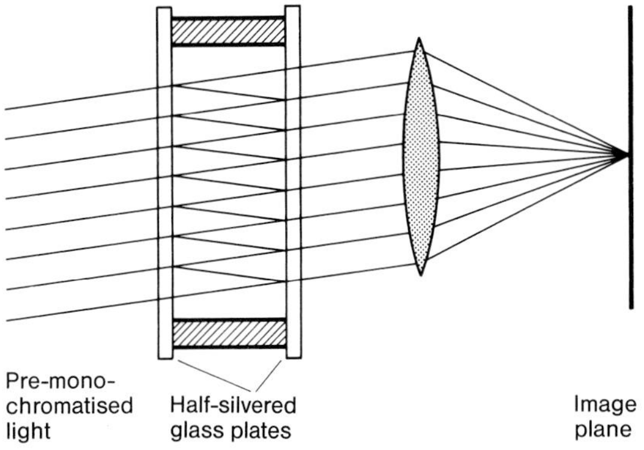

Fig. 22.1. Fabry-Pérot interferometer. The interference fringes at a given angle which are produced between the two plane-parallel glass plates yield a system of concentric rings in the image plane. The usable range of wavelengths is very small; for this reason, the light must be pre-monochromatised with a prism or a grating monochromator

ferences of high order m occur. If the spacing of the plates is 2.5 cm, then the path length difference between neighbouring rays is equal to 5 cm. For light of wavelength 500 nm, this corresponds to an order $m = 10^5$, since $m = (5 \cdot 10^{-2}\,\mathrm{m})/(5 \cdot 10^{-7}\,\mathrm{m})$. The number of rays N which can interfere is limited by the reflectivity of the mirrors and is of order 10. The resolving power is thus about 10^6 according to the formula $\lambda/\Delta\lambda = m \cdot N$.

With such spectrographs, the line shapes or the splitting behaviour in a magnetic field are investigated for individual spectral lines. For this reason, usually a simple monochromator to pre-select the light is used in front of the Fabry-Pérot interferometer. In the image plane of the interferometer, a system of concentric rings is observed. For monochromatic light, these are the neighbouring interference orders, e.g. $m = 1000000$, 1000001, 1000002, etc. The intensity distribution in the interference pattern may be recorded photographically or photoelectrically. The wavelength range may be increased somewhat by changing the spacing of the plates or the gas pressure between the plates.

In the past 10 years, completely new methods of ultra-high-resolution optical spectroscopy have become possible as a result of developments in laser technology. In the following sections, we shall treat these methods in more detail.

22.2 Quantum Beats

In this section, we shall treat a modern spectroscopic method which has come to be known by the name "quantum beats". This method permits the resolution of closely spaced neighbouring levels. To understand the principle, let us imagine an atom with three levels, which we shall denote by the indices 0, 1, and 2 (Fig. 22.2). The energy levels 1 and 2 are supposed to be closely spaced, with a frequency spacing denoted by $\omega_{12} = (E_2 - E_1)/\hbar$. The atom is at first taken to be in the ground state 0 and is then excited by means of a brief light pulse. The pulse length τ is chosen so that it fulfils the relation $\tau \leqq 1/\omega_{12}$. The centre frequency ("carrier frequency") ω_0 of the pulse corresponds to the transition frequency from level 0 to one of the levels 1 or 2. If we suppose the pulse to be expanded in sine waves, these have frequencies for the most part in the range $-\omega_{12} + \omega_0$ to $+\omega_{12} + \omega_0$. This light pulse can cause an electron to undergo a transition from the ground state into one of the two states 1 or 2, or, more generally, into a superposition of the two states. A superposition of two wavefunctions may be interpreted, as we saw in Sect. 16.1.4, as an oscillation, for example of a dipole, with the frequency ω_{12}. If the electron now radiates light, the exponential decay of the

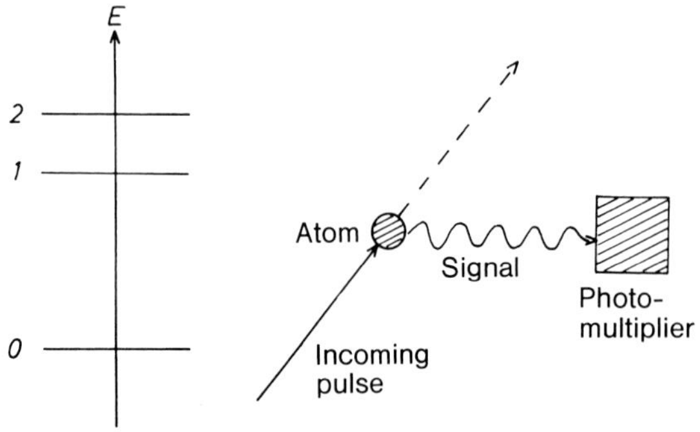

Fig. 22.2. Quantum beats. *Left:* Schematic energy level diagram. *Right:* Experimental set-up. The atom is excited by an incoming light pulse. The emitted radiation is measured in another direction with a photomultiplier. Typical experimental results are shown in Fig. 22.3

radiated intensity from spontaneous emission is modulated at the above oscillation frequency ω_{12}. Applying the statistical interpretation of quantum mechanics, this phenomenon is often explained as follows: the excited electron, when it happens to be in state 1, can emit a photon with the frequency $\omega_{10} \equiv (E_1 - E_0)/\hbar$; in state 2, it can emit a photon of frequency ω_{20}. To these two light *quanta*, there corresponds a *beat* frequency $\omega_{12} \equiv \omega_{20} - \omega_{10}$; thus we arrive at the name "quantum beats".

We now give a more exact quantum mechanical treatment. Using quantum theory, we must take as the wavefunction of the excited electron at time $t = 0$

$$\psi(\boldsymbol{r}, 0) = \alpha_1 \phi_1(\boldsymbol{r}) + \alpha_2 \phi_2(\boldsymbol{r}) , \tag{22.1}$$

where ϕ_i is the wavefunction of the electron in the level $i = 1$ or 2. The α's are coefficients which are determined by the initial excitation process. In the time that follows, the electron will undergo a transition to the ground state, accompanied by the emission of a photon; in the process, the occupation probability in the levels 1 and 2 decreases exponentially with a decay constant 2Γ, while the ground state becomes populated. The corresponding wavefunction is denoted by ϕ_0. This implies the state in which the electron is in the ground level and, additionally, a photon is present. The total wavefunction therefore takes the form

$$\psi(\boldsymbol{r}, t) = \alpha_1 \exp(-\mathrm{i}E_1 t/\hbar - \Gamma t)\, \phi_1(\boldsymbol{r}) + \alpha_2 \exp(-\mathrm{i}E_2 t/\hbar - \Gamma t)\, \phi_2(\boldsymbol{r}) + \alpha_0(t)\, \phi_0(\boldsymbol{r}) . \tag{22.2}$$

The coefficient $\alpha_0(t)$, which is especially interesting for the present application, can be quantum mechanically determined. Unfortunately, we cannot, for reasons of space, give the Wigner-Weisskopf theory, which is applicable to this determination, here. We therefore simply show the result, according to which the coefficient α_0 consists of the two parts

$$\alpha_0(t) = c_1(t) + c_2(t) , \tag{22.3}$$

in which the coefficients c_j have, in principle, the form

$$c_j(t) = \alpha_j \boldsymbol{e} \cdot \boldsymbol{\theta}_{0j}\, a\, \mathrm{e}^{-\mathrm{i}E_j t/\hbar} (\mathrm{e}^{-\Gamma t} - \mathrm{e}^{\mathrm{i}(\omega_{j0} - \omega)t}) . \tag{22.4}$$

Thus the coefficient c_j depends on the initial coefficient α_j, as well as on the polarisation $\boldsymbol{e}$ of the emitted photon and on the dipole moment $\boldsymbol{\theta}_{0j}$ which connects the ground state with the excited state j; a is a proportionality factor which need not interest us further here. As may be immediately seen, the coefficient c_j vanishes at $t = 0$, but reaches a maximum value at later times. The probability of finding the electron in the ground state grows continuously until it reaches 1. The probability of finding a photon present is given by $|\alpha_0|^2$. If we use the sum (22.3) and the explicit form (22.4), we can see that $|\alpha_0|^2$ has the form

$$|\alpha_0|^2 = \mathrm{e}^{-\Gamma t} [A + B \cos \omega_{12}(t + \phi)] . \tag{22.5}$$

Here, A and B are time-independent constants. Through the superposition of the emission from the states 1 and 2, an oscillating contribution with the frequency $\omega_{12} = (E_2 - E_1)/\hbar$ is produced. If (22.5) is plotted as a function of time, the modulated

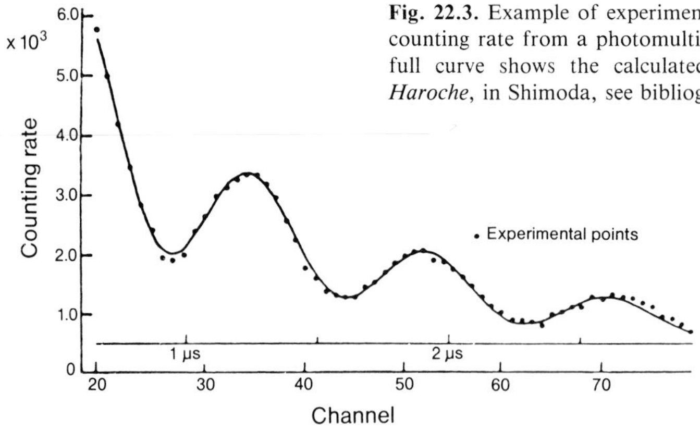

Fig. 22.3. Example of experimental data for quantum beats. The counting rate from a photomultiplier is plotted against time. The full curve shows the calculated quantum-beat function (after *Haroche*, in Shimoda, see bibliography for this chapter

decay curve shown in Fig. 22.3 is obtained. The frequency ω_{12}, and thus the energy difference of the two excited levels, can be determined from experiment, which is used to obtain precision measurements of this energy difference.

22.3 Doppler-free Saturation Spectroscopy

As we have already seen in Sect. 16.2, the spectral lines from gas atoms are broadened because of the Doppler effect. Since this Doppler broadening is, in general, much greater than the natural linewidth, it represents a considerable obstacle to the measurement of narrow spectral lines. With the help of lasers, it has been found possible to avoid Doppler broadening. To understand how, we recall the chapter on laser theory, considering in particular the gas laser. There we saw that because of laser action, the originally inverted populations of the atoms (i.e. $N_2 - N_1 > 0$) become less inverted.

Now imagine that we can distinguish the individual atoms in a gas laser according to their velocities in the axial direction of motion. Each group of atoms with velocity v has a particular occupation number $N_{1,v}$ or $N_{2,v}$ in the lower or in the upper state, respectively. If we multiply the occupation number $N_{2,v}$, which is supposed to be the same for all groups of atoms v, by the Maxwell velocity distribution (16.73), we obtain a profile like that shown in Fig. 22.4. If we now imagine that the laser light is produced with a particular frequency ω, it will only be able to interact with those groups of atoms v for which the Doppler shift is

$$|\omega - \omega_0 + \omega_0 v/c| < \gamma\,, \tag{22.6}$$

where γ represents the order of magnitude of the natural linewidth of the atomic transition, and ω_0 is the transition frequency of the atoms at rest. Due to this interaction, the inversion of this particular group v of atoms will be lowered, i.e. the number of atoms in the upper level is reduced. In this manner, the originally unperturbed population density has a "hole burned" in it (Fig. 22.5).

Because of the laser tube, we are dealing with a standing wave, which may be thought of as being composed of two travelling waves. The wave travelling in the one direction is then in resonance with atoms having a velocity $v = c(1 - \omega/\omega_0)$, the other

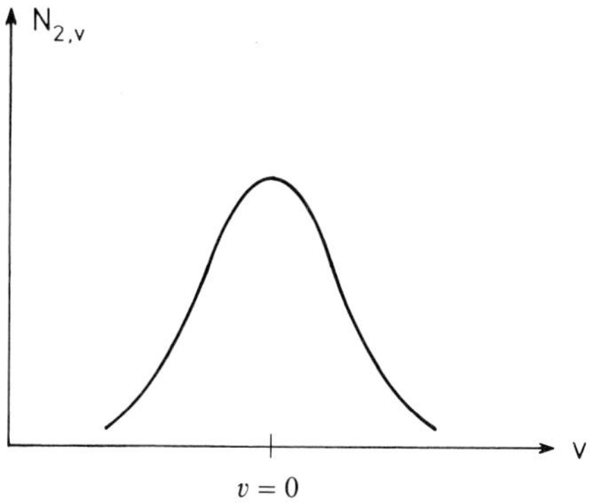

Fig. 22.4. The number of excited atoms N_2 with velocity v is represented as a function of v by a Gaussian curve, according to the Maxwell distribution

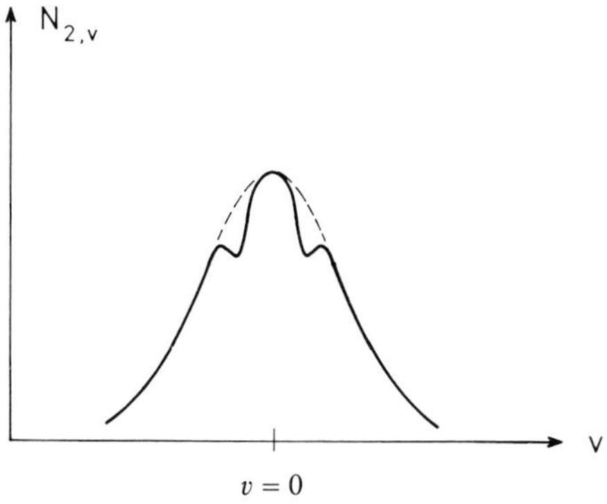

Fig. 22.5. This figure shows the effect of a standing laser wave on the distribution of excited atoms. If we imagine the standing wave to be composed of two travelling waves, each of these components interacts with a group of atoms whose Doppler frequency just coincides with the laser-wave frequency. Therefore, in general, two holes are burned in the distribution. If these two holes overlap, the notch in the distribution is especially deep

is in resonance with atoms having a velocity $v = -c(1-\omega/\omega_0)$, where v is the velocity component along the axis of the laser tube. In this way, *two* holes are burned in the population inversion distribution. If the laser frequency ω exactly agrees with the frequency of the atomic transition ω_0, the two holes coincide and a particularly deep notch is cut out of the distribution. This is clearly the case for those atoms whose velocity component along the laser axis is zero, or, more precisely, for which $|\omega_0 v/c| < \gamma$. The intensity emitted by the laser naturally depends upon the population inversion; thus, the intensity goes through a minimum when the laser light is exactly in resonance with the transition frequency of the atoms at rest, $v \cong 0$. This effect is known as the "Lamb dip". It was, by the way, independently and simultaneously predicted theoretically by *Haken* and *Sauermann*. Using the Lamb dip, the position of a spectral line can clearly be much more precisely determined, namely to the order of magnitude of the natural linewidth. To be sure, this method is limited to media which are themselves laser active. Furthermore, in the Lamb dip an additional source of broadening, called power broadening, plays a rôle; we shall not discuss it in detail here.

A decisive improvement was suggested by *Letokhov*: the substance which is to be investigated is brought, in gas form, between one of the mirrors and the laser-active medium. When the gas atoms, which once again have a Maxwell velocity distribution, are illuminated by the intense laser beam, those which fulfil condition (22.6) absorb strongly. Thereby, an appreciable fraction of these atoms is pumped into the excited state(s), which decreases the population in the ground state and thus reduces the absorption. We have burned a hole in the absorption line. If we now suppose that the laser light and the central frequency for the atomic transition are in resonance, we find the picture shown in Fig. 22.6. Above the Doppler broadened line, a very narrow line is

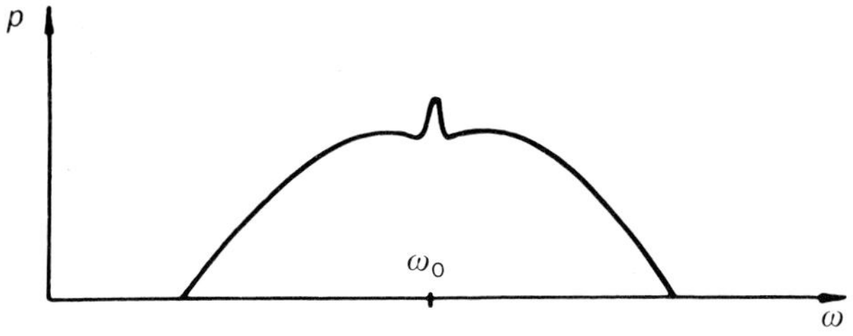

Fig. 22.6. Burning a hole in the absorption line produces a peak in the emitted radiation, which is shown here as a function of the frequency of the laser field. The width of the peak corresponds to the natural linewidth

produced, with a width of the order of the natural linewidth of the material investigated. In general, the frequencies of the laser light and the absorbing material will not be exactly the same. But even then, there is a sharp line of enhanced total emission to be observed above the Doppler broadened line.

Figure 22.7 is a diagram of an experimental arrangement. With such methods it has been possible to observe Lamb shifts, such as the one mentioned in Sect. 12.11, optically (refer to the spectrum shown in Fig. 12.24). There are a number of variations of the method discussed here, but we must refer the reader to the literature on them (see the list of references for this chapter at the end of the book).

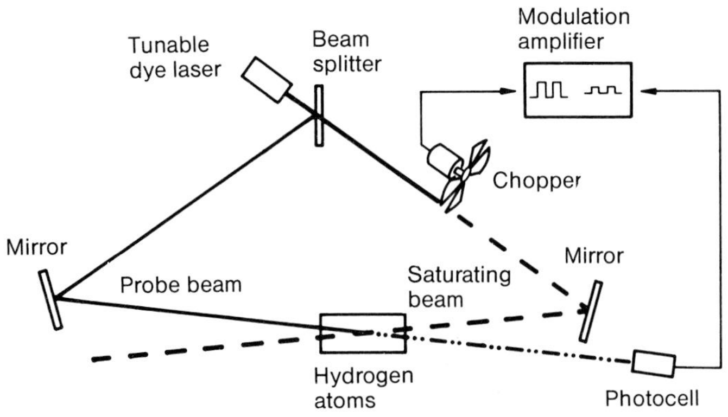

Fig. 22.7. Diagram of an apparatus for saturation spectroscopy. The light from a tunable dye laser is divided into an intense saturation beam and a weaker test beam. These pass in opposite directions through a vessel filled with hydrogen atoms (according to *Hänsch*, *Schawlow* and *Series*

22.4 Doppler-free Two-Photon Absorption

In an intense field of laser light, it is possible that an atom will absorb two photons. In this case the energy difference ΔE between the two levels is exactly twice the photon energy, i.e.

$$\Delta E = 2h\nu \equiv 2\hbar\omega . \tag{22.7}$$

An atom can also absorb photons with different frequencies ω_1 and ω_2, where the conservation of energy requires that

$$\Delta E = \hbar\omega_1 + \hbar\omega_2 . \tag{22.8}$$

In addition, we must take into account that the selection rules for two-photon absorption are different from those governing one-photon absorption. For example, transi-

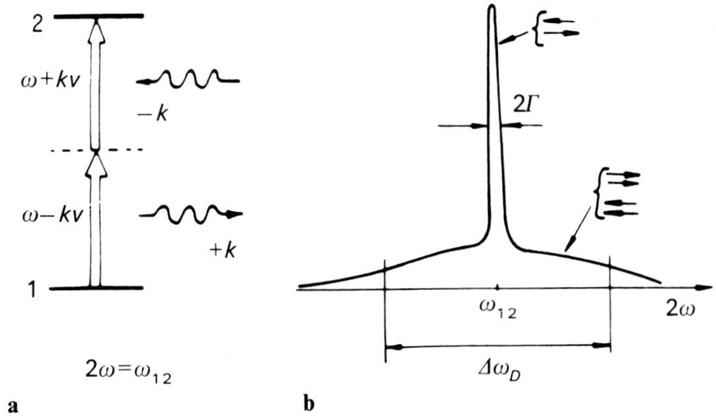

Fig. 22.8. a) Two-photon absorption scheme. We think of the standing laser wave as being broken down again into two waves moving in opposite directions. The part moving to the right interacts with atoms at a Doppler-shifted transition frequency $\omega - kv$. The wave moving to the left, however, interacts with the same atom at a Doppler frequency of $\omega + kv$. When the two waves add, the velocity components cancel out. **b)** The shape of the resonance line in two-photon absorption. $\Delta\omega_D$ is the Doppler width, and 2Γ the total width of the resonance line

tions from one s state to another are now allowed. This can be justified, within the framework of perturbation theory, but we cannot undertake it here.

We now imagine that we pass intense laser light, which forms a standing wave in a resonator, through a gas of the atoms being studied. If we pick out an atom with the velocity v in the axial direction, the frequency of a photon (ω) moving in one direction will appear, in the moving frame of reference of our atom, to be $\omega_1 = \omega(1 + v/c)$. The frequency of a photon moving in the opposite direction will be, in the atom's frame of reference, $\omega_2 = \omega(1 - v/c)$. Obviously the atom can absorb both photons, obeying the law of conservation of energy, if

$$\Delta E = \hbar\omega_1 + \hbar\omega_2 = \hbar\omega(1 + v/c) + \hbar\omega(1 - v/c) = 2\hbar\omega\,. \tag{22.9}$$

Since the velocity v of the chosen atom cancels out of this equation, all the atoms can carry out this two-photon process, so that a very intense absorption line with a line-width equal to the natural linewidth can form (Fig. 22.8). In this way the Doppler broadening can be cancelled out. Figure 22.9 is a diagram of an experimental arrangement.

In the last chapter, especially, it was clear that atomic physics is still an area in which much research is being done. The new methods of high-resolution spectroscopy have been made possible by the laser, and they offer the possibility of examining atomic structure more closely and of carrying out entirely new kinds of experiments. The development is quite certainly still in progress.

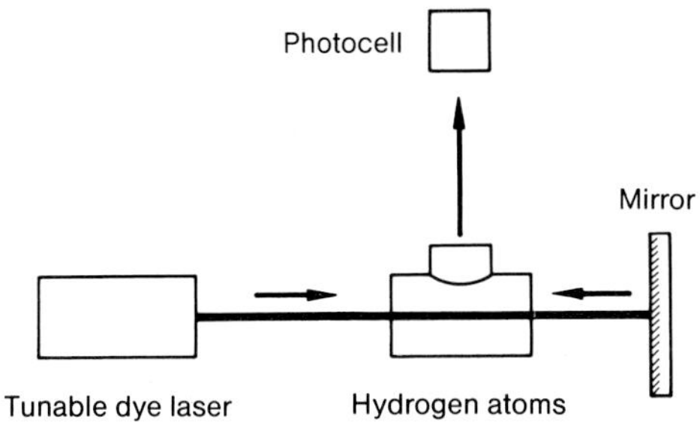

Fig. 22.9. Diagram of an experimental arrangement for two-photon spectroscopy. Two laser beams moving in opposite directions along the optical axis are passed through the gas being studied

22.5 Level-Crossing Spectroscopy and the Hanle Effect

The coherent excitation of two different excitation states of an atom, which was introduced in the section on quantum beats, is also the basis of another spectroscopic technique. The latter, however, can be done with a conventional light source or electron bombardment, and was first used in 1959 by *Franken*. It is called *level-crossing spectroscopy*.

If atomic terms are split in an external magnetic field $\boldsymbol{B}_0$ (Chap. 13), then in the range of weak or medium fields, levels can cross. At certain field strengths $\boldsymbol{B}_c$, therefore, two different Zeeman terms coincide energetically. Such crossing occurs when there is already fine structure splitting in the absence of an external field $\boldsymbol{B}_0$, or when there are hyperfine multiplets with the same J value (Fig. 22.10). Let us consider two resonance transitions from the term splitting in Fig. 22.10. Both end at the common ground state, and both are excited simultaneously by the Doppler broadened line of the light source. Each excited state returns to the ground state by emitting resonant light. The intensity and polarisation of the emitted light can be calculated from the selection rules given in Chaps. 12 and 13. The sum of the individual intensities is observed by the photodetector. If we assume, for example that each state by itself emits a linearly polarised wave with amplitude $a_1 \cos \omega_1 t$ or $a_2 \cos \omega_2 t$, then the total intensity of the observed resonance light when there is no level crossing is proportional to $a_1^2 + a_2^2$, on a time average.

If the two levels cross, however, the transition frequencies are the same. Both levels are coherently excited, and the light waves emitted from them interfere. The observed intensity is then proportional to $(a_1 + a_2)^2$. There is thus a change in the spatial distribution of the emitted radiation, and in the direction of observation established by the arrangement of the photodetector there is generally a change in the observed total intensity whenever the external field takes on a value $\boldsymbol{B}_c$ at which crossing occurs between two levels. An experimental observation is reproduced at the bottom of Fig. 22.10.

The crossing fields $\boldsymbol{B}_c$, and thus the dependence of the terms on the field strength $\boldsymbol{B}_0$, can be deduced from such changes in intensity (Fig. 22.10, upper part). The measurement of these term changes is an important experimental aid in the determination of fine structure and hyperfine constants.

If linearly polarised light is used for excitation, it is possible to excite coherently two terms with the same transition frequency if the selection rule for one, for example is $\delta m = +1$, while the rule for the other is $\delta m = -1$. Both are fulfilled, because linearly polarised light can be decomposed into a σ^+ and a σ^- component. The m-values of the level crossings are indicated in Fig. 22.10.

The significance of the method for spectroscopic analysis of atomic excited states results in particular from the high spectral resolution. The precision with which the $\boldsymbol{B}_c$ values can be measured, i.e. the "linewidth" of level-crossing spectroscopy, is not limited by the Doppler width of the spectral lines, since here the coherence of the excitation within an atom is utilised. Rather the linewidth is limited essentially by the lifetime of the excited state and is, in practice, at least several MHz. One thus obtains the same high spectral resolution as with high-frequency spectroscopy. As an example, the fine structure splitting $2^2P_{3/2} - 2^2P_{1/2}$ in the ^{6}Li atom could be determined to be 10052.76 MHz.

In a similar manner, the *Hanle effect*, discovered in 1924, can be thought of as a zero-field level crossing. Terms which are not split without an applied magnetic field "cross" in zero field, when the field $\boldsymbol{B}_0$ is varied from negative values through zero to

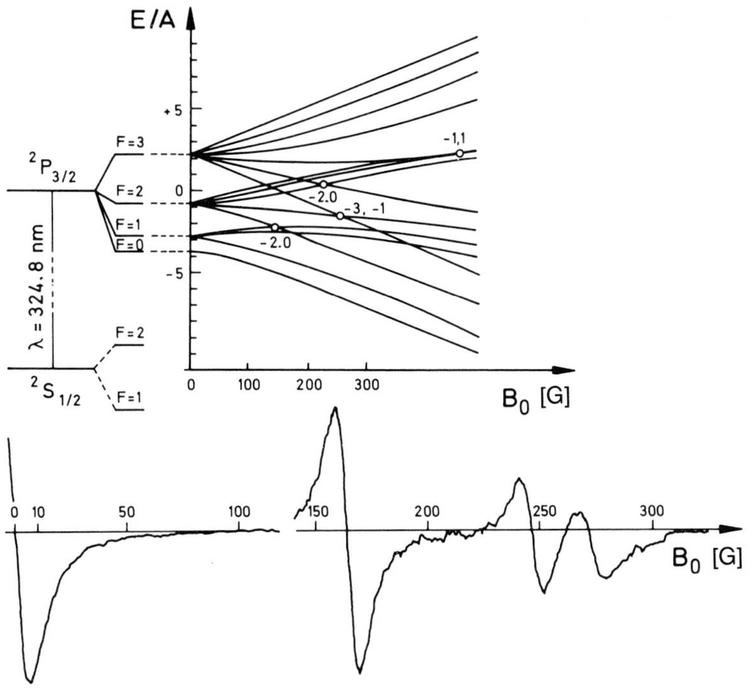

Fig. 22.10. Level-crossing spectroscopy and Hanle effect on the transition $^2S_{1/2}-{}^2P_{3/2}$ in Cu, $I = 3/2$ (from *Bucka* et al.). In the upper portion of the figure, the energy level diagram is shown; the splitting of the excited state $^2P_{3/2}$ is given in units of the hyperfine structure constant A. The crossing points which are accessible to level-crossing spectroscopy are denoted by their m_F-values. The strength of the applied field $\boldsymbol{B}_0$ is plotted along the abscissa. In the lower portion of the figure, experimental curves for the Hanle effect (*left*) and for the level crossings (*right*) are shown. For experimental reasons, the derivative of the intensity w.r.t. magnetic field dI/dB_0 is plotted rather than the intensity itself. The plots therefore are dispersion curves and the level-crossing fields $\boldsymbol{B}_c$ correspond to the zero-crossings of the curve, i.e. the points where it crosses the abscissa. At the left, one can see the Hanle effect at $\boldsymbol{B}_0 = 0$; on the right, the 3 level crossings, corresponding to the crossing points $(-2, 0)$, $(-2, 0)$ and $(-3, -1)$ in the term diagram. To convert the linewidth in Gauss to the value in MHz, (13.7) is to be used

positive values. In zero field, degenerate atoms may be coherently excited for the same reasons as in level-crossing spectroscopy, assuming validity of the same selection rules. The coherence of the excitation and the interference of the emitted resonance light are

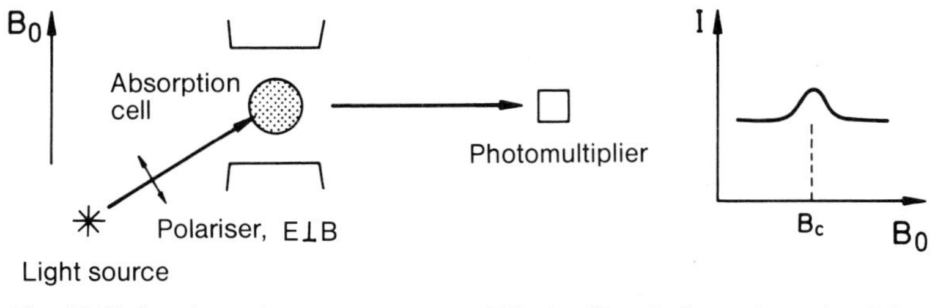

Fig. 22.11. Level-crossing spectroscopy and Hanle effect. Left, a schematic of the experimental setup. The atoms in the gas cell are excited by polarised light from a conventional light source and the resonance light (with or without polarisation analyser) is observed as a function of the magnetic field $\boldsymbol{B}_0$. At level-crossing fields $\boldsymbol{B}_c$, the intensity and the polarisation of the resonance light detected by the photomultiplier change, in general. This is indicated on the right

removed when the degeneracy of the terms is lifted through the application of a magnetic field. Figure 22.10 also shows an example of a measurement of the Hanle effect. It is an important experimental method for determining the lifetimes of excited atomic states. A schematic of the experimental arrangement for observation of the Hanle effect and of level crossings is shown in Fig. 22.11.

23. Fundamentals of the Quantum Theory of Chemical Bonding

23.1 Introductory Remarks

As we know from chemistry, many atoms can combine to form particular molecules, e.g. chlorine and sodium atoms form NaCl molecules. But atoms of the same type can also form bonds, as, for example in the case of hydrogen, H_2. Before the development of quantum theory, the explanation of chemical bonding was a puzzle to chemists and physicists alike. Bonding between ions, as in the negatively charged chlorine ion and the positively charged sodium ion, could, to be sure, be understood in the light of the Coulomb attraction between oppositely charged bodies; it remained, however, inexplicable that two similar atoms, which are electrically neutral (as, for example two hydrogen atoms) could form a bound state (homopolar bonding). Here it only became possible with the aid of quantum mechanics to attain a fundamental understanding. Even in the case of ionic bonding (also called heteropolar bonding), basic new insights have been obtained through quantum theory. For example, it must be understood why the ions form in the first place, and why the electron which is transferred from sodium to chlorine thus finds an energetically more favourable state.

In the following, we wish to develop some important basic ideas for the quantum theory of chemcial bonding. It should be pointed out at the outset, however, that physics and chemistry are still far removed from a complete solution to these problems. This is a result of the fact that to understand chemical bonding, the interactions of several particles must in each case be taken into account: given n atomic nuclei and m electrons, one would have to find the complete wavefunction and the corresponding energies of the total system. In order to find suitable approaches to a solution in such problems, one must trust to intuition to a considerable extent at first, in order to gradually develop more systematic methods. In attacking this problem it is useful to keep in mind that the nuclear masses are much greater than those of the electrons. This fact makes it reasonable as a first step to treat the nuclear masses as though they were infinite. Then we may ignore the motion of the nuclei and treat them as fixed. In atomic physics, we were able to obtain much information from spectroscopic observations and could direct our attention to both the ground states and to the excited states; in the study of chemical bonding, the determination of the wavefunction of the ground state of the particular molecule plays a more important rôle. We will therefore direct our attention in the following to this question, which is of basic importance for chemistry. Let us now turn to concrete problems.

23.2 The Hydrogen-Molecule Ion H_2^+

Certainly the simplest case of chemical bonding occurs in the hydrogen-molecule ion H_2^+. This species is observed as a bound state in gas discharges in a hydrogen

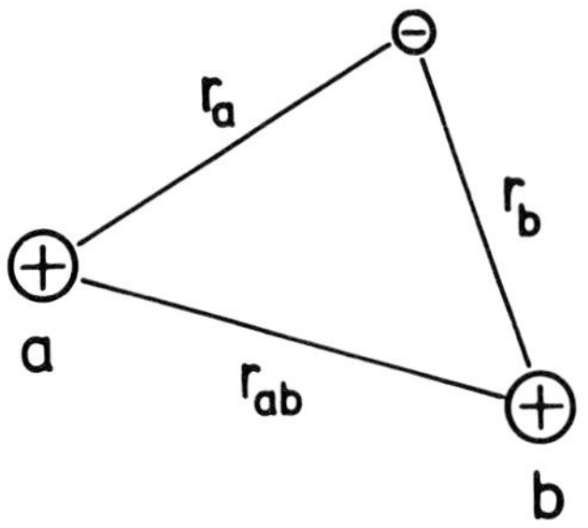

Fig. 23.1. Overview of the hydrogen-molecule ion. The two nuclei are denoted as a and b, their distance apart as r_{ab}. r_a and r_b give the distance of the electron from nucleus a and from nucleus b, respectively

atmosphere; in such a gas discharge, the hydrogen molecule may have an electron torn out. The bonding energy, equivalent to the dissociation energy, has been determined to be 2.65 eV for H_2^+. Here, we are dealing with two hydrogen nuclei, i.e. two protons but only one electron. The two nuclei may be denoted by the indices a and b (Fig. 23.1). If the nuclei are far removed from one another, we can imagine that the electron is localised on one nucleus or the other. The wavefunctions are then those of the hydrogen atomic ground state. In the following, we will denote the distance of the electron from nucleus a or b by r_a or r_b. If we call the wavefunction of the hydrogen ground state belonging to nucleus a ϕ_a, the latter must satisfy the Schrödinger equation

$$\underbrace{\left(-\frac{\hbar^2}{2m_0}\nabla^2-\frac{e^2}{4\pi\varepsilon_0 r_a}\right)}_{\mathscr{H}_a}\phi_a(r_a)=E_a^0\phi_a(r_a)\,, \tag{23.1}$$

and correspondingly for the wavefunction ϕ_b, with the energies E_a^0 and E_b^0 being equal:

$$E_a^0=E_b^0=E^0\,. \tag{23.2}$$

If we now let the nuclei approach one another, the electron, which was, for example, at first attached to nucleus a, will experience the attractive Coulomb force of nucleus b. Conversely, an electron which was at first bound to nucleus b will experience the attractive Coulomb force of nucleus a. We must therefore set up a Schrödinger equation which contains the Coulomb potentials of *both* nuclei (Fig. 23.2). Furthermore, in order to calculate the total energy of the system, we must take into account the Coulomb repulsion between the two nuclei. If we call the distance between these two nuclei r_{ab}, the additional energy is $e^2/(4\pi\varepsilon_0 r_{ab})$. Since this additional energy is not directly related to the energy of the electron, it will only produce a constant shift of all the energy eigenvalues. We will therefore leave this constant out of the calculations at first, and will reintroduce it at the end of our considerations.

This leads us to the Schrödinger equation

$$\left(-\frac{\hbar^2}{2m_0}\nabla^2-\frac{e^2}{4\pi\varepsilon_0 r_a}-\frac{e^2}{4\pi\varepsilon_0 r_b}\right)\psi=E\psi\,, \tag{23.3}$$

in which the wavefunction ψ and the corresponding energy E are yet to be calculated.

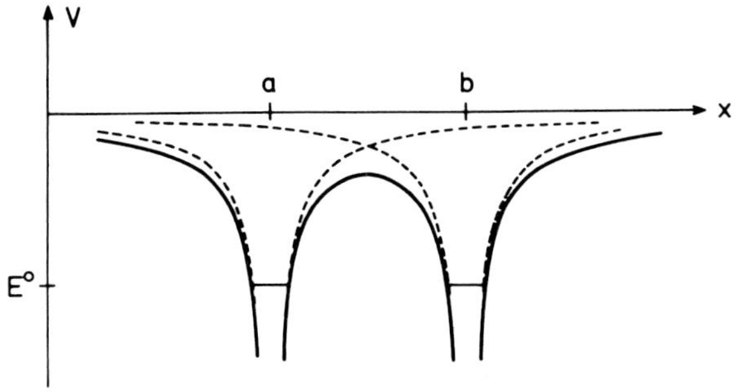

Fig. 23.2. Hydrogen-molecule ion. The potential energy V which acts on the electron due to the attractive forces from nuclei a and b is plotted. The dashed curves show the energy of an electron in the field of nucleus a or nucleus b alone. The full curve is the total potential energy. The abscissa is the position coordinate x. The binding energy of the electron in the field of a single nucleus, E^0, is also indicated

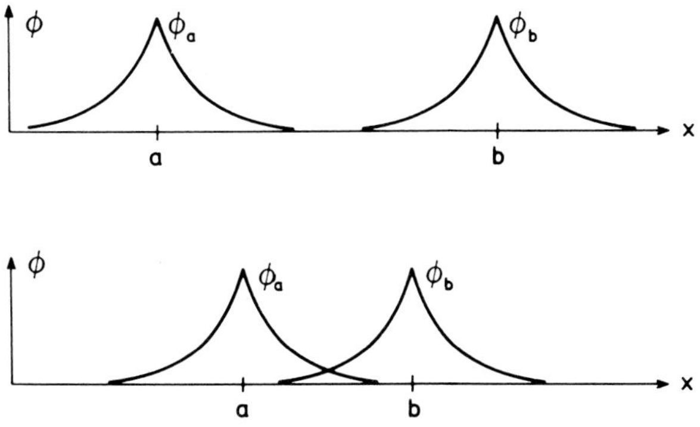

Fig. 23.3. *Upper portion:* The wavefunction ϕ_a of an electron which is moving in the field of nucleus a and the corresponding wavefunction ϕ_b for the electron in the neighbourhood of nucleus b. *Lower portion:* If the distance between nuclei a and b is reduced, the two wavefunctions ϕ_a and ϕ_b begin to overlap

We will now determine the wavefunction ψ approximately. For this purpose we will utilise an idea which is borrowed from perturbation theory with degeneracy (Sect. 15.2.3). In principle, the electron could be in the neighbourhood of nucleus a or of nucleus b (Fig. 23.3), with the same energy in each case, as in (23.1) and (23.2). These two states ϕ_a and ϕ_b are thus degenerate. Now however, the other nucleus, from which the electron is by chance more distant, acts as a perturbation to the electronic state. We can thus expect that the degeneracy will be lifted by this perturbation. Exactly as in perturbation theory in the presence of degeneracy, we form as a solution to (23.3) a linear combination:

$$\psi = c_1\phi_a + c_2\phi_b\,, \tag{23.4}$$

where the two coefficients c_1 and c_2 are to be determined. To find their values, we proceed in the usual manner. We first insert the trial solution (23.4) into (23.3) and obtain

$$\begin{aligned}&\underbrace{\left(-\frac{\hbar^2}{2m_0}\nabla^2 - \frac{e^2}{4\pi\varepsilon_0 r_a}\right.}_{\mathcal{H}_a}\left.-\frac{e^2}{4\pi\varepsilon_0 r_b}\right)c_1\phi_a\\ &+\underbrace{\left(-\frac{\hbar^2}{2m_0}\nabla^2 - \frac{e^2}{4\pi\varepsilon_0 r_b}\right.}_{\mathcal{H}_b}\left.-\frac{e^2}{4\pi\varepsilon_0 r_a}\right)c_2\phi_b = E(c_1\phi_a + c_2\phi_b)\,.\end{aligned} \tag{23.5}$$

We have collected the terms in the two brackets in (23.5) so that the operator $\mathcal{H}_a$ operates on ϕ_a and $\mathcal{H}_b$ on ϕ_b. We can simplify these expressions immediately with the help of (23.1) and the corresponding equation for $\mathcal{H}_b$, by writing $E_a^0\phi_a$ in place of $\mathcal{H}_a\phi_a$, for example.

If we now bring the right-hand side of (23.5) to the left side, we obtain

$$\Bigg(\underbrace{E^0-E}_{-\Delta E}-\frac{e^2}{4\pi\varepsilon_0 r_b}\Bigg)c_1\phi_a+\Bigg(\underbrace{E^0-E}_{-\Delta E}-\frac{e^2}{4\pi\varepsilon_0 r_a}\Bigg)c_2\phi_b=0\,. \tag{23.6}$$

While ϕ_a and ϕ_b are functions of position, the coefficients c_1 and c_2 are independent of position. To find a position-independent equation for the c's, we multiply (23.6) with ϕ_a^* or ϕ_b^*, as we are accustomed from perturbation theory, and integrate over the electronic coordinates. For this purpose we assume in the following that the functions ϕ_a and ϕ_b are real, as is the case for the hydrogen-atom ground state wavefunctions. We must take the fact into account that the functions ϕ_a and ϕ_b are not orthogonal, i.e. that the integral

$$\int\phi_a\phi_b dV=S \tag{23.7}$$

does not vanish. If we multiply (23.6) by ϕ_a and then integrate over electronic coordinates, we obtain expressions which have the form of matrix elements, namely the integrals

$$\int\phi_a(r_a)\left(\frac{e^2}{4\pi\varepsilon_0 r_b}\right)\phi_a(r_a)dV=C\,, \tag{23.8}$$

$$\int\phi_a(r_a)\left(\frac{e^2}{4\pi\varepsilon_0 r_a}\right)\phi_b(r_b)dV=D\,. \tag{23.9}$$

which we have abbreviated with the letters C and D. The meaning of the first integral becomes immediately clear if we recall that $-e|\phi_a|^2$ stands for the charge density of the electron. Expression (23.8) is then nothing other than the Coulomb interaction energy between the electronic charge density and the nuclear charge e of nucleus b (Fig. 23.4). In the integral (23.9), by contrast, instead of the electronic charge density, the expression $-e\phi_a\phi_b$ occurs. This means that the electron in some sense is partly in state a and partly in state b, or, in other words, an exchange between these two states takes place. For this reason, $\phi_a\phi_b$ is referred to as the *exchange density*, and integrals in which such exchange densities $\phi_a\phi_b$ occur as *exchange integrals* (Fig. 23.5). These integrals represent a particular quantum mechanical effect.

If we multiply (23.6) by ϕ_b instead of by ϕ_a and integrate, we obtain expressions which are quite similar to (23.8) and (23.9), except that the indices a and b are exchanged. Since, however, the problem is fully symmetric with respect to the indices a and b, the new integrals must have the same values as the old. If we collect all the terms resulting from multiplying by ϕ_a and integrating, (23.6) becomes

$$(-\Delta E+C)c_1+(-\Delta E\cdot S+D)c_2=0\,, \tag{23.10}$$

and correspondingly we obtain after multiplying by ϕ_b and integrating:

$$(-\Delta E\cdot S+D)c_1+(-\Delta E+C)c_2=0\,. \tag{23.11}$$

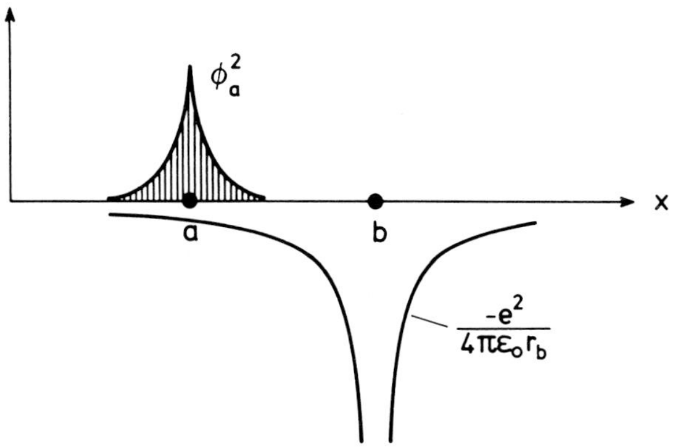

Fig. 23.4. Intuitive explanation of the integral (23.8). This integral represents the Coulomb interaction energy of an electron density cloud with the probability density distribution ϕ_a^2 in the Coulomb field of the nucleus. The density distribution is shown as a shaded area, while the full curve gives the potential energy of a point charge in the Coulomb field of nucleus b. To calculate the integral, at each point in space the value of ϕ_a^2 must be multiplied with the corresponding value of $-e^2/(4\pi\varepsilon_0 r_b)$ and the products must be added (integrated) over the entire volume

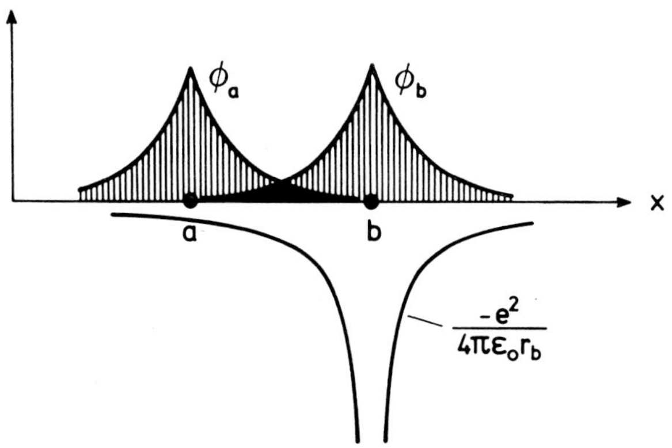

Fig. 23.5. Intuitive explanation of the integral (23.9). The three functions ϕ_a, ϕ_b, and $-e^2/(4\pi\varepsilon_0 r_b)$ which occur in the integral are plotted. Since the product of these three functions occurs, nonvanishing contributions are only possible when the wavefunctions ϕ_a and ϕ_b overlap, as is indicated by the heavily shaded area. The integral is obtained by multiplying the functional values of ϕ_a, ϕ_b, and $-e^2/(4\pi\varepsilon_0 r_b)$ at each point in space and adding up over the whole volume

These are two quite simple algebraic equations for the unknown coefficients c_1 and c_2. In order that these equations have a nontrivial solution, the determinant must vanish, that is

$$(-\Delta E+C)^2-(-\Delta E\cdot S+D)^2=0\,. \tag{23.12}$$

This is a quadratic equation for the energy shift ΔE, which we can solve immediately:

$$E=E^0+\Delta E=E^0+\frac{C\pm D}{1\pm S}\,, \tag{23.13}$$

where the two possible signs are due to the square root entering the solution of (23.12). If we insert ΔE in (23.10), for example, we see that the upper sign corresponds to a solution with

$$c_2=-c_1\equiv -c\,. \tag{23.14}$$

In this case, we obtain for the total wavefunction

$$\psi=c(\phi_a-\phi_b)\,. \tag{23.15}$$

The constant c still has to be determined through the normalisation of the total wavefunction. This wavefunction is represented in Fig. 23.6. If we use the lower sign in (23.13), we obtain the coefficients $c_2=c_1=c$ and, with these, the total wavefunction

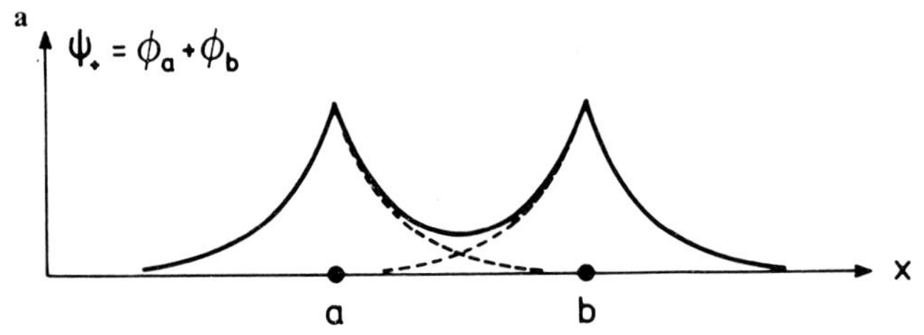

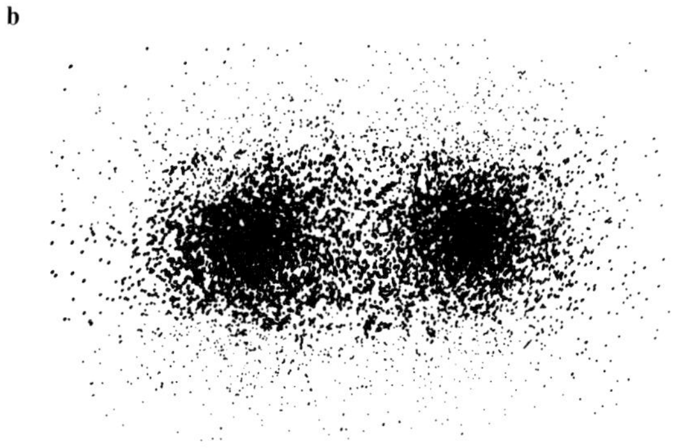

Fig. 23.6. a) The symmetric wavefunction ψ_+ is formed by superposing the wavefunctions ϕ_a and ϕ_b. Because of the overlap between ϕ_a and ϕ_b, the occupation probability for ψ_+ between the two nuclei is increased. **b)** Illustration of the density distribution of the electron in the ψ_+ state

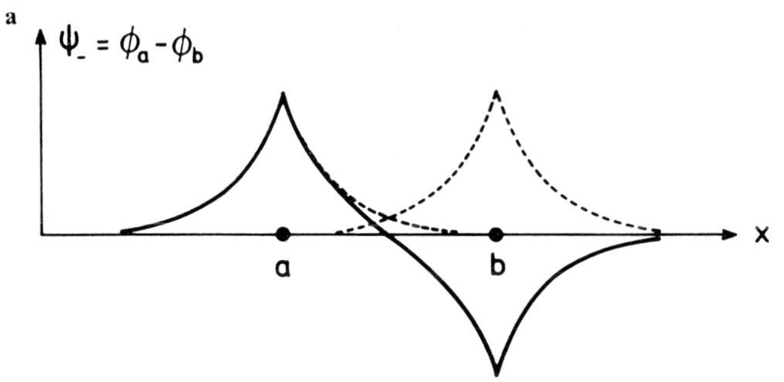

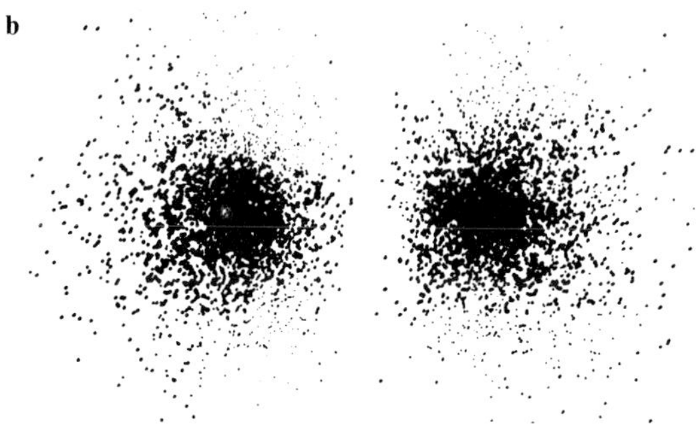

Fig. 23.7. a) The antisymmetric wavefunction ψ_- is formed by taking the difference of ϕ_a and ϕ_b. The occupation probability is clearly zero in the plane of symmetry between the two nuclei. **b)** Illustration of the density distribution of the electron in the hydrogen-molecule ion in the case of the antisymmetric wavefunction ψ_-

$$\psi = c(\phi_a + \phi_b) \tag{23.16}$$

(Fig. 23.7).

As may be seen in Figs. 23.4 and 23.5, the quantities S, C, and D depend on the internuclear distance r_{ab}. If the two nuclei approach one another, the electronic energy E splits up in a way depending on whether we are dealing with a symmetric state (23.16) or with an antisymmetric state (23.15). In the symmetric case the energy is reduced; here, we may speak of a bonding state or bonding orbital. In the case of the antisymmetric wavefunction, the energy is increased and we may speak of an antibonding state (Fig. 23.8).

In order to determine whether the hydrogen-molecule ion is, in fact, bound, we must add the Coulomb repulsion energy between the two nuclei a and b to the electronic energy E. We then find for the bonding state that the total energy depends on the internuclear distance as shown in Fig. 23.8. One can recognise from this figure that there is a certain internuclear distance at which the total energy of the system has a minimum; thus, a bound state exists.

As may be seen in Fig. 23.6, in the bound state the probability density of the electron is relatively large between the two nuclei. The electron can thus, from an energetic point of view, profit from the Coulomb attraction of both nuclei, which

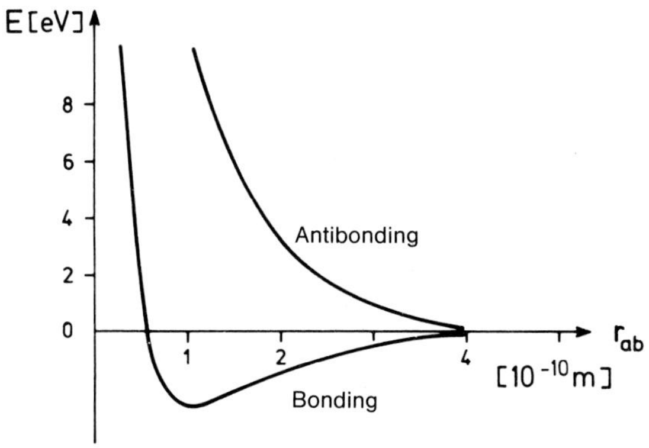

Fig. 23.8. The energy E of the hydrogen-molecule ion, taking the Coulombic repulsion between the two nuclei into consideration. The energy curves are plotted to show their dependence on the internuclear distance r_{ab} for the bonding and the antibonding states

lowers the potential energy of the total system. In the antibonding state (Fig. 23.6), the probability density of the electron between the two nuclei is small, in the middle in fact 0, which means that the electron only experiences the attractive force of one nucleus at a time.

23.3 The Tunnel Effect[1]

Using the example of the hydrogen-molecule ion, we may get an excellent description of an effect which is known as the "tunnel effect" and which represents a typically quantum mechanical phenomenon. Let us first consider the behaviour of a classical particle in a potential field, shown in Fig. 23.2. If the particle is first in the region of the left potential well with a total energy as represented by the horizontal line in the figure, it cannot pass the potential barrier between the left and right regions; it remains forever trapped in the left portion of the potential. We will now show that this is *not* true in quantum theory: a particle which is originally in the left-hand region of the potential can pass over into right-hand region after a time.

In order to simplify the discussion somewhat, we will assume that the overlap integral S is small ($S \ll 1$) and can be neglected. The energy of the electron can then be given in the form

$$E = E^0 + C \pm D\,. \tag{23.17}$$

We consider the solutions of the time-dependent Schrödinger equation, which in general have the form

$$\psi(\boldsymbol{r},t) = \psi(\boldsymbol{r})\,\mathrm{e}^{-\mathrm{i}Et/\hbar}\,. \tag{23.18}$$

For $\psi(\boldsymbol{r})$ we will use the approximate wavefunctions (23.15, 16), and for E the corresponding energies (23.17). To simplify the notation, we introduce instead of the energies the frequencies, related by $E = \hbar\omega$:

[1] The reader who is interested only in chemical bonding may wish to skip this section

$$(E^0 + C)/\hbar = \omega_0 \,, \tag{23.19}$$

$$D/\hbar = \omega_T/2 \,. \tag{23.20}$$

In (23.20) the factor 1/2 and the subscript T were introduced for a good reason, as we will shortly see. According to the different solutions (23.15) and (23.16) we obtain the solutions of the time-dependent Schrödinger equation in the form

$$\psi_+ = (\phi_a + \phi_b)\frac{1}{\sqrt{2}}\,e^{-i(\omega_0 - \omega_T/2)t} \tag{23.21}$$

and

$$\psi_- = (\phi_a - \phi_b)\frac{1}{\sqrt{2}}\,e^{-i(\omega_0 + \omega_T/2)t} \,. \tag{23.22}$$

From such time-dependent solutions we can construct so-called wave packets by superposition of the two functions (23.21) and (23.22):

$$\psi(\boldsymbol{r}, t) = d_1\psi_+ + d_2\psi_- \,, \tag{23.23}$$

where the coefficients d_1 and d_2 are time-dependent constants. By inserting (23.23) into the time-dependent Schrödinger equation

$$\mathscr{H}\psi = i\hbar\dot{\psi} \,, \tag{23.24}$$

where the Hamiltonian $\mathscr{H}$ is the same as in the Schrödinger equation (23.3), it is possible to convince oneself that the trial solution (23.23) is, in fact, an approximate solution to (23.24). We now choose d_1 and d_2 in such a way that at the initial time $t = 0$, the electron is in state ϕ_a on atom a; we do this by setting $d_1 = d_2$ and choosing both equal to $1/\sqrt{2}$ for reasons of normalisation. Then the wavefunction is given by

$$\psi(\boldsymbol{r}, t) = \tfrac{1}{2}[\underbrace{(\phi_a + \phi_b)}_{\propto \psi_+} + \underbrace{(\phi_a - \phi_b)}_{\propto \psi_-}\exp(-i\omega_T t)]\exp(-i\omega_0 t + i\omega_T t/2) \,. \tag{23.25}$$

If we allow the time t to vary, the relative phase between the states ψ_+ and ψ_- changes, as can be seen from (23.25).

We now recall a relation from mathematics, namely

$$\exp(i\omega_T t) = \cos\omega_T t + i\sin\omega_T t \,.$$

If we choose a time $t = t_0$ such that $\omega_T t_0 = \pi$, the cosine will be equal to -1 and the sine to zero. For this t_0 we obtain instead of (23.25)

$$\begin{aligned}\psi(\boldsymbol{r}, t) &= \tfrac{1}{2}[(\phi_a + \phi_b) - (\phi_a - \phi_b)]\exp(-i\omega_0 t_0 + i\omega_T t_0/2)\\ &\equiv \phi_b \exp(-i\omega_0 t_0 + i\omega_T t_0/2) \,.\end{aligned}$$

The electron is thus now in the state ϕ_b on atom b! It has tunnelled through the potential barrier. After the twofold time $2t_0$ it has tunnelled back to atom a. The circular frequency for the tunnelling process is thus given by $2\pi/2t_0$, which is just the frequency introduced in (23.20) as ω_T. According to (23.20), ω_T is, aside from constant factors, given by the exchange integral (23.9). The larger the overlap between the wavefunctions ϕ_a and ϕ_b, the more rapidly the electron tunnels back and forth.

The tunnel effect plays an important role in many areas of modern physics. An especially well-known example is the α decay of atomic nuclei, where the α particle is able to overcome the attractive nuclear potential barrier by means of tunnelling.

23.4 The Hydrogen Molecule H_2

We will now turn to the problem of chemical bonding when more than one electron is involved. Before we take up the simplest concrete example, that of the hydrogen molecule, we will make a general remark which is of fundamental importance for many problems in quantum theory.

We meet up again and again with the task of solving a Schrödinger equation of the general type

$$\mathcal{H}\psi = E\psi\,, \tag{23.26}$$

which is however often not possible in closed form. As well as the methods of solution which we have treated up to now, for example perturbation theory, there is another, fundamentally different method, which is based on a variational principle.

To explain this method, let us imagine the Schrödinger equation (23.26) to be multiplied by ψ^* and integrated over all coordinates on which ψ depends. We then obtain

$$E = \frac{\int \psi^* \mathcal{H} \psi\, dV_1 \ldots dV_n}{\int \psi^* \psi\, dV_1 \ldots dV_n}\,. \tag{23.27}$$

Since the Hamiltonian $\mathcal{H}$ is the operator for the total energy, we have here nothing other than the expectation value for the energy, which in the present case is the same as the energy eigenvalue of the Schrödinger equation. What will happen now, if instead of the true solution of the Schrödinger equation, we substitute another function for ψ? Then expression (23.27) still has the dimensions of an energy, but it must no longer represent a correct eigenvalue of the Schrödinger equation, which is what we are seeking. Mathematically, one can now prove an extremely important relation: if we in fact use some function other than the correct eigenfunction ψ for the ground state, the energy expectation values which we obtain will always be *higher* than the eigenvalues of the solution to (23.26). In this sense, we can define a criterion for the quality of our approximate wavefunctions: they are better and better, the lower the corresponding energy expectation values.

We shall use this criterion repeatedly in later sections. Now we will return to the problem of determining approximately the wavefunction and the energy of the hydrogen molecule in the ground state.

The two atomic nuclei (protons) will be denoted by the indices a and b, the electrons by 1 and 2. Since the Coulomb force acts between all the various particles, we must introduce the corresponding distances, which are defined in Fig. 23.9. In order to write

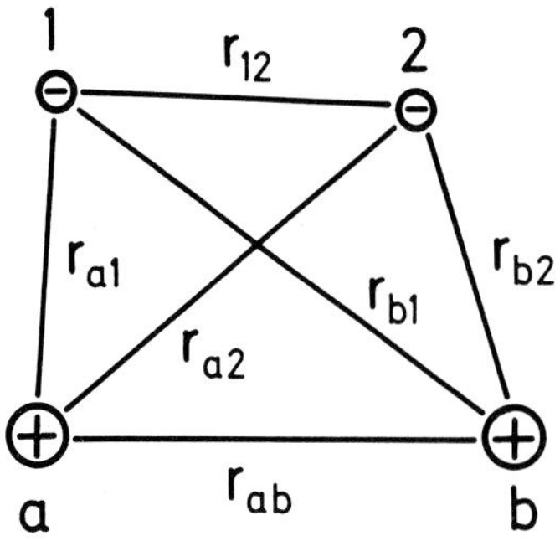

Fig. 23.9. Overview of the hydrogen molecule. The two nuclei are denoted by indices a and b, the two electrons by indices 1 and 2. The distances between the nuclei, between the electrons, and between electrons and nuclei, and the notations used for them, are shown in the sketch

the Hamiltonian, we recall again the energy balance of classical physics: we are concerned here with the kinetic energies of electrons 1 and 2, as well as with the various contributions of the Coulomb energy. If we express the kinetic energies as quantum mechanical operators, we obtain for the Hamiltonian the expression:

$$\mathcal{H} = \underbrace{-\frac{\hbar^2}{2m_0}\nabla_1^2 - \frac{e^2}{4\pi\varepsilon_0 r_{a1}}}_{\mathcal{H}_1} \underbrace{-\frac{\hbar^2}{2m}\nabla_2^2 - \frac{e^2}{4\pi\varepsilon_0 r_{b2}}}_{\mathcal{H}_2}$$
$$-\frac{e^2}{4\pi\varepsilon_0 r_{b1}} - \frac{e^2}{4\pi\varepsilon_0 r_{a2}} + \frac{e^2}{4\pi\varepsilon_0 r_{ab}} + \frac{e^2}{4\pi\varepsilon_0 r_{12}}\,. \tag{23.28}$$

∇_1^2 and ∇_2^2 are Laplace operators (19.3) which refer to the coordinates of electrons 1 and 2, respectively. We once again assume that the nuclei are infinitely massive.

We have the task of solving the Schrödinger equation

$$\mathcal{H}\psi(\boldsymbol{r}_1,\boldsymbol{r}_2) = E\psi(\boldsymbol{r}_1,\boldsymbol{r}_2) \tag{23.29}$$

with the Hamiltonian (23.28). If the nuclei were infinitely far apart, it would suffice to treat each atom independently, i.e. to solve the equations

$$\left(-\frac{\hbar^2}{2m_0}\nabla_1^2 - \frac{e^2}{4\pi\varepsilon_0 r_{a1}}\right)\phi_a(\boldsymbol{r}_1) = E_0\phi_a(\boldsymbol{r}_1) \tag{23.30}$$

and

$$\left(-\frac{\hbar^2}{2m_0}\nabla_2^2 - \frac{e^2}{4\pi\varepsilon_0 r_{b2}}\right)\phi_b(\boldsymbol{r}_2) = E_0\phi_b(\boldsymbol{r}_2)\,. \tag{23.31}$$

Since we are concerned here with a two-electron problem, we must take the *Pauli principle* into consideration (Sect. 17.2), that is we must take special account of the fact that the electrons possess spins. If the two hydrogen atoms did not interact with one another, we could immediately write down the total wavefunction with the help of the two wavefunctions ϕ_a and ϕ_b which occur in (23.30) and (23.31). As may be seen by substituting into the Schrödinger equation with $\mathcal{H} = \mathcal{H}_1 + \mathcal{H}_2$,

$$\phi_a(r_1)\,\phi_b(r_2) \tag{23.32}$$

would be a solution.

To take account of the spins of the electrons, we must multiply this solution with spin functions. The reader who is not familiar with the spin formalism need not despair at this point, since we will use only a very few properties of the spin functions and then can ignore them in the further course of the calculation.

We denote the function which represents an electron with its spin up by α; a spin wavefunction of this type was called $\phi_\uparrow$ in Sect. 14.2.2. If the electron in question is electron 1, we will call the spin function $\alpha(1)$. If both spins point in the same direction (upwards), our wavefunction is then given by

$$\phi_a(r_1)\,\phi_b(r_2)\,\alpha(1)\,\alpha(2)\,. \tag{23.33}$$

This, however, does not satisfy the Pauli Principle, which in its mathematical interpretation requires that a wavefunction be antisymmetric in all the coordinates of the electrons, or, in other words, that the wavefunction reverse its sign when we exchange the indices 1 and 2 everywhere. The wavefunction (23.33) does not have this property. By contrast, the wavefunction

$$\psi = \phi_a(r_1)\,\alpha(1)\,\phi_b(r_2)\,\alpha(2) - \phi_a(r_2)\,\alpha(2)\,\phi_b(r_1)\,\alpha(1) \tag{23.34}$$

does. If we factor out $\alpha(1)$ and $\alpha(2)$, this wavefunction is reduced to the simple form

$$\psi = \alpha(1)\,\alpha(2)\,\underbrace{[\phi_a(r_1)\,\phi_b(r_2) - \phi_a(r_2)\,\phi_b(r_1)]}_{\psi_\mathrm{u}}\,, \tag{23.35}$$

that is a product of a spin function and a function which depends only on spatial coordinates. [In quantum theory, wavefunctions which are symmetric with respect to exchange of the electronic spatial coordinates are denoted as "even" (symbol "g", from the German word "gerade"), while those which are antisymmetric are called "odd" ("u", from the German word "ungerade")].

As we know from the study of atoms, the energetically lowest states, of the lighter atoms at least, are constructed in a many-electron atom by filling the states from the lowest upwards with electrons of antiparallel spins. We therefore expect, and the expectation is confirmed by calculation, that the wavefunction ψ in (23.35) does not represent the energetically lowest state, since here the spins are parallel. We must instead consider a wavefunction in which the spins are antiparallel, in which one electron is described by a "spin up" function α and the other by a "spin down" function β. Here there are a number of various possibilities obtained by combining expressions of the form (23.33). One such possibility would be

$$\phi_a(r_1)\,\phi_b(r_2)\,\alpha(1)\,\beta(2)\,. \tag{23.36}$$

Other trial functions can be obtained from (23.36) by exchanging the coordinates r_1 and r_2 or the arguments of α and β, i.e. 1 and 2, or by exchanging all the coordinates with one another. None of these combinations is itself antisymmetric. We shall try to find a combination of (23.36) and similar trial functions of the sorts just mentioned which will be antisymmetric and permit the factorisation of the wavefunction into a spin part and a spatial part, as in (23.35). This is, in fact (as one discovers after a bit of trial and error), possible, and one finds a trial function

$$\psi = \underbrace{[\phi_a(r_1)\,\phi_b(r_2) + \phi_a(r_2)\,\phi_b(r_1)]}_{\psi_\mathrm{g}}\,[\alpha(1)\,\beta(2) - \alpha(2)\,\beta(1)]\,. \tag{23.37}$$

The spin function is here visibly antisymmetric, while the spatial part is symmetric. If we exchange all the spatial and spin coordinates of the two electrons simultaneously, we find in agreement with the Pauli principle an antisymmetric wavefunction. The spin functions were just an aid to help us take the overall symmetry of the wavefunction into account. Since, however, no operators occur in the Hamiltonian (23.28) which in any way act on the spin, we can completely neglect these spin functions in our following calculations of the energy. This means that in the approximation used here, the interaction of the spins with each other (spin-spin interaction) and the interaction of the spins with the spatial functions (spin-orbit interaction) are not taken into account. We will concern ourselves from now on only with the spatial wavefunctions ψ_u and ψ_g.

Following the basic suggestion of *Heitler* and *London*, we imagine these wavefunctions ψ_u and ψ_g as approximate solutions of the Schrödinger equation with the Hamiltonian (23.28), which contains *all* the Coulomb interactions between electrons and protons, and assume that using them, the exact energy is approached with the aid of (23.27). We thus have the task of calculating the energy expectation value for these wavefunctions.

As a first step in the calculation of the expectation value, let us consider the normalisation integral which occurs in the denominator of (23.27). It has the form

$$\begin{aligned}&\int\!\!\int |\psi(r_1,r_2)|^2 dV_1\,dV_2\\&\quad= \int\!\!\int [\phi_a(r_1)\,\phi_b(r_2) \pm \phi_a(r_2)\,\phi_b(r_1)]\\&\qquad\times [\phi_a(r_1)\,\phi_b(r_2) \pm \phi_a(r_2)\,\phi_b(r_1)]\,dV_1\,dV_2\,,\end{aligned} \tag{23.38}$$

where we have assumed ϕ_a, ϕ_b real.

After multiplying out the terms we obtain

$$\begin{aligned}&\int |\phi_a|^2 dV_1 \int |\phi_b|^2 dV_2 + \int |\phi_a|^2 dV_2 \int |\phi_b|^2 dV_1\\&\quad\pm \int \phi_a(r_1)\,\phi_b(r_1)\,dV_1 \int \phi_a(r_2)\,\phi_b(r_2)\,dV_2\\&\quad\pm \int \phi_a(r_2)\,\phi_b(r_2)\,dV_2 \int \phi_b(r_1)\,\phi_a(r_1)\,dV_1\,.\end{aligned} \tag{23.39}$$

The first two expressions can be reduced because of the normalisation of the wavefunctions ϕ_a and ϕ_b to

$$\int |\phi_a|^2 dV_1 = \int |\phi_b|^2 dV_2 = 1\,, \tag{23.40}$$

while the remaining terms are squares of the overlap integral

$$\int \phi_a(r_1)\,\phi_b(r_1)\,dV_1 = S\,. \tag{23.41}$$

With these results, we can write the normalisation integral (23.38) in the simple form

$$2(1 \pm S^2) . \tag{23.42}$$

In evaluating the numerator of the energy expectation value (23.27), we encounter similar expressions, of which we give two examples here.

We begin with the expression

$$\iint \phi_a(r_1)\,\phi_b(r_2)\left(\mathscr{H}_1 + \mathscr{H}_2 - \frac{e^2}{4\pi\varepsilon_0 r_{b1}} - \frac{e^2}{4\pi\varepsilon_0 r_{a2}} + \frac{e^2}{4\pi\varepsilon_0 r_{ab}} + \frac{e^2}{4\pi\varepsilon_0 r_{12}}\right)$$

$$\times\, \phi_a(r_1)\,\phi_b(r_2)\,dV_1\,dV_2 . \tag{23.43}$$

Since the Hamiltonian $\mathscr{H}_1$ in (23.43) operates only on ϕ_a, we can use the fact that ϕ_a is a solution of the Schrödinger equation (23.30) for the further evaluation of (23.43). Evaluating the terms with $\mathscr{H}_2$ in similar fashion, we arrive at the expression

$$\iint |\phi_a(r_1)|^2 |\phi_b(r_2)|^2 \Bigg(\underbrace{2E_0}_{(1)} - \underbrace{\frac{e^2}{4\pi\varepsilon_0 r_{b1}}}_{(2)} - \underbrace{\frac{e^2}{4\pi\varepsilon_0 r_{a2}}}_{(3)} + \underbrace{\frac{e^2}{4\pi\varepsilon_0 r_{ab}}}_{(4)} + \underbrace{\frac{e^2}{4\pi\varepsilon_0 r_{12}}}_{(5)}\Bigg) dV_1\,dV_2 . \tag{23.44}$$

It is useful in the following to investigate the meanings of the individual terms (1 – 5) separately.

1) $2E_0$ is the energy of the infinitely separated hydrogen atoms. (23.45)

2) $$\int |\phi_a(r_1)|^2 \left(\frac{-e^2}{4\pi\varepsilon_0 r_{b1}}\right) dV_1 = A < 0 . \tag{23.46}$$

This expression represents the Coulomb interaction energy of electron 1 in state a with nucleus b.

3) $$\int |(\phi_b(r_2)|^2 \left(\frac{-e^2}{4\pi\varepsilon_0 r_{a2}}\right) dV_2 = A < 0 . \tag{23.47}$$

This integral represents the Coulomb interaction energy of electron 2 in state b with nucleus a. From the symmetry of the problem it follows that the two integrals (2) and (3) are equal.

4) Because of the normalisation of the wavefunctions ϕ_a and ϕ_b, we obtain here

$$\frac{e^2}{4\pi\varepsilon_0 r_{ab}} . \tag{23.48}$$

This is the energy of Coulomb repulsion of the two nuclei.

5) $$\iint |\phi_a(r_1)|^2 |\phi_b(r_2)|^2 \frac{e^2}{4\pi\varepsilon_0 r_{12}} dV_1\,dV_2 = E_{\rm int} . \tag{23.49}$$

This integral represents the repulsive Coulomb energy between the two electron clouds.

Summarising the contributions (23.45 – 49), we find a term from the energy expectation value of (23.43):

$$\hat{E}/2 = 2E_0 + 2A + E_{\mathrm{int}} + \frac{e^2}{4\pi\varepsilon_0 r_{ab}} . \tag{23.50}$$

This is, however, still not the total energy expectation value, because on substituting ψ_{u} and ψ_{g} into (23.27), we find the occurrence also of exchange terms of the type

$$\iint \phi_b(\boldsymbol{r}_1)\,\phi_a(\boldsymbol{r}_2)(\ldots)\,\phi_b(\boldsymbol{r}_2)\,\phi_a(\boldsymbol{r}_1)\,dV_1\,dV_2 \tag{23.51}$$

in which the parenthetical expression (...) is the same as that in (23.43). Explicitly, (23.51) is given by

$$\pm \iint \phi_b(\boldsymbol{r}_1)\,\phi_a(\boldsymbol{r}_2)\,\phi_a(\boldsymbol{r}_1)\,\phi_b(\boldsymbol{r}_2) \times \Bigg(\underbrace{2E_0}_{(1)} - \underbrace{\frac{e^2}{4\pi\varepsilon_0 r_{b1}}}_{(2)} - \underbrace{\frac{e^2}{4\pi\varepsilon_0 r_{a2}}}_{(3)} + \underbrace{\frac{e^2}{4\pi\varepsilon_0 r_{ab}}}_{(4)} + \underbrace{\frac{e^2}{4\pi\varepsilon_0 r_{12}}}_{(5)} \Bigg)\,dV_1\,dV_2 . \tag{23.52}$$

The terms have the following forms and meanings:

1) $\pm 2E_0 S^2$ (23.53)

is the energy of the two separated hydrogen atoms multiplied by the square of the overlap integral S, which we have already met in (23.7).

2) $\pm \int \phi_a(\boldsymbol{r}_2)\,\phi_b(\boldsymbol{r}_2)\,dV_2 \underbrace{\int \phi_b(\boldsymbol{r}_1)\left(-\frac{e^2}{4\pi\varepsilon_0 r_{b1}}\right)\phi_a(\boldsymbol{r}_1)\,dV_1}_{D} .$ (23.54)

This exchange integral is a product of the overlap integral S with the single electron exchange integral D (23.9).

3) $\pm SD$. (23.55)

The result is the same as in (23.54), except that the indices of the electrons and the atoms are interchanged.

4) $\pm S^2 \dfrac{e^2}{4\pi\varepsilon_0 r_{ab}}$. (23.56)

The square of the overlap integral is multiplied by the energy of the Coulomb interaction between the two nuclei.

5) $$\pm \iint \phi_b(r_1)\,\phi_a(r_2)\,\frac{e^2}{4\pi\varepsilon_0 r_{12}}\,\phi_a(r_1)\,\phi_b(r_2)\,dV_1\,dV_2 = \pm E_{CE}\,. \tag{23.57}$$

This integral represents the Coulomb interaction between the two electrons, in which however the normal charge density does not occur, but, instead, the exchange density. It is therefore referred to as the Coulombic exchange interaction.

The total contribution resulting from (23.53 – 57) is thus

$$\tilde{E}/2 = \pm 2E_0 S^2 \pm 2DS \pm E_{CE} \pm S^2 \frac{e^2}{4\pi\varepsilon_0 r_{ab}}\,. \tag{23.58}$$

At this point we recall our original goal, which was to calculate the numerator of (23.27) using the wavefunctions ψ_u and ψ_g. If we multiply out all the terms in ψ_u and ψ_g, we obtain integrals of the type (23.43) twice, and integrals of the type (23.51) twice. Finally, we must divide the result by the normalisation integral; we then obtain the total energy of the hydrogen molecule:

$$E_{g,u} = \frac{\hat{E} + \tilde{E}}{2(1 \pm S^2)}\,, \tag{23.59}$$

in which, for the wavefunction ψ_g or ψ_u the upper or the lower sign is to be chosen:

$$E_g = 2E_0 + \frac{2A + E_{int}}{1 + S^2} + \frac{2DS + E_{CE}}{1 + S^2} + \frac{e^2}{4\pi\varepsilon_0 r_{ab}}\,, \tag{23.60}$$

$$E_u = 2E_0 + \frac{2A + E_{int}}{1 - S^2} - \frac{2DS + E_{CE}}{1 - S^2} + \frac{e^2}{4\pi\varepsilon_0 r_{ab}}\,. \tag{23.61}$$

In order to determine whether the molecule is chemically bound or not, the integrals in (23.60) and (23.61) must be numerically evaluated, since different effects are in competition and it is not immediately clear which of them is predominant. We have partly negative and partly positive contributions from the Coulomb interactions of the electrons between themselves, the nuclei between themselves, and the electrons with the nuclei. In addition, typically quantum mechanical effects occur, such as exchange interactions, which may be combined in the expression

$$K = 2DS + E_{CE} < 0\,. \tag{23.62}$$

It should not be our goal here to carry out the numerical evaluation of the integrals. Such an evaluation shows that the exchange contributions, for example (23.62), are negative. This produces the effect that the even wavefunctions are lower in energy than the odd wavefunctions. Furthermore, it is found that the net effect of the various Coulomb interactions is to lower the total energy of the hydrogen molecule relative to that of two free hydrogen atoms; this lowering, together with the exchange effects (23.62), comes about because the electrons can both be between the two nuclei simultaneously. They thus can profit from the attractive Coulombic potentials of both nuclei, and in such a way, that the Coulombic repulsion of the electrons for each other

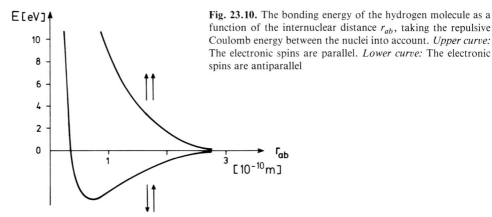

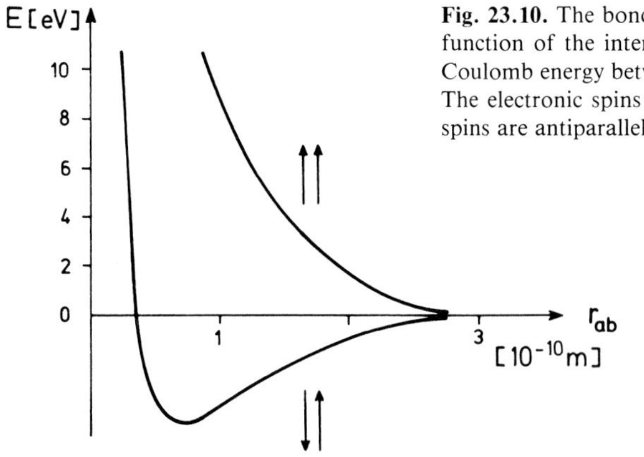

Fig. 23.10. The bonding energy of the hydrogen molecule as a function of the internuclear distance r_{ab}, taking the repulsive Coulomb energy between the nuclei into account. *Upper curve:* The electronic spins are parallel. *Lower curve:* The electronic spins are antiparallel

and of the nuclei for each other are to some extent compensated. The net energy lowering depends on the internuclear distance r_{ab}.

At very small distances, the positive repulsive energy of the two nuclei finally becomes dominant. Therefore, the total energy exhibits a minimum at a certain value of r_{ab} (Fig. 23.10). The dissociation energy, which is given by the difference between the minimum energy at the equilibrium distance and the energy at a distance $r_{ab}=\infty$, is, according to the calculation, 3.14 eV. The observed bonding energy, which is equal to the dissociation energy, is, by contrast, 4.48 eV, whereby it must still be taken into account that the nuclei contribute some energy through their kinetic energy. If this contribution, whose value we will not calculate here, is subtracted, the bonding energy would become even larger, 4.75 eV. We thus see that a considerable discrepancy exists between the calculated and the measured bonding energies. This means that the wavefunctions of the Heitler-London model are still relatively inaccurate. While they show us that the hydrogen molecule is, in fact, bound, they can give the shape of the wavefunctions only very approximately. To improve these wavefunctions several other effects must be taken into account more precisely; we will discuss only one typical example, which is known as covalent-ionic resonance.

23.5 Covalent-Ionic Resonance

In the previous section we employed a wavefunction for the two electrons in which the one electron is always in the neighbourhood of the opposite nucleus to the other electron. In this case, known as "covalent bonding", the wavefunction was of the type

$$\psi_{\rm cov}=N[\phi_a(r_1)\,\phi_b(r_2)+\phi_a(r_2)\,\phi_b(r_1)]\,, \qquad (23.63)$$

where N is a normalisation factor.

It is of course possible that – at least with a certain probability – one electron is localised at the same time as the other on the same atom, in which case the wavefunction would be of the type

$$\phi_a(r_1)\,\phi_a(r_2)\,. \qquad (23.64)$$

Since the two atomic nuclei are equivalent, both electrons could just as well be near atom b, which would lead to the wavefunction

$$\phi_b(\boldsymbol{r}_1)\,\phi_b(\boldsymbol{r}_2)\,. \tag{23.65}$$

The functions (23.64, 65) describe states in which a negatively charged hydrogen atom is present; for this reason, they are called "ionic". Since the states (23.64, 65) are energetically degenerate, we must form a linear combination from them, which we will do in the symmetric form:

$$\psi_{\text{ion}} = N'\,[\phi_a(\boldsymbol{r}_1)\,\phi_a(\boldsymbol{r}_2) + \phi_b(\boldsymbol{r}_1)\,\phi_b(\boldsymbol{r}_2)] \tag{23.66}$$

so that (23.66) has the same symmetry as (23.63). Now we must expect that in nature neither wavefunction (23.63) nor wavefunction (23.66) occurs uniquely, since the electrons avoid each other for part of the time, but can still be in the neighbourhood of the same nucleus at other times. Since both possibilities exist, we must – according to the basic rules of quantum mechanics – construct the wavefunction which best fits the observations by taking a linear combination of (23.63) and (23.66):

$$\psi = \psi_{\text{cov}} + c\,\psi_{\text{ion}}\,, \tag{23.67}$$

where the constant c is still a parameter which must be determined by using the condition that the energy expectation value belonging to (23.67) be minimised.

23.6 The Hund-Mulliken-Bloch Theory of Bonding in Hydrogen

As well as the Heitler-London method, which we have discussed above, a second technique is often used in molecular physics; to be sure, it does not yield such good results for the bond energies as the Heitler-London method, but it allows the wavefunction of the individual electrons to be more closely described. This is of particular interest in molecular spectroscopy, where in general only a single-electron state is changed by an observed transition and one would like to treat just this change theoretically.

In this method, the fact that two electrons are involved in the problem is at first ignored. We consider instead the motion of a single electron in the field of the atomic nuclei, or, in other words, we start from the solution to the problem of the hydrogen-molecule ion. This solution was derived in Sect. 23.2; the result had the form

$$\psi_{\text{g}}(\boldsymbol{r}) = N[\phi_a(\boldsymbol{r}) + \phi_b(\boldsymbol{r})]\,. \tag{23.68}$$

The idea is now to insert *both* of the electrons of the hydrogen molecule one after the other into this state (23.68). As a trial solution of the Schrödinger equation with the Hamiltonian (23.28) for both electrons we therefore use

$$\psi(\boldsymbol{r}_1,\boldsymbol{r}_2) = \psi_{\text{g}}(\boldsymbol{r}_1)\,\psi_{\text{g}}(\boldsymbol{r}_2)\cdot\text{spin function}\,, \tag{23.69}$$

in which we will restrict ourselves to the case that the spins are antiparallel; the spin function is antisymmetric, and thus has the form

$$\text{spin function} = \frac{1}{\sqrt{2}}[\alpha(1)\beta(2) - \alpha(2)\beta(1)] \; . \tag{23.70}$$

The total wavefunction (23.69) may be seen to be antisymmetric with respect to the electronic space and spin coordinates. With the trial solution (23.69), we can again calculate the expectation value of the total energy; the result is not quite as accurate as that of the Heitler-London method. The procedure which we are describing is called the "*Linear Combination of Atomic Orbitals*" method, abbreviated LCAO. A linear combination such as that given in (23.68) represents the wavefunction of a single electron in the molecule and is also referred to, for this reason, as a *molecular orbital*.

This technique may be extended to other more complicated molecules, in which case the wavefunction is first constructed for each pair of centers (atomic nuclei) and then filled with a pair of electrons. The method requires some modifications for many molecules, some of the most important and widely applied of which we will describe in the following.

23.7 Hybridisation

A particularly important case, which is of special interest in organic chemistry, is *hybridisation*. In treating it, we will be led to atoms with several electrons. In the formation of chemical bonds and molecules, the electrons of the closed inner shells are not strongly affected. Chemical bonding takes place through the outer electrons (valence electrons), which are more loosely bound to the nuclei. In the case of the carbon atom, of the six electrons, two are in the $1s$ state, two in the $2s$ state, and the remaining two in two of the three states $2p_x$, $2p_y$, and $2p_z$ (Sect. 19.2). The degeneracy of the state with principal quantum number $n = 2$, which is familiar from the *hydrogen atom*, is lifted here; to be sure, the energy difference between the $2s$ and the $2p$ states, 4 eV, is not too large, and there is in fact an excited state of the carbon atom in which an electron is excited out of the $2s$ state into the $2p$ level (Sect. 19.2). Then the states $2s$, $2p_x$, $2p_y$, and $2p_z$ are all occupied. Let us now consider these states, each of which contains a single electron, more carefully, while letting external forces act on the electrons, for example by bringing a hydrogen atom into the neighbourhood of the carbon atom. Then the energy splitting which still exists between the $2s$ and the $2p$ states may be more or less compensated by the external forces, so that the $2s$ and $2p$ states are practically degenerate.

As we know from perturbation theory in the presence of degeneracy, we must, in such a situation, form linear combinations of the old, degenerate functions. For example, instead of the $2s$ and the $2p_x$ functions, we can construct two new wavefunctions of the form

$$\begin{aligned} \psi_+ &= \phi_s + \phi_{p_x} \, , \\ \psi_- &= \phi_s - \phi_{p_x} \, . \end{aligned} \tag{23.71}$$

As we have shown earlier, linear combinations of this type mean that the centre of charge is shifted as compared to the s function (Fig. 23.11). Just such a phenomenon occurs in hybridisation. Let us discuss some types of hybridisation, beginning with the most well-known case, that of methane, CH_4, where the carbon atom is surrounded by

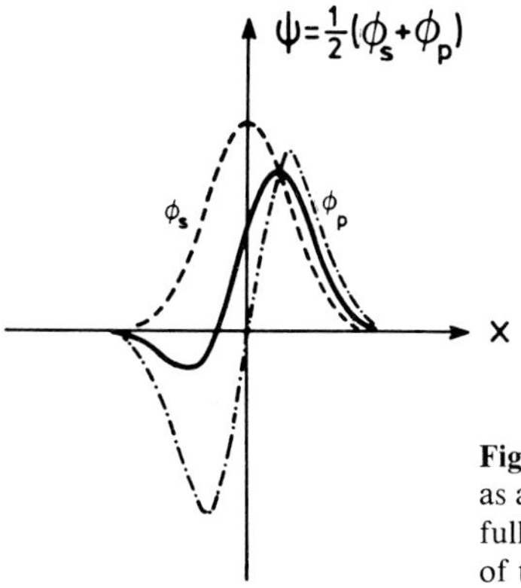

Fig. 23.12. *Left:* The density distributions of the four electrons in the case of tetragonal (sp^3-) hybridisation of carbon. *Right:* Exploded view

Fig. 23.11. Shape of the wavefunctions in diagonal hybridisation (sp hybridisation). The s function is shown as a dashed line, the p function as a dot-dashed line, and the function resulting from their superposition as a full curve. The figure shows clearly how the centre of gravity is displaced to the right due to the superposition of the wavefunctions ϕ_s and ϕ_p

four hydrogen atoms. It is known experimentally that the carbon atom sits in the centre of a tetrahedron whose vertices are occupied by the hydrogen atoms. Interestingly, the four degenerate wavefunctions with principal quantum number $n = 2$ in the carbon atom can form four linear combinations, whose centres of charge are displaced in the directions of these four vertices. If we now recall the fact that the wavefunctions of the p state have the form $f(r)x$, $f(r)y$, and $f(r)z$, we can see in elementary fashion that the following linear combinations generate the charge displacements mentioned above:

$$\begin{aligned}
\psi_1 &= \tfrac{1}{2}(\phi_s + \phi_{p_x} + \phi_{p_y} + \phi_{p_z})\,, \\
\psi_2 &= \tfrac{1}{2}(\phi_s + \phi_{p_x} - \phi_{p_y} - \phi_{p_z})\,, \\
\psi_3 &= \tfrac{1}{2}(\phi_s - \phi_{p_x} + \phi_{p_y} - \phi_{p_z})\,, \\
\psi_4 &= \tfrac{1}{2}(\phi_s - \phi_{p_x} - \phi_{p_y} + \phi_{p_z})\,.
\end{aligned} \tag{23.72}$$

With these new linear combinations, the electrons of the carbon atom are "adjusted" to the tetrahedral environment. Each one of the four wavefunctions (23.72) can take part in forming a bond with the corresponding hydrogen atom (Fig. 23.12). For example, we select the direction of vertex 1 and denote the wavefunction ψ_1 of (23.72) in the carbon atom more exactly as ψ_{C1}, and that in the hydrogen atom as ϕ_{H1}. Similarly to the hydrogen molecule, according to the LCAO method, a wavefunction for each of the electrons which participates in the bonding can be written in the form

$$\psi(r) = \psi_{C1}(r) + \alpha\phi_{H1}(r)\,. \tag{23.73}$$

Since the carbon atom and the hydrogen atom are different, the coefficient α will not be equal to 1 (in contrast to the hydrogen molecule) and must be determined by a variational method.

In the present case, we have guided our considerations by the experimental evidence, which indicates that four hydrogen atoms are located at the vertices of a tetrahedron. It is tempting to ask whether, in a theoretical treatment of this problem, the wavefunctions (23.72) are to be regarded as given in advance, and the hydrogen atoms then attach themselves to the vertices of the (pre-existing) tetrahedron, or whether the hydrogen atoms first place themselves on the vertices of a tetrahedron and the carbon wavefunctions then adjust themselves accordingly. From the viewpoint of quantum

Fig. 23.13. *Left:* The density distributions of the three electrons in the case of trigonal (sp^2-) hybridisation of carbon. *Right:* Exploded view

mechanics, such considerations are superfluous. The position of the hydrogen atoms and the orientation of the wavefunctions mutually influence each other. The overall configuration is chosen by the CH_4 molecule in such a way as to minimise the total energy.

The tetragonal hybridisation discussed above, i.e. arrangement of the wavefunctions in tetrahedral symmetry, is not the only possible form of hybridisation in the carbon atom. A further form has already been anticipated, namely diagonal hybridisation, which is expressed in the wavefunctions (23.71) (Fig. 23.11).

In carbon, a third type of hybridisation is possible, the trigonal form, in which the wavefunctions of s-, p_x- and p_y types generate preferred orientations in three directions within a plane by forming suitable linear combinations. In order to give the reader an impression of the form which such linear combinations take, we will give them explicitly here (Fig. 23.13):

$$\begin{aligned} \psi_1 &= \sqrt{\tfrac{1}{3}}(\phi_s + \sqrt{2}\,\phi_{p_x}) \,, \\ \psi_2 &= \sqrt{\tfrac{1}{3}}(\phi_s + \sqrt{\tfrac{3}{2}}\,\phi_{p_y} - \sqrt{\tfrac{1}{2}}\,\phi_{p_x}) \,, \\ \psi_3 &= \sqrt{\tfrac{1}{3}}(\phi_s - \sqrt{\tfrac{3}{2}}\,\phi_{p_y} - \sqrt{\tfrac{1}{2}}\,\phi_{p_x}) \,. \end{aligned} \tag{23.74}$$

Clearly, the original fourth wavefunction, p_z, is not used in these three wavefunctions. It plays an additional rôle in chemical bonding, as we will see: we consider the case of ethene, C_2H_4. Here the two carbon atoms each form a trigonal configuration with two hydrogen atoms. The hydrogen-carbon bonds are given by wavefunctions of the type (23.73), where, e.g. for ψ_{C1} we now take ψ_2 from (23.74). A carbon-carbon bond is formed by the first of these wavefunctions, with each carbon atom contributing one electron. The electrons in the p_z orbitals remain unused; these orbitals can be employed to form linear combinations following the Hund-Mulliken-Bloch method in analogy to the hydrogen molecule, leading to an additional bond between the two carbon atoms. We thus finally see that we are dealing with a double carbon-carbon bond (Fig. 23.14).

Fig. 23.14. Density distribution of the hybridised electrons of carbon in ethene C_2H_4. *Left:* The two carbon atoms are located at the two nodes and form, together with their respective bound hydrogen atoms, a trigonal configuration. *Right:* The p_z functions of the two carbon atoms, which are oriented perpendicularly to the plane of the trigonal bonds, form a second carbon-carbon bond

23.8 The π Electrons of Benzene, C_6H_6

To conclude our brief excursion into the quantum mechanics of chemical bonding, we will treat another typical case, that of benzene, C_6H_6. As is experimentally known, this molecule is planar. The hydrogen atoms lie in the same plane as the carbon atoms which are joined to form a six-membered ring (Fig. 23.15). If we look at a particular carbon atom, we find again a trigonal arrangement with respect to the projecting hydrogen atom and the two neighbouring carbon atoms. We see that, in a quite similar fashion to ethene, one p_z orbital per carbon atom, containing one electron, is left over. All such p_z orbitals on the six different carbon atoms are equivalent energetically. An electron can thus in principle be in any one of the orbitals.

Fig. 23.15. Electron density distribution in benzene, C_6H_6. *Left:* A series of trigonal arrangements in the six-membered ring of benzene (σ bonds). *Right:* The density distribution of the π states, which are formed from the p_z orbitals of the six carbon atoms

Let us now apply the fundamental idea of the LCAO method, i.e. the method of linear combinations of atomic orbitals. It tells us to seek the wavefunction for a single electron in the field of all the atomic nuclei, here in particular in the field of the six carbon nuclei. We then have to deal with a generalisation of the hydrogen molecule, where, however, the electron is not shared between two, but rather among six different positions. If we denote the p_z orbitals at the six carbon positions as π_1, π_2, $\pi_3, \dots, \pi_6$, we thus arrive at a wavefunction of the form

$$\psi = c_1\pi_1 + c_2\pi_2 + \dots + c_6\pi_6 \,. \tag{23.75}$$

To determine the coefficients c_j, we can utilise the symmetry of the problem. That is, if we were to move each carbon atom in the ring one step ahead, the overall potential field which acts on the electron we are considering would remain unchanged. We have an example of a rotational symmetry, in which the potential field remains unchanged when we rotate the molecule by 60°, or a multiple thereof, about an axis perpendicular to the plane of the molecule. At this point, we can use the results of Sect. 16.1.2. There we found that the wavefunctions in an atom must have a certain form when the Hamiltonian possesses rotational symmetry. It may be shown, as one can verify using the discussion in that section, that the wavefunction (23.75) must be multiplied by a factor of the form

$$\mathrm{e}^{2\mathrm{i}m\pi/6} \tag{23.76}$$

when we rotate the molecule by an angle of 60°. Here the quantum number m can take on the values $m = 0, 1, \dots, 5$. The relation

$$\psi_{\text{rotated}} = \mathrm{e}^{2\mathrm{i}m\pi/6}\,\psi_{\text{unrotated}} \tag{23.77}$$

results in relations between the coefficients c_j in (23.75): on rotation of the molecule by 60°, the π wavefunctions are permuted among themselves:

$$\pi_2 \to \pi_1,\ \pi_3 \to \pi_2, \dots, \pi_1 \to \pi_6 \,.$$

We then obtain from (23.77)

$$c_1\pi_6 + c_2\pi_1 + c_3\pi_2 + \ldots + c_6\pi_5 = e^{2im\pi/6}(c_1\pi_1 + c_2\pi_2 + \ldots + c_6\pi_6) \,. \tag{23.78}$$

Since the wavefunctions π_j are linearly independent of one another, (23.78) can only be fulfilled if

$$c_2 = e^{2im\pi/6}c_1 \,, \qquad c_3 = e^{2im\pi/6}c_2 \,, \qquad \text{etc.} \tag{23.79}$$

holds. The solutions of these equations are

$$c_j = e^{2\pi imj/6} \,, \tag{23.80}$$

with which we have determined the coefficients in (23.75) up to a normalisation factor. If we examine the result of these considerations, we see that in benzene, there are not only electrons which are localised between C and H and between C and C atoms, but also there are electrons which can move about the entire molecule. Since the phase factors in (23.78) represent a wave, each of the π electrons which we refer to here can propagate around the benzene molecule like a wave.

The methods of quantum mechanics which we have introduced here are not only applicable to still more complex molecules – they also form the basis for the quantum theory of electronic states in solids. In particular, the wavefunctions for electrons in the conduction band of a crystal may be regarded as direct generalisations of the benzene wavefunctions (23.75) and (23.80).

Problems

23.1 The problem of the hydrogen molecule ion can be simplified by following the motion of the electron in only one dimension and representing the attractive potential of the nuclei by $-\beta\delta(x-x_a)-\beta\delta(x-x_b)$. Calculate the wavefunctions for the (two) bonding states and discuss their symmetry properties. How does the binding energy depend on the distance?

Hint: Proceed as in Problem 9.3. For the definition of the δ-function cf. Appendix A.

23.2 Repeat the calculation of Problem 23.1 for the case when the attractive forces β of the two nuclei are different and discuss how this affects the probability distribution of the position of the electron.

23.3 To arrive at a model for the π electrons of benzene, one can think of six δ potentials of strength β, periodically arranged. To imitate the ring, it is required in the one-dimensional calculation that the wavefunction $\psi(x)$ repeat itself after six lengths a between the potentials:

$$\psi(x+6a) = \psi(x) \,.$$

Calculate $\psi(x)$ and E for the states with $E<0$.

Hint: Use Problem 23.1 and (23.77).

Appendix

A. The Dirac Delta Function and the Normalisation of the Wavefunction of a Free Particle in Unbounded Space

The English physicist *Dirac* introduced a function which is extremely useful for many purposes of theoretical physics and mathematics. Precisely speaking, it is a generalised function which is only defined under an integral. We shall first give its definition, and then discuss its uses. The delta function (δ function) is defined by the following properties (x is a real variable, $-\infty \leqq x \leqq \infty$):

1) $\delta(x) = 0 \quad \text{for} \quad x \neq 0\,,$ (A.1)

2) $\int_a^b \delta(x)\,dx = 1 \quad \text{for} \quad a < 0 < b\,.$ (A.2)

The δ function thus vanishes for all values of $x \neq 0$, and its integral over every interval which contains $x = 0$ has the value 1. The latter property means, speaking intuitively, that the δ function must become infinitely large at $x = 0$. The unusual properties of the δ function become more understandable when we consider it as the limiting case of functions which are more familiar. Such an example is given by the function

$$\frac{1}{\sqrt{\pi}u}\,\mathrm{e}^{-x^2/u^2}\,, \tag{A.3}$$

which is shown in Fig. A.1.

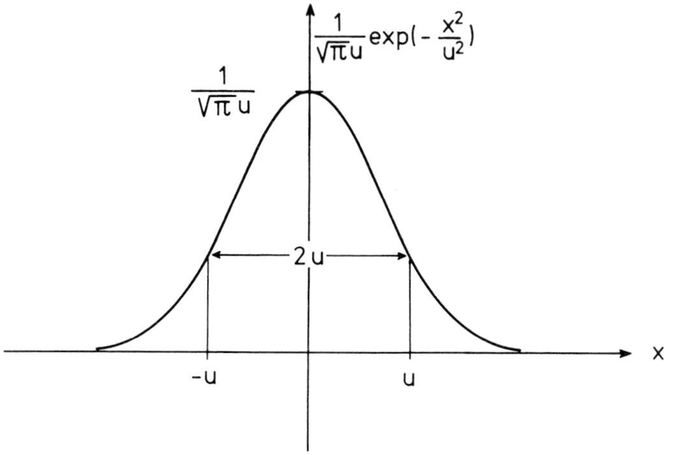

Fig. A.1. The function $(1/\sqrt{\pi}u)\exp(-x^2/u^2)$ plotted against the variable x. If we let the parameter u become smaller, the value of the function at $x = 0$ gets larger and larger and the decrease to both sides gets steeper, until the function has finally pulled itself together into a δ function

If we allow u to go to zero, the function becomes narrower and higher until it is finally just a "vertical line". We thus have

$$\text{for} \quad x \neq 0: \lim_{u \to 0} \frac{1}{\sqrt{\pi} u} e^{-x^2/u^2} = 0 . \tag{A.4}$$

On the other hand, one can find in any integral table the following fact:

$$\int_{-\infty}^{+\infty} \frac{1}{\sqrt{\pi} u} e^{-x^2/u^2} dx = 1 , \tag{A.5}$$

independently of the value of u. If the limit $u \to 0$ is calculated it becomes clear that because of (A.4), we can write the integral (A.5) with any finite limits a and b with $a < 0 < b$ without changing its value. This is just the relation (A.2).

In many practical applications in quantum mechanics, the δ function occurs as the following limit:

$$\delta(x) = \lim_{u \to \infty} \frac{1}{\pi} \frac{\sin ux}{x} . \tag{A.6}$$

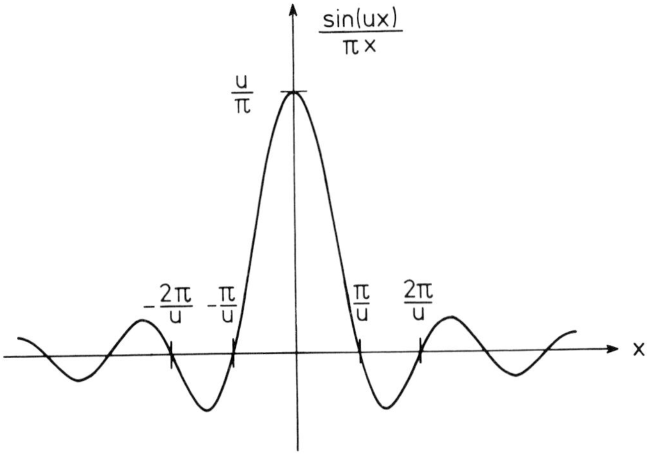

Fig. A.2. The function $\sin(ux)/\pi x$ plotted against x. If we allow u to go to the limit infinity, the value of the function at $x = 0$ becomes larger and larger. At the same time, the position of the zero-crossing moves towards $x = 0$

The property (A.2) is found to be fulfilled when we take into account that

$$\int_{-\infty}^{+\infty} \frac{1}{\pi} \frac{\sin(ux)}{x} dx = 1 . \tag{A.7}$$

The property (A.1) is not so obvious. To demonstrate it, one has to consider that for $u \to \infty$, $x \neq 0$, $\sin(ux)/x$ oscillates extremely rapidly back and forth, so that when we average the function over even a small region, the value of the function averages out to zero (Fig. A.2).

The δ function has, in particular, the following properties:

for a continuous function $f(x)$,

$$\int_a^b f(x)\,\delta(x-x_0)\,dx = f(x_0)\,, \qquad a < x_0 < b \qquad \text{is valid}\,. \tag{A.8}$$

For a function $f(x)$ which is n times continuously differentiable,

$$\int_a^b f(x)\,\delta^{(n)}(x-x_0)\,dx = (-1)^n f^{(n)}(x_0)\,, \qquad a < x_0 < b \tag{A.9}$$

holds. Here $f^{(n)}$ and $\delta^{(n)}$ mean the nth derivatives w.r.t. x. The proof of (A.8) follows immediately from (A.1, 2). The proof of (A.9) is obtained by n-fold partial integration. Furthermore,

$$\delta(cx) = \frac{1}{|c|}\,\delta(x)\,, \qquad c \text{ real} \tag{A.10}$$

is valid. The relation (A.1) is seen to be fulfilled on both sides. If we insert (A.10) in (A.2), we find

$$\int_a^b \delta(cx)\,dx\,,$$

which, after changing variables using $cx = x'$, becomes

$$\int_{a'}^{b'} \frac{1}{c}\,\delta(x')\,dx'\,, \qquad a' \lessgtr b' \qquad \text{for} \qquad c \gtrless 0\,,$$

which is thus, according to (A.2), equal to $1/|c|$.

We now turn to the question of the normalisation of wavefunctions in unbounded space, where we can limit ourselves to the one-dimensional case without missing any essentials. We start with wavefunctions which are normalised in the interval L,

$$\psi_k(x) = (1/\sqrt{L})\,e^{ikx}\,, \tag{A.11}$$

for which we have the normalisation integral

$$\int_{-L/2}^{L/2} |\psi(x)|^2 dx = 1\,. \tag{A.12}$$

If we furthermore assume that $\psi(x)$ is periodic, $\psi(x+L) = \psi(x)$, the k's must have the form

$$k = \frac{2\pi n}{L}\,, \qquad n = 0, \pm 1, \pm 2, \dots\,. \tag{A.13}$$

It is easy to convince oneself that

$$\int_{-L/2}^{L/2} \psi_k^*(x)\,\psi_{k'}(x)\,dx = \delta_{k,k'} \tag{A.14}$$

$$= \begin{cases} 1 & \text{for} \quad k = k' \\ 0 & \text{for} \quad k \neq k' . \end{cases} \tag{A.15}$$

To prove this relation, the integral must be computed, taking account of (A.13). The integral yields

$$\frac{1}{L}\int_{-L/2}^{L/2} \mathrm{e}^{-\mathrm{i}kx+\mathrm{i}k'x}dx = \frac{1}{\mathrm{i}L(k'-k)}(\exp[\mathrm{i}(k'-k)L/2] - \exp[-\mathrm{i}(k'-k)L/2]) . \tag{A.16}$$

If we now abbreviate $k'-k$ with ξ and $L/2$ with u, we may write (A.16) in the form

$$\sin(\xi u)/\xi u . \tag{A.17}$$

This is, however, apart from the factor $2\pi/L$, just the function which appears in (A.6) on the right under the limit, if we identify ξ with x. If we thus divide (A.16) by $2\pi/L$ and form $\lim\limits_{L\to\infty}$, we obtain on the left side of (A.16)

$$\frac{1}{2\pi}\lim_{L\to\infty}\int_{-L/2}^{L/2} \exp(-\mathrm{i}kx+\mathrm{i}k'x)\,dx , \tag{A.18}$$

which we may also write somewhat differently:

$$\int_{-\infty}^{+\infty}\left[\frac{1}{\sqrt{2\pi}}\mathrm{e}^{\mathrm{i}kx}\right]^*\left[\frac{1}{\sqrt{2\pi}}\mathrm{e}^{\mathrm{i}k'x}\right]dx . \tag{A.19}$$

The right-hand side of (A.16), using (A.17) and (A.6), goes to $\delta(k'-k)$. We thus finally obtain

$$\int_{-\infty}^{+\infty} \psi_k^*(x)\,\psi_{k'}(x)\,dx = \delta(k'-k) , \tag{A.20}$$

where

$$\psi_k(x) = \frac{1}{\sqrt{2\pi}}\mathrm{e}^{\mathrm{i}kx} . \tag{A.21}$$

Equation (A.20) with (A.21) generalises the relation (A.14) [with (A.11)] to the case of wavefunctions without finite boundary conditions and thus to the corresponding case of continuous k values. As may be seen in all practical applications, the δ function in (A.20) always occurs under further integrals over k or k' (or both), so that we have found a self-consistent formalism.

Let the wavefunctions in (A.21) depend not upon k, but upon $p = \hbar k$; then we must observe (A.10). In order to normalise the new wavefunctions

$$\psi_p(x) = N\mathrm{e}^{\mathrm{i}px/h}$$

correctly, we must set N equal to $(1/\sqrt{\hbar})\,(1/\sqrt{2\pi}) = (1/\sqrt{h})$. The normalised wavefunction is now given by

$$\psi_p(x) = \frac{1}{\sqrt{h}}\,\mathrm{e}^{\mathrm{i}px/\hbar}\,.$$

B. Some Properties of the Hamiltonian Operator, Its Eigenfunctions and Its Eigenvalues

We write the time-independent Schrödinger equation in the form

$$\mathscr{H}\psi_n = E_n\psi_n \tag{B.1}$$

with the Hamiltonian

$$\mathscr{H} = -\frac{\hbar^2}{2m_0}\nabla^2 + V(\boldsymbol{r})\,, \qquad V(\boldsymbol{r})\ \text{real}\,.$$

The $\psi_n(\boldsymbol{r})$ are square-integrable eigenfunctions with the eigenvalues E_n. Here, $\psi_n = 0$ is excluded. The eigenvalues E_n may be discrete or they may be continuous.

In the following, we denote by ψ_μ and ψ_ν the wavefunctions on which the operator $\mathscr{H}$ can act. We can now easily read off the following properties:

a) $\mathscr{H}$ is a linear operator, i.e. the relation

$$\mathscr{H}(c_\mu\psi_\mu + c_\nu\psi_\nu) = c_\mu\mathscr{H}\psi_\mu + c_\nu\mathscr{H}\psi_\nu$$

holds, where c_μ and c_ν are some complex numbers. In particular, it follows from this that every linear combination of eigenfunctions of $\mathscr{H}$ with the same eigenvalue E is itself an eigenfunction of $\mathscr{H}$ with the eigenvalue E.

b) $\mathscr{H}$ is Hermitian, i.e. the equation

$$\int \psi_\mu^*(\boldsymbol{r})\,[\mathscr{H}\psi_\nu(\boldsymbol{r})]\,dV = \int [\mathscr{H}\psi_\mu(\boldsymbol{r})]^*\,\psi_\nu(\boldsymbol{r})\,dV \tag{B.2}$$

is valid. It follows from (B.2) that for the operator of the potential energy, $V^*(\boldsymbol{r}) = V(\boldsymbol{r})$. For the kinetic energy operator, (B.2) can be proved by double partial integration, taking into account the fact that the wavefunctions vanish at infinity.

c) The eigenvalues E_n are real. This is a consequence of (B.2), if one inserts for ψ_μ and ψ_ν the same *eigenfunction* ψ_n and utilises (B.1).

d) Eigenfunctions with different eigenvalues are orthogonal.

We take the following scalar products (different eigenvalues belong to the functions ψ_m and ψ_n):

$$\int \psi_m^*(\mathbf{r}) [\mathcal{H} \psi_n(\mathbf{r})] \, dV = E_n \int \psi_m^*(\mathbf{r}) \, \psi_n(\mathbf{r}) \, dV \,, \tag{B.3}$$

$$\int [\mathcal{H} \psi_m(\mathbf{r})]^* \, \psi_n(\mathbf{r}) \, dV = E_m^* \int \psi_m^*(\mathbf{r}) \, \psi_n(\mathbf{r}) \, dV \,. \tag{B.4}$$

We subtract (B.4) from (B.3) and use (B.2) and the property that the eigenvalues are real:

$$0 = \underbrace{(E_n - E_m)}_{\neq 0} \underbrace{\int \psi_m^*(\mathbf{r}) \, \psi_n(\mathbf{r}) \, dV}_{0} \,.$$

The second bracket indicates the orthogonality of the wavefunctions.

Furthermore, it can be shown that eigenfunctions with the same eigenvalue may always be *chosen* to be orthogonal by using appropriate linear combinations.

Solutions to the Problems

2.1 b) $N_A = 6.25 \cdot 10^{23}\,\text{mol}^{-1}$

2.2 $r_{He} = 2.08 \cdot 10^{-10}\,\text{m} = 2.08\,\text{Å}$

2.3

Pressure p [mbar]	1	10^{-2}	10^{-4}
Number of collisions	1706	17	0.17

2.4 Helium: $r_{He} = 1.33\,\text{Å}$; mercury: $r_{Hg} = 1.19\,\text{Å}$

2.5 The Bragg Law leads to a lattice-plane spacing of $d = 2.85\,\text{Å}$

2.6 Debye-Scherrer arrangement; Bragg's Law gives $r = 16.1$ cm in first order ($n = 1$) and $r = 33.1$ cm in second order ($n = 2$).

2.7 Bragg reflection in first order: $\theta = 3.63°$. The neutron wavelength is $\lambda = 2.023 \cdot 10^{-11}$ m.

3.1 The orbital radius of a charged particle:

$$r = \frac{p}{qB} \rightarrow \text{momentum filter for } q, B \text{ constant.}$$

With $p = \sqrt{2mE}$ (classical energy-momentum relationship), we have

$$r = \frac{\sqrt{2E}}{qB}\sqrt{m} \rightarrow \text{mass filter for } q, B \text{ and } E \text{ constant.}$$

3.2

	$^1H^+$	$^2H^+$	$^3H^+$
Mass m [kg]	$1.67 \cdot 10^{-27}$	$3.35 \cdot 10^{-27}$	$5.02 \cdot 10^{-27}$
Radius r of the circular orbit [cm]	9.14	12.9	15.8
Deflection on the screen [cm]	14.6	9.39	7.46

3.3 Parabola method:

$$Y = AX^2 \tag{1}$$

	H^+	H_2^+
Mass m [kg]	$1.673 \cdot 10^{-27}$	$3.347 \cdot 10^{-27}$
A in Eq. (1) [1/m]	65.27	130.58
v (1000 V) [m/s]	$4.376 \cdot 10^5$	$3.094 \cdot 10^5$
X (1000 V) [m]	$1.751 \cdot 10^{-2}$	$1.238 \cdot 10^{-2}$
Y (1000 V) [m]	$2.000 \cdot 10^{-2}$	$2.000 \cdot 10^{-2}$
v (4000 V) [m/s]	$8.752 \cdot 10^5$	$6.188 \cdot 10^5$
X (4000 V) [m]	$8.753 \cdot 10^{-3}$	$6.188 \cdot 10^{-3}$
Y (4000 V) [m]	$5.000 \cdot 10^{-3}$	$5.000 \cdot 10^{-3}$

If both positive and negative particles are injected, one obtains parabolas in the $(+x, +y)$ and $(-x, -y)$ quadrants.

3.4 Abundance ratio P_n of ^{235}U to ^{238}U after the nth separation step:

$$P_n = P_0 q^n \quad \text{where} \quad q = \frac{P_1}{P_0} = \frac{0.754/99.246}{0.72/99.28} \tag{1}$$

Amount of ^{235}U	50%	99%
P_n	1	99
n in Eq. (1)	106	205

4.1 When the aluminium foil is replaced by a gold foil of the same thickness, Z and N in the Rutherford scattering formula (4.20) are changed, i.e.

$$\frac{\left(\frac{dn}{dt}\right)_{Al}}{Z_{Al}^2 N_{Al}} = \frac{\left(\frac{dn}{dt}\right)_{Au}}{Z_{Au}^2 N_{Au}} \,.$$

One finds $\left(\frac{dn}{dt}\right)_{\mathrm{Au}} = 36.11 \cdot \left(\frac{dn}{dt}\right)_{\mathrm{Al}}$, meaning that 36110 particles/second are measured.

4.2 From the Rutherford scattering formula (4.20) it follows that

$$\left(\frac{dn}{dt}\right)(\theta) = \frac{\sin^4\left(\frac{10°}{2}\right)}{\sin^4\left(\frac{\theta}{2}\right)}\left(\frac{dn}{dt}\right)(10°)$$

$$= \frac{57.7\,\frac{1}{\mathrm{s}}}{\sin^4\left(\frac{\theta}{2}\right)}$$

$\theta[°]$	10	45	90	135	180
$\left(\frac{dn}{dt}\right)\left[\frac{1}{\mathrm{s}}\right]$	10^6	2690.5	230.8	79.2	57.7

4.3 The kinetic energy of the proton is converted completely to potential energy as a result of the Coulomb interaction. The distance of closest approach a is given by

$$a = \frac{e^2 Z}{4\pi\varepsilon_0 E_0} = \frac{1.138\cdot 10^{-7}\,\mathrm{eVm}}{E_0}\ .$$

$E_0 = 1$ MeV: $a = 1.14\cdot 10^{-13}$ m
$E_0 = 10$ MeV: $a = 1.14\cdot 10^{-14}$ m

The radius of a gold nucleus (cf. Sect. 4.2.5):

$$R = (1.3\pm 0.1)\,A^{1/3}\cdot 10^{-15}\,\mathrm{m} = 7.6\cdot 10^{-15}\,\mathrm{m}$$

i.e. the proton "touches" the nucleus for $E_0 = 15$ MeV.

4.4 It follows from (4.12) that $\theta = 2\,\mathrm{arccot}\frac{bmv_0^2}{k} = 12.49°$.

4.5 It follows from (4.12) that $b = 2.16\cdot 10^{-13}$ m.

4.6 Sketch:

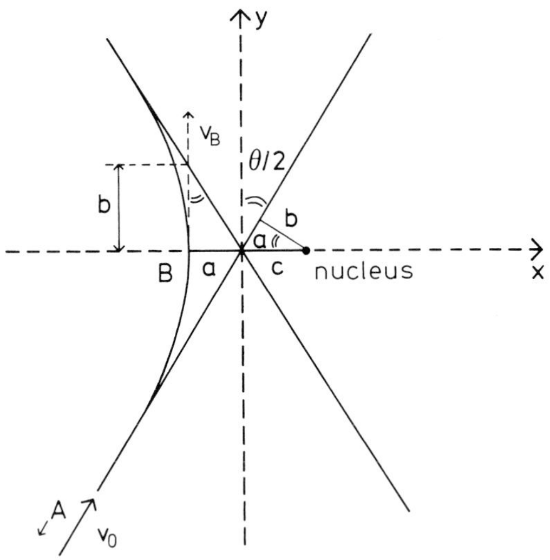

a) Integration of (4.8) from infinity (Point A) to Point B, the closest approach, with

$$\phi_{\mathrm{B}} = \frac{180° - \theta}{2} \quad \text{and} \quad v_{\perp\mathrm{B}} = \frac{v_0 b}{r_{\mathrm{B}}}\cos\phi_{\mathrm{B}}$$

yields the distance

$$r_{\mathrm{B}} = \frac{m v_0 \cos\phi_{\mathrm{B}}}{k(1-\cos\phi_{\mathrm{B}})}\,b^2\ ; \qquad k = \frac{27\,e^2}{4\pi\varepsilon_0}\ .$$

Equation (4.12) leads to

$$r_{\mathrm{B}} = \frac{k}{2E_0}\left(1 + \frac{1}{\sin\theta/2}\right) = r_{\mathrm{Al}} + r_\alpha\ .$$

The radius of the Al nucleus is thus $r_{\mathrm{Al}} = 2.7\cdot 10^{-15}$ m.

b) From energy conservation,

$$E_0 = \frac{1}{2}m v_{\mathrm{B}}^2 + \frac{k}{r_{\mathrm{B}}}\ ; \qquad (1)$$

from momentum conservation, $v_0 b = r_{\mathrm{B}} v_{\mathrm{B}}$ (2)

Equation (4.12) $b = \frac{k}{2E_0}\cot\theta/2\ .$ (3)

Elimination of v_{B} and b from (1 – 3) yields

$$r_B = \frac{k}{2E_0}\left(1 + \frac{1}{\sin\theta/2}\right)$$

c) hyperbolic orbits $\frac{x^2}{a^2} - \frac{y^2}{b^2} = 1$.

From the sketch, we see that $r_B = a + c = a + \sqrt{a^2 + b^2}$; with $b = (2Ze^2/4\pi\varepsilon_0 m v_0^2)\cot\theta/2$ (Equation 4.12) and $a = b\tan\theta/2$ we find

$$r_B = \frac{k}{2E_0}\left(1 + \frac{1}{\sin\theta/2}\right)$$

4.7 a) The Rutherford scattering formula for protons (cf. 4.20):

$$\left(\frac{dn}{n\,d\Omega}\right) = \frac{Z^2 e^4 DN}{(16\pi\varepsilon_0)^2 E_0^2 \sin^4(\theta/2)} \quad (1)$$

Integrating over the angle ϕ ($d\Omega = \sin\theta\, d\theta\, d\phi$) gives

$$\left(\frac{dn}{n\,d\theta}\right) = \frac{Z^2 e^4 DN\cos(\theta/2)}{64\pi\varepsilon_0^2 E_0^2 \sin^3(\theta/2)} \quad (2)$$

i.e. $E_0 = 5.62 \cdot 10^{-13}$ J = 3.51 MeV.

b) Setting

$$\frac{dn/n}{d\Omega} = ND\frac{da(\theta)}{d\Omega}$$

in (1), we obtain the differential cross section

$$\frac{da}{d\Omega} = 1.052 \cdot 10^{-27}\,\mathrm{m}^2 = 10.52\ \mathrm{barn} .$$

c) From (4.12), with $k = k_p = Ze^2/4\pi\varepsilon_0$, we obtain

$$b = 2.81 \cdot 10^{-14}\,\mathrm{m} .$$

5.1 $E_{photon} = hc/\lambda = m_{ph}c^2 \rightarrow m_{ph} = h/\lambda c$

5.2 $p_{photon} = mc = h/\lambda = E_{photon}/c$

$E_{photon} = 1\ \mathrm{eV} \rightarrow p_{photon} = 5.34 \cdot 10^{-28}\ \mathrm{kg\,m\,s^{-1}}$

$\lambda = 12400\ \text{Å} = 1.24 \cdot 10^{-6}\,\mathrm{m}$

5.3 Mass reduction: $m = E/c^2 = Pt/c^2$ $= 3.504 \cdot 10^{-8}$ kg

5.4 a) $E_{kin,e^-} = E_{kin,e^+} = \frac{1}{2}E_\gamma - m_{e^-}c^2 = 0.489$ MeV $= 7.83 \cdot 10^{-14}$ J

b) $E = mc^2 = \dfrac{m_0 c^2}{\sqrt{1 - v^2/c^2}}$

$$\rightarrow \frac{v}{c} = \sqrt{1 - \left(\frac{m_0 c^2}{E}\right)^2} = 0.86$$

5.5 $E_\gamma = hc/\lambda = 1.136 \cdot 10^{-18}$ J = 7.09 eV.

This corresponds to the binding energy of the O_2 molecule.

5.6 Photon energy:

$$E_\gamma = hc/\lambda = 3.31 \cdot 10^{-19}\ \mathrm{J} = 2.07\ \mathrm{eV} .$$

Photon flux for a radiant power of $P = 1.8 \cdot 10^{-18}$ W:

$$\frac{dN}{dt} = \frac{P}{E_\gamma} = 5.4\ \mathrm{s}^{-1}$$

5.7

Frequency ν [Hz]	10^3	$2.42 \cdot 10^{21}$
Photon energy $h\nu$ [eV]	$4.14 \cdot 10^{-12}$	10^7
Number of photons per m^2 and second	$1.51 \cdot 10^{30}$	$6.2 \cdot 10^{11}$

5.8 Photon momentum $p_\gamma = m_\gamma c = E_\gamma/c$.

The radiation pressure for perpendicular incidence and complete absorption:

$$P_{rad} = \frac{\Delta p_\gamma}{\Delta t\,\Delta A} = \frac{\Delta E_\gamma}{c\,\Delta t\,\Delta A} = 4.67 \cdot 10^{-6}\ \mathrm{kg/ms^2}$$

$$\rightarrow \frac{P_{rad}}{P_{atm}} = 4.61 \cdot 10^{-11} .$$

The force on an area of 1 m^2:

$$F = P_{rad}A = 4.67 \cdot 10^{-6}\ \mathrm{N} .$$

Upon complete reflection, the momentum transfer is twice as large; P_{rad} and F are doubled.

5.9 a) Momentum conservation:

$$\frac{E_\gamma}{c} = \frac{h\nu}{c} = \sqrt{2 M_{atom} E_{atom}} \rightarrow E_{atom} = \frac{h^2}{2 M \lambda^2} \ .$$

b) Mercury:

$$E_{atom} = 1.02 \cdot 10^{-29}\,\mathrm{J} = 6.4 \cdot 10^{-11}\,\mathrm{eV} \ .$$

From (7.29) the energy-time uncertainty relation:
$\Delta E \cdot \Delta t \approx h$.
With $\tau = \Delta t = 10^{-8}$ s, we have
$\Delta E \approx 6.6 \cdot 10^{-26}\,\mathrm{J} \gg E_{atom}$.

→ No effect on the position of the line, reabsorption of the photon by other mercury atoms is possible.

c) Nickel:

$$E_{atom} = 2.59 \cdot 10^{-18}\,\mathrm{J} = 16.2\,\mathrm{eV} \ .$$

With $\tau = \Delta t = 10^{-14}$ s, we have
$\Delta E \approx 6.6 \cdot 10^{-20}\,\mathrm{J} \ll E_{atom}$.

→ Noticeable line shift, reabsorption of the γ quantum by other (stationary) Ni atoms is not possible.

5.10 The Stefan-Boltzmann law gives the power radiated by the sphere:

$$P = \pi d^2 \sigma T^4 \ , \quad T = 487\,\mathrm{K} \quad \text{or} \quad 214\,°\mathrm{C} \ .$$

Annual loss of mass

$$\Delta m = \frac{\Delta E}{c^2} = \frac{Pt}{c^2} = 3.5 \cdot 10^{-8}\,\mathrm{kg} \ .$$

5.11 The intensity of radiation at the sun's surface I_S is

$P = I_E 4\pi r^2 = I_S 4\pi R^2$ and therefore
$I_S = 6.4 \cdot 10^7\,\mathrm{W/m^2}$.

From the Stefan-Boltzmann law (5.3), we have for the temperature

$$T = \sqrt[4]{\frac{I_S}{\sigma}} = 5800\,\mathrm{K} \ .$$

Integration of the spectral energy density $u(\nu, T)\,d\nu$ from (5.6) gives the total energy density in the interior of a black body:

$$U = \int_0^\infty u(\nu)\,d\nu = \frac{8\pi^5 k^4}{15 c^3 h^3} T^4 = 0.854\,\mathrm{J/m^3} \ .$$

The energy radiated from the sun is generated in the sun's core and transported outwards. Therefore, I_S gets larger and larger towards the interior, and T and U increase strongly.

5.12 Using Wien's displacement law (5.4)

$$\lambda_{max} T = 0.29\,\mathrm{cm\,K} \ , \quad \lambda_{max} = 9.67\,\mu\mathrm{m} \ .$$

Monochromatic energy density (5.6)

$$u(\nu, T) = \frac{8\pi h \nu^3}{c^3} \frac{1}{e^{h\nu/kT} - 1} = 1.3 \cdot 10^{-19}\,\mathrm{Js/m^3} \approx 0.8\,\mathrm{eV\,s/m^3} \ .$$

5.13 $h\nu = mv^2/2 + eV_A = 4\,\mathrm{eV}$, $\lambda = c/\nu = 310\,\mathrm{nm}$

5.14 a) Braking voltage from (5.28):

$$V_{max} = \frac{h\nu}{e} - V_A = \frac{hc}{\lambda e} - V_A = 1.2\,\mathrm{V} \ .$$

b) $E_{kin}^{max} = eV_{max} = 1.2\,\mathrm{eV} = 1.92 \cdot 10^{-19}\,\mathrm{J}$,

$$v^{max} = \sqrt{\frac{2E_{kin}}{m}} = 6.5 \cdot 10^5\,\mathrm{m/s} \ .$$

5.15 a) Assuming that each photon releases an electron, the number of electrons per s and area j_s is

$$j_s = \frac{\Delta N}{\Delta t \cdot A} = \frac{I\lambda}{hc} = 6.04 \cdot 10^9\,(\mathrm{s\,m^2})^{-1} \ .$$

b) The energy absorbed per unit of surface area and unit time is

$$\frac{\Delta W}{\Delta t A} = \frac{\Delta N \cdot eV_A}{\Delta t \cdot A} = j_s e V_A$$
$$= 1.208 \cdot 10^{10}\,\mathrm{eV/m^2 s} \ .$$

c) $E_{kin} = h\nu - eV_A = 1.77 \cdot 10^{-19}\,\mathrm{J} = 1.1\,\mathrm{eV}$.

5.16 Table of values:

λ [nm]	ν [10^{14} Hz]	V [V]
366	8.20	1.48
405	7.41	1.15
436	6.88	0.93
492	6.10	0.62
546	5.50	0.36
579	5.18	0.24

Equation (5.27): $V = \frac{h\nu}{e} - V_A$.

Fitting using the least-squares method gives

$$\frac{h}{e} = (4.111 \pm 0.012) \cdot 10^{-15}\,\mathrm{Js/C}$$
$$V_A = (1.8939 \pm 0.0076)\,\mathrm{V} \ .$$

a) $V = 0$: $\quad \nu_{lim} = \frac{V_A}{h/e} = (4.607 \pm 0.032) \cdot 10^{14}\,\mathrm{Hz}$

b) $W_A = eV_A = (3.034 \pm 0.012) \cdot 10^{-19}\,\mathrm{J}$

c) $h/e = (4.111 \pm 0.012) \cdot 10^{-15}\,\mathrm{Js/C}$.

5.17 Na D lines: $\lambda_D \approx 589\,\mathrm{nm}$

$$\rightarrow h\nu = \frac{hc}{\lambda} = eV \quad \text{with} \quad V = 2.11\,\mathrm{V}$$
$$\rightarrow h/e = \frac{V\lambda}{c} = 4.14 \cdot 10^{-15}\,\mathrm{Js/C} \ .$$

5.18 Calculation of the kinetic energy of an electron from the orbital data:

$$evB = m_0 \frac{v^2}{R} \rightarrow E_{e^-} = \frac{1}{2} m_0 v^2 = \frac{e^2 B^2 R^2}{2\,m_0}$$
$$\approx 7.9\,\mathrm{keV} \ . \quad (1)$$

From Problem 5.19 we have

$$E_{e^-} = \Delta E_\gamma = E\frac{\xi(1-\cos\theta)}{1+\xi(1-\cos\theta)} \ ; \ \xi = \frac{E}{m_0 c^2} \ . \quad (2)$$

From (1) and (2) it follows that

$$\xi = \frac{A + \sqrt{A^2 + 8A}}{2} \approx 0.184 \ ; \quad A = \frac{e^2 B^2 R^2}{2\,m_0^2 c^2} \ .$$

$\rightarrow$ Energy and wavelength of the incident photon are

$$E = m_0 c^2 \xi = 94\,\mathrm{keV} \ , \quad \lambda = hc/E = 0.132\,\text{Å} \ .$$

5.19 Energy and momentum conservation:

$$E - E' = E_e \quad (1)$$
$$\frac{E}{c} - \frac{E'}{c}\cos\theta = p_e \cos\Theta \quad (2)$$
$$\frac{E'}{c}\sin\theta = p_e \sin\Theta \ . \quad (3)$$

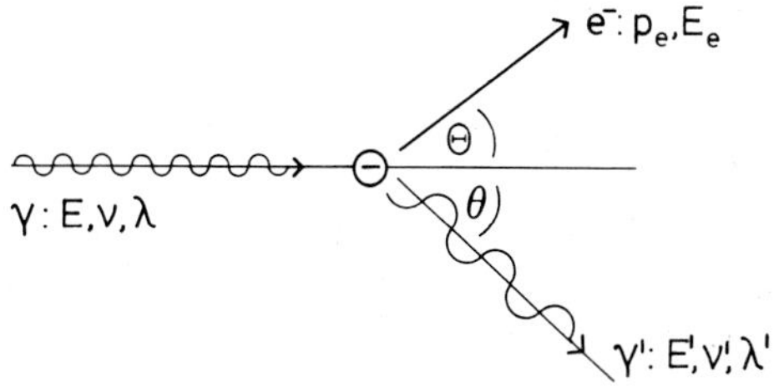

Relativistic energy-momentum relation:

$$p_e^2 = \frac{1}{c^2}(E_e^2 + 2\,m_0 c^2 E_e) \ . \quad (4)$$

a) $(1)^2 + (2)^2 - (1)^2/c^2$ and (4) yield

$$E' = \frac{E}{1+\xi(1-\cos\theta)} \ ; \quad \xi = \frac{E}{m_0 c^2} \quad (5)$$

thus, the energy loss of the photon is

$$\Delta E_\gamma = E - E' = E\frac{\xi(1-\cos\theta)}{1+\xi(1-\cos\theta)} = 96.8\,\mathrm{eV}$$

and the change in its frequency and wavelength is

$$\Delta\nu = \frac{\Delta E_\gamma}{h} = 2.34 \cdot 10^{16}\,\mathrm{Hz} \ ,$$

$$\Delta\lambda = \frac{h}{m_0 c}(1-\cos\theta) = 1.21 \cdot 10^{-12}\,\mathrm{m} \ .$$

b) $(2)^2 + (3)^2$ and (5) yield

$$p_e^2 = m_0^2 c^2 \frac{\xi^2(1-\cos\theta)\,[2+\xi(2+\xi)(1-\cos\theta)]}{(1+\xi(1-\cos\theta))^2} \ .$$

This gives for the electron's momentum $p_e = 5.31 \cdot 10^{-24}$ kg m/s, and for its kinetic energy,

$$E_e = \Delta E_\gamma = 96.8\,\mathrm{eV} \ .$$

Equations (2), (3) and (5) give

$$\cot\theta = (\xi+1)\sqrt{\frac{1-\cos\Theta}{1+\cos\Theta}} \ ,$$

leading to $\theta = 59.5°$.

5.20 (Compare Problem 5.19)

a) $\Delta\lambda = \dfrac{h}{m_0 c}(1-\cos\theta) = 0.0243\,\text{Å}$.

b) $E_{\mathrm{kin}} = \Delta E_\gamma = E\dfrac{(E/m_0 c^2)(1-\cos\theta)}{1+(E/m_0 c^2)(1-\cos\theta)}$;

$E = \dfrac{hc}{\lambda}$; $\rightarrow E_{\mathrm{kin}} = 293.7\,\mathrm{eV}$.

c) $\dfrac{\Delta E_\gamma}{E}(1\,\text{Å}, 90°) = \dfrac{293.7\,\mathrm{eV}}{12409\,\mathrm{eV}} = 2.37\%$

d) $\dfrac{\Delta E_\gamma}{E}(0.1\,\text{Å}, 90°) = 19.5\%$.

6.1 a) For levitation equilibrium

$$F\!\uparrow\, = F_{\mathrm{electr.}} = e\frac{V}{d} = \frac{4}{3}\pi r^3 \varrho g = F_{\mathrm{grav.}} = F\!\downarrow \ , \quad (1)$$

giving

$$r = \sqrt[3]{\frac{3\,\mathrm{eV}}{4\pi d\varrho g}} = 2.79 \cdot 10^{-7}\,\mathrm{m} \ .$$

Motion after field reversal

$$F\!\uparrow\, = F_{\mathrm{visc.}} = 6\pi\eta r v = e\frac{V}{d} + \frac{4}{3}\pi r^3 \varrho g$$
$$= F_{\mathrm{electr.}} + F_{\mathrm{grav.}} = F\!\downarrow \ , \quad (2)$$

giving

$$v = \frac{1}{6\pi\eta r}\left(e\frac{V}{d} + \frac{4}{3}\pi r^3 \varrho g\right)$$
$$= 1.67 \cdot 10^{-5}\,\mathrm{m/s} \ .$$

b) The correction introduces an additional term $1/(1+A\lambda/r)$ on the left side of (2). Eliminating r from (1) and (2) yields

$$e = \frac{\pi d\varrho g}{6V}\left(-A\lambda + \sqrt{A^2\lambda^2 + \frac{9\eta v}{\varrho g}}\right)^3 \ .$$

6.2 Radius of the flight path $r = 1.668$ m

Mass of the particle $m = \dfrac{q^2 B^2 r^2}{2E}$

$= 1.86 \cdot 10^{-28}$ kg

Rest energy $E = mc^2 = 104.5$ MeV .

The particle is most probably a muon.

6.3 b) The electrons leave the apparatus without deflection when the midpoints of both condensers are reached just at the moment of opposite zero-crossings of the high frequency field.

Possible frequencies:

$$\nu_n = \sqrt{\frac{e}{m}}\left(n+\frac{1}{2}\right)\frac{\sqrt{2V}}{l}$$

c) $\nu_{\min} = \nu_0 = \sqrt{2V\left(\dfrac{e}{m}\right)}\,\dfrac{1}{2l} = 66.31\,\mathrm{MHz}$.

6.4 $E_{ges} = E_0 + E_{kin} = \dfrac{m_0 c^2}{\sqrt{1 - v^2/c^2}}$;

$E_{kin} = E_0 = m_0 c^2$

$\rightarrow v = \dfrac{\sqrt{3}}{2} c \approx 0.866\, c$

6.5 $\dfrac{m}{m_0} = \dfrac{E_0 + E_{kin}}{E_0} = 1 + \dfrac{E_{kin}}{m_0 c^2} = 2.96$

6.6 $\lambda = \dfrac{h}{p} = \dfrac{h}{m_0 c} \dfrac{\sqrt{1 - v^2/c^2}}{v/c} = 1.818 \cdot 10^{-12}\,\mathrm{m}$

6.7 Relativistic relationship $\lambda(E_{kin})$:

$$\lambda = \frac{hc}{E_0} \frac{1}{\sqrt{K^2 + 2K}} \;;\quad K = \frac{E_{kin}}{E_0} .$$

Scattering angle α in the 1st order from Bragg's Law:

$\lambda = 2d \sin\dfrac{\alpha}{2}$.

E_{kin} [eV]	1	10^2	10^3	10^5
λ [m]	$1.227 \cdot 10^{-9}$	$1.227 \cdot 10^{-10}$	$3.877 \cdot 10^{-11}$	$3.702 \cdot 10^{-12}$
α [°]	–	33.15	10.35	0.99

For scattering angle $\alpha = 30°$: $E_{kin} = 121.4\,\mathrm{eV}$.

6.8 $\bar{E}_{kin}(298\,\mathrm{K}) = \frac{3}{2}\,kT = 6.17 \cdot 10^{-21}\,\mathrm{J} = 38.5\,\mathrm{meV}$

$$\lambda = \frac{h}{\sqrt{2m\bar{E}_{kin}}} = 1.458 \cdot 10^{-10}\,\mathrm{m}$$

Bragg reflection in 1st order at $\theta = 15.0°$.

6.9 $\lambda = \dfrac{h}{p} = \dfrac{h}{\sqrt{2mE_{kin}}}$ (nonrelativistic)

$\lambda(a = 1\,\mathrm{m}) = 3.232 \cdot 10^{-5}\,\mathrm{m}$

$\lambda(a = 0.5 \cdot 10^{-10}\,\mathrm{m}) = 2.287 \cdot 10^{-10}\,\mathrm{m}$

7.1 The normalisation constant N is $N = (2\Delta k)^{-1/2}\pi^{-3/4}$.

After Fourier transformation, one obtains the spatial wavefunction of the free particle for $t > 0$:

$$\psi(x,t) = \sqrt{\frac{m_0 \Delta k}{\sqrt{\pi}\,(m_0 + \mathrm{i}\hbar t \Delta k^2)}} \cdot \exp\left\{-\frac{x^2 m_0 \Delta k^2}{2(m_0 + \mathrm{i}\hbar t \Delta k^2)}\right\} .$$

From it, the probability density for the particle can be calculated to be:

$$|\psi(x,t)|^2 = \frac{m_0 \Delta k}{\sqrt{\pi (m_0^2 + \hbar^2 t^2 \Delta k^4)}} \cdot \exp\left\{-\frac{x^2 m_0^2 \Delta k^2}{(m_0^2 + \hbar^2 t^2 \Delta k^4)}\right\} .$$

The normalisation remains valid for all times, since every solution of a time-dependent Schrödinger equation remains normalised. This can be demonstrated in general by calculating the time dependence of the normalisation integral and using the fact that the wavefunctions obey the Schrödinger equation. If the wavefunction is known as an explicit function of x and t, one could naturally calculate the normalisation integral for any arbitrary time; this method, however, demands considerably more calculational effort. The probability density is a Gaussian function having a width at arbitrary time t of:

$$\Delta x(t) = \frac{1}{m_0 \Delta k} \sqrt{m_0^2 + \hbar^2 t^2 \Delta k^4} \;;$$

thus, the wave packet spreads out with time.
The factor $N^2 = \exp(-(k/\Delta k)^2)$ is the probability density in k-space, i.e. it denotes the probability of finding the particle with a momentum $= \hbar k$ as a result of a momentum measurement.

7.2 $\Delta x = \sqrt{\dfrac{1}{(\Delta k)^2} + \dfrac{\hbar^2 t^2 (\Delta k)^2}{m_0^2}}$.

For $t = 0$ we have: $\Delta k = 10^8\,\mathrm{cm}^{-1}$

$$t = \frac{m_0}{\hbar \Delta k} \sqrt{(\Delta x)^2 - \frac{1}{(\Delta k)^2}}$$

$t \approx m_0 \cdot 1.5 \cdot 10^{32}\,\mathrm{s}$

$\rightarrow t \approx 1.3 \cdot 10^4\,\mathrm{s}$ for an electron.

7.3 In three dimensions, we use a trial wavefunction for the particle consisting of the product of the three one-dimensional wavefunctions from Problem 7.1:

$$\Psi(r, t) = \psi(x, t)\,\psi(y, t)\,\psi(z, t) \ .$$

This is correct for arbitrary times, since the Hamiltonian is the sum of three one-dimensional operators, each of which commutes with the others:

$$\mathscr{H} = -\frac{\hbar^2}{2m_0}\left(\frac{d^2}{dx^2} + \frac{d^2}{dy^2} + \frac{d^2}{dz^2}\right) .$$

The three-dimensional Schrödinger equation thus is decomposed into three mutually independent, one-dimensional Schrödinger equations, whose solutions were already found in Problem 7.1. Thus, the wave packet spreads out in each of the three spatial directions:

$$\Delta x(t) = \frac{1}{m_0 \Delta k_x} \sqrt{m_0^2 + \hbar^2 t^2 \Delta k_x^4} \ ,$$

$$\Delta y(t) = \frac{1}{m_0 \Delta k_y} \sqrt{m_0^2 + \hbar^2 t^2 \Delta k_y^4} \ ,$$

$$\Delta z(t) = \frac{1}{m_0 \Delta k_z} \sqrt{m_0^2 + \hbar^2 t^2 \Delta k_z^4} \ .$$

8.1 $E_{\mathrm{kin}} = 1.387 \cdot 10^{-26}\,\mathrm{J} = 8.657 \cdot 10^{-8}\,\mathrm{eV}$

$v = 4.07\,\mathrm{ms}^{-1}$

8.2

λ [Å]	$\tilde{\nu}$ [cm^{-1}]	$\tilde{\nu}/R_{\mathrm{H}}$	n
3669.42	27252.3	0.24848	>20
3770.06	26524.8	0.24184	>10
3835.40	26072.9	0.23772	9
3970.07	25188.5	0.22966	7
4340.47	23039.0	0.21006	5

8.3 Practically all the atoms are in their ground states before the absorption; one obtains the lines of the Lyman series

$$E_\gamma = hc\tilde{\nu} = hcR_{\mathrm{H}}\left(1 - \frac{1}{n^2}\right) ,$$

$$\lambda = \frac{1}{\tilde{\nu}} = \frac{1}{R_{\mathrm{H}}\left(1 - \frac{1}{n^2}\right)} \ ; \quad n = 2, 3, \ldots .$$

8.4 a) Diffraction of mth order by a grating: $\lambda = (d/m)\sin\theta$ gives the wavelength of the Balmer line, $\lambda = 486.08$ nm and the initial state level $n = 4$.

b) The Balmer lines from the $n = 32$ and $n = 31$ states must still be able to be separated, giving the necessary resolution A:

$$A = \frac{\tilde{\nu}(n=32)}{\tilde{\nu}(n=32) - \tilde{\nu}(n=31)} = 3890 = m \cdot p \ ,$$

where m is the diffraction order and p the number or rulings of the grating; the latter must thus be at least 3890.

8.5 From Sect. 8.4 we have

$$E_{\mathrm{kin}} = \frac{a}{2r_n} \ ; \quad E_{\mathrm{pot}} = -\frac{a}{r_n} \ ; \quad a = \frac{Ze^2}{4\pi\varepsilon_0} \ ;$$

therefore $E_{\mathrm{kin}} \neq |E_{\mathrm{pot}}|$.

Total energy $E_n = E_{\mathrm{pot}} + E_{\mathrm{kin}} = -(a/2r_n)$.
This amount of energy is released, e.g. by the emission of photons, when a Bohr atom is formed.

$$\frac{E_{\mathrm{pot}}}{E_{\mathrm{kin}}} = \frac{-a/r_n}{a/2r_n} = -2 \quad \text{independently of } n.$$

8.6 a) $F_z = m\dfrac{v^2}{r} = G\dfrac{Mm}{r^2}$ (1)

b) $E_{\mathrm{kin}} = \dfrac{1}{2}mv^2 = \dfrac{1}{2}G\dfrac{Mm}{r}$

c) $E_{\mathrm{pot}} = -\dfrac{GMm}{r}$

d) $E_{tot} = E_{pot} + E_{kin} = -G\dfrac{Mm}{2r}$

e) Quantisation of orbital angular momentum

$$L = mvr = n\hbar \qquad (2)$$

f) (1) yields

$$v^2 r = GM \qquad (3)$$

(2) and (3) lead to

$$v_n = \frac{GMm}{\hbar} \cdot \frac{1}{n} = 9.654 \cdot 10^{-34}\,\mathrm{m/s} \cdot \frac{1}{n}$$

$$r_n = \frac{\hbar^2}{GMm^2} \cdot n^2 = 1.199 \cdot 10^{29}\,\mathrm{m} \cdot n^2 \ .$$

8.7 For a ring current:

$$I = -\frac{e\omega}{2\pi} = -1.055 \cdot 10^{-3}\,\mathrm{A} \cdot \frac{1}{n^3}$$

and the magnetic moment

$$|\mu| = \frac{1}{2} e\omega r^2 = \frac{e}{2m_0}|l| = \frac{e\hbar}{2m_0} n = \mu_B n$$

(compare Sect. 12.2).

n	I [A]	$\lvert\mu\rvert$ [Am²]
1	$1.055 \cdot 10^{-3}$	$9.274 \cdot 10^{-24}$
2	$1.318 \cdot 10^{-4}$	$1.855 \cdot 10^{-23}$
3	$3.905 \cdot 10^{-5}$	$2.782 \cdot 10^{-23}$

8.8

$$\frac{\omega}{2\pi} = \frac{me^4}{64\pi^3 \varepsilon_0^2 \hbar^3} \cdot \frac{1}{n^3} = 3.288 \cdot 10^{15}\,\mathrm{s}^{-1} \cdot \frac{1}{n^3}$$

$$r = \frac{8\pi\varepsilon_0 \hbar^2}{me^2} n^2 = 1.059 \cdot 10^{-10}\,\mathrm{m} \cdot n^2$$

$$E_n = -\frac{me^4}{64\pi^2 \varepsilon_0^2 \hbar^2} \cdot \frac{1}{n^2} = -6.80\,\mathrm{eV} \cdot \frac{1}{n^2}$$

where r is the distance between the e^- and e^+. (For the calculation, see Sect. 8.4.)

8.9 a) $E_n(\mu) = -\dfrac{Z^2 e^4 m_\mu}{32\pi^2 \varepsilon_0^2 \hbar^2} \cdot \dfrac{1}{n^2} = -2813\,\mathrm{eV} \cdot \dfrac{Z^2}{n^2}$

b) $r_n = \dfrac{4\pi\varepsilon_0 \hbar^2}{Ze^2 m_\mu} n^2 = 2.56 \cdot 10^{-3}\,\text{Å} \cdot \dfrac{n^2}{Z}$

c) $h\nu = E_2(\mu) - E_1(\mu) = 2110\,\mathrm{eV} \cdot Z^2$

8.10 With ω_n from (8.9) and $\Delta t = 10^{-8}\,\mathrm{s}$, we have

$$N = \frac{\Delta t}{2\pi} \cdot \omega = 6.583 \cdot 10^7 \cdot \frac{1}{n^3} \ .$$

a) $n = 2$ gives $N = 8.228 \cdot 10^6$;

$$K = \frac{4.5 \cdot 10^9}{N} = 547$$

b) $n = 15$ gives $N = 1.950 \cdot 10^4$;

$K = 230\,800$.

8.11 Pickering series: $\tilde{\nu}_p = 4\,R_{He}\left(\dfrac{1}{4^2} - \dfrac{1}{n^2}\right)$

Rydberg constant: $R_{He} = \dfrac{R_\infty}{1 + \dfrac{m_0}{M_{He}}}$;

1st line: $n = 5$; $\Delta\tilde{\nu} = 0.443\,\mathrm{cm}^{-1}$;

$\Delta E = 5.5 \cdot 10^{-5}\,\mathrm{eV}$

3rd line: $n = 7$; $\Delta\tilde{\nu} = 0.828\,\mathrm{cm}^{-1}$;

$\Delta E = 1.03 \cdot 10^{-4}\,\mathrm{eV}$.

8.12 Hydrogen-like atoms:

$$\tilde{\nu} = R\left(\frac{1}{n'^2} - \frac{1}{n^2}\right) ; \quad n' < n \ .$$

$R_H = 109\,677.581\,\mathrm{cm}^{-1}$;

$R_{He} = 109\,722.398\,\mathrm{cm}^{-1}$.

We find $400\,\mathrm{nm} < \lambda < 700\,\mathrm{nm}$ for

Hydrogen	with $n = 3,4,5,6$	i.e. $n' = 2$
$Helium^+$	with $n = 4$	$n' = 3$
	$n = 6 \ldots 13$	$n' = 4$
	$n = 12 \ldots \infty$	$n' = 5$.

8.13 The relative deviation γ between the Sommerfeld energy formula $E_{n,k}$ (8.29) and the Bohr energy formula E_n in the hydrogen atom for $n = 2$:

$$\gamma = \frac{E_{2,k} - E_2}{E_2} = \frac{\alpha^2}{4}\left(\frac{2}{k} - \frac{3}{4}\right)$$

k	1	2
γ	$1.67 \cdot 10^{-5}$	$3.33 \cdot 10^{-6}$

8.14 Wavelength λ_2 of the second laser

$$\lambda_2 = \frac{hc}{E_1(\mathrm{H})\left(\frac{1}{n^2} - 1\right) - 11.5\ \mathrm{eV}}$$

n_2	E_{n_2} [eV]	r_{n_2} [Å]	λ_2 [nm]
20	$-3.40 \cdot 10^{-2}$	212	603.5
30	$-1.51 \cdot 10^{-2}$	477	598.0
40	$-0.85 \cdot 10^{-2}$	848	596.1
50	$-0.54 \cdot 10^{-2}$	1325	595.3

Required linewidth $\Delta E \approx E_{51} - E_{50} \approx 2 \cdot 10^{-4}$ eV.

8.15 a) $\nu_n = \dfrac{\omega_n}{2\pi} = \dfrac{e^4 m_0}{32\pi^3 \varepsilon_0^2 \hbar^3 n^3} = \dfrac{6.58 \cdot 10^{15}\ \mathrm{s}^{-1}}{n^3}$

b) $$\nu_{n\to n-1} = \frac{e^4 m_0}{32\pi^3 \varepsilon_0^2 \hbar^3 n^3}\left(\frac{1 - \frac{1}{2n}}{1 - \frac{2}{n} + \frac{1}{n^2}}\right)$$

c) $\lim\limits_{n\to\infty} \nu_{n\to n-1} = \nu_n$

(compare Sect. 8.11: correspondence principle)

8.16 The motion of the nucleus can be taken into account by replacing the electron's mass m_0 with the reduced mass

$\mu = m_0 M/(m_0 + M)$ (M = nuclear mass) .

$$E_n(A, Z) = \frac{E_1(\mathrm{H})}{n^2}\,\frac{1}{1 + \frac{1}{A}\cdot\frac{m_0}{m_\mathrm{p}}}\,Z^2\ ;$$

$$\frac{m_0}{m_\mathrm{p}} = \frac{1}{1836.15}$$

Atom	$\dfrac{\Delta E(Z,n)}{E(Z,n)}$
^{1}H	$5.45 \cdot 10^{-4}$
^{2}H	$2.75 \cdot 10^{-4}$
^{3}H	$1.82 \cdot 10^{-4}$
$^4He^+$	$1.36 \cdot 10^{-4}$
$^7Li^{2+}$	$0.78 \cdot 10^{-4}$

8.17 $n = 1 \to n = 2$; $n = 1 \to n = 3$

9.1 The Schrödinger equation of the force-free particle is:

$$\mathrm{i}\hbar\frac{\partial}{\partial t}\psi(x,t) = -\frac{\hbar^2}{2m_0}\frac{\partial^2}{\partial x^2}\psi(x,t)\ .$$

Both sides of the equation can be calculated by exchanging the differentiations with the integrations in k-space. Substituting the given dispersion relation for the free particle then demonstrates the equality of the left and the right sides.

9.2 a) By substituting $\psi(r, t)$ into (9.32) .

b) We require $\int \psi^* \psi dV = 1$.

Since $\int \phi_j^* \phi_k dV = \delta_{jk}$,

$|C_1|^2 + |C_2|^2 = 1$ must hold ,

or, in general, $\sum\limits_j |C_j|^2 = 1$.

9.3 From the requirement that the wavefunction be normalisable, we find from the time-independent Schrödinger equation:

$$\psi(x) = \begin{cases} N\exp(kx)\ , & x \le 0 \\ N\exp(-kx)\ , & x \ge 0 \end{cases},$$

with $k = \sqrt{-\dfrac{2m_0E}{\hbar^2}}$.

The normalisation factor of the exponential function must be the same on both sides, in order for the wavefunction to remain continuous at $x = 0$. The jump condition is

$$\psi'(\varepsilon) - \psi'(-\varepsilon) = -\frac{2m_0\beta}{\hbar^2}\psi(0) \ .$$

From this, one finally obtains $k = m_0\beta/\hbar^2$, and the energy eigenvalue $E = -(m_0\beta^2/2\hbar^2)$. The normalisation factor is $N = \sqrt{k}$.

9.4
$$\begin{aligned} \psi_{\mathrm{I}} &= A_1 e^{\kappa x} , && x < -L \\ \psi_{\mathrm{II}} &= A_2 \cos(kx) + B_2 \sin(kx) , && -L \le x \le L \\ \psi_{\mathrm{III}} &= B_3 e^{-\kappa x} , && x > L \end{aligned}$$

$$\kappa = \sqrt{\frac{-2m_0E}{\hbar^2}} \ , \quad k = \sqrt{\frac{2m_0(E+V_0)}{\hbar^2}} \ .$$

The requirements that the solutions be continuous and differentiable at $x = \pm L$ yield:

$\kappa = k\tan(kL) \rightarrow E(L, V_0)$ (graphically or numerically) ,

and the two types of solutions:

$A_1 = B_3$, $B_2 = 0$ symmetric solution;

$A_1 = -B_3$, $B_2 \neq 0$ antisymmetric solution .

9.5
$$a = \frac{1}{1 - \dfrac{m_0\beta}{\mathrm{i}\hbar^2 k}} - 1$$

$$b = \frac{1}{1 - \dfrac{m_0\beta}{\mathrm{i}\hbar^2 k}} \ , \quad \text{with} \quad k = \sqrt{\frac{2m_0E}{\hbar^2}} \ .$$

The meanings of a and b:

$|a|^2$ = reflection coefficient,

$|b|^2$ = transmission coefficient.

$$|a|^2 = \frac{1}{\dfrac{\hbar^4k^2}{m_0^2\beta^2} + 1}$$

$$|b|^2 = \frac{1}{\dfrac{m_0^2\beta^2}{\hbar^4k^2} + 1}$$

$|a|^2$ and $|b|^2$ are independent of the sign of β!

9.6 The wall is located at the position $x = 0$. Since $\psi(x)$ must be continuous at $x = 0$, we find $\psi(0) = 0$.
For the wavefunctions, the result is:

$$\psi(x) = \begin{cases} A\sin(k, x) \ , & x \le 0 \\ 0 & x \ge 0 \end{cases} ,$$

with $k = \sqrt{\dfrac{2m_0E}{\hbar^2}}$,

where E is the energy of the particle.

9.7
$$\langle x \rangle = \int \psi^*(x,t)\, x\, \psi(x,t)\, dx = 0$$

$$\langle p \rangle = \int \psi^*(x,t)\left(-\mathrm{i}\hbar\frac{d}{dx}\right)\psi(x,t)\, dx = 0$$

$$\langle E_{\mathrm{kin}} \rangle = \frac{\langle p^2 \rangle}{2m_0} = \int \psi^*(x,t)\left(-\mathrm{i}\hbar\frac{d}{dx}\right)^2 \psi(x,t)\, dx = \frac{\hbar^2(\Delta k)^2}{4m_0}$$

$$\langle x^2 \rangle = \int \psi^*(x,t)\, x^2 \psi(x,t)\, dx = \frac{1}{2(\Delta k)^2} + \frac{\hbar(\Delta k)^2}{2m_0^2}t^2$$

$\langle x^2 \rangle$ illustrates directly the spreading out of the wave packet.

9.8
$$\bar{E} = \frac{1}{\sqrt{\pi}\,\Delta k}\int_{-\infty}^{+\infty} dk \frac{\hbar^2k^2}{2m_0}\exp\{-k^2/(\Delta k)^2\}$$

9.9
$$\begin{aligned}[\hat{l}_x, \hat{l}_y] &= -\hbar^2\left\{\left(y\frac{\partial}{\partial z} - z\frac{\partial}{\partial y}\right)\left(z\frac{\partial}{\partial x} - x\frac{\partial}{\partial z}\right) - \left(z\frac{\partial}{\partial x} - x\frac{\partial}{\partial z}\right)\left(y\frac{\partial}{\partial z} - z\frac{\partial}{\partial y}\right)\right\} \\ &= -\hbar^2\left\{y\frac{\partial}{\partial x} - x\frac{\partial}{\partial y}\right\} = \mathrm{i}\hbar\hat{l}_z\end{aligned}$$

$[\hat{l}_y, \hat{l}_z]$ and $[\hat{l}_z, \hat{l}_x]$ analogously.

$$[\hat{l}^2, \hat{l}_x] = [\hat{l}_x^2, \hat{l}_x] + [\hat{l}_y^2, \hat{l}_x] + [\hat{l}_z^2, \hat{l}_x]$$
$$= \hat{l}_y[\hat{l}_y, \hat{l}_x] + [\hat{l}_y, \hat{l}_x]\hat{l}_y + \hat{l}_z[\hat{l}_z, \hat{l}_x] + [\hat{l}_z, \hat{l}_x]\hat{l}_z$$
$$= -\mathrm{i}\hbar\hat{l}_y\hat{l}_z - \mathrm{i}\hbar\hat{l}_z\hat{l}_y + \mathrm{i}\hbar\hat{l}_z\hat{l}_y + \mathrm{i}\hbar\hat{l}_y\hat{l}_z$$

$[\hat{l}^2, \hat{l}_y]$ and $[\hat{l}^2, \hat{l}_z]$ analogously.

9.10 $[\hat{l}_x, x] = 0$ and $[\hat{l}_x, V(r)] = 0$

9.11 a) $(ab)^* = a^* b^*$ with x and dx real.

b)
$$\int_{-\infty}^{\infty} \psi_1^* \frac{\hbar}{\mathrm{i}} \frac{\partial}{\partial x} \psi_2 dx = \left[\frac{\hbar}{\mathrm{i}} \psi_1^* \psi_2\right]_{-\infty}^{\infty} - \int_{-\infty}^{\infty} \psi_2 \frac{\hbar}{\mathrm{i}} \frac{\partial}{\partial x} \psi_1^* dx = \left(\int_{-\infty}^{\infty} \psi_2^* \frac{\hbar}{\mathrm{i}} \frac{\partial}{\partial x} \psi_1 dx\right)^*$$

c)
$$\int_{-\infty}^{\infty} \psi_1^* \frac{\partial^2}{\partial x^2} \psi_2 dx = \left[\psi_1^* \frac{\partial}{\partial x} \psi_2\right]_{-\infty}^{\infty} - \int_{-\infty}^{\infty} \left(\frac{\partial}{\partial x}\psi_1^*\right)\left(\frac{\partial}{\partial x}\psi_2\right) dx$$
$$= \left[-\frac{\partial}{\partial x}(\psi_1^*)\psi_2\right]_{-\infty}^{\infty} + \int_{-\infty}^{\infty} \psi_2 \frac{\partial^2}{\partial x^2} \psi_1^* dx = \left(\int_{-\infty}^{\infty} \psi_2^* \frac{\partial^2}{\partial x^2} \psi_1 dx\right)^*$$

9.12
$$\frac{d}{dt}\langle x \rangle = \int \dot{\psi}^* x \psi\, dx + \int \psi^* x \dot{\psi}\, dx$$

with
$$\dot{\psi} = \frac{1}{\mathrm{i}\hbar}\left(-\frac{\hbar^2}{2m_0}\frac{d^2}{dx^2} + V\right)\psi$$

Substituting:

$$\frac{\mathrm{i}\hbar}{2m_0}\int dx \left(\psi^* x \frac{d^2}{dx^2}\psi - \left(\frac{d^2}{dx^2}\psi^*\right) x\psi\right)$$

partial integration:

$$\frac{\mathrm{i}\hbar}{2m_0}\int dx \left(\psi \frac{d}{dx}\psi^* - \left(\frac{d}{dx}\psi\right)\psi^*\right)$$

partial integration:

$$\frac{1}{m_0}\int dx\, \psi^* p \psi = \frac{1}{m_0}\langle p \rangle$$

$$\frac{d}{dt}\langle p \rangle = -\mathrm{i}\hbar\int \dot{\psi}^* \frac{d}{dx}\psi\, dx - \mathrm{i}\hbar\int \psi^* \frac{d}{dx}\dot{\psi}\, dx$$

Substitute in the Schrödinger equation and integrate by parts two times:

$$\frac{d}{dt}\langle p \rangle = \int dx\, \psi^* \left(-\frac{d}{dx}V\right)\psi = -\left\langle \frac{dV}{dx} \right\rangle.$$

In three dimensions, we have for each component i:

$$m_0\frac{d}{dt}\langle x_i \rangle = \langle p_i \rangle\ , \quad \frac{d}{dt}\langle p_i \rangle = -\left\langle \frac{dV}{dx_i} \right\rangle.$$

9.13 The potential is $V(x) = \frac{1}{2}m_0\omega^2(x - x_0)^2 - \varepsilon_0$,

with $x_0 = \frac{k_0}{k}$ and $\varepsilon_0 = \frac{1}{2}\frac{k_0^2}{k}$.

The time-independent Schrödinger equation is then

$$\left[-\frac{\hbar^2}{2m_0}\frac{d^2}{dx^2} + \frac{1}{2}m_0\omega^2(x - x_0)^2 - \varepsilon_0\right]\psi(x) = E\psi(x)\ .$$

Employing the transformations $y = x - x_0$ and $\hat{E} = E + \varepsilon_0$, we obtain the well-known form

$$\left[-\frac{\hbar^2}{2m_0}\frac{d^2}{dy^2} + \frac{1}{2}m_0\omega^2 y^2\right]\psi(y) = \hat{E}\psi(y)\ .$$

The solutions of this equation are transformed back, and we find for the eigenfunctions:

$$\psi_n(x) = \left(\frac{m_0\omega}{\hbar}\right)^{1/4} \exp\left[-\frac{m_0\omega}{2\hbar}(x - x_0)^2\right]$$

$$\cdot H_n\left[\sqrt{\frac{m_0\omega}{\hbar}}(x-x_0)\right] ,$$

with eigenvalues:

$$E_n=\left(n+\frac{1}{2}\right)\hbar\omega-\frac{1}{2}\frac{k_0^2}{k} .$$

9.14 $b=\frac{1}{\sqrt{2}}\left(\frac{\partial}{\partial\xi}+\xi\right)$; $b^+=\frac{1}{\sqrt{2}}\left(-\frac{\partial}{\partial\xi}+\xi\right)$;

$$\xi=\sqrt{\frac{m_0\omega}{\hbar}}x$$

$$[b,b^+]=\frac{1}{2}\left\{\left(\frac{\partial}{\partial\xi}+\xi\right)\left(-\frac{\partial}{\partial\xi}+\xi\right)-\left(-\frac{\partial}{\partial\xi}+\xi\right)\left(\frac{\partial}{\partial\xi}+\xi\right)\right\}$$

$$=\frac{1}{2}\left\{-\frac{\partial^2}{\partial\xi^2}-\xi\frac{\partial}{\partial\xi}+\frac{\partial}{\partial\xi}\xi+\frac{\partial^2}{\partial\xi^2}-\xi\frac{\partial}{\partial\xi}+\frac{\partial}{\partial\xi}\xi+\xi^2-\xi^2\right\}$$

$$=-\xi\frac{\partial}{\partial\xi}+\frac{\partial}{\partial\xi}\xi=-\xi\frac{\partial}{\partial\xi}+\xi\frac{\partial}{\partial\xi}+1=1$$

9.15 $\psi(t)=\exp\left(-\mathrm{i}\frac{\omega}{2}t\right)[\psi_0+\psi_1\exp(-\mathrm{i}\omega t)]$

Since ψ_0 and ψ_1 are real, we obtain

$$|\psi(t)|^2=\psi_0^2+\psi_1^2+2\psi_0\psi_1\cos(\omega t) .$$

If one chooses e.g. $t_n=2\pi n/\omega$, with $n=0,1,2,\dots$, the result is

$$|\psi(t_n)|^2=(\psi_0+\psi_1)^2 ,\quad \text{and}$$

$$|\psi(t_n+\pi/\omega)|^2=(\psi_0-\psi_1)^2 .$$

The wavefunction thus oscillates periodically between the two forms.

9.16 a) $\int_{-\infty}^{\infty}\left(\frac{\partial u}{\partial\xi}\right)v(\xi)d\xi=[u(\xi)v(\xi)]_{-\infty}^{\infty}-\int_{-\infty}^{\infty}u(\xi)\left(\frac{\partial v}{\partial\xi}\right)d\xi$.

For functions which vanish at infinity, then,

$$\int_{-\infty}^{\infty}\left(\frac{\partial u}{\partial\xi}\right)v(\xi)\,d\xi=\int_{-\infty}^{\infty}u(\xi)\left(-\frac{\partial v}{\partial\xi}\right)d\xi .$$

b) $\int(b^+\phi_n)^*(b^+\phi_n)d\xi=\int\phi_n^*bb^+\phi_nd\xi$

$$=\int\phi_n^*(1+b^+b)\phi_nd\xi$$

$$=(1+n)\int\phi_n^*\phi_n\,d\xi$$

c) $\int\phi_{n+1}^*\phi_{n+1}d\xi=\frac{1}{n+1}\int(b^+\phi_n)^*(b^+\phi_n)d\xi$

with (b), $=\int\phi_n^*\phi_n\,d\xi$

d) $\phi_n=\frac{1}{\sqrt{n!}}(b^+)^n\phi_0$ follows immediately from (c).

e) With $\phi_{n+1}=\frac{1}{\sqrt{n+1}}b^+\phi_n$ we have

$$b^+\phi_n=\sqrt{n+1}\,\phi_{n+1} ;$$

from $\phi_n=\frac{1}{\sqrt{n}}b^+\phi_{n-1}$ we have

$$b\phi_n=\frac{1}{\sqrt{n}}bb^+\phi_{n-1}=\frac{1}{\sqrt{n}}(1+b^+b)\phi_{n-1}$$

$$=\frac{1+n-1}{\sqrt{n}}\phi_{n-1} ;$$

and therefore:

$$b\phi_n=\sqrt{n}\,\phi_{n-1} .$$

f) Assuming

$$K_n=[b,(b^+)^n]=n(b^+)^{n-1}$$

is valid for a particular n, then it follows that

$$K_{n+1}=[b,(b^+)^{n+1}]=b(b^+)^n b^+-(b^+)^{n+1}b$$
$$=((b^+)^n b+K_n)b^+ -(b^+)^{n+1}b$$
$$=(b^+)^n bb^+ +n(b^+)^{n-1}b^+ -(b^+)^{n+1}b$$
$$=(b^+)^n(b^+b+1)+n(b^+)^n -(b^+)^{n+1}b$$
$$=(n+1)(b^+)^n \ .$$

Thus, the assumption is also valid for $n+1$. Since it is valid for $n=1$, it is valid for all n! Analogously:

$$Q_n=[b^+,b^n]=-nb^{n-1}$$

is supposed to be valid for a particular n. Then we also have

$$Q_{n+1}=b^+b^{n+1}-b^{n+1}b^+$$
$$=(b^n b^+ +Q_n)b-b^{n+1}b^+$$
$$=b^n b^+ b-nb^n-b^{n+1}b^+$$
$$=b^n(bb^+-1)-nb^n-b^{n+1}b^+$$
$$=-(n+1)b^n \ .$$

Thus, the assumption is also valid for $n+1$. Since it is valid for $n=1$, it is also valid for all n!

9.17 Momentum $p\sim(b-b^+)/\mathrm{i}$, position $x\sim(b+b^+)$ with

$$\int\phi_n^* b\phi_n\,d\xi=\sqrt{n}\int\phi_n^*\,\phi_{n-1}\,d\xi=0 \quad \text{and}$$
$$\int\phi_n^*\,b^+\,\phi_n\,d\xi=(\sqrt{n+1}\,)^{-1}\int\phi_n^*\,\phi_{n+1}\,d\xi=0$$

it follows that

$$\int\phi_n^* p\phi_n\,d\xi=\int\phi_n^* x\phi_n\,d\xi=0.$$

Kinetic energy:

$$\tfrac{1}{2}p^2\sim-\tfrac{1}{4}(b-b^+)^2$$
$$=-\tfrac{1}{4}(bb+b^+b^+-bb^+-b^+b)$$
$$=-\tfrac{1}{4}(b^+b^++bb-2\,b^+b-1) \ .$$

$$\int\phi_n^*\frac{p^2}{2}\phi_n\,d\xi \quad \text{is then} \quad \int\phi_n^*\left(\frac{1}{2}b^+b+\frac{1}{4}\right)\phi_n\,d\xi$$
$$=\tfrac{1}{2}(n+\tfrac{1}{2}) \ .$$

Analogously for the potential energy:

$$\tfrac{1}{2}x^2\sim\tfrac{1}{4}(b+b^+)^2=\tfrac{1}{4}(b^+b^++bb+2\,b^+b+1)$$
$$=\int\phi_n^*\frac{x^2}{2}\phi_n\,d\xi\to\frac{1}{2}\left(n+\frac{1}{2}\right) \ .$$

9.18 To be shown:

$$\int\phi_n^*\phi_m\,d\xi=\delta_{n,m} \ .$$

ϕ_n and ϕ_m are solutions of the Schrödinger equations

$$b^+b\phi_n=n\phi_n \quad \text{and} \quad b^+b\phi_m=m\phi_m \ .$$

Multiplying by ϕ_m or ϕ_n, we obtain

$$\int\phi_n^* b^+ b\phi_n\,d\xi=n\int\phi_n^*\phi_n\,d\xi \quad \text{and} \qquad (1)$$
$$\int(b^+b\phi_m)^*\phi_n\,d\xi=m\int\phi_m^*\phi_n\,d\xi$$
$$=\int\phi_m^* b^+b\phi_n\,d\xi$$
$$=m\int\phi_m^*\phi_n\,d\xi \ . \qquad (2)$$

Taking the difference between (1) and (2), we find

$$0=(n-m)\int\phi_m^*\phi_n\,d\xi \ ,$$

i.e. the eigenfunctions are orthogonal and – with Problem 9.16c – are normalised.

9.19 a) as in Problem 9.16e
b) as in Problem 9.18
c), d) as in Problem 9.17.

10.1 a) $\bar{E}_{\text{kin}}=\dfrac{1}{2}\dfrac{m_0e^4}{(4\pi\varepsilon_0)^2\hbar^2}$ $\quad \bar{E}_{\text{pot}}=-\dfrac{m_0e^4}{(4\pi\varepsilon_0)^2\hbar^2}$

b) $\bar{E}_{\text{kin}}=\dfrac{1}{8}\dfrac{m_0e^4}{(4\pi\varepsilon_0)^2\hbar^2}$ $\quad \bar{E}_{\text{pot}}=-\dfrac{1}{4}\dfrac{m_0e^4}{(4\pi\varepsilon_0)^2\hbar^2}$

10.2 a) $\boldsymbol{D}=(0,0,0)$

b) 1. $\boldsymbol{D}=(0,0,0)$

2. $\boldsymbol{D}=d/\sqrt{3}\,(0,0,1)$

3. $\boldsymbol{D}=\frac{1}{3}\sqrt{(3/2)}\,d(-1,\pm\mathrm{i},0)$

$$d=\frac{256}{81}\frac{1}{\sqrt{6}}\frac{\hbar^2}{m_0e^2}4\pi\varepsilon_0$$

10.3 $\kappa_1 = [0.529 \cdot 10^{-8}\,\text{cm}]^{-1}$ $E_1 = -13.55\,\text{eV}$

$\kappa_2 = [1.058 \cdot 10^{-8}\,\text{cm}]^{-1}$ $E_2 = -3.39\,\text{eV}$

$\kappa_3 = [1.587 \cdot 10^{-8}\,\text{cm}]^{-1}$ $E_3 = -1.51\,\text{eV}$

10.4 a) $N = (2/\pi)^{3/4}(r_0)^{-3/2}$

$$r_{0\,\text{min}} = 6\pi^{3/2}\varepsilon_0\hbar^2/(\sqrt{2}e^2 m_0)$$

$$\bar{E}_{\text{min}} = -e^4 m_0/(12\hbar^2\pi^3\varepsilon_0^2) = \frac{8}{3\pi}E_0$$

b) $N = (\pi r_0^3)^{-1/2}$

$$r_{0\,\text{min}} = 4\pi\varepsilon_0\hbar^2/(m_0 e^2)$$

$$\bar{E}_{\text{min}} = -e^4 m_0/(32\hbar^2\pi^2\varepsilon_0^2) = E_0$$

E_0: exact ground-state energy.

10.5 Differential equation for $g(x)$:

$$\frac{d^2}{dx^2}g(x) + 2\left(\frac{\sigma}{x} - \sqrt{\varepsilon}\right)\frac{d}{dx}g(x) + \frac{g(x)}{x}(\tilde{c}_1 - 2\sigma\sqrt{\varepsilon}) = 0$$

Power series expansion for $g(x)$ as trial solution:

$$g(x) = \sum_{n=0}^{\infty} a_n x^n$$

Recursion formula for the expansion coefficients a_n:

$$a_{n+1} = a_n(2n\sqrt{\varepsilon} - \tilde{c}_1 + 2\sigma\sqrt{\varepsilon})/[(n+1)(n+2\sigma)]$$

Energy eigenvalues:

$$E_n = -\frac{m_0 c_1^2}{2\hbar^2}\left[n + \frac{1}{2} + \frac{1}{2}\sqrt{1 + 8m_0 c_2/\hbar^2}\right]^{-2}$$

11.1

	$E_{n,l}$ [eV]	$l = 0$	$l = 1$	$l = 2$
Li:	$n = 2$	−5.31	−3.54	–
	$n = 3$	−2.01	−1.55	−1.51
	$n = 4$	−1.05	−0.87	−0.85
Na:	$n = 3$	−5.12	−3.03	−1.52
	$n = 4$	−1.97	−1.40	−0.85
	$n = 5$	−1.03	−0.80	−0.55

11.2 $E_{n,l} = -\dfrac{R_{\text{alkali}}\,hc}{(n - \Delta(l))^2}$;

$\Delta(0) = 0.41$; $\Delta(1) = 0.04$

Transition	$\tilde{\nu}$ [cm^{-1}]	λ [nm]	Region
$3p \to 2s$	31 000	323	UV
$3p \to 3s$	3 800	2619	far IR
$3s \to 2p$	12 200	820	IR
$2p \to 2s$	14 900	671	visible

11.3

		3^2D	3^2P
Principal quantum number	n	3	3
Spin quantum number	s	1/2	1/2
Angular momentum quantum number	l	2	1

Three lines result from this transition:

$3^2D_{5/2} \to 3^2P_{3/2}$

$3^2D_{3/2} \to 3^2P_{3/2}$

$3^2D_{3/2} \to 3^2P_{1/2}$.

The new index represents the total angular momentum quantum number $J = L \pm 1/2$.

12.1 Electron:

$$\omega_{\text{L}} = 2\pi\nu_L = \frac{2\mu_{\text{B}}B}{\hbar} = 3.52 \cdot 10^6\,\text{s}^{-1}$$

Proton:

$$\omega_L = 2\pi\nu_L = \frac{5.585\,\mu_{\text{N}}B}{\hbar} = 5345\,\text{s}^{-1}$$

with $\mu_{\text{N}} = \mu_{\text{B}}/1836$.

12.2 Result:

$$\mu_z = \frac{m\,d\,v^2}{2\,\partial B_z/\partial z\;l_1(l_2 + l_1/2)} = 9.32 \cdot 10^{-24}\,\text{Am}^2\,.$$

The nuclear spin contributes to the deflection by an amount which is 3 orders of magnitude smaller and can thus be neglected.

12.3 With $l = n\hbar$ (cf. Bohr model) we have from Sect. 12.8

$$B_l = \frac{e\mu_0}{4\pi r^3 m_0}\,l = 12.54\,\text{T}\,.$$

12.4 In general (compare Sect. 12.2):

$\mu_l = \mu_B\sqrt{l(l+1)}\,; \quad \mu_{l,z} = \mu_B m_l\,.$

Electron: $\mu_B = \dfrac{e\hbar}{2m_0} = 9.274\cdot 10^{-24}\,\mathrm{Am}^2$

Muon: $\mu_B(\text{muon}) = \mu_B/207 = 4.480\cdot 10^{-26}\,\mathrm{Am}^2\,.$

In the case of positronium, due to the opposite charges and equal masses, no current flows and the magnetic moment is zero.

12.5
$$V_{l,s} = \frac{V_0}{n^3 l(l+\frac{1}{2})(l+1)}\cdot(j(j+1)-l(l+1)-s(s+1))$$

with
$$V_0 = \frac{Z^4 e^2\mu_0\hbar^2}{16\pi m_0^2 a_0^3} = 5.81\cdot 10^{-23}\,\mathrm{J} = 3.63\cdot 10^{-4}\,\mathrm{eV}$$

n	l	j	$\frac{V_{l,s(n,l,j)}}{V_0}$	$V_{l,s(n,l,j)}$ [eV]
2	1	1/2	−1/12	$-3.03\cdot 10^{-5}$
		3/2	1/24	$1.51\cdot 10^{-5}$
3	1	1/2	−2/81	$-8.96\cdot 10^{-6}$
		3/2	1/81	$4.48\cdot 10^{-6}$
	2	3/2	−1/135	$-2.69\cdot 10^{-6}$
		5/2	2/405	$1.79\cdot 10^{-6}$
⋮				
30	1	1/2	−1/40500	$-8.96\cdot 10^{-9}$
		3/2	1/81000	$4.48\cdot 10^{-9}$
	⋮			
	29	28.5	$-4.33\cdot 10^{-8}$	$-1.57\cdot 10^{-11}$
		29.5	$4.19\cdot 10^{-8}$	$1.52\cdot 10^{-11}$

12.6 The energy difference ΔE between the states:

$$\Delta E = hc\left(\frac{1}{\lambda_0} - \frac{1}{\lambda_0+\Delta\lambda}\right) = 1.101\cdot 10^{-20}\,\mathrm{J} = 0.0687\,\mathrm{eV}$$

with $\lambda_0 = 852.1$ nm, $\quad \Delta\lambda = 422$ nm.

The fine-structure interaction energy (cf. Sect. 12.8):

$$V_{l,s} = -\boldsymbol{\mu}_s\cdot\boldsymbol{B}_l = g_s\mu_B\sqrt{s(s+1)}\,B_l\cos(\sphericalangle(\boldsymbol{s},\boldsymbol{B}_l))$$
$$= \frac{a}{2}(j(j+1)-l(l+1)-s(s+1))$$

(For the calculation of $\cos(\sphericalangle(\boldsymbol{s},\boldsymbol{B}_l))$ see Sect. 12.8.)

$\rightarrow \Delta E = V_{l,s}(j=3/2) - V_{l,s}(j=1/2)$

$\rightarrow B_l = 559.5$ T; $a = 7.339\cdot 10^{-21}\,\mathrm{J} = 0.0458$ eV .

12.7 1) Energy states of the hydrogen atom as described by the Dirac theory (cf. Sect. 12.10):

$$E_{n,j} = E_n + \Delta E_{n,j} = E_n + \frac{E_n\alpha^2}{n}\left(\frac{1}{j+1/2} - \frac{3}{4n}\right)$$

with $E_n = -hcR_H/n^2$.

2) The Lamb shift lifts the degeneracy of the states with the same value of n, so that states with smaller l values have slightly higher energies.

3) Selection rules for the possible transitions:

$\Delta l = \pm 1$; $\Delta j = 0, \pm 1$.

A total of 18 lines occur; see Fig. 12.20.

12.8 a)
$$-\frac{E_{FS}\,n^3}{hcRZ^4\alpha^2} = \left(\frac{1}{j+1/2} - \frac{3}{4n}\right) > \frac{1}{j_{max}+1/2} - \frac{3}{4n} = \frac{1}{4n} > 0$$

with $j_{max} = n - 1/2$. Then, for all n and j, we have $E_{FS} < 0$.

b) In hydrogen-like atoms, the energies of the electronic states depend only on n and j. For a given n, there are just n different j values: 1/2, 3/2, ..., $n-1/2$. Thus, the terms for $n=3$ and $n=4$ are split into 3 or 4 levels, respectively.

c)
$$\Delta\tilde{\nu} = \frac{\Delta E_{FS}}{hc} = -\frac{\alpha^2 RZ^4}{n^3}\left(\frac{1}{j+1/2} - \frac{3}{4n}\right)$$

n	3	3	3	4	4	4	4
j	1/2	3/2	5/2	1/2	3/2	5/2	7/2
$-\Delta\tilde{\nu}$ [cm^{-1}]	0.162	0.054	0.018	0.074	0.029	0.013	0.006

d) Selection rules: $\Delta l = \pm 1$; $\quad \Delta j = 0, \pm 1$.

The following table shows the possible transitions and the deviations in the positions of the lines as a result of the fine-structure interaction.

$n=3$ \ $n=4$	$^2f_{7/2}$	$^2f_{5/2}$	$^2d_{5/2}$	$^2d_{3/2}$	$^2p_{3/2}$	$^2p_{1/2}$	$^2s_{1/2}$
$^2d_{5/2}$	0.20	0.08			−0.17		
$^2d_{3/2}$		0.66			0.41	−0.32	
$^2p_{1/2}$				2.14			1.41
$^2p_{3/2}$			0.66	0.41			−0.32
$^2s_{1/2}$					2.14	1.41	

12.9 Fine structure (spin-orbit) interaction:

$$V_{L,S}=\frac{A}{2}[J(J+1)-L(L+1)-S(S+1)]$$

The fine-structure constant A is a constant for a particular multiplet (L, S = constant).

Selection rules: $\Delta L=(0),\ \pm 1 \quad \Delta J=0,\ \pm 1$
(for the case of $\Delta L=0$ see Sect. 17.3.2)
(see figure on the *right*)

12.10 a) A magnetic dipole always produces a magnetic field $\boldsymbol{B}(\boldsymbol{r})$ in its neighbourhood:

$$\boldsymbol{B}(\boldsymbol{r})=\frac{-\mu_0}{4\pi}\operatorname{grad}\left(\frac{\boldsymbol{\mu}\cdot\boldsymbol{r}}{r^3}\right)=\frac{\mu_0}{4\pi}\left(\frac{3(\boldsymbol{\mu}\cdot\boldsymbol{r})\boldsymbol{r}-\boldsymbol{\mu}r^2}{r^5}\right).$$

Thus, for the interaction between two dipoles,

$$E=-\boldsymbol{\mu}_2\cdot\boldsymbol{B}_1(\boldsymbol{r})=\frac{\mu_0}{4\pi}\left[\frac{\boldsymbol{\mu}_1\cdot\boldsymbol{\mu}_2}{r^3}-3\frac{(\boldsymbol{\mu}_1\cdot\boldsymbol{r})(\boldsymbol{\mu}_2\cdot\boldsymbol{r})}{r^5}\right]$$

E is always zero when $\boldsymbol{\mu}_2\perp\boldsymbol{B}_1(\boldsymbol{r})$.

b) If $\boldsymbol{\mu}_1$ and $\boldsymbol{\mu}_2$ are parallel and θ is the angle between $\boldsymbol{\mu}_1$ or $\boldsymbol{\mu}_2$ and $\boldsymbol{r}$, we have

$$E=\frac{\mu_0}{4\pi}\frac{\mu_1\mu_2}{r^3}(1-3\cos^2\theta)\,.$$

Extreme values for fixed $|\boldsymbol{r}|=r$:

$$\theta=0,\pi:\quad E=-\frac{\mu_0}{2\pi}\frac{\mu_1\mu_2}{r^3}$$

$$\theta=\pi/2:\quad E=\frac{\mu_0}{4\pi}\frac{\mu_1\mu_2}{r^3}\,.$$

c) Electron-electron:

$$\mu_s=g_s\mu_B\sqrt{s(s+1)};\ s=1/2;\ \mu_B=\frac{e\hbar}{2m_0};\ g_s=2$$

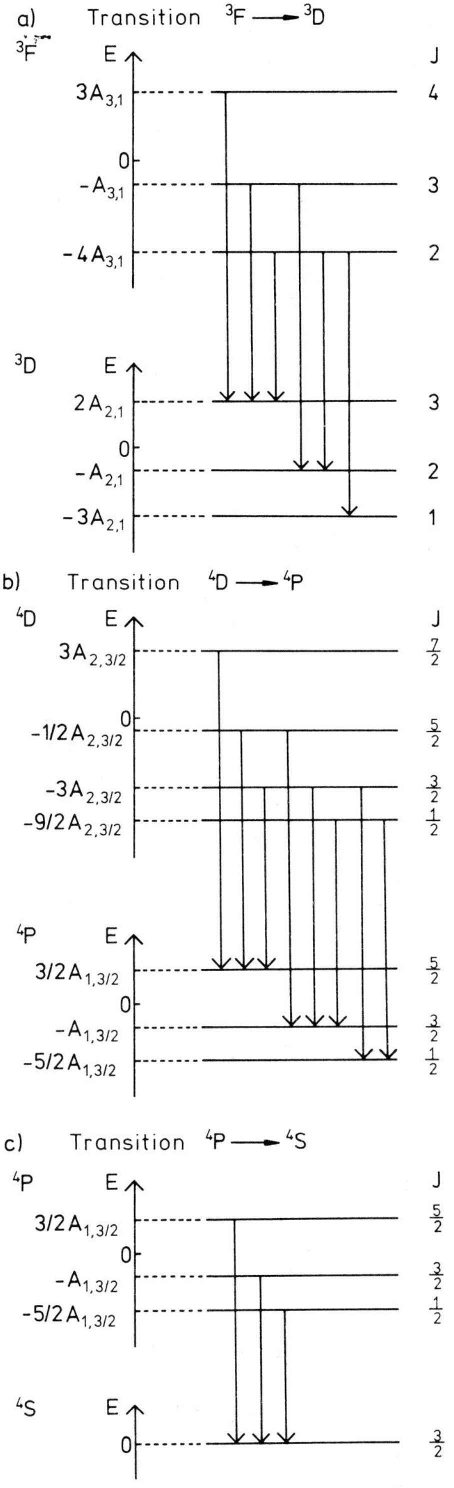

Diagram for Solution **12.9**

Angle θ	Energy E	Magnetic field $\|B_1(r)\|$
$0, \pi$	$-6.451 \cdot 10^{-24}$ J	0.402 T
$\pi/2$	$3.226 \cdot 10^{-24}$ J	0.201 T

Proton-proton:

$$\mu_p = g_p \mu_N \sqrt{I(I+1)}; \quad I = 1/2; \quad \mu_N = \frac{\mu_B}{1836.15};$$

$g_P = 5.585$

Angle θ	Energy E	Magnetic field $\|B_1(r)\|$
$0, \pi$	$-1.492 \cdot 10^{-29}$ J	$6.11 \cdot 10^{-4}$ T
$\pi/2$	$7.46 \cdot 10^{-30}$ J	$3.05 \cdot 10^{-4}$ T

13.1 $\nu = \frac{2\mu_B B_z}{h} = 2.8 \cdot 10^9\,\text{Hz} = 2.8\,\text{GHz}$

13.2 $^4D_{1/2}$: $L = 2$, $S = 3/2$, $J = 1/2$

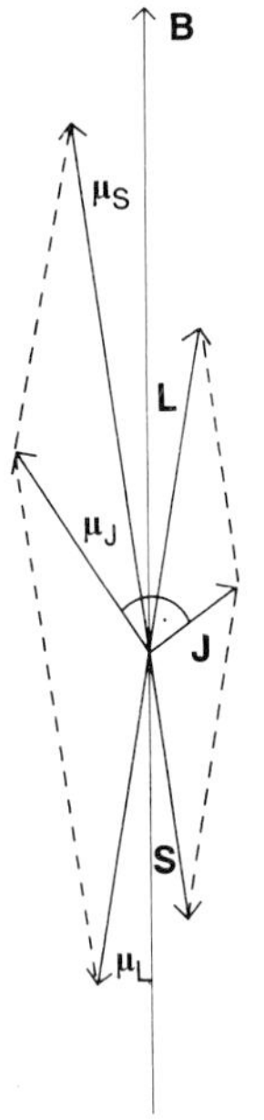

Vector model, cf. Fig. 13.9

The resultant vector μ_J for the magnetic moment is perpendicular to J and precesses around it. Only the component $(\mu_J)_J$ which is parallel to J is measurable; however, it does not occur here (cf. Fig. 13.9).

Therefore, the $^4D_{1/2}$ term does not split in a magnetic field.

13.3 $\phi = \sphericalangle(L, J) = 50.77°$

The condition $V_{1/2}(\frac{1}{2}, 1) - V_{-3/2}(\frac{3}{2}, 1) = \Delta E$ gives

$$B_0 = \frac{3}{7}\frac{hc}{\mu_B}\left(\frac{1}{\lambda_1} - \frac{1}{\lambda_2}\right) = 15.6\,\text{T}$$

b) D_1 line: $\Delta\nu = \frac{8}{3}\frac{\mu_B B}{h} = 3.73 \cdot 10^{10}$ Hz

D_2 line: $\Delta\nu = \frac{10}{3}\frac{\mu_B B}{h} = 4.67 \cdot 10^{10}$ Hz.

13.4 a) The energy difference between the $^2P_{3/2}$ and $^2P_{1/2}$ states without a magnetic field is

$$\Delta E = \frac{hc}{\lambda_1} - \frac{hc}{\lambda_2}.$$

Additional energy in a magnetic field:

$^2P_{1/2}$: $l = 1, j = 1/2 \quad g_j = 2/3$

$$V_{mj}(\tfrac{1}{2}, 1) = \tfrac{2}{3}\mu_B B_0 m_j \tag{1}$$

$^2P_{3/2}$: $l = 1, j = 3/2 \quad g_j = 4/3$

$$V_{mj}(\tfrac{3}{2}, 1) = \tfrac{4}{3}\mu_B B_0 m_j \tag{2}$$

(cf. Sect. 13.3.4 and 5).

13.5 The extra term in the energy for the anomalous Zeeman effect:

$\Delta E_B = g_j \mu_B B m_j$;

$$g_j = 1 + \frac{j(j+1) + s(s+1) - l(l+1)}{2j(j+1)}$$

Landé factors:

$g_j(^2D_{5/2}) = 6/5$; $g_j(^2D_{3/2}) = 4/5$; $g_j(^2P_{3/2}) = 4/3$; $g_j(^2P_{1/2}) = 2/3$.

Selection rules: $\Delta l = \pm 1$, $\Delta j = 0, \pm 1$, $\Delta m_j = 0, \pm 1$.

1) $^2D_{5/2} \leftrightarrow {}^2P_{3/2}$ transition:

$$\frac{\Delta(h\nu)}{\mu_B B} = g_j(^2D_{5/2}) \cdot m_j(^2D_{5/2}) - g_j(^2P_{3/2}) \cdot m_j(^2P_{3/2})$$

$^2P_{3/2}$ \ $^2D_{5/2}$	$m_j = 5/2$	$m_j = 3/2$	$m_j = 1/2$	$m_j = -1/2$	$m_j = -3/2$	$m_j = -5/2$
$m_j = 3/2$	1	$-1/5$	$-7/5$			
$m_j = 1/2$		17/15	$-1/15$	$-19/15$		
$m_j = -1/2$			19/15	1/15	$-17/15$	
$m_j = -3/2$				7/5	1/5	-1

2) $^2D_{3/2} \leftrightarrow {}^2P_{3/2}$-transition:

$$\frac{\Delta(h\nu)}{\mu_B B} = g_j(^2D_{3/2}) \cdot m_j(^2D_{3/2}) - g_j(^2P_{3/2}) \cdot m_j(^2P_{3/2})$$

$^2P_{3/2}$ \ $^2D_{3/2}$	$m_j = 3/2$	$m_j = 1/2$	$m_j = -1/2$	$m_j = -3/2$
$m_j = 3/2$	$-4/5$	$-8/5$		
$m_j = 1/2$	8/15	$-4/15$	$-16/15$	
$m_j = -1/2$		16/15	4/15	$-8/15$
$m_j = -3/2$			8/5	4/5

3) $^2D_{3/2} \leftrightarrow {}^2P_{1/2}$-transition:

$$\frac{\Delta(h\nu)}{\mu_B B} = g_j(^2D_{3/2}) \cdot m_j(^2D_{3/2}) - g_j(^2P_{1/2}) \cdot m_j(^2P_{1/2})$$

$^2P_{1/2}$ \ $^2D_{3/2}$	$m_j = 3/2$	$m_j = 1/2$	$m_j = -1/2$	$m_j = -3/2$
$m_j = 1/2$	13/15	1/15	$-11/15$	
$m_j = -1/2$		11/15	$-1/15$	$-13/15$

13.6 a) No magnetic field: Each line in the Balmer series, neglecting the Lamb shift, consists of five individual lines.

b) Weak magnetic field: Anomalous Zeeman effect.

Transition	F−S Energy $-\frac{\Delta(h\nu)_{F-S}}{E_1\alpha^2}$	Number of Zeeman lines	Total Zeeman splitting in $\mu_B B$
$n\,^2D_{5/2} \rightarrow 2^2P_{3/2}$	0.01254	12	2.8
$n\,^2D_{3/2} \rightarrow 2^2P_{3/2}$	0.006366	10	3.2
$n\,^2D_{3/2} \rightarrow 2^2P_{1/2}$	0.06887	6	1.73
$n\,^2P_{3/2} \rightarrow 2^2S_{1/2}$	0.06887	6	3.33
$n\,^2P_{1/2} \rightarrow 2^2S_{1/2}$	0.05035	4	2.67
$n\,^2S_{1/2} \rightarrow 2^2P_{1/2}$	0.05035	4	2.67
$n\,^2S_{1/2} \rightarrow 2^2P_{3/2}$	−0.01215	6	3.33

(cf. also Problem 13.5).

c) Strong magnetic field: In the Paschen-Back effect, three lines separated by $\mu_B B$ are found in the spectrum. The central line corresponds to the transition energy without spin-orbit coupling.

Magnetic moments [from (13.17)]

State	$\lvert(\mu_j)_j\rvert/\mu_B$	$\lvert(\mu_j)_j\rvert$[Am^2]
$^2S_{1/2}$	$\sqrt{3}$	$1.61 \cdot 10^{-23}$
$^2P_{1/2}$	$\frac{1}{\sqrt{3}}$	$5.35 \cdot 10^{-23}$
$^2P_{3/2}$	$2\sqrt{\frac{5}{3}}$	$2.40 \cdot 10^{-23}$
$^2D_{3/2}$	$2\sqrt{\frac{3}{5}}$	$1.44 \cdot 10^{-23}$
$^2D_{5/2}$	$3\sqrt{\frac{7}{5}}$	$3.29 \cdot 10^{-2}$

Estimate of the transition from the Zeeman to the Paschen-Back region:

$$\Delta E_{F-S} \approx hc \cdot 0.1\ \text{cm}^{-1} \approx \mu_B B_{\text{lim}}$$

thus, $B_{\text{lim}} \approx 0.2$ T.

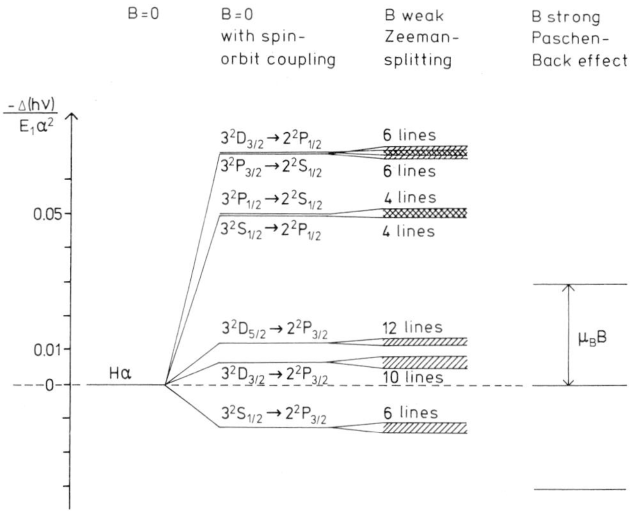

13.7 a) According to Problem 13.6, the transition from the Zeeman to the Paschen-Back region occurs around $B = 0.2$ T, i.e. here we are in the region of the Paschen-Back effect.

b) $\Delta E_{m_l, m_s} = -\mu_{l,z} B_0 - \mu_{s,z} B_0 = \mu_B B_0 (m_l + 2m_s)$.

Note that $\Delta E_{m_l, m_s}$ is the deviation of the energy from that of the same state without fine-structure splitting.

Selection rules: $\Delta m_l = 0;\ \pm 1;\ \Delta m_s = 0$.

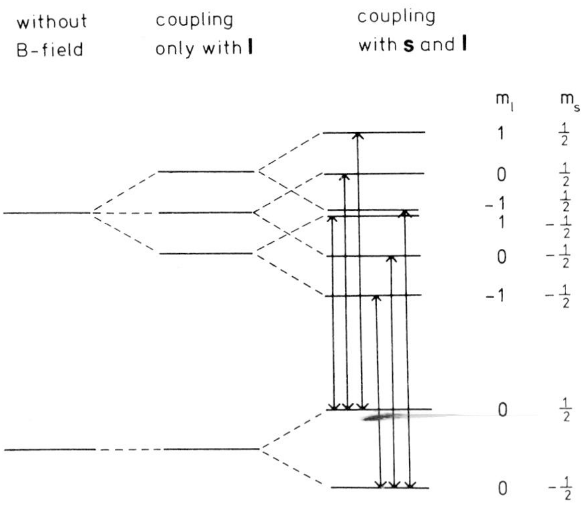

In the spectrum, one observes three lines with a splitting of $\Delta \tilde{\nu} = 2.098\ \mathrm{cm}^{-1}$.

c) Let ΔE be the energy splitting between the levels in the case of the Paschen-Back effect:

$$\Delta E = \mu_B B_0 = \frac{e\hbar}{2m_0} B_0 = h\nu$$

then $$\left(\frac{e}{m_0}\right) = \frac{4\pi\nu}{B_0} = 1.76 \cdot 10^{11}\ \mathrm{C/kg}$$

(Literature value: $e/m_0 = 1.7588 \cdot 10^{11}$ C/kg).

d) Since the energy-level splitting due to the Paschen-Back effect is independent of n, the splitting of the first line in the Lyman series is the same in wavenumbers as that of the H_α line. Thus

$$\Delta\lambda(\text{Lyman}) < \lambda(\mathrm{H}_\alpha)\,.$$

14.1 The Hamiltonian is

$$\mathscr{H} = \frac{1}{2m_0}\left(p + \frac{e}{c}A\right)^2$$
$$= \frac{1}{2m_0}\left(-\hbar^2\left(\frac{\partial^2}{\partial x^2} + \frac{\partial^2}{\partial y^2}\right) + \frac{2e}{c}Bx\frac{\hbar}{\mathrm{i}}\frac{\partial}{\partial y} + \frac{e^2}{c^2}B^2x^2\right).$$

Result:

$$E_n = \hbar\,\omega_L\,(n + \tfrac{1}{2})$$

$$\omega_L = \frac{|e|\,|B|}{m_0 c}\,.$$

14.2 By substituting in the corresponding matrices, the given relations can be immediately verified.

14.3 $$\sqrt{c^2p^2 + m_0^2c^4} = m_0c^2\sqrt{1 + p^2/m_0^2c^2} \approx m_0c^2(1 + \tfrac{1}{2}p^2/m_0^2c^2)$$

(Power series expansion of the square root terminated after the second term.)

14.4 $(\Box^2 - (m_0c/\hbar)^2)\,\psi = 0$ (Klein-Gordon Equation).

From this:

$$\psi^*(\Box^2 - (m_0c/\hbar)^2)\,\psi = 0\,.$$

The complex conjugate expression is

$$\psi(\Box^2 - (m_0c/\hbar)^2)\,\psi^* = 0\,.$$

Taking the difference:

$$\psi^*(\Box^2 - (m_0c/\hbar)^2)\,\psi - \psi(\Box^2 - (m_0c/\hbar)^2)\,\psi^* = 0,$$

and

$$\psi^*\Box^2\psi - \psi\Box^2\psi^* = 0\,.$$

With

$$\Box^2 = \left(\nabla^2 - \frac{1}{c^2}\frac{\partial^2}{\partial t^2}\right) \quad \text{we have}$$

$$\psi^*\nabla^2\psi - \psi\nabla^2\psi^* - \frac{1}{c^2}\left(\psi^*\frac{\partial^2\psi}{\partial t^2} - \psi\frac{\partial^2\psi^*}{\partial t^2}\right) = 0\,.$$

From this we obtain

$$\frac{1}{c^2}\frac{\partial}{\partial t}\left(\psi\frac{\partial\psi^*}{\partial t} - \psi^*\frac{\partial\psi}{\partial t}\right) + \operatorname{div}(\psi^*\nabla\psi - \psi\nabla\psi^*) = 0\,.$$

Multiplying the preceding expression by $\hbar/2\mathrm{i}m_0$, we obtain

$$\frac{\partial}{\partial t}\left[\frac{\mathrm{i}\hbar}{2m_0c^2}\left(\psi^*\frac{\partial\psi}{\partial t} - \psi\frac{\partial\psi^*}{\partial t}\right)\right] + \operatorname{div}\left[\frac{\hbar}{2\mathrm{i}m_0}(\psi^*\nabla\psi - \psi\nabla\psi^*)\right] = 0\,,$$

or

$$\frac{\partial}{\partial t}\varrho+\operatorname{div}\boldsymbol{j}=0\,.$$

14.5 $$(\mathrm{i}\hbar)^2\frac{\partial^2}{\partial t^2}\psi=\mathrm{i}\hbar\frac{\partial}{\partial t}\mathscr{H}\psi=\mathscr{H}\mathrm{i}\hbar\frac{\partial}{\partial t}\psi=\mathscr{H}^2\psi\,,$$

$$=(-\hbar^2c^2\nabla^2+m_0^2c^4)\,\psi \quad \text{or}$$

$$\left(\nabla^2-\frac{1}{c^2}\frac{\partial^2}{\partial t^2}\right)\psi=\frac{m_0^2c^2}{\hbar^2}\psi\,.$$

14.6 Substitution into the Dirac equation yields

$$\begin{pmatrix} -E+m_0c^2 & 0 & c\hbar k & 0\\ 0 & -E+m_0c^2 & 0 & -c\hbar k\\ c\hbar k & 0 & -E-m_0c^2 & 0\\ 0 & -c\hbar k & 0 & -E-m_0c^2\end{pmatrix}\cdot\begin{pmatrix}\Psi_1\\ \Psi_2\\ \Psi_3\\ \Psi_4\end{pmatrix}=0\,.$$

Using the trial solutions $(\Psi_1, 0, \Psi_3, 0)$ and $(0, \Psi_2, 0, \Psi_4)$, one can decompose this system of equations. The corresponding energies are

$$E=\pm\sqrt{m_0^2c^4+c^2\hbar^2k^2}\,.$$

15.1 Electric field strength $= F$

$n=1$ level:

$$\psi=\phi_1-\tilde F\phi_0+\sqrt{2}\,\tilde F\phi_2$$

$$E^{(1)}=0,\quad E^{(2)}=\frac{-e^2F^2}{2m_0\,\omega^2}\,.$$

$n=2$ level:

$$\psi=\phi_2-\sqrt{2}\,\tilde F\phi_1+\sqrt{3}\,\tilde F\phi_3$$

$$E^{(1)}=0,\,E^{(2)}=\frac{e^2F^2}{2m_0\,\omega^2}\,,\,\tilde F=-eF/\sqrt{2m_0\,\omega^3\hbar}$$

15.2 Eigenfunctions and eigenvalues:

$$\theta_n(\phi)=\frac{1}{\sqrt{2\pi}}\mathrm{e}^{\mathrm{i}n\phi}\qquad n=0,\,\pm1,\,\pm2,\,\pm3\,\dots$$

$$E_n^0=\frac{\hbar^2n^2}{M_0r_0^2}\,.$$

First-order perturbation theory:

$$E_n^1=0\quad\text{for}\quad n\neq\pm1$$

$$\text{for } n=\pm1\text{: } E_+^1=\frac{1}{2}a,\quad \theta_+=\frac{1}{\sqrt{4\pi}}(\mathrm{e}^{\mathrm{i}\phi}+\mathrm{e}^{-\mathrm{i}\phi})$$

$$E_-^1=-\frac{1}{2}a,\ \theta_-=\frac{\mathrm{i}}{\sqrt{4\pi}}(\mathrm{e}^{\mathrm{i}\phi}-\mathrm{e}^{-\mathrm{i}\phi})$$

15.3 $$\mathscr{H}_{\text{field}}\,\Phi=\sum_\lambda\hbar\omega_\lambda(b_\lambda^+b_\lambda+\tfrac{1}{2})\,\Phi=E\,\Phi \tag{1}$$

with $\Phi_{\lambda'}=b_{\lambda'}^+\Phi_0$, we obtain

$$\sum_\lambda\hbar\omega_\lambda(b_\lambda^+b_\lambda b_{\lambda'}^+\Phi_0+\tfrac{1}{2}b_{\lambda'}^+\Phi_0)=E\,b_{\lambda'}^+\Phi_0\,.$$

Making use of $b_\lambda b_{\lambda'}^+-b_{\lambda'}^+b_\lambda=\delta_{\lambda\lambda'}$, it follows that

$$\hbar\omega_{\lambda'}b_{\lambda'}^+\Phi_0+\sum_\lambda\hbar\omega_\lambda b_\lambda^+b_{\lambda'}^+b_\lambda\Phi_0+\sum_\lambda\hbar\omega_\lambda(\tfrac{1}{2})b_{\lambda'}^+\Phi_0=E\,b_{\lambda'}^+\Phi_0\,.$$

Since $b_\lambda\Phi_0=0$, we have

$$\hbar\omega_{\lambda'}b_{\lambda'}^+\Phi_0+\sum_\lambda\hbar\omega_\lambda(\tfrac{1}{2})\,b_{\lambda'}^+\Phi_0=E\,b_{\lambda'}^+\Phi_0\,.$$

Then $\Phi_{\lambda'}$ obeys expression (1) under the condition that $E=\hbar\omega_{\lambda'}+\frac{1}{2}\sum_\lambda\hbar\omega_\lambda$.

15.4 $$N\sum_\lambda\hbar\omega_\lambda b_\lambda^+b_\lambda(b_1^+)^{n_1}(b_2^+)^{n_2}\dots(b_N^+)^{n_N}\Phi_0+(N/2)\sum_\lambda\hbar\omega_\lambda(b_1^+)^{n_1}(b_2^+)^{n_2}\dots(b_N^+)^{n_N}\Phi_0=EN(b_1^+)^{n_1}(b_2^+)^{n_2}\dots(b_N^+)^{n_N}\Phi_0\,, \tag{1}$$

with $N=1/\sqrt{n_1!\,n_2!\dots n_N!}$.

Taking into account the commutation relations given, we obtain:

$$N\hbar\omega_1b_1^+n_1(b_1^+)^{n_1-1}(b_2^+)^{n_2}\dots(b_N^+)^{n_N}\Phi_0$$
$$+N\hbar\omega_2b_2^+n_2(b_1^+)^{n_1}(b_2^+)^{n_2-1}\dots(b_N^+)^{n_N}\Phi_0$$
$$+\dots N\hbar\omega_Nb_N^+n_N(b_1^+)^{n_1}(b_2^+)^{n_2}\dots(b_N^+)^{n_N-1}\Phi_0+N\sum_\lambda\hbar\omega_\lambda b_\lambda^+(b_1^+)^{n_1}(b_2^+)^{n_2}$$
$$\dots(b_N^+)^{n_N}b_\lambda\Phi_0+(N/2)\sum_\lambda\hbar\omega_\lambda(b_1^+)^{n_1}\times(b_2^+)^{n_2}\dots(b_N^+)^{n_N}\Phi_0$$
$$=EN(b_1^+)^{n_1}(b_2^+)^{n_2}\dots(b_N^+)^{n_N}\Phi_0\,.$$

With $b_\lambda \Phi_0 = 0$ and $b_\lambda^+ (b_\kappa^+)^n - (b_\kappa^+)^n b_\lambda^+ = 0$, it follows that

$$N \sum_\lambda \hbar\omega_\lambda n_\lambda (b_1^+)^{n_1} (b_2^+)^{n_2} \dots (b_N^+)^{n_N} \Phi_0$$
$$+ (N/2) \sum_\lambda \hbar\omega_\lambda (b_1^+)^{n_1} (b_2^+)^{n_2} \dots (b_N^+)^{n_N} \Phi_0$$
$$= NE (b_1^+)^{n_1} (b_2^+)^{n_2} \dots (b_N^+)^{n_N} \Phi_0 .$$

Then Φ obeys relation (1) with the condition

$$E = \sum_\lambda \hbar\omega_\lambda n_\lambda + (\tfrac{1}{2}) \sum_\lambda \hbar\omega_\lambda .$$

15.5
$$\int \psi_n^*(r) \hat{p}^2 \psi_n(r)\, dr$$
$$= \sum_n \int \psi_n^*(r) \hat{p}_r \psi_{n'}(r)\, dr \int \psi_{n'}^*(r') \hat{p}_{r'} \psi_n(r')\, dr'$$
$$= \int [\psi_n^*(r) \hat{p}_r \hat{p}_{r'} \psi_n(r') \sum_{n'} \psi_{n'}(r) \psi_{n'}^*(r')]\, dr'\, dr$$
$$= \int [\psi_n^*(r) \hat{p}_r \int \hat{p}_{r'} \psi_n(r') \delta(r - r')\, dr']\, dr$$
$$= \int \psi_n^*(r) \hat{p}_r \hat{p}_r \psi_n(r)\, dr$$
$$= \int \psi_n^*(r) \hat{p}^2 \psi_n(r)\, dr$$

15.6 Rearrangements: $\sum_{n'} \langle n' | \hat{p} | n \rangle^2 (E_{n'}^0 - E_n^0)$
$$= \sum_n \langle n | \hat{p} | n' \rangle \langle n' | \hat{p} | n \rangle (E_{n'}^0 - E_n^0)$$
$$= \tfrac{1}{2} \sum_{n'} \{ \langle n | \hat{p} | \mathcal{H}_{\mathrm{el}} n' \rangle \langle n' | \hat{p} | n \rangle$$
$$+ \langle n | \hat{p} | n' \rangle \langle \mathcal{H}_{\mathrm{el}} n' | \hat{p} | n \rangle - \langle \mathcal{H}_{\mathrm{el}} n | \hat{p} | n' \rangle$$
$$\times \langle n' | \hat{p} | n \rangle + \langle n | \hat{p} | n' \rangle \langle n' | \hat{p} | \mathcal{H}_{\mathrm{el}} n \rangle \}$$
$$= \tfrac{1}{2} \sum_{n'} \{ \langle n | [\hat{p}, \mathcal{H}_{\mathrm{el}}] | n' \rangle \langle n' | \hat{p} | n \rangle$$
$$- \langle n | \hat{p} | n' \rangle \langle n' | [\hat{p}, \mathcal{H}_{\mathrm{el}}] | n \rangle \}$$
$$= -\tfrac{1}{2} \langle n | [\hat{p}, [\hat{p}, \mathcal{H}_{\mathrm{el}}]] | n \rangle$$

15.7 Taylor expansion:

$$\langle \psi_n(r)\, V(r+s)\, \psi_n(r) \rangle = \langle \psi_n(r)\, V(r)\, \psi_n(r) \rangle$$
$$+ \frac{1}{2} \sum_{i,j} \frac{e}{4\pi\varepsilon_0} \left\langle \psi_n(r)\, s_i s_j \left(\frac{\partial}{\partial x_i} \frac{\partial}{\partial x_j} \frac{1}{|r|} \right) \psi_n(r) \right\rangle$$
$$+ \dots .$$

Furthermore, we have

$$\langle s_i s_j \rangle = \delta_{i,j} \langle s_i s_i \rangle = \tfrac{1}{3} \delta_{i,j} \langle s \cdot s \rangle .$$

We thus have for the second-order term:

$$\varepsilon^{(2)} = \frac{1}{3} \frac{e}{8\pi\varepsilon_0} \left\langle \psi_n(r)\, s \cdot s \left(\Delta \frac{1}{|r|} \right) \psi_n(r) \right\rangle$$
$$= \frac{1}{6\varepsilon_0} \langle s \cdot s \rangle \int \psi_n(r)^2 \delta(r)\, d^3r$$
$$= \frac{1}{6\varepsilon_0} \langle s \cdot s \rangle\, \psi_n(0)^2 .$$

17.1

n	2	3	4
$-E/hcR$	17/4	37/9	65/16
E [eV]	−57.76	−55.87	−55.21
E_{exp} [eV]	−58.31	−56.01	−55.25

With increasing n, the regions of high probability density of the two electrons move apart, so that the inner electron behaves like an electron in a He^+ atom ($Z = 2$) and the outer electron behaves like that of a hydrogen atom ($Z = 1$). Then the theoretical and experimental values agree more closely.

17.2 $N = (2L+1)(2S+1)$ is the number of different quantum-mechanical states of an atom with orbital angular momentum L and total spin S. In $L-S$ coupling with $S \leqq L$, we have

$$\sum (2J+1) = \sum_{k=-S}^{S} (2(L+k)+1) = (2S+1)(2L+1)$$
$$= N$$
$$J = L - S \dots L + S .$$

That is, the number of states is conserved in $L-S$ coupling. (The case $L < S$ can be obtained by exchanging L and S.)

17.3 a) $L-S$ coupling:

$l_1 = 1$, $l_2 = 2$, $L = 1, 2, 3$
$s_1 = 1/2$, $s_2 = 1/2$, $S = 0, 1$

Possible states:

S \ L	1	2	3
0	$^1P_1$	$^1D_2$	$^1F_3$
1	$^3P_0\,{}^3P_1\,{}^3P_2$	$^3D_1\,{}^3D_2\,{}^3D_3$	$^3F_2\,{}^3F_3\,{}^3F_4$

Each term includes $(2J+1)$ states; thus, there are altogether 60 states.

b) $j-j$ coupling:

$$l_1 = 1,\ s_1 = 1/2 \qquad j_1 = 1/2,\ 3/2$$
$$l_2 = 2,\ s_2 = 1/2 \qquad j_2 = 3/2,\ 5/2$$

Possible states:

$$\binom{1/2}{3/2} \rightarrow J = 1;\ 2 \qquad \binom{3/2}{3/2} \rightarrow J = 0;\ 1;\ 2;\ 3$$
$$\binom{1/2}{5/2} \rightarrow J = 2;\ 3 \qquad \binom{3/2}{5/2} \rightarrow J = 1;\ 2;\ 3;\ 4$$

The same abundances of the different J values are found, and thus the same total number of states.

17.4 a) Electron configuration of the excited C atom:

$$\underbrace{\mathrm{C}: 1s^2\ 2s^2}\quad 2p \quad 3d \text{ — } l_2 = 2,\quad s_2 = 1/2$$
$$2p \text{ — } l_1 = 1,\quad s_1 = 1/2$$
$$1s^2\,2s^2 \text{ — } L_{\text{core}} = S_{\text{core}} = 0.$$

Possible L and S values: $L = 1, 2, 3;\ S = 0, 1$
Possible J values: $J = 0, 1, 2, 3, 4$
Possible terms: ${}^1P_1\ {}^1D_1\ {}^1F_3$ singlet $S = 0$
${}^3P_0\ {}^3D_1\ {}^3F_2$ triplet $S = 1$
$\quad\ \ 1 \quad\ 2 \quad\ 3$
$\quad\ \ 2 \quad\ 3 \quad\ 4$

b) Occupation of the $3d$ subshell due to Hund's rules (cf. Sect. 19.2):

$3d$	↑	↑	↑		
$m_l =$	-2	-1	0	1	2

$\rightarrow L = 3;\ S = 3/2;\ J = 3/2 \rightarrow {}^4F_{3/2}$

Magnetic moment [cf. (13.17)]:

$$(\mu_J)_J = \sqrt{\tfrac{3}{5}}\,\mu_B$$

c) Occupation according to Hund's rule (cf. Sect. 19.2) gives for the ground state in

Y: $L = 2,\quad S = 1/2,\quad J = 3/2 \rightarrow {}^2D_{3/2}$
Zr: $L = 3,\quad S = 1,\quad J = 2 \rightarrow {}^3F_2$.

d) Electron configuration of Mn:

Mn: $1s^2\,2s^2\,2p^6\,3s^2\,3p^6\,4s^2\,3d^5$

$3d$	↑	↑	↑	↑	↑
$m_l =$	-2	-1	0	1	2

Ground state: $L = 0,\ S = 5/2,\ J = 5/2 \rightarrow {}^6S_{5/2}$.

17.5 a) General procedure:

1) Determination of L and S from the term symbol;
2) Determination of J from the number m of particle beams in the Stern-Gerlach experiment: $m = 2J+1$;
3) Maximum magnetic moment in the direction of the magnetic field:

$$\mu_{z,\max} = g_J J \mu_B;$$
$$g_J = 1 + \frac{J(J+1) + S(S+1) - L(L+1)}{2J(J+1)}$$

Atom	Term symbol	L	S	J	g_J	$\mu_{z,\max}/\mu_B$
Vanadium	${}^4F$	3	$\frac{3}{2}$	$\frac{3}{2}$	$\frac{2}{5}$	$\frac{3}{5}$
Manganese	${}^6S$	0	$\frac{5}{2}$	$\frac{5}{2}$	2	5
Iron	${}^5D$	2	2	4	$\frac{3}{2}$	6

b) For $S = 0$, the energy level splitting in a magnetic field is given by:

$$\Delta E = E(m_L = L) - E(m_L = -L) = 2L\mu_B B;$$

thus,

$$L = \frac{\Delta E}{2\mu_B B} = \frac{hc\tilde{\nu}}{2\mu_B B} = 3, \quad \text{giving } {}^1F_3.$$

18.1 $\lambda_{\min} = 3.09 \cdot 10^{-11}\ \mathrm{m} = 0.309\ \text{Å}$

18.2 $\Delta E_{\mathrm{Co}} = hc/\lambda = 1.114 \cdot 10^{-15}\ \mathrm{J} = 6952\ \mathrm{eV}$

In comparison to the hydrogen atom, we have

$$\Delta E_H = hcR_H\left(\frac{1}{1^2} - \frac{1}{2^2}\right) = 10.21\ \mathrm{eV} \approx \frac{\Delta E_{\mathrm{Co}}}{(Z_{\mathrm{Co}} - 1)^2}$$

(cf. Moseley's rule, Sect. 18.4).

18.3 $\lambda_{K_\alpha}(\mathrm{Cu}) = 1.55 \cdot 10^{-10}\ \mathrm{m} = 1.55\ \text{Å}$

18.4 Manganese ($Z = 25$)

18.5 Linear absorption coefficient μ:

$$I(x) = I_0 e^{-\mu x}; \qquad \mu = \frac{\ln(I_0/I(x))}{x}.$$

I_0 and $I(x)$ are the intensities of the incident beam and of the beam after the distance x. From the experimental data it follows that

$\bar{\mu} = (131.7 \pm 0.3)\ \mathrm{m}^{-1}$.

18.6 $I = I_0 \mathrm{e}^{-\mu x} \rightarrow x = \dfrac{\ln I_0/I}{\mu}$

a) $h\nu$ [MeV]	0.05	0.3	1
μ [1/m]	8000	500	78
x [m]	$2.88 \cdot 10^{-4}$	$4.61 \cdot 10^{-3}$	$2.95 \cdot 10^{-2}$
b) $h\nu$ [MeV]	0.05	0.3	1
I/I_0	$4.25 \cdot 10^{-18}$	$8.21 \cdot 10^{-2}$	0.677

18.7 The half-absorption length of a material:

$$I = I_0 \mathrm{e}^{-\mu d_{1/2}} = \tfrac{1}{2} I_0; \quad \text{thus, } d_{1/2} = \frac{\ln 2}{\mu}.$$

When $I_0/I(x)$ is known, it follows for the thickness x that

$$x/d_{1/2} = \frac{\ln(I_0/I)}{\ln 2}.$$

I_0/I	16	20	200
$x/d_{1/2}$	4.00	4.32	7.64

18.8 a) Photon energies of the K series:

$$\varepsilon_n = h\nu_n = hcR_Z(Z-1)^2\left(\frac{1}{1^2} - \frac{1}{n^2}\right).$$

Energy levels:

Shell	K	L	M	N
Energy [eV]	−69660	−10620	−2280	−390

b) In order to excite the L series, an electron must be removed from the L shell, i.e. one requires at least the energy

$-E_L = -10.62$ keV.

The wavelength of the L_α line is $\lambda_{L_\alpha} = 1.49$ Å.

18.9 $E_{\mathrm{kin}} = E_1 - 2E_2 = 5.57 \cdot 10^{-15}\ \mathrm{J} = 34.8$ keV.

The velocity of the electron is calculated from the relativistic relation to be

$$v = c\sqrt{1 - \left(\frac{m_0 c^2}{E_{\mathrm{tot}}}\right)^2} = 1.05 \cdot 10^8\ \mathrm{m/s}.$$

19.1 a) 15 electrons: phosphorous
b) 46 electrons: palladium

19.2 Consider an atom with several valence electrons having the same n, l quantum numbers, and neglect the spin-orbit interaction and interactions of the electrons among themselves. Then the total state of the valence electrons can be represented as a product of one-electron states. A one-electron state is characterized by the quantum numbers n, l, m_l, and m_s.

$$\Psi_{\mathrm{tot}} = \Psi_{m_{l_1}, m_{s_1}}(1) \cdot \Psi_{m_{l_2}, m_{s_2}}(2) \cdot \ldots \quad (1)$$

The operators $\hat{L}_Z$ and $\hat{S}_Z$ for the Z components of the total orbital angular momentum and the total spin are

$$\hat{L}_Z = \hat{l}_{z_1} + \hat{l}_{Z_2} + \ldots; \quad \hat{S}_Z = \hat{s}_{Z_1} + \hat{s}_{Z_2} + \ldots \quad (2)$$

If a shell n, l is fully occupied, all of the possible one-electron states (m_l, m_s) are filled just once, according to the Pauli principle. From (1), the following wavefunction is obtained:

$$\Psi_{\mathrm{tot}} = \Psi_{L,+1/2}(1) \cdot \Psi_{L,-1/2}(2) \cdot \ldots$$
$$\cdot\, \Psi_{-L,+1/2}(4L+1) \cdot \psi_{-L,-1/2}(4L+2).$$

Ψ_{tot} is an eigenfunction of $\hat{L}_Z$ and of $\hat{S}_Z$. The eigenvalues are $m_L\hbar = 0$ and $m_S\hbar = 0$.
Therefore, the total orbital angular momentum and the total spin must vanish, i.e. $L = S = 0$.

Note: the function Ψ_{tot} in (1) does not yet fulfill the Pauli principle, since it is not antisymmetric with respect to exchange of two electrons. Strictly speaking, one would have to set up a determinant as in (19.18); this has no effect on the conclusions drawn above.

19.3 Pauli Principle:
The state of an electron in an atom can be specified by four quantum numbers. If the spin-orbit coupling and the interaction with other electrons which may be present are neglected, the quantum

numbers n, l, m_l, and m_s may be used. The Pauli principle then requires that in a system containing several electrons, no two electrons may have exactly the same four quantum number values. This principle is based on the indistinguishability of electrons.

If one uses the quantum numbers n, l, m_l, and m_s, then, taking the Pauli principle into account, the following two-electron states are found ($\times$) for an np^2 configuration:

Table 1

$(m_l, m_s)_2$ \ $(m_l, m_s)_1$	(1, +)	(1, −)	(0, +)	(0, −)	(−1, +)	(−1, −)
(1, +)		×	×	×	×	×
(1, −)			×	×	×	×
(0, +)				×	×	×
(0, −)					×	×
(−1, +)						×
(−1, −)						

In the following, we denote such a two-particle state by the expression $[(m_{l_1}, m_{s_1})(m_{l_2}, m_{s_2})]$ with $m_l = +1, 0, -1$ and $m_s = \pm(1/2)$.

The coupling of two angular momenta:

When two arbitrary angular momenta $\boldsymbol{J}_1$ and $\boldsymbol{J}_2$ are coupled to a total angular momentum $\boldsymbol{J} = \boldsymbol{J}_1 + \boldsymbol{J}_2$, the quantum number J can take on the values $J = J_1 + J_2, \ldots, |J_1 - J_2|$. For each J value-there are the m_J values $m_J = -J, \ldots, +J$.

The corresponding total angular momentum state (J, m_J) can be represented as a linear combination of the product states $(J_1, m_{J_1}) \cdot (J_2, m_{J_2})$, where $m_J = m_{J_1} + m_{J_2}$ is always fulfilled.

$$(J, m_J) = \sum_{\substack{m_{J1}, m_{J2} \\ m_{J1} + m_{J2} = m_J}} C_{m_{J1}, m_{J2}} (J_1, m_{J1}) \cdot (J_2, m_{J2}) \qquad (1)$$

After this coupling, m_{J_1} and m_{J_2} are no longer valid quantum numbers. An angular momentum state is specified uniquely by the quantum numbers J_1, J_2, J, and m_J. In both $j-j$ and in $L-S$ coupling, one arrives at the total angular momentum J of the system by means of such a two-stage angular momentum coupling, starting from the non-interacting particle model.

a) *$j-j$ Coupling:*

In the case of $j-j$ coupling, in the first step, the orbital angular momentum l and the spin s of each particle (electron) are coupled to a single-particle angular momentum j. Table 2 shows the ordering of all the states which are consistent with the Pauli principle from Table 1, according to the new single-particle quantum number m_j.

Table 2

m_{j2} \ m_{j1}	3/2	1/2	−1/2	−3/2
3/2		$m_J = 2$	$m_J = 1$	$m_J = 0$
1/2	[(1, +)(1, −)] [(1, +)(0, +)]	[(1, −)(0, +)]		$m_J = -1$
−1/2	[(1, +)(0, −)] [(1, +)(−1, +)]	[(1, −)(0, −)] [(0, +)(0, −)] [(1, −)(−1, +)] [(0, +)(−1, +)]	[(0, −)(−1, +)]	$m_J = -2$
−3/2	[(1, +)(−1, −)]	[(1, −)(−1, −)] [(0, +)(−1, −)]	[(0, −)(−1, −)] [(−1, +)(−1, −)]	

In the second step, the single-particle angular momenta are coupled to give the total angular momentum J ($j-j$ coupling). From (1), only those states contribute to a state with quantum number m_J (belonging to J) for which $m_{j_1} + m_{j_2} = m_J$. The number of such states is conserved, so the abundance of the various possible m_J values may be calculated.

m_J	2	1	0	−1	−2
Abundance	2	3	5	3	2

Here, the following quantum numbers J occur:

J	2	1	0
Abundance	2	1	2

b) *$L-S$ Coupling:*

In the first step, the orbital angular momenta l_1 and l_2 couple to give the total orbital angular momentum L and the spins s_1 and s_2 couple to the total spin S. Table 3 gives the ordering of the states which are consistent with the Pauli principle from Table 1, according to the new quantum numbers m_S and m_L.

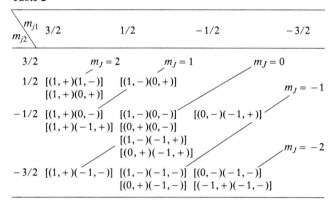

Table 3

m_L \ m_S	1	0	−1	
		$m_J = 2$		$m_J = 1$
2		[(1, +)(1, −)]		
				$m_J = 0$
1	[(1, +)(0, +)]	[(1, +)(0, −)] [(1, −)(0, +)]	[(1, −)(0, −)]	
				$m_J = -1$
0	[(1, +)(−1, +)]	[(1, +)(−1, −)] [(0, +)(0, −)] [(1, −)(−1, +)]	[(1, −)(−1, −)]	
				$m_J = -2$
−1	[(0, +)(−1, +)]	[(0, +)(−1, −)] [(0, −)(−1, +)]	[(0, −)(−1, −)]	
−2		[(−1, +)(−1, −)]		

For the $L-S$ coupling to a total angular momentum J we find, similarly to part (a):

m_J	2	1	0	−1	−2
Abundance	2	3	5	3	2

and from this:

J	2	1	0
Abundance	2	1	2

Thus, the same J values occur with the same abundances.

19.4 The given electron configurations are considered in the light of $L-S$ coupling. If several electrons have the same n and l values, the Pauli principle must be observed (cf. Problem 19.3).

a) Electron configuration ns:
Possible term: $^2S_{1/2} \rightarrow 2$ states

b) Electron configuration np^3:
One finds the following abundances for the various (m_L, m_S) pairs:

m_L \ m_S	3/2	1/2	−1/2	−3/2
2	0	1	1	0
1	0	2	2	0
0	1	3	3	1
−1	0	2	2	0
−2	0	1	1	0

leading to the possible terms:

L	S	Terms	Number of states
2	1/2	$^2D_{3/2}\ ^2D_{5/2}$	10
1	1/2	$^2P_{1/2}\ ^2P_{3/2}$	6
0	3/2	$^4S_{3/2}$	4

c) Electron configuration $np^2n's$:
Possible terms:

L	S	Terms	Number of states
2	1/2	$^2D_{3/2}\ ^2D_{5/2}$	10
1	3/2	$^4P_{1/2}\ ^4P_{3/2}\ ^4P_{5/2}$	12
1	1/2	$^2P_{1/2}\ ^2P_{3/2}$	6
0	1/2	$^2S_{1/2}$	2

d) Electron configuration np^5:

Possible terms: $L = 1$; $S = 1/2$: $^2P_{1/2}\ ^2P_{3/2}$; 6 states

e) Electron configuration $nd^2\ n'p$:
Possible terms:

L	S	Terms	Number of states
5	1/2	$^2H_{9/2}\ ^2H_{11/2}$	22
4	1/2	2x ($^2G_{7/2}\ ^2G_{9/2}$)	36
	3/2	$^4G_{5/2}\ ^4G_{7/2}\ ^4G_{9/2}\ ^4G_{11/2}$	36
3	1/2	3x ($^2F_{5/2}\ ^2F_{7/2}$)	42
	3/2	$^4F_{3/2}\ ^4F_{5/2}\ ^4F_{7/2}\ ^4F_{9/2}$	28
2	1/2	3x ($^2D_{3/2}\ ^2D_{5/2}$)	30
	3/2	2x ($^4D_{1/2}\ ^4D_{3/2}\ ^4D_{5/2}\ ^4D_{7/2}$)	40
1	1/2	3x ($^2P_{1/2}\ ^2P_{3/2}$)	18
	3/2	$^4P_{1/2}\ ^4P_{3/2}\ ^4P_{5/2}$	12
0	1/2	$^2S_{1/2}$	2
	3/2	$^4S_{3/2}$	4

f) Electron configuration $nd\,n'd$:
Possible terms:

L	S	Terms	Number of states
4	0	$^1G_4$	9
	1	$^3G_3\ ^3G_4\ ^3G_5$	27
3	0	$^1F_3$	7
	1	$^3F_2\ ^3F_3\ ^3F_4$	21
2	0	$^1D_2$	5
	1	$^3D_1\ ^3D_2\ ^3D_3$	15
1	0	$^1P_1$	3
	1	$^3P_0\ ^3P_1\ ^3P_2$	9
0	0	$^1S_0$	1
	1	$^3S_1$	3

19.5

Element	Electron configuration	Ground state			
		L	S	J	Term
Silicon	$1s^2\,2s^2\,2p^6\,3s^2\,3p^2$	1	1	0	$^3P_0$
Chlorine	$1s^2\,2s^2\,2p^6\,3s^2\,3p^5$	1	$\frac{1}{2}$	$\frac{3}{2}$	$^2P_{3/2}$
Arsenic	$1s^2\,2s^2\,2p^6\,3s^2$ $3p^6\,3d^{10}\,4s^2\,4p^3$	0	$\frac{3}{2}$	$\frac{3}{2}$	$^4S_{3/2}$

19.6

Element	Electron configuration	Number of unpaired e^-	Ground state
Sulfur	$1s^2\,2s^2\,2p^6\,3s^2\,3p^4$	2	$^3P_2$
Calcium	$1s^2\,2s^2\,2p^6\,3s^2\,3p^6\,4s^2$	0	$^1S_0$
Iron	$1s^2\,2s^2\,2p^6\,3s^2$ $3p^6\,3d^6\,4s^2$	4	$^5D_4$
Bromine	$1s^2\,2s^2\,2p^6\,3s^2\,3p^6$ $3d^{10}\,4s^2\,4p^5$	1	$^2P_{3/2}$

19.7 Electron configuration: nd^2
Possible terms (cf. Problems 19.3 and 19.4):

L	S	Terms	Number of states
4	0	$^1G_4$	9
3	1	$^3F_2\ ^3F_3\ ^3F_4$	21
2	0	$^1D_2$	5
1	1	$^3P_0\ ^3P_1\ ^3P_2$	9
0	0	$^1S_0$	1

According to Hund's rules, for the ground state term we have $L=3$, $S=1$, $J=2$; i.e. $^3F_2$.

19.8 Proof of relation (19.25):

$$\Sigma_z\,\phi_\uparrow(1)\,\phi_\uparrow(2) = (\sigma_{z,1}+\sigma_{z,2})\,\phi_\uparrow(1)\,\phi_\uparrow(2) = (\tfrac{1}{2}\hbar+\tfrac{1}{2}\hbar)\,\phi_\uparrow(1)\,\phi_\uparrow(2).$$

The proofs of relations (19.26) and (19.32) are analogous. To prove relations (19.35–38), we require the following identities:

$$\Sigma^2 = \sigma_1^2 + 2\sigma_1\sigma_2 + \sigma_2^2;$$

$$\sigma_1\cdot\sigma_2 = \sigma_{z,1}\sigma_{z,2} + \sigma_{x,1}\sigma_{x,2} + \sigma_{y,1}\sigma_{y,2};$$

$$\sigma_{x,1}\sigma_{x,2} + \sigma_{y,1}\sigma_{y,2} = \tfrac{1}{2}(\sigma_{+,1}\sigma_{-,2} + \sigma_{-,1}\sigma_{+,2}).$$

With these, it follows that

$$\Sigma^2 = \sigma_1^2 + \sigma_2^2 + 2\sigma_{z,1}\sigma_{z,2} + \sigma_{+,1}\sigma_{-,2} + \sigma_{-,1}\sigma_{+,2}$$

and the following relations are valid:

$$\sigma_1^2\phi_\uparrow(1) = \tfrac{3}{4}\hbar\,\phi_\uparrow(1),\ \sigma_1^2\phi_\downarrow(1) = \tfrac{3}{4}\hbar\,\phi_\downarrow(1);$$

$$\sigma_{+,1}\phi_\downarrow(1) = 0,\ \sigma_{+,1}\phi_\downarrow(1) = \hbar\,\phi_\uparrow(1);$$

$$\sigma_{-,1}\phi_\downarrow(1) = 0,\ \sigma_{-,1}\phi_\uparrow(1) = \hbar\,\phi_\downarrow(1)$$

$$\sigma_{z,1}\phi_\uparrow(1) = \hbar\,\phi_\uparrow(1),\ \sigma_{z,1}\phi_\downarrow\phi_\downarrow(1) = -\hbar\,\phi_\downarrow(1)$$

(similar relations hold for the second electron.) With these, one can prove (19.35–38)!

19.9 When the quantum numbers q_1 and q_2 are identical, no triplet state exists.
[See (19.19, 20 and 30)]

20.1 The formula for hyperfine structure (20.10) is

$$\Delta E_F = \frac{a}{2}[F(F+1) - I(I+1) - J(J+1)]$$

The interval rule (20.12) states that

$$\Delta E_{F+1} - \Delta E_F = a(F+1).$$

From this, F_{max} and a may be determined to be

$F_{max} = 7 \quad a = (0.0789 \pm 0.0006)\ \mathrm{cm}^{-1}$,
and $F = 2, \ldots, 7$ so that $I = 9/2$.

F	2	3	4	5	6	7
$\Delta E_F/a$	−13.75	−10.75	−6.75	−1.75	4.25	11.25

20.2 The hyperfine structure formula (20.10) is

$$\Delta E_F = \frac{a}{2}[F(F+1) - I(I+1) - J(J+1)];$$

$$a = \frac{g_I \mu_N B_J}{\sqrt{J(J+1)}}.$$

In the hydrogen atom, (20.11) is valid:

$$a = \frac{2}{3}\mu_0 g_e \mu_B g_I \mu_N |\psi(0)|^2; |\psi(0)|^2 = \frac{1}{\pi r_H^3 n^3}.$$

(Here, r_H is the first Bohr radius.) Then

$$B_J = \tfrac{2}{3}\mu_0 g_e \mu_B |\psi(0)|^2 \sqrt{J(J+1)}.$$

According to the interval rule, the energy difference between parallel ($F=1$) and antiparallel ($F=0$) orientations of the electronic and the proton spins is just a.

n	$\lvert\psi(0)\rvert^2$ [1/m³]	B_J [T]	a [cm^{-1}]
1	$2.15 \cdot 10^{30}$	29.0	0.0475
2	$2.69 \cdot 10^{29}$	3.63	0.00594
3	$7.96 \cdot 10^{28}$	1.07	0.00176

20.3 $a/hc = (5.85 + 0.52) \cdot 10^{-4}\,\text{cm}^{-1}$
$B_J = (3.03 \pm 0.27)$ T

20.4 Zeeman energy of the electron:

$$E_{e^-} = g_e \mu_B B_0 m_s = 5.57 \cdot 10^{-24}\,\text{J} \cdot m_s$$
$$= 3.48 \cdot 10^{-5}\,\text{eV} \cdot m_s = hc \cdot 0.281\,\text{cm}^{-1} \cdot m_s.$$

Hyperfine interaction energy with $a = 0.0475\,\text{cm}^{-1}$:

$$E_{HFS} = a m_I m_s = hc \cdot 0.0475\,\text{cm}^{-1} \cdot m_I m_s$$
$$= 5.89 \cdot 10^{-6}\,\text{eV} \cdot m_I m_s$$
$$= 9.44 \cdot 10^{-25}\,\text{J} \cdot m_I m_s.$$

Zeeman energy of the nucleus ($g_I = 5.585$):

$$E_p = -g_I \mu_N B_0 m_I = -8.46 \cdot 10^{-27}\,\text{J} \cdot m_I$$
$$= -5.28 \cdot 10^{-8}\,\text{eV} \cdot m_I = hc \cdot 4.26 \cdot 10^{-4}\,\text{cm}^{-1} \cdot m_I.$$

20.5 $\gamma = 6.624 \cdot 10^7\,\text{m}^2/\text{Vs}^2$

$g_I = 1.38$

$$\frac{\mu_{z,\max}}{\mu_N} = 3.45$$

20.6 According to (20.15), in the case of a "strong" applied magnetic field we have

$$\Delta E_{HFS} = g_J \mu_B m_J B_1 + a m_I m_J.$$

In the Stern-Gerlach experiment, the beam splits into $(2J+1)$ $(2I+1)$ sub-beams; thus, we have $I(^{23}\text{Na}) = 3/2$.

In the case of a "weak" magnetic field, the beam is also split into 8 sub-beams, which, however, have a different geometrical arrangement.

21.1 $W = 1.224 \cdot 10^{-12}\,\text{s}^{-1}$

21.2

$L \rightarrow$	1 cm	10 cm
$R = 99\%$	$t_0 = 3.336 \cdot 10^{-9}$ s	$t_0 = 3.336 \cdot 10^{-8}$ s
$R = 90\%$	$t_0 = 3.336 \cdot 10^{-10}$ s	$t_0 = 3.336 \cdot 10^{-9}$ s
$R = 10\%$	$t_0 = 3.706 \cdot 10^{-11}$ s	$t_0 = 3.706 \cdot 10^{-10}$ s

$L \rightarrow$	100 cm
$R = 99\%$	$t_0 = 3.336 \cdot 10^{-7}$ s
$R = 90\%$	$t_0 = 3.336 \cdot 10^{-8}$ s
$R = 10\%$	$t_0 = 3.706 \cdot 10^{-9}$ s

21.3 $R = 90\%$, $L = 100$ cm:

The number of photons decreases to $1/e$ of its initial value after a time $T = 3.336 \cdot 10^{-8}$ s.

21.4 The laser condition is

$$\frac{N_2 - N_1}{V} < \frac{8\pi\nu^2 \Delta\nu\tau}{c^3 t_0}.$$

The formula

$$\frac{1}{t_0} = \frac{c}{L}(1-R)$$

gives the lifetime of the photons in the resonator.

For the length of the resonator, we assume e.g. 10 cm, and thus obtain the value $1.2 \cdot 10^{18}/\text{cm}^3$ for the critical inversion density.

23.1 Transform the coordinates according to $x_b = -x_a$, $x_a > 0$.

$$\psi(x) = A\,e^{\kappa x} \qquad x \le -x_a$$

$$\psi(x) = B\,e^{\kappa x} + C\,e^{-\kappa x} \qquad -x_a \le x \le -x_a$$

$\psi(x) = D\,\mathrm{e}^{-\kappa x} \qquad x \geq x_a$

$\kappa = \sqrt{2 m_0 |E|/\hbar^2}$

Symmetric state: $\psi(x) = \psi(-x)$.

$\kappa - m_0\beta/\hbar^2 = \mathrm{e}^{-2\kappa x_a} m_0\beta/\hbar^2$

$B = C,\ A = D$

Antisymmetric state: $\psi(x) = -\psi(-x)$.

$\kappa - m_0\beta/\hbar^2 = -\mathrm{e}^{-2\kappa x_a} m_0\beta/\hbar^2$

$B = -C,\ A = -D$

23.2 Transform the coordinates, so that $x_b = -x_a$, $x_a > 0$.

$\psi(x) = A\,\mathrm{e}^{\kappa x} \qquad x \leq -x_a$

$\psi(x) = B\,\mathrm{e}^{\kappa x} + C\,\mathrm{e}^{-\kappa x} \qquad -x_a \leq x \leq -x_a$

$\psi(x) = D\,\mathrm{e}^{-\kappa x} \qquad x \geq x_a$

$\kappa = \sqrt{2 m_0 |E|/\hbar^2}$

$(\kappa - m_0\beta_a/\hbar^2)(\kappa - m_0\beta_b/\hbar^2) = \mathrm{e}^{-4\kappa x_a}\beta_a\beta_b m_0^2/\hbar^4$

$(C/B)^2 = \beta_a(\kappa - m_0\beta_b/\hbar^2)/(\beta_b(\kappa - m_0\beta_a/\hbar^2))$

23.3 Wavefunction:

$\psi_k(x) = \mathrm{e}^{\mathrm{i}nka}(A_k \mathrm{e}^{\kappa(x-na)} + B_k\,\mathrm{e}^{-\kappa(x-na)})$
for $na \leq x \leq (n+1)a$

$A_k = B_k(\mathrm{e}^{\mathrm{i}ka} - \mathrm{e}^{-\kappa a})(\mathrm{e}^{\kappa a} - \mathrm{e}^{\mathrm{i}ka})$

$\kappa = \sqrt{2 m_0 |E|/\hbar^2}$

$ka = \frac{2\pi}{6}, 2\frac{2\pi}{6}, \ldots, 6\frac{2\pi}{6}$

Auxiliary equation for κ:

$$\cos ka = \cosh \kappa a - \frac{\sinh \kappa a}{\kappa} m_0\beta/\hbar^2 .$$

Bibliography of Supplementary and Specialised Literature

1. General Physics Textbooks

M. Alonso, E. J. Finn: *Fundamental University Physics* (Addison-Wesley, Reading 1972)

R. P. Feynman, R. B. Leighton, M. Sands: *Lectures on Physics* (Addison-Wesley, New York 1980)

D. Halliday, R. Resnick: *Physics I and II*, 3rd ed. (Wiley, New York 1978)

J. N. H. Hume, D. G. Ivey: *Physics 1 and 2* (Wiley, New York 1974)

K. S. Krane: *Modern Physics* (Wiley, New York 1983)

2. Atomic Physics Textbooks

M. Born: *Atomic Physics* (Hafner, New York 1966)

B. H. Bransden, C. J. Joachain: *Physics of Atoms and Molecules* (Longman, Inc., New York 1983)

B. Cagnac, J.-C. Pebay-Peyroula: *Modern Atomic Physics: Quantum Theory and Its Applications* (MacMillan, London 1975)

R. Eisberg, R. Resnick: *Quantum Physics of Atoms, Molecules, Solids and Particles* (Wiley, New York 1974)

U. Fano, L. Fano: *The Physics of Atoms and Molecules; an Introduction to the Structure of Matter* (University of Chicago Press, 1972)

T. A. Littlefield, N. Thorley: *Atomic and Nuclear Physics: an Introduction* (Van Nostrand Reinhold, New York 1979)

R. L. Sproull, W. A. Phillips: *Modern Physics: the Quantum Physics of Atoms, Solids, and Nuclei* (Wiley, New York 1980)

3. Quantum Mechanics Textbooks

J. Avery: *Quantum Theory of Atoms, Molecules, and Photons* (McGraw-Hill, New York 1972)

H. A. Bethe, E. F. Salpeter: *The Quantum Mechanics of One- and Two-Electron Atoms* (Plenum, New York 1977)

W. A. Blanpied: *Modern Physics: An Introduction to Its Mathematical Language* (Holt Rinehart Winston, New York 1971)

A. Böhm: *Quantum Mechanics* (Springer, Berlin, Heidelberg, New York 1979)

C. Cohen-Tannoudji, B. Diu, F. Laloë: *Quantum Mechanics I and II*, 2nd ed. (Wiley, New York 1977)

L. D. Landau, E. M.Lifshitz: *Quantum Mechanics*, 3rd ed. (Pergamon, Oxford 1982)

J. L. Martin: *Basic Quantum Mechanics* (Oxford University Press, Oxford 1981)

E. Merzbacher: *Quantum Mechanics*, 2nd ed. (Wiley, New York 1970)
A. Messiah: *Quantum Mechanics I and II* (Wiley, New York 1958)
R. Shankar: *Principles of Quantum Mechanics* (Plenum, New York 1980)
E. H. Wichmann: *Quantum Physics* (Vol. IV of *Berkeley Physics Course*) (McGraw-Hill, New York 1971)

4. Specialised Literature

Chapter 1

J. L. Heilbron: *Elements of Early Modern Physics* (University of California Press, Berkeley 1982)
F. Hund: *The History of Quantum Theory* (Hanup, London 1974)
B. L. van der Waerden (ed.): *Sources of Quantum Mechanics* (Dover, New York 1967)

Chapter 2

S. G. Brush: *Statistical Physics and the Theory of Atomic Spectra* (Princeton University Press, Princeton 1983)
J. W. Gibbs: *Statistical Mechanics* (Oxbow Press, Woodbridge 1981)
J. P. Holman: *Thermodynamics*, 3rd Ed. (McGraw-Hill, New York 1980)
Sir J. Jeans: *Introduction to the Kinetic Theory of Gases* (Cambridge University Press, Cambridge 1982)
C. Kittel, H. Kroemer: *Thermal Physics* (Freeman, San Francisco 1980)

Chapter 3

P. Richard (ed.): *Atomic Physics: Accelerators* (Academic, New York 1980)
S. Villani (ed.): *Uranium Enrichment* (Springer, Berlin, Heidelberg, New York 1979)

Chapter 4

M. A. Preston, R. K. Bhaduri: *Structure of the Nucleus*, 2nd ed. (Addison-Wesley, New York 1975)
E. Segrè: *Nuclei and Particles*, 2nd ed. (Benjamin, Reading 1977)

Chapter 5

F. Bitter, H. A. Medicus: *Fields and Particles: an Introduction to Electromagnetic Wave Phenomena and Quantum Physics* (Amer. Elsevier, New York 1973)
R. P. Feynman, R. B. Leighton, M. Sands: *Lectures on Physics* (Addison-Wesley, New York 1980)
L. D. Landau, E. M. Lifshitz: *Statistical Physics*, 3rd ed. (Pergamon, Oxford 1982)
E. M. Lifshitz, L. P. Pitaevskii: *Statistical Physics*, 3rd Ed. (Pergamon, Oxford 1980)

Chapter 6 Atomic Physics Textbooks

Chapter 7 Quantum Mechanics Textbooks

Chapter 8

R. W. Dixon: *Spectroscopy and Structure* (Methuen, London and Wiley, New York 1965)

J. L. Heilbron: *Elements of Early Modern Physics* (University of California Press, Berkeley 1982)
W. R. Hindmarsh: *Atomic Spectra* (Pergamon, Oxford 1967)
H. G. Kuhn: *Atomic Spectra*, 3rd ed. (Longmans, London 1969)
A. Sommerfeld: *Atomic Structure and Spectral Lines*, 3rd ed. (Methuen, London 1934)
B. L. van der Waerden (ed.): *Sources of Quantum Mechanics* (Dover, New York 1967)
H. E. White: *An Introduction to Atomic Spectra* (McGraw-Hill, New York 1954)

Chapters 9, 10 Quantum Mechanics Textbooks

Chapters 11, 12

H. G. Kuhn: *Atomic Spectra*, 3rd ed. (Longmans, London 1969)
M. Weissbluth: *Atoms and Molecules* (Academic Press, New York 1978)
H. E. White: *An Introduction to Atomic Spectra* (McGraw-Hill, New York 1954)

Chapters 13, 14

A. Abragam: *The Principles of Nuclear Magnetism* (Oxford University Press, Oxford 1961)
R. McWeeny: *Spins in Chemistry* (Academic, New York 1970)
C. P. Slichter: *Principles of Magnetic Resonance*, Springer Ser. Solid-State Sci., Vol. 1, 2nd ed. (Springer, Berlin, Heidelberg, New York 1978)
J. D. Bjorken, S. D. Drell: *Relativistic Quantum Mechanics* (Mc Graw-Hill, New York 1964)

Chapter 15

R. P. Feynman: *Quantum Electrodynamics* (Benjamin, New York 1961)
W. Heitler: *The Quantum Theory of Radiation*, 3rd ed (Clarendon Press, Oxford 1954)
N. F. Ryde: *Atoms and Molecules in Electric Fields* (Almqvist and Wicksell, Stockholm 1976)

Chapter 16

E. U. Condon, G. H. Shortley: *The Theory of Atomic Spectra* (Cambridge University Press, Cambridge 1979)
I. I. Sobelman: *Atomic Spectra and Radiative Transitions*, Springer Ser. Chem. Phys., Vol. 1 (Springer, Berlin, Heidelberg, New York 1979)

Chapter 17

R. D. Cowan: *The Theory of Atomic Structure and Spectra* (UC Press, Berkely 1981)
G. Herzberg: *Atomic Spectra and Atomic Structure* (Dover, New York 1944)
H. G. Kuhn: *Atomic Spectra*, 3rd ed. (Longmans, London 1969)
H. E. White: *An Introduction to Atomic Spectra* (McGraw-Hill, New York 1954)

Chapter 18

B. K. Agarwal: *X-Ray Spectroscopy*, Springer Ser. Opt. Sci., Vol. 15 (Springer, Berlin, Heidelberg, New York 1979)
C. R. Bründle, A. D. Baker (eds.): *Electron Spectroscopy; Theory, Techniques and Applications I – VI* (Academic, London 1977 – 1984)

E. E. Koch (ed.): *Handbook on Synchrotron Radiation* (North-Holland, Amsterdam 1983)

C. Kunz: *Synchroton Radiation*, Topics Curr. Phys., Vol. 10 (Springer, Berlin, Heidelberg, New York 1979)

Chapter 19

M. Cardona, L. Ley (eds.): *Photoemission in Solids I, II*, Topics Appl. Phys., Vols. 26, 27 (Springer, Berlin, Heidelberg, New York 1978, 1979)

Chapter 20

A. Abragam: *The Principles of Nuclear Magnetism* (Oxford University 1961)

A. J. Freeman, R. B. Frankel: *Hyperfine Interactions* (Academic, New York 1967)

H. Kopfermann: *Nuclear Moments* (Academic, New York 1958)

C. P. Slichter: *Principles of Magnetic Resonance*, Springer Ser. Solid-State Sci., Vol. 1, 2nd ed. (Springer, Berlin, Heidelberg, New York 1978)

Chapter 21

H. Haken: *Laser Theory* (Springer, Berlin, Heidelberg, New York, Tokyo 1984); *Light II, Laser Light Dynamics* (North-Holland, Amsterdam 1985)

D. C. O'Shea, W. R. Callen, W. T. Rhodes: *Introduction to Lasers and Their Applications* (Addison-Wesley, Reading 1978)

Chapter 22

Chi H. Lee (ed.): *Picosecond Optoelectronic Devices* (Academic, New York 1984)

A. Corney: *Atomic and Laser Spectroscopy* (Oxford University Press, Oxford 1977)

W. Demtröder: *Laser Spectroscopy*, Springer Ser. Chem. Phys., Vol. 5 (Springer, Berlin, Heidelberg, New York 1982)

D. C. Hanna, M. A. Yuratich, D. Cotter: *Nonlinear Optics of Free Atoms and Molecules*, Springer Ser. Opt. Sci., Vol. 17 (Springer, Berlin, Heidelberg, New York 1979)

V. S. Letokhov, V. P. Chebotayev: *Nonlinear Laser Spectroscopy*, Springer Ser. Opt. Sci., Vol. 4 (Springer, Berlin, Heidelberg, New York 1977)

D. C. O'Shea, W. R. Callen, W. T. Rhodes: *Introduction to Lasers and Their Applications* (Addison-Wesley, Reading 1978)

K. Shimoda (ed.): *High-Resolution Laser Spectroscopy*, Topics Appl. Phys., Vol. 13 (Springer, Berlin, Heidelberg, New York 1976)

S. Stenholm: *Foundations of Laser Spectroscopy* (Wiley, New York 1984)

H. Walther (ed.): *Laser Spectroscopy of Atoms and Molecules*, Topics Appl. Phys., Vol. 2 (Springer, Berlin, Heidelberg, New York 1975)

H. Walther, K. W. Rothe (eds.): *Laser Spectroscopy IV*, Springer Ser. Opt. Sci., Vol. 21 (Springer, Berlin, Heidelberg, New York 1979)

Chapter 23

L. Pauling: *The Nature of the Chemical Bond* (Cornell University Press, Ithaca 1960)

K. S. Pitzer: *Quantum Chemistry* (Prentice-Hall, Englewood Cliffs 1961)

F. O. Rice, E. Teller: *The Structure of Matter* (Wiley, New York 1961)

M. Karplus, R. M. Porter: *Atoms and Molecules: an Introduction for Students of Physical Chemistry* (Benjamin, New York 1970)

Subject Index